Unsupervised Learning Algorithms

M. Emre Celebi • Kemal Aydin

Editors

Unsupervised Learning Algorithms

 Springer

Editors
M. Emre Celebi
Department of Computer Science
Louisiana State University in Shreveport
Shreveport, LA, USA

Kemal Aydin
North American University
Houston, TX, USA

ISBN 978-3-319-79590-4   ISBN 978-3-319-24211-8   (eBook)
DOI 10.1007/978-3-319-24211-8

Springer Cham Heidelberg New York Dordrecht London
© Springer International Publishing Switzerland 2016
Softcover reprint of the hardcover 1st edition 2016

Printed on acid-free paper

Springer International Publishing AG Switzerland is part of Springer Science+Business Media (www.
springer.com)

Preface

With the proliferation of massive amounts of unlabeled data, unsupervised learning algorithms–which can automatically discover interesting and useful patterns in such data–have gained popularity among researchers and practitioners. These algorithms have found numerous applications including pattern recognition, market basket analysis, web mining, social network analysis, information retrieval, recommender systems, market research, intrusion detection, and fraud detection. The difficulty of developing theoretically sound approaches that are amenable to objective evaluation has resulted in the proposal of numerous unsupervised learning algorithms over the past half-century.

The goal of this volume is to summarize the state of the art in unsupervised learning. The intended audience includes researchers and practitioners who are increasingly using unsupervised learning algorithms to analyze their data.

This volume opens with two chapters on anomaly detection. In "Anomaly Detection for Data with Spatial Attributes," P. Deepak reviews anomaly detection techniques for spatial data developed in the data mining and statistics communities. The author presents a taxonomy of such techniques, describes the most representative ones, and discusses the applications of clustering and image segmentation to anomaly detection.

In "Anomaly Ranking in a High Dimensional Space: The Unsupervised TreeRank Algorithm," Clémençon et al. describe a computationally efficient anomaly ranking algorithm based on the minimization of the mass-volume criterion. This algorithm does not involve any sampling; therefore, it is especially suited for large and high-dimensional data.

Clustering is undoubtedly the most well-known subfield of unsupervised learning. The volume continues with 12 chapters on clustering. In "Genetic Algorithms for Subset Selection in Model-Based Clustering," Scrucca describes a genetic algorithm that maximizes the Bayesian information criterion (BIC) to select a subset of relevant features for model-based clustering. In particular, the criterion is based on the BIC difference between a candidate clustering model for the given subset and a model, which assumes no clustering for the same subset. The implementation of this algorithm uses the facilities available in the GA package for the open-source statistical computing environment R.

In "Clustering Evaluation in High-Dimensional Data," Tomašev and Radovanović investigate the performance of popular cluster quality assessment indices on synthetically generated high-dimensional Gaussian data. Extensive experiments reveal that dimensionality and degree of cluster overlap can affect both the mean quality score assigned by an index and the stability of its quality estimation. The authors also discover that appropriate treatment of hub points may improve the quality assessment process.

In "Combinatorial Optimization Approaches for Data Clustering," Festa presents an overview of clustering algorithms with particular emphasis on algorithms based on combinatorial optimization. The author first reviews various mathematical programming formulations of the partitional clustering problem and some exact methods based on them. She then provides a brief survey of partitional clustering algorithms based on heuristics or metaheuristics.

In "Kernel Spectral Clustering and Applications," Langone et al. present a survey of the recently proposed kernel spectral clustering (KSC) algorithm. The authors describe the basic KSC algorithm as well as its probabilistic, hierarchical, and sparse extensions. They also provide an overview of the various applications of the algorithm such as text clustering, image segmentation, power load clustering, and community detection in big data networks.

In "Uni- and Multi-Dimensional Clustering via Bayesian Networks," Keivani and Peña discuss model-based clustering using Bayesian networks. For unidimensional clustering, the authors propose the use of the Bayesian structural clustering (BSC) algorithm, which is based on the celebrated expectation-maximization algorithm. For the multidimensional case, the authors propose two algorithms, one based on a generalization of the BSC algorithm and the other based on multidimensional Bayesian network classification. The former algorithm turns out to be computationally demanding. So, the authors provide a preliminary evaluation of the latter on two representative data sets.

In "A Radial Basis Function Neural Network Training Mechanism for Pattern Classification Tasks," Niros and Tsekouras propose a novel approach for designing radial basis function networks (RBFNs) based on hierarchical fuzzy clustering and particle swarm optimization with discriminant analysis. The authors compare the resulting RBFN classifier against various other classifiers on popular data sets from the UCI machine learning repository.

In "A Survey of Constrained Clustering," Dinler and Tural provide an in-depth overview of the field of constrained clustering (a.k.a. semi-supervised clustering). After giving an introduction to the field of clustering, the authors first review unsupervised clustering. They then present a survey of constrained clustering, where the prior knowledge comes from either labeled data or constraints. Finally, they discuss computational complexity issues and related work in the field.

In "An Overview of the Use of Clustering for Data Privacy," Torra et al. give a brief overview of data privacy with emphasis on the applications of clustering to data-driven methods. More specifically, they review the use of clustering in masking methods and information loss measures.

In "Nonlinear Clustering: Methods and Applications," Wang and Lai review clustering algorithms for nonlinearly separable data. The authors focus on four approaches: kernel-based clustering, multi-exemplar-based clustering, graph-based clustering, and support vector clustering. In addition to discussing representative algorithms based on each of these approaches, the authors present applications of these algorithms to computer vision tasks such as image/video segmentation and image categorization.

In "Swarm Intelligence-Based Clustering Algorithms: A Survey," İnkaya et al. present a detailed survey of algorithms for hard clustering based on swarm intelligence (SI). They categorize SI based clustering algorithms into five groups: particle swarm optimization based algorithms, ant colony optimization based algorithms, ant based sorting algorithms, hybrid algorithms, and miscellaneous algorithms. In addition, they present a novel taxonomy for SI based clustering algorithms based on agent representation.

In "Extending Kmeans-Type Algorithms by Integrating Intra-Cluster Compactness and Inter-Cluster Separation," Huang et al. propose a framework for integrating both intra-cluster compactness and inter-cluster separation criteria in k-means-type clustering algorithms. Based on their proposed framework, the authors design three novel, computationally efficient k-means-type algorithms. The performance of these algorithms is demonstrated on a variety of data sets, using several cluster quality assessment indices.

In "A Fuzzy-Soft Competitive Learning Approach for Grayscale Image Compression," Tsolakis and Tsekouras propose a novel, two-stage vector quantization algorithm that combines the merits of hard and soft vector quantization paradigms. The first stage involves a soft competitive learning scheme with a fuzzy neighborhood function, which can measure the lateral neuron interaction phenomenon and the degree of neuron excitations. The second stage improves the partition generated in the first stage by means of a codeword migration strategy. Experimental results on classic grayscale images demonstrate the effectiveness and efficiency of the proposed algorithm in comparison with several state-of-the-art algorithms.

This volume continues with two chapters on the applications of unsupervised learning. In "Unsupervised Learning in Genome Informatics," Wong et al. review a selection of state-of-the-art unsupervised learning algorithms for genome informatics. The chapter is divided into two parts. In the first part, the authors review various algorithms for protein-DNA binding event discovery and search from sequence patterns to genome-wide levels. In the second part, several algorithms for inferring microRNA regulatory networks are presented.

In "The Application of LSA to the Evaluation of Questionnaire Responses," Martin et al. investigate the applicability of Latent Semantic Analysis (LSA) to the automated evaluation of responses to essay questions. The authors first describe the nature of essay questions. They then give a detailed overview of LSA including its historical and mathematical background, its use as an unsupervised learning system, and its applications. The authors conclude with a discussion of the application of LSA to automated essay evaluation and a case study involving a driver training system.

This volume concludes with two chapters on miscellaneous topics regarding unsupervised learning. In "Mining Evolving Patterns in Dynamic Relational Networks," Ahmed and Karypis present various practical algorithms for unsupervised analysis of the temporal evolution of patterns in dynamic relational networks. The authors introduce various classes of dynamic patterns, which enable the identification of hidden coordination mechanisms underlying the networks, provide information on the recurrence and stability of its relational patterns, and improve the ability to predict the relations and their changes in these networks.

Finally, in "Probabilistically Grounded Unsupervised Training of Neural Networks," Trentin and Bongini present a survey of probabilistic interpretations of artificial neural networks (ANNs). The authors first review the use of ANNs for estimating probability density functions. They then describe a competitive neural network algorithm for unsupervised clustering based on maximum likelihood estimation. They conclude with a discussion of probabilistic modeling of sequences of random observations using a hybrid ANN/hidden Markov model.

We hope that this volume, focused on unsupervised learning algorithms, will demonstrate the significant progress that has occurred in this field in recent years. We also hope that the developments reported in this volume will motivate further research in this exciting field.

Shreveport, LA, USA   M. Emre Celebi
Houston, TX, USA   Kemal Aydin

Contents

Anomaly Detection for Data with Spatial Attributes

P. Deepak

Abstract The problem of detecting spatially-coherent groups of data that exhibit anomalous behavior has started to attract attention due to applications across areas such as epidemic analysis and weather forecasting. Earlier efforts from the data mining community have largely focused on finding outliers, individual data objects that display deviant behavior. Such point-based methods are not easy to extend to find groups of data that exhibit anomalous behavior. Scan statistics are methods from the statistics community that have considered the problem of identifying regions where data objects exhibit a behavior that is atypical of the general dataset. The spatial scan statistic and methods that build upon it mostly adopt the framework of defining a character for regions (e.g., circular or elliptical) of objects and repeatedly sampling regions of such character followed by applying a statistical test for anomaly detection. In the past decade, there have been efforts from the statistics community to enhance efficiency of scan statistics as well as to enable discovery of arbitrarily shaped anomalous regions. On the other hand, the data mining community has started to look at determining anomalous regions that have behavior divergent from their neighborhood. In this chapter, we survey the space of techniques for detecting anomalous regions on spatial data from across the data mining and statistics communities while outlining connections to well-studied problems in clustering and image segmentation. We analyze the techniques systematically by categorizing them appropriately to provide a structured birds-eye view of the work on anomalous region detection; we hope that this would encourage better cross-pollination of ideas across communities to help advance the frontier in anomaly detection.

1 Introduction

Anomaly may be broadly defined as something that departs from what is generally regarded as *normal* (i.e., common). Thus, a group of objects can be considered as an anomaly if their collective behavior deviates from what is regarded as common.

P. Deepak (✉)
IBM Research - India, Bangalore, India
e-mail: deepaksp@acm.org

© Springer International Publishing Switzerland 2016
M.E. Celebi, K. Aydin (eds.), *Unsupervised Learning Algorithms*,
DOI 10.1007/978-3-319-24211-8_1

Fig. 1 Example grid dataset

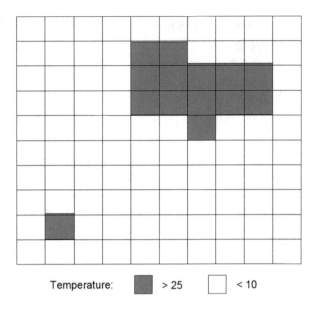

Temperature: ▮ > 25   ☐ < 10

For example, a county where schools record an average of 80 % pass in a test where the national average is 40 % would be classified as an anomaly under this definition. On data that have a spatial positioning, anomalous groups that are coherent on the spatial attributes (e.g., objects that fall within a county, as in the example) are often the anomalies of interest. Figure 1 illustrates a geographic region gridded into squares, each cell in the grid colored according to the temperature recorded within it. For simplicity, we consider that each cell takes a temperature from either of two ranges, the higher range being greater than 25 units, and the lower range set at less than 10 units. While most of the cells in the grid record the lower range, two regions, a large set of contiguous cells in the top and a single cell region at the bottom left are seen to record the higher range. According to the definition of anomaly, both these regions would intuitively be classified as anomalies due to being high-temperature regions within a largely cooler grid.

Outlier is a related but more fine-grained concept in that it quantifies uncommon behavior at the level of individual objects. Without going into the technical details of what would be an outlier, the cell in the lower left region of the grid in Fig. 1 would be regarded as an outlier since it is seen to be warm while both its local neighborhood as well as the entire grid are significantly cooler. This brings us to finer details of the definition of an outlier, where there is no consensus on what should be chosen as the baseline to contrast a candidate outlier with. This, however, is often due to good reason since some applications may inherently warrant a comparison with the local neighborhood whereas others may be more concerned about the level of contrast with the global behavior. Under the global comparison paradigm, each cell in the warm region in the top-half of the grid would be deemed to be an outlier. However, the local comparison would unveil a different story; most of the cells in the warm region have warm cells as neighbors, except perhaps the bottom-most warm cell that protrudes into a cooler neighborhood. Thus, the local comparison approach would classify most of the cells in the warm region as non-outliers.

Terminology Throughout this paper, we will use the term *spatial data* as a shorthand to describe data that has spatial attributes in addition to other attributes (like temperature in the example above); in particular, it is necessary to mention here that we exclude purely spatial data, those which have only spatial attributes, from our definition of spatial data, for this chapter.

Roadmap In this chapter, we intend to survey techniques from across diverse disciplines that could be used towards identifying anomalous regions on data with spatial attributes. We focus on unsupervised methods and restrict the discussion to simple objects that may be described by a set of values, one per attribute; for example, we would not consider anomaly identification in multimedia such as videos. We organize the content into seven sections, as described below.

- **Problem Definition and Taxonomy of Techniques:** We start with a section outlining the problem of anomaly detection in full generality, and present a taxonomy of approaches that address the problem. The rest of the sections are organized so as to mostly focus on specific areas within the taxonomy.
- **Object Anomalies:** In this section, we survey the various approaches for anomaly detection at the object level. We will consider the applicability of general outlier detection methods to the task of identifying anomalous regions on spatial data, as well as describe a method from literature that was specifically proposed for the task.
- **Region Anomalies—Global:** This section describes approaches that seek to identify globally anomalous regions, i.e., regions comprising groups of objects that are divergent from global behavior. This encompasses methods from statistics such as spatial scan as well as a few mining-based approaches.
- **Region Anomalies—Local:** Here, we consider approaches that look for regions that are anomalous when contrasted against behavior in the local neighborhood. These include the localized homogeneous anomaly (LHA) detection method as well as image segmentation methods that address a stricter version of the problem from a different domain.
- **Region Anomalies—Grouping:** In this section, we will survey techniques that target to group objects in such a way that groups are coherent while being mutually divergent. Techniques in this category group objects such that each group is divergent from other groups in the output; some of the groups in the output may be regarded as anomalies based on the level of contrast. We will specifically look at grouping methods for spatial data as well as clustering-based anomaly detection methods from intrusion detection literature.
- **Discussion:** In this discussion section, we will explore other ways of using clustering methods in the anomaly detection pipeline. In this context, we will briefly describe density-based clustering algorithms.
- **Directions for Future Work:** In this section, as the name suggests, we outline promising directions for future work in the area of anomaly detection.

2 Problem Definition and Taxonomy of Techniques

2.1 Problem Definition

We now define the problem of anomaly detection on spatial data at a level of generality so that most anomaly detection techniques would be applicable. Consider a set of objects $\mathscr{D} = \{d_1, d_2, \ldots, d_n\}$ where each object d_i may be represented by the values it takes for a set of attributes from a pre-defined schema:

$$d_i = [s_{i1}, s_{i2}, \ldots, s_{im_s}, v_{i1}, v_{i2}, \ldots, v_{im_v}] \tag{1}$$

The first m_s attributes are designated as spatial attributes, whereas the remaining m_v are non-spatial attributes; we will call the non-spatial attributes as value attributes. In our example from Fig. 1, each cell is a data object with the spatial attributes being spatial location defined by the x and y co-ordinates, and a single value attribute denoting the temperature. In the case of studies on epidemics, data points may be people with their spatial attribute being their geographic co-ordinates and their value attribute being a single boolean attribute indicating whether they are diseased or not. In case of weather modeling, temperature, humidity, etc., may form different value attributes.

Additionally, anomaly identification methods use a notion of proximity between data points, defined based on their spatial attributes. For data in the form of grid cells, the proximity measure may simply be the adjacency matrix that could then be used to determine whether a set of cells is connected or not. In the case of data in the form of points, this may be a similarity measure defined on the spatial attributes.

The problem of anomaly detection is to identify a set of anomalies $\mathscr{A} = \{A_1, A_2, \ldots, A_k\}$ such that each set $A_i \subseteq \mathscr{D}$ adheres to the following criteria:

- **Spatial Coherence:** The set of objects in A_i satisfy a coherence condition defined on the *spatial attributes*; examples are that they are the only objects in a circular or elliptical region within the space defined by the spatial attributes, or that they form a set of connected cells in a gridded dataset.
- **Contrast from Context:** A model built over objects in A_i using their *value attributes*, contrasts well with those built over objects in $C(A_i)$, the *context* of A_i. The model could simply be the *mean* of temperatures over cells and the context could be the set of cells adjoining A_i (i.e., local) or the set of all cells in the dataset (i.e., global). Typical models used in anomaly detection are simple in construction, with the mean being among the most commonly used models.

Some anomaly detection methods, especially, those that come from the clustering family, use a different construction to handle the second condition; we will elaborate on those aspects in respective sections. There are different possibilities to outline the spatial coherence condition; so is the case with building the value attribute model for each A_i, defining the context and specifying the contrast condition between models. Different anomaly detection techniques differ in the specifications used for each of the above phases.

Applications Beyond Spatial Data Though we will discuss algorithms and techniques designed for and motivated by scenarios involving spatial data where geographic attributes form spatial attributes, the problem itself is much more general. There are motivating scenarios for anomaly detection where metadata attributes replace the role of spatial attributes. For example, an enhanced rate of incidence of disease among teens in a specific county could be an anomalous pattern; the methods for spatial anomaly detection would be able to find such anomalies if age is modeled as a "spatial" attribute. Similarly, there are other motivating anomaly detection scenarios where ordinal attributes such as weight, height, education level, etc. may be modeled as "spatial" attributes followed by application of spatial anomaly detection methods. In short, the problem we are looking at in this chapter as well as techniques for addressing it, are applicable beyond data having conventional spatial attributes.

2.2 Taxonomy of Techniques

Figure 2 illustrates a taxonomy of methods that have been explored in anomaly detection on spatial data. The greyed boxes in the taxonomy represent methods for anomaly detection that do not differentiate between attributes as spatial and value attributes; thus, these are not readily applicable to the problem of anomaly

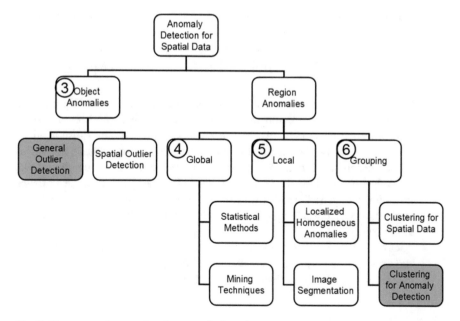

Fig. 2 Taxonomy of approaches for anomaly detection

detection on spatial data. However, we include those in our taxonomy since they may be adapted to the problem in question. We now briefly describe each class in the taxonomy, leading up to a detailed treatment in subsequent sections. Figure 2 also indicates the section number where each sub-tree of the taxonomy is described.

- **Object Anomalies:** Techniques under this category enforce that the output anomalies are all made up of singleton sets, i.e., $\forall i, |A_i| = 1$, and are thus also called outlier detection techniques.

  - **General Outlier Detection:** General outlier detection methods,[1] the ones that we will survey under this class, do not differentiate attributes as spatial or value attributes. However, they frequently use concepts such as neighbors and neighborhood which are more meaningful in the context of spatial attributes; thus, these may be seen as techniques that do not use value attributes, i.e., those that assume $m_v = 0$. They mostly estimate the outlier-ness of an object based on whether there is disparity between density of objects around them and the density of objects around their neighbors. We will also comment on how such methods can be adapted to the problem of outlier detection on spatial data.
  - **Spatial Outlier Detection:** Unlike general outlier detection methods, outlier detection applied to spatial data is relatively less studied. Spatial outliers are those that take on values different from those of their spatial neighbors. Among the few methods that have been proposed under this category, we will take a detailed look at SLOM [7].

- **Region Anomalies:** As the name suggests, methods for region anomaly detection seek to identify anomalies that could potentially comprise multiple objects, i.e., $|A_i| \geq 1$, in a spatially coherent region as outlined in Sect. 2.1. We classify these techniques based on the construction of the context for estimating anomalousness.

  - **Global:** Under the global context setup, models built across objects in a candidate anomalous region are compared against those built over objects across the entire dataset. In a weather dataset, a hot desert surrounded by warm plains would be regarded as an anomaly by these methods regardless of whether its temperature differs substantially from its local neighborhood. There has been a large body of work under this category from the statistics community, e.g., SaTScan [20] and its numerous variants; we will survey such methods as well as techniques from the data mining community that fall under this category.
  - **Local:** Techniques that use local context estimate the qualify of a candidate anomaly by comparing the model against that built over *nearby* objects, with the notion of nearness defined over spatial attributes. In our example, the hot

---

[1]We use the prefix *general* to differentiate these from spatial outlier detection methods, that we will see shortly.

desert wont qualify as an anomaly as long as the contrast in warmth is not good enough with the surrounding plain, regardless of whether it is much warmer than the average temperature in the whole area under consideration. Under this category, we will describe the LHA anomaly detection method [33] and also comment on how image segmentation techniques are applicable for local anomaly detection.

– **Grouping:** Unlike anomaly detection methods that specify a context for comparison crisply, methods under this category use an indirect form of context comparison for anomaly estimation. These techniques, largely inspired from literature on data clustering, group objects into clusters in such a way that the clusters are mutually divergent on the value attributes (as well as, sometimes, on the spatial attributes). Since the notion of mutual divergence implies that each cluster can be different from the models corresponding to its neighboring clusters as well, these methods are closer in spirit to the *local* methods than to the *global* methods discussed above. While some methods are only for forming groups and do not perform anomaly detection, they can be easily extended to the problem of anomaly detection by employing a test for anomalous-ness on the groups so that the subset of clusters that qualify be labeled as anomalies. There has been a lot of work, mostly from network intrusion detection literature, on clustering for anomaly detection; however, these, being designed for intrusion detection scenarios where there are no spatial considerations, treat all attributes on an equal footing without differentiating them as spatial or value attributes. We will survey such methods too, in addition to clustering methods specifically designed for spatial datasets such as those we are interested in.

In the following sections, we will describe techniques under each category in the taxonomy in a greater amount of detail.

3   Object Anomalies: Techniques for Outlier Detection

Outlier detection methods, as observed earlier, address the problem of finding individual objects in the dataset that are anomalous, i.e., by using the constraint:

$$\forall\, i, |A_i| = 1$$

This is a field that has attracted tremendous attention in the data mining community for the last two decades. Among the popular definitions of outliers is one from Hawkins [16]: "An outlier is an observation which deviates so much from other observations as to arouse suspicions that it was generated by a different mechanism". The vast majority of techniques that have been proposed for outlier detection consider the general case, where attributes are not classified into categories such as value and spatial; though not strictly the case, these may be regarded to consider data as purely spatial attributes (i.e., $m_v = 0$) since they make use of notions such as *neighborhood* and *density* that are more meaningful in spatial data.

Due to the large volume of literature around general outlier detection, we will take a closer look at those before considering methods for spatial outlier detection.

3.1 General Outlier Detection

The problem of outlier detection on general datasets, those where attributes are not classified into two classes as in the case of data that we have been considering in this chapter, has attracted plentiful attention from the data mining community, notably since [4]. Though these methods do not address our problem directly, they could be adapted to our problem of anomaly detection on spatial data. Given the large amount of work in general outlier detection, and the consequent maturity of techniques, we will consider those techniques in some detail in this section. We start with a framework for general outlier detection methods proposed in [28], and then take a closer look at LOF [4]. We will then comment on how general outlier detection methods can be adapted to the problem of outlier detection on spatial data. A comprehensive and excellent survey of outlier detection techniques appears in [28], for further reading on general outlier detection methods.

3.1.1 General Framework for Outlier Detection

The survey paper by Schubert et al. [28] proposes a five phase framework that covers most well-known methods for general outlier detection. The task may be seen as that of associating each data object d_i in the database with a score that is directly related to the estimated outlier-ness of the object; objects can then be sorted based on this score and those that pass a threshold of outlier-ness may be output. While the framework uses terms such as proximity and neighborhood as if the data has only spatial attributes, the methods are applicable for non-spatial attributes too. It may be noted that these methods handle all attributes equally and are not designed to handle spatial and non-spatial attributes differently. The five phase approach for outlier-ness estimation is given below:

- **Context Estimation:** The context of an object is typically defined as a subset of the data that are proximal to the object under consideration. The simplest and most widely used model for context is the collection of k nearest neighbors of the object in question. Another intuitive model is the set of objects that fall within a distance threshold. In certain cases, however, the entire database is used as the context.
- **Model Building:** This step summarizes the set of objects in the context of the object in question, to form an object-specific model. The model typically summarizes the distance of the considered object from objects in its context using a metric such as the sum of the distances. The simplest models could simply

count the number of objects in the context, whereas sophisticated ones try to fit a distribution such as Gaussian.

- **Reference Estimation:** While the context determines the objects that would feed into building a model, the reference of an object is a subset of the dataset that would be used for comparing the object's model against. Like the context, the reference could range from simply choosing the k nearest neighbors to considering the entire dataset.
- **Comparison:** This phase does the first part of the outlier-ness estimation wherein the model of the object is compared with the models of the objects in its reference. In the case of simple models like object count or sum of distances, the comparison could simply be the quotient of the object model with those of objects in the reference; these separate quotients (one for each object in the reference) may then be simply averaged to obtain a single score denoting outlier-ness.
- **Normalization:** The normalization step, which is not used in many methods, would normalize the object-specific outlier scores obtained from the previous step. A simple normalization approach would be to divide the outlier score of each object by the sum of scores across objects so that an outlier probability distribution is obtained across all objects in the dataset.

Having outlined this framework, any general outlier detection approach may be simply specified by the way each of these steps are instantiated. It may be noted that another smaller category of general outlier detection methods, called *global outlier detection*, do not fall into the above framework. Some of them regard objects that do not have at least a fixed fraction of database objects within a specified distance threshold as global outliers (e.g., [19]). Clustering-based outlier detection methods leverage clustering algorithms that do not force every data instance to be assigned to a cluster (e.g., DBSCAN [10], ROCK [15], RGC [2]); a simple heuristic, such as that used in FindOut [35] would be to demarcate objects that do not belong to any cluster as outliers. To contrast from such methods, the methods under the framework described above are some times referred to as *local outlier detection* methods. We will now turn our attention to LOF, an early method for local outlier detection.

3.1.2 LOF

LOF [4] is among the earliest methods for outlier detection. This draws upon ideas from density-based clustering approaches such as DBSCAN [10]. LOF uses the set of k nearest neighbors to each object as the context of the object. In the model building phase, LOF calculates a smoothened version of pairwise distance between each object d and every object c in its context; this is called the *reachability distance* and is calculated as follows:

$$RD_p(d, c \in context(d)) = max\{dist(d, pthNN(d)), dist(d, c)\} \qquad (2)$$

where $pthNN(d)$ denotes the pth nearest neighbor of d based on a distance measure defined on the attributes. This is a smoothened version of the distance since objects far away from d would have their $RD_p(d, .)$ equal to their actual distance, whereas the $pthNN(d)$ would dominate in the case of objects closer to d. It may be noted that if $p = k$ (as is set in many cases), $RD_p(d, .)$ would be equal to the kNN distance for all objects in the context of d. The object-specific model used by LOF is called local reachability density that is defined as follows:

$$lrd(d) = 1 \Bigg/ \frac{\sum_{c \in context(d)} RD_p(d, c)}{|context(d)|} \tag{3}$$

Thus, $lrd(d)$, the model for d, is the reciprocal of the average reachability distance of d to objects in its context. Moving on to the third phase in the framework, LOF uses the reference of an object as its k nearest neighbors itself, exactly as in the case of the context. In the model comparison phase, the model of d is compared to those built for objects in its reference space to obtain an outlier factor called the local outlier factor.

$$LOF(d) = average \left\{ \frac{lrd(r)}{lrd(d)} \middle| r \in reference(d) \right\} \tag{4}$$

LOF does not use a global normalization phase, and thus outputs objects with high LOF values as outliers. Informally, $lrd(.)$ captures the density of the space around an object, and those that are in a sparser neighborhood (i.e., with high distance to nearest neighbors) as compared to the neighborhoods of the objects in their context would end up with large LOF values and thus be considered outliers.

3.1.3 Adapting for Spatial Outlier Detection

The task of spatial outlier detection is that of finding objects whose values for the value attributes contrast well with those in its specified spatial context. We could use general outlier detection methods for anomaly detection on spatial data by using the following approach:

1. *Partitioning:* Partition the dataset based on solely the value attributes such that the objects within each partition are reasonably homogeneous with respect to the *value attributes*. This could be achieved by clustering the dataset on the value attributes using any well-known partitional clustering algorithm [5], so that objects within each cluster has similar values for the value attributes. In simpler cases of datasets with just one value attribute such as temperature, one could create buckets on the range of values of the attribute to create partitions; each partition would then comprise the subset of data that have objects having temperature within the range of the respective bucket.
2. *Partition-wise Outlier Detection:* In this phase, each partition is *separately* fed to the general outlier detection methods which will then work on finding outliers based on only the *spatial attributes*. These would then identify partition-specific outliers as those objects whose neighborhood within that partition is sparse.

Given the nature of our partitioning step, the sparse neighborhood of the outlier is likely to be caused due to many objects in close proximity to the identified outlier having been assigned to other partitions. Thus, this satisfies our condition of spatial outliers as those that contrast well with local neighborhood on the value attributes. It needs to be noted, however, that there could be false positives under this approach. The sparse neighborhood of the partition-specific outlier could be caused also due to the general sparsity of the full dataset (i.e., across partitions) as against the objects in the neighborhood being assigned to other partitions; a post-processing step would need to be designed to weed out cases where the partition-specific outlier is induced by the former factor than the latter. It may also be noted that this phase treats all objects in a partition as if they are identical on the value attributes; this could lead to missing some legitimate outliers.

Though adaptations of general outlier detection to spatial data have not been considered much, a recent approach adapting LOF to spatial data [28] is worthy of attention.

3.2 Spatial Outlier Detection

Spatial outlier detection methods, as observed earlier, look to identify objects whose values for the value attributes contrast well with those in its specified spatial context. The common approach used by spatial outlier detection methods is to estimate a deviation score, where the value attributes of an object are compared to the value attributes of its spatial neighbors [30], with the spatial neighbors forming the context. Among the relatively few methods that have been proposed for the same, we will look at SLOM [7] in this section.

3.2.1 SLOM

SLOM [7] starts by defining a context for each object that would be used for comparison later on. The context of each object in SLOM is simply the set of spatial neighbors in gridded or tessellated data; in the case of a non-gridded dataset of multi-dimensional point objects, the context may be defined as the set of kNN objects. SLOM then calculates the trimmed mean for the object, which is its average distance with those on its context, distance measured on the value attributes only; the context object with the largest distance is ignored to reduce sensitivity to noise, thus giving this measure the adjective *trimmed*. In addition to calculating the trimmed mean, denoted by $\tilde{d}$, of the object itself, the trimmed mean is also calculated for all objects in the context, and its average is estimated:

$$avgtm(d) = average\{\tilde{d}(c)|c \in \{context(d) \cup d\}\} \tag{5}$$

This is used to define an oscillation measure β that estimates the extent of fluctuation of the trimmed mean around the object in question.

$$\beta(d) = \frac{|\{c \in econtext(d) \wedge \tilde{d}(c) > avgtm(d)\}| - |\{c \in econtext(d) \wedge \tilde{d}(c) < avgtm(d)\}|}{|econtext(d) - 2|} \tag{6}$$

where $econtext(d)$ indicates the set of objects in $context(d)$ and itself. In essence, β captures the normalized asymmetry of the distribution of the trimmed means around $avgtm(d)$; if there are equal number of objects on either side, β evaluates to 0. The SLOM measure for an object is then defined as follows:

$$SLOM(d) = \frac{\tilde{d}(d) \times \beta(d)}{1 + avg\{\tilde{d}(c)|c \in context(d)\}} \tag{7}$$

Thus, objects with a large $\tilde{d}(.)$ score with respect to those in their context would then be good SLOM outliers. As seen from the construction, higher β values also favor labeling an object as an outlier. To position SLOM on the framework for general outlier detection introduced earlier, estimation of $\tilde{d}(d)$ may be seen as the model building phase. The reference is the same neighborhood as used for the context, and the estimation of β followed by the SLOM score form part of the comparison phase. As in the case of LOF, there is no global normalization in SLOM as well.

4 Region Anomalies: Global

Turning our attention to anomalous region detection techniques where the sets of anomalies output could comprise more than one point, we will now look at methods that quantify anomalous-ness by contrasting the behavior of the set of objects in question against a global behavior model estimated across all the objects in the dataset. Such methods may be classified into two groups: the larger group is that consisting of statistical methods that build upon scan statistics such as SaTScan [20], whereas methods in the second group explore data mining approaches for the task. We will look at these two kinds of techniques in separate sub-sections herein.

4.1 Statistical Approaches: Spatial Scan Statistics

Among the earliest statistical techniques for globally divergent anomaly detection is the spatial scan statistic proposed by Kulldorff [20]. Numerous extensions have been proposed to extend the basic spatial scan statistic while preserving the core framework. We will consider the basic spatial scan statistic as well as one of its extensions, in this section.

While the basic spatial scan statistic is general and can be applied across domains, we illustrate it by means of an example application domain. Consider the problem of identifying spatial clusters with high incidence of a particular disease that is under study; these clusters would be anomalies due to the disparity between the cluster disease rate and the global rate. With coarse-grained modeling where we record statistics at the county level, each data object d_i could be county, with its population denoted by a_i, and the number of diseased people denoted by y_i. The core parameter of interest is the disease rate (or more generally, the *response rate*),

$$r_i = \frac{y_i}{a_i} \tag{8}$$

Due to this construction, response rate varies within the range $[0, 1]$. When the cell is denoted using a variable such as d, we will use r_d to denote the response rate in the cell. The spatial scan statistic is designed to identify *hotspots* (i.e., anomalies) that are spatial regions (e.g., county) with an elevated (or decreased) response rate when assessed against the rate in the entire dataset. These techniques usually do not allow to identify irregular closed shapes such as county boundaries, but restrict themselves to the identification of regular-shaped spatial anomalies (e.g., circular regions). Hotspots are regions of cells that satisfy at least the following two properties:

- **Spatial Coherence:** The initial version of the spatial scan statistic enforced that hotspots be circular in shape, i.e., a hotspot be a region defined by a circle, intuitively comprising of all cells that fall within the circle. In the case of non-gridded data where each observation is a point object, all objects that fall within the circle would then be considered. There have been extensions that have allowed for elliptically shaped [21] hotspots, and even arbitrarily shaped regions that are connected [25].
- **Limited Size:** Hotspots should not be excessively large with respect to the entire dataset. For example, if the region were to be more than half the size of the dataset, it could be argued that the region, and not its exterior, should be considered as the background. The initial version of the spatial scan statistics, and most of its variants, impose a constraint that a hotspot not be more than 50 % of the size of the entire dataset.

In addition to the above properties that are agnostic to the response rate, we would want to ensure that the response rate within the candidate hotspot be high (low) before it be declared as a hotspot. The spatial scan statistic uses a hypothesis testing approach to assess this hotness. The null hypothesis assumes that the response rate is the same for all regions in the dataset, and thus, consequently, that there is no hotspot in the dataset:

$$H_0 : \forall d \in \mathcal{D}, r_d = p_a \tag{9}$$

The alternative hypothesis, on the other hand, assumes that there is a hotspot in the dataset, and is outlined as follows:

$$H_1 : \exists p_0, p_1, Z \text{ such that} \tag{10}$$

$$0 \le p_0 < p_1 \le 1$$

$$Z \subset \mathcal{D} \wedge Z \ne \phi$$

$$\forall d \in Z, r_d = p_1$$

$$\forall d \in (\mathcal{D} - Z), r_d = p_0$$

Informally, H_1 assumes that there is a region Z such that the response rate is a high value, p_1, for cells within it, and is a low value, p_0, for other cells in the dataset $\mathcal{D}$. In addition to the response rate conditions as outlined above, the alternative hypothesis also requires that Z, the hotspot region, adheres to the spatial coherence and size restrictions as outlined earlier. H_1, as outlined above, assumes that the hotspot exhibits an elevated response rate; the analogous hypothesis for reduced response rate hotspots may be obtained by simply inverting the inequality between p_0 and p_1 so that p_1 be lesser than p_0.

As may be seen, there are three unknowns in the alternative (i.e., hotspot) hypothesis, viz., Z, p_0, and p_1. When all of these are known, one can estimate the likelihood of the data being generated under the hotspot hypothesis, by assuming that the data is distributed according to one of the well-known models such as Bernoulli or Poisson distributions. Let the likelihood of the combination of parameters be denoted by $L(Z, p_0, p_1)$; now, the problem is to find values $\hat{Z}, \hat{p_0}, \hat{p_1}$ that maximize the likelihood $L(Z, p_0, p_1)$ for a given dataset. For a given value of Z and an assumed distribution of data, it is easy to determine the values of p_0 and p_1 that maximize the likelihood, as follows:

$$\{\widehat{p_{0Z}}, \widehat{p_{1Z}}\} = \arg\max_{p_0, p_1} L(Z, p_0, p_1) \tag{11}$$

where $\widehat{p_{iZ}}$ denotes the optimal value of the parameter for a fixed Z. The difficult part, however, is estimating the value of $\hat{Z}$ that maximizes $L(.)$ since the space of possible values of Z, $Dom(Z)$ is typically large. The optimization problem we are talking about is the following:

$$\hat{Z} = \arg\max_{Z \in Dom(Z)} L(Z, \widehat{p_{0Z}}, \widehat{p_{1Z}}) \tag{12}$$

For example, any circular region in the dataset that satisfies the size restriction is a candidate value of Z, and it is not difficult to see that there could be infinitely many circles. Strategies to do this optimization fall under two main categories:

- **Parameter Space Reduction:** Instead of looking for all candidate values across $Dom(Z)$, parameter space reduction strategies look to find optimal values of Z

within a sub-space $R(Z)$. The reduced space may be defined by a tessellation of the space followed by a topological sort, as used in ULS Scan [25]. However, this approach could potentially lead to sub-optimal result since $\hat{Z}$ may not always be contained within the search space $R(Z)$.

- **Stochastic Optimization Methods:** Stochastic optimization methods such as genetic algorithms and simulated annealing start from an initialization of Z, and scan the space through perturbations of the value of Z. These could miss $\hat{Z}$ and discover a sub-optimal value, but, are expected to converge to the optima under certain assumptions. An example of a simulated annealing approach for spatial cluster detection appears in the technique presented in [9].

It is not just enough to have a high likelihood under H_1; we would need to compare with the likelihood under the competing hypothesis, H_0. Having determined the $\hat{Z}$ that maximizes the likelihood for H_1, the spatial scan statistic uses a likelihood ratio test to determine the number of times the data is more likely under H_1 as compared to the null hypothesis, H_0. To do this, we would need to also estimate the likelihood of the data under the null hypothesis. It may be observed that H_0 is the limit of H_1 as $p_1 \rightarrow p_0$. Thus, the likelihood under the null hypothesis may simply be written as $L(Z, p_a, p_a)$. Further, it is easy to observe that Z is not identifiable in the limit[2]; any value of Z would serve just as fine since the response rate within and outside are assumed to be equal. Thus, we can omit Z and simply denote the likelihood under the null hypothesis as L_0, a shorthand for $L(Z, p_a, p_a)$. The likelihood ratio is then:

$$\lambda = \frac{L(\hat{Z})}{L_0} \qquad (13)$$

We have omitted the parameters p_0 and p_1 from the numerator for simplicity. Beyond just estimating the value of the test statistic, λ, we additionally would like to see whether it is high enough under an expected distribution of the test statistic. For example, even a synthetic dataset of $|\mathcal{D}|$ objects generated under the null hypothesis may have a yield a high value of λ due to localized regions that have a high response rate. Monte Carlo simulation methods [34] may now be used to generate a large number of synthetic datasets, each of $|\mathcal{D}|$ objects and overall response rate (i.e., p_a) equal to the dataset under consideration. Consider 10,000 such generated datasets, each of which are subjected to the same analysis to yield a separate λ value; now, if the λ value for the dataset in question (i.e., the non-synthetic dataset) is higher than 9500 of the 10,000 generated λ values, the test statistic may be adjudged to be significant at a p-value of <0.05. Similarly, if the test statistic is within the range of the top 100 values, the test statistic may be deemed to be significant at a p-value of <0.01. If the hotspot hypothesis is indeed

---

[2]An intuitive likelihood estimates the chances of generating the data points as against the expected probability, and aggregates it across the objects. Under the condition that the expected probability is the same for objects within and outside Z, any value of Z would yield the same likelihood.

statistically significant, the corresponding region, $\hat{Z}$, is regarded as a hotspot or an anomaly. There could be many such regions—i.e., values of Z—in the dataset that yield a statistically significant λ; in cases where one would like to identify multiple hotspots, the likelihood ratio provides an intuitive ranking.

4.1.1   ULS Scan Statistic

Having described the basic approach, which is also the core framework that most spatial scan statistics variants operate upon, we will now look at the upper level set scan statistic [25] which is interesting due to being able to detect arbitrarily shaped anomalies (i.e., hotspots). Most of the details of the likelihood and test statistic computation remain similar as in the spatial scan statistic; we will focus on the construction of the candidate hotspots and computational aspects therein, which are interesting from a data mining perspective. Spatial scan statistics that restrict hotspots to be of regular shapes (e.g., circular and elliptic) would be unable to detect clusters in geographic regions that are culturally and politically coherent while being rather elongated geographically such as a settlement on a valley. Many epidemiological patterns are generated due to policies determined at a country/province level, neither of which need to be circular or regular or even convex in shape. The ULS Scan Statistic may be applied in such scenarios.

ULS Scan Statistic works on datasets in a tessellated space where the presence or absence of a shared boundary between regions is well-defined. ULS can admit hotspots that are composed of multiple connected cells, i.e., a set of cells such that there exists a pairwise path between any two cells that is entirely confined to within cells in the set. The ULS Scan Statistic works by first constructing a data structure called the ULS Tree that is composed of cells as nodes, and the edges determined using a combination of connectedness and response rates.

Figure 3 illustrates the working of the ULS on a sample dataset and the construction of the ULS Tree. Figure 3a shows a tessellated dataset with different regions colored with an intensity that is proportional to their response rate. The cells are numbered and the corresponding response rates are seen to satisfy the following:

$$r_1 < r_3 < r_2 < r_5 < r_4 \tag{14}$$

The y-co-ordinate of each cell in Fig. 3b is equivalent to the response rate; thus, the cell 4 appears at the top whereas cell 1 appears at the bottom. The x-co-ordinate in the same figure does not have any meaning beyond visualization ease, and cells could be arbitrarily moved around on the X-axis. The construction of the ULS Tree starts from the top and proceeds downwards, and follows the following simple rules:

- If the next cell (the cell with the highest response rate among those yet to be considered) is not spatially adjacent to any higher cell, do not introduce any edges. Since this step allows to add nodes that are not connected to any existing

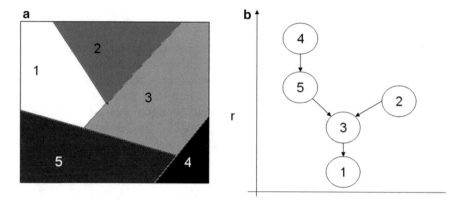

Fig. 3 Illustration of ULS Tree construction (**a**) dataset, (**b**) ULS Tree

node, it allows to create multiple components that are mutually disconnected, across steps; we will refer to each such component as a tree.
- Else, in the case that the next cell is spatially adjacent to a higher cell, add an edge from the lowest node in the tree containing the latter cell. If there are multiple trees containing cells that are spatially adjacent to the next cell, add edges from each of them.

This construction process ensures a unique entry point for each cell in the tree, as may be seen in the example below.

Tree Construction in Our Example: In our example, 4 starts off as a singleton node. The next cell, 5, is seen to be adjacent to 4, and this induces the link between them. The third cell, 2, is not adjacent to any cell considered so far, and thus starts another tree on its own. When the algorithm processes 3, it is seen to be spatially adjacent to 4, 5, and 2. 4 and 5 appear in the same sub-tree necessitating just one edge for both, whereas an additional edge is introduced to connect to 2. 1 is then added as the lowermost cell, completing the tree construction process.

Identifying hotspots under ULS Scan: Each node in the ULS Tree defines a candidate region that comprises all cells above it in its sub-tree. Thus, the region at node 5 is to be understood to stand for the set of cells $\{4, 5\}$, whereas the set at 3 is $\{4, 5, 2, 3\}$. The tree construction method ensures that each such region is a set of connected cells. As is obvious, the bottom-most node is the set of all cells in the dataset; this would be excluded from consideration as a hotspot due to the size constraint that excludes regions that cover more than 50 % of the dataset. It is likely, but not necessary, that the hotspots with high likelihood scores would be towards the upper part of the tree. All candidate regions are then processed using the statistical machinery from that of the spatial scan statistic introduced earlier. ULS Scan is computationally efficient since the number of candidate hotspots (i.e., nodes in the tree) is intuitively bounded by the number of cells in the dataset.

4.1.2 Other Extensions

As observed earlier, there have been numerous extensions of the basic spatial scan statistic. The Bayesian spatial scan statistic [24] proposes a Bayesian method that is shown to be easier to incorporate prior information and can provide much better response time by avoiding the randomization testing phase. Among the most active areas of work have been to relax the shape restriction of hotspots in the original framework to ellipses [21], and arbitrary shapes [9, 32]. Readers interested in scan statistics may find the book on scan statistics [13] useful.

4.2 Mining Approaches

Unlike the statistics community, the data mining community has paid lesser attention to the problem of identifying globally divergent anomalies. The two intuitive directions for the mining community towards this problem are top-down or bottom-up. Top-down approaches could start with the entire dataset and progressively zoom-in on candidates for anomalies, whereas bottom-up approaches could start with small anomalous regions and merge regions while keeping track of how they fare with respect to anomalous-ness. We will look at Bump Hunting [12], a top-down approach, in this section.

4.2.1 Bump Hunting

Bump Hunting [12] is a top-down approach for spatial cluster detection designed for datasets with only one value attribute. It starts with the entire spatial dataset in one spatial *box*, B. The algorithm iteratively peels of spatial sub-boxes from B, in accordance with the following greedy strategy:

$$B = B - \underset{b \in C(b)}{\arg \max} \, average\{v_{il} | d_i \in B - b\} \qquad (15)$$

$C(b)$ denotes the set of spatial sub-boxes that are available to be peeled off. v_{il} is the only value attribute of the object d_i. Thus, this strategy peels off the spatial sub-box such that the average value of the value attribute among objects in the remainder of B is maximized. The peeling process stops when the number of objects in B is small enough that further peeling would violate a support threshold. To partially offset for the sub-optimality in B due to the greedy strategy employed, an additional phase of pasting is applied that would enlarge the box by pasting sub-boxes to it, using considerations similar to that employed in the peeling phase.

As is obvious from the description, this addresses a special case of globally divergent anomaly detection. The approach would need to be adapted to address the problem of detecting multiple anomalies. Extending the technique to work with datasets having multiple value attributes is possible, but is not straightforward. As may be inferred from the description, the basic approach can only identify anomalies where the average value of the value attribute is higher than the average value in the dataset; similar deviations to the lower side of the average may also be deemed anomalous and can be identified by changing the arg max condition to arg min.

5 Region Anomalies: Local

We now consider techniques addressing the problem of discovering anomalies that exhibit differences from their local neighborhood. We will first describe a recent data mining algorithm which identifies homogeneous regions that contrast well with their generalized local neighborhood as anomalies. This will be followed by a discussion on how the task of image segmentation relates to the problem.

5.1 Localized Homogeneous Anomalies

LHA detection [33] addresses the problem of discovering regions that are spatially coherent and homogeneous on the value attributes, while also contrasting well on the value attributes with their *generalized local neighborhood*. This is designed to work on gridded or tessellated data, where the adjacency/neighborhood relation is well-defined. Homogeneity is measured on the value attributes; thus, a region with people of similar incomes would be considered homogeneous, while a city with luxury residences interspersed with slums would not satisfy the homogeneity constraint on the income attribute. The generalized neighborhood condition is best illustrated by means of the example in Fig. 4 which depicts two plots of spatial datasets that use a single value attribute, with each point colored according to the intensity of the value attribute (e.g., temperature) at them. In each, the middle circle is the region under consideration and the ring-shaped enclosing region bounded by the dotted lines form the local neighborhood. The circle in Fig. 4a is not considered as an LHA due to being of medium intensity, which would not differ much from the average intensity of its neighborhood (which is composed of two halves of low and high intensity). Telang et al. [33] propose that it be considered a non-anomalous *transitional* region of intermediate values between the low values in the left and the high values in the right. On the other hand, Fig. 4b is clearly anomalous since its left and right neighborhoods contain low and intermediate values, respectively, both contrasting well with the high values in the central circle. Thus, despite both the circles contrasting well with their local neighborhoods separately, the one in (a) does not differ much from the generalized neighborhood since any measure of central tendency (recall response rate in scan statistics) estimated over the objects in the neighborhood would be quite close to the values within the circle.

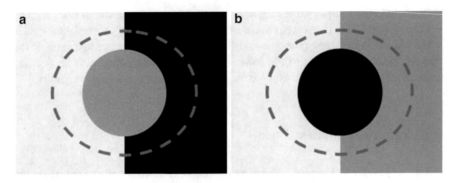

Fig. 4 Generalized neighborhood example (**a**) transition, (**b**) anomaly

The algorithm proceeds in two phases, discovering homogeneous clusters in the first phase, followed by a phase where non-anomalous clusters are filtered out. The cluster detection algorithm starts by initializing all cells as unclustered, followed by picking up each unclustered cell and trying to grow a cluster from them. A set of cells, C, is grown iteratively by greedily adding cells that keep the gini co-efficient,[3] *measured on the value attributes*, as low as possible.

$$C = C \cup \underset{c \in C.neighbors}{\arg\min} \ Gini\{C \cup \{c\}\} \tag{16}$$

Each cluster is grown as long as there are neighbors whose inclusion would retain the gini co-efficient to within a threshold; if no such neighbors exist, the cluster is marked as completed, and the algorithm proceeds to the next unclustered cell. In the second phase, each cluster C is compared against its generalized neighborhood where the neighborhood is defined as a set of cells:

$$N_C = \{d | d \in \mathscr{D} : \exists d' \in C, dist(d, d') \leq \rho\} \tag{17}$$

Informally, the neighborhood of C comprises all cells that are at-most ρ hops away from any cell in C; thus, the neighborhood of a circular cluster would be a ρ-width ring just enclosing the cluster. Conceptually, the neighborhood forms a set of cells that form a sheet of width ρ around the cluster C. Each cluster is then assessed for anomalous-ness using the likelihood ratio test on the value attributes under the following hypotheses:

$$H_0 : \forall d \in C \cup N_C, r_d = p_a \tag{18}$$

$$H_1 : \exists p_0, p_1 \ \text{such that} \tag{19}$$

---

[3]http://en.wikipedia.org/wiki/Gini_coefficient.

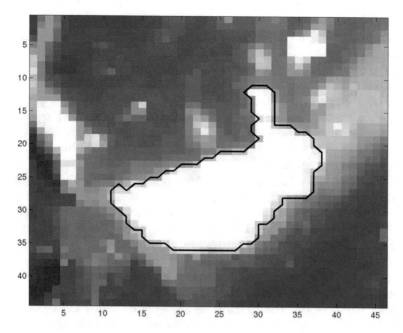

Fig. 5 Example LHA anomaly

$$\forall d \in C, r_d = p_1$$
$$\forall d \in N_C, r_d = p_0$$
$$0 \leq p_0, p_1 \leq 1$$
$$p_0 \neq p_1$$

Basically, H_1 proposes that the response rate is different between the interior and exterior of C whereas H_0 assumes uniform response rate across the dataset. The response rate r_d above may be replaced by the average temperature in scenarios such as climate modeling where the value attribute is the temperature. The likelihood ratio, the number of times the data is likely under H_1 as compared to H_0, is compared against the χ^2 value to a desired statistical significance. All clusters that qualify are output as anomalies, whereas the others are discarded. Figure 5 illustrates, using a black border line, an anomaly detected using the LHA identification approach.

5.2 Image Segmentation

We will now consider the applicability of image segmentation methods towards the problem of identifying region anomalies that contrast with their local neighborhood. Image segmentation is the task of partitioning a digital image into multiple segments, i.e., sets of pixels, such that pixels within each set share certain characteristics. Since the segments are typically expected to stand for objects, pixels within each segment are usually expected to be connected and to have similar colors. To exemplify the connection of such methods to our problem, consider the pixels to be akin to data points in a spatial dataset, where the positioning of the pixel is determined by the spatial attributes and the color of the pixel being determined by the value attributes. In the case of an image, there are typically only two spatial attributes and three value attributes (i.e., RGB values); however, most image segmentation methods can be easily adapted to deal with more or fewer spatial and value attributes.

To illustrate the working of simple methods for image processing, consider the example in Fig. 1 again. It is not hard to see that the simplest possible cluster (i.e., anomaly) membership propagation algorithm that could identify the anomaly could be similar to flood-fill [29]. Under flood-fill, the cluster is grown on all directions, with the growth restricted such that boundaries on value attributes be not crossed. This criteria of stopping the cluster growth upon seeing a significant difference in value attributes at the frontier implicitly ensures that the cluster contrasts with the local neighborhood on the value attributes. In simple painting tools, the boundary would be a cell that has a different color from that of the interior of the cluster; simplistically, image segmentation algorithms such as blob detection may be thought of as a relaxation of the boundary condition of flood-fill (this is indeed an over-simplification). In our case, the boundary detection piece could look for deviation on the value attributes (i.e., temperature in the case of the example in Fig. 1). Advanced image segmentation methods such as blob detection techniques use a variety of methods such as histograms [3], graph partitioning [14, 31], and region growing [11, 27].

Differences from LHA Despite being a locally contrasting method, clusters discovered by image segmentation methods have striking differences from LHA anomalies. LHA anomalies classify regions as anomalies based on whether they differ enough from the *generalized* local neighborhood. On the other hand, image segmentation-style approaches do not differentiate between clusters that differ from the local neighborhood in piecewise fashion and those that differ from the generalized local neighborhood. Thus, the central circle in Fig. 4 that would be discarded as a transition region by LHA would be identified as an anomaly by an image segmentation-style method.

Adaptations Though image segmentation methods are probably the most mature family of techniques from another domain that can handle spatial and value attributes differently and in accordance with the semantics as necessary for our problem, they need to be adapted to be used for anomaly detection meaningfully. We list some possible adaptations:

- **Data Sparsity:** Unlike the case of digital images where every possible combination of (x, y) values would be taken by a pixel, spatial data are often sparse. For example, in the case of disease or weather modeling, we need to model data points corresponding to humans or weather sensors, respectively; intuitively, these objects are far fewer in number as compared to the number of possible combinations of [*lat, long*] values. One simple way of handling this case is to coarsen the space by gridding it so that each grid cell has data points within it, with the value attributes of the grid cell estimated as the mean of the value attributes of the data points within it. The image segmentation technique can then be applied on the dataset comprising cells of the grid. This introduces a parameter, i.e., the width of the grid cell. Devising better ways of transforming sparse data so that image segmentation methods could be applied would be an interesting direction for future work.

- **Anomaly Sizes:** Image segmentation methods, when applied on Fig. 1, could produce the entire white region as a segment. Similarly, they could identify even small regions with very few cells as objects too. There are intuitive reasons to exclude very large as well as very small anomalies. For example, if the candidate anomaly is larger than half of the dataset, it might be argued that the background, and not the region, be considered an anomaly; the 50 % size constraint in scan statistics is motivated by such concerns. A lower bound on the anomaly size might be necessary to avoid a lot of very small and non-noteworthy anomalies in the output.

- **Scoring:** The segments generated out of the image segmentation methods do not have an intuitive scoring. For scenarios where a user, due to time or screen size constraints, is interested in only the top-k anomalies, there needs to be a mechanism to rank the segments based on their anomalous-ness. An intuitive idea would be to score segments based on the difference in value attributes between objects in the segment and those in their immediate local neighborhood.

6 Region Anomalies: Grouping

Clustering is probably the most widely studied task in unsupervised learning. Given that clustering targets to find groups of data that have similar behavior, it is only obvious that clustering techniques be adapted to the problem of finding anomalous, i.e., out-of-the-common behavior. General clustering algorithms target to partition the data into clusters so that the intra-cluster object similarity is maximized while the inter-cluster object similarity is minimized. There have been relatively few efforts to adapting clustering to form clusters in spatial data, i.e., to form clusters as regions that are spatially coherent while differing from other regions based on the value attributes of the component objects. However, using clustering to address the problem of finding outliers in general data, i.e., where the attributes are not differentiated into value and spatial, has been the subject of much study; most efforts have been motivated from use cases in network intrusion detection. In this section, we will look at two classes of techniques and possible adaptations to our problem:

- Clustering methods for spatial data and how they can be adapted to identify anomalies.
- Clustering-inspired anomaly detection methods on general data, and their suitability to anomaly detection on spatial data.

In particular, we do not consider how clustering methods for general data can be used for the task of anomaly detection on spatial data in this section.

6.1 Clustering for Spatial Data

We will now consider techniques for clustering spatial data in a way that the clusters are *spatially contiguous* while being coherent on the value attributes. Thus, these methods use different considerations for spatial and value attributes. While contiguity is typically correlated with coherence (i.e., similarity), these considerations are markedly different in semantics. As an example, no matter how similar an object is, to a cluster, on the value attributes, it cannot be part of the cluster unless it is spatially contiguous with the cluster; no such membership constraint exists on the value attributes. The clustering techniques that we discuss in this section output clusters from the data. Depending on the usage scenario, the output clusters may need to be filtered using constraints on size, homogeneity, or contrast, in order to find a subset of clusters that would be regarded as anomalies. In this section, we consider two methods for clustering spatial data.

6.1.1 HAC-A

This is a simple adaptation of hierarchical agglomerative clustering [36], proposed in [33] as a technique to identify spatially contiguous clusters. Hierarchical agglomerative clustering works by initially assigning all objects to singleton clusters, followed by merging the most similar pair of clusters iteratively, until only the desired number of clusters remains. In HAC-A, the search for pairs of clusters is directed to find only such pairs that are spatially adjacent:

$$PairToMerge = \underset{c_1, c_2 \in Clusters, isAdjacent(c_1, c_2) = true}{\arg\max} Similarity(c_1, c_2) \qquad (20)$$

Similarity is assessed using the value attributes, while the *isAdjacent(., .)* relation is determined using the spatial attributes. This, as may be obvious, is applicable to scenarios where the spatial adjacency relation is well-defined, such as tessellated or gridded data. Extending this to work with datasets of multi-dimensional points such as sensors distributed across a spatial region could lead to interesting future work.

6.1.2 Clustering Ensuring Spatial Convexity

Deepak et al. [8] proposes a method that treats spatial and value attributes differently in clustering. In particular, it exploits the property of the K-Means algorithm [22] in generating clusters with spatially contiguous convex shapes. The method operates in two phases as below:

- **All Attribute Clustering:** In this phase, any clustering algorithm may be used to cluster the data objects across all attributes, without differentiating between spatial and value attributes.
- **K-Means:** The means of the clusters from the first phase are considered, and these are then projected to the spatial attributes alone (i.e., the value attributes are excluded from the cluster means). Now, each object is assigned to the closest mean, similarity measured on just the spatial attributes:

$$cluster(d_i) = \arg\max_{c \in clusters} sim(\pi(d_i), \pi(mean(c))) \qquad (21)$$

where $\pi(.)$ denotes the projection on the spatial attributes.

The second phase ensures that the clusters are spatially contiguous; this is due to the property of the K-Means assignment in generating disjoint and convex shaped clusters. This is in contrast with the first phase clustering that assigns data objects to clusters based on their similarity on all attributes. However, the second phase can significantly alter the cluster memberships from that obtained in the first phase. Evidently, there is room for future work to build a principled technique where convexity and homogeneity conditions are considered together.

6.2 Clustering-Based Anomaly Detection

This section covers methods for clustering-based anomaly detection for general data where attributes are not differentiated as value and spatial attributes. Similar to the structure of Sect. 3.1 on general outlier detection, we will describe methods without considering how they can be adapted to spatial data. We will then comment on how they can be adapted to our problem of anomaly detection on spatial data.

Most techniques for clustering-based anomaly detection have been designed keeping in mind the problem of intrusion detection in networks. It is easy to think of these methods as being instantiations of a two-phase framework as follows:

- **Clustering:** Cluster the data so that the data is partitioned into multiple groups.
- **Anomaly Scoring:** Use statistical measures based on the clustering produced in the first step to identify anomalous groups of data.

The various algorithms for clustering-based anomaly detection differ in the kind of clustering process in the first step, and in the way in which clusters are scored as anomalous in the second step. We will look at a few of them now.

CLAD [23] is an approach for clustering-based anomaly detection. In the first phase, it uses fixed-width clustering to cluster the objects in the dataset to form clusters. CLAD allows an object to be a member of multiple clusters. Fixed-width clustering, which has been quite popular in network intrusion detection, works by traversing the dataset object by object. It starts by setting the first object to the center of a cluster. Every subsequent object would then be added to those existing clusters whose centers are at most w unit of distance away from it; if there are no such clusters, a new cluster is formed with the object in question as the center. As a final phase, CLAD reconsiders every object and adds it to all clusters whose centers are not more than w distance away. In the second phase, every cluster is assigned an inter-cluster distance (ICD) with the ICD for cluster c defined as:

$$ICD_c = \frac{1}{|C| - 1} \sum_{c' \in C, c \neq c'} distance(c, c') \tag{22}$$

where C is the set of clusters and $distance(.,.)$ is a distance function between clusters. CLAD considers clusters that are distant and either sparse or dense, as anomalous. Thus, a cluster c is flagged as anomalous if both the below conditions hold:

- If ICD_c is more than one standard deviation away from the ICDs of clusters in C.
- Number of objects in c is numerically away from the average number of objects across clusters in C by at least one median absolute deviation.

This completes the anomaly detection process in CLAD.

Another approach, proposed in [18], uses an adaptation of K-Means clustering in the first phase. In the second phase, a graph is formed with clusters as nodes and distance between cluster centers acting as the edge weight. A minimal spanning tree is built out of this graph; the longest edge is then removed from the spanning tree, and the clusters in the smaller tree are labeled as outliers. He et al. [17] also uses a two-phase method, where any clustering method is used to identify clusters. This is followed by determining a threshold such that clusters having more and less objects than the threshold are called large and small clusters, respectively. Each object d is then scored as follows:

$$CBLOF(d) = \begin{cases} |d.cluster| \times dist(d, d.cluster) & \text{if } d.cluster \text{ is small} \\ |d.cluster| \times min\{dist(d, l) | l \in large\ clusters\} & \text{if } d.cluster \text{ is large} \end{cases} \tag{23}$$

where $d.cluster$ denotes the cluster to which d belongs. $CBLOF$ functions as an outlier-ness score, and tends to be high for ds that is far away from the center of low-cardinality clusters. Though this score is assigned at an object level, and is hence more close to outlier detection, whole clusters could be marked as anomalous by virtue of their small size. Yet another approach [26] that uses clustering uses a scoring that prefers to label small clusters as anomalous. KD-trees have been used instead of clustering in an anomaly detection method proposed in [6].

6.2.1 Adapting for Anomaly Detection on Spatial Data

Similar to the discussion in Sect. 3.1.3, a straightforward method of using general clustering-based anomaly detection methods for spatial data would be to partition data using value attributes, followed by feeding them separately to the anomaly detection method with just the spatial attributes. Since the techniques targeting network intrusion that we saw in the previous section mostly employ clustering algorithms that produce contiguous convex clusters in the attribute space they work with, using them on data with just spatial attributes would produce spatially contiguous clusters (a desirable property). The partitioning approach, however, could pose more serious issues for region anomaly detection than outlier detection since region anomalies comprise multiple objects. Consider the case of anomalies that straddle the value attributes-based partitioning boundary; these could go undetected under the partitioning-based adaptation. Partitioning data using fuzzy clustering methods could be an interesting direction to pursue to get to a more principled approach to leverage clustering-based anomaly detection for our task.

7 Discussion

We have so far seen many methods for anomaly detection on spatial data. Additionally, we also considered applications of methods proposed for general data (i.e., where attributes are not classified into value and spatial) such as outlier detection and clustering-based anomaly detection. Another potential avenue for consideration is that of usage of clustering algorithms for general data, for anomaly detection in spatial data.

The various clustering approaches differ in how they operate; while density-based clustering algorithms (as well as fixed-width clustering as seen in Sect. 6.2) use a bottom-up style approach by aggregating data objects to form clusters, other popular algorithms like K-Means [22] use a more global strategy by assigning data points to clusters based on their proximity to various cluster models. The former class of methods, especially, DBSCAN [10] and variants such as OPTICS [1] have been shown to be useful for detecting patterns in spatial data. One may recollect that many of the region anomaly methods do not have a very sophisticated approach for determining candidate anomalies; while most spatial scan statistics use a completely data agnostic approach to determine candidates (e.g., for the basic spatial scan statistics, any circular region, regardless of the sparsity or density of data within it, was regarded as a candidate), the LHA method for finding homogeneous regions is applicable only for gridded data. Since we are ultimately interested in characterizing groups of objects, rather than spatial regions, it is intuitive to think of modifying the candidate discovery phase by ideas from density-based clustering algorithms such as DBSCAN. Specifically, one could run DBSCAN on the whole dataset using the spatial attributes only; the clusters generated therein can then used as candidate regions for anomalies which would then be scored by statistical methods

Fig. 6 DBSCAN example
(from http://commons.
wikimedia.org/wiki/File:
DBSCAN-Illustration.svg)

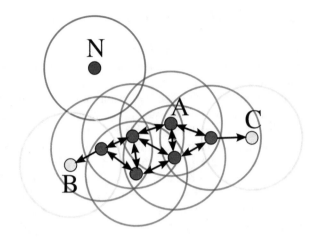

such as those used in SaTScan and LHA. Motivated by this application, we will briefly introduce density-based clustering algorithms in this section, to complete the discussion on anomaly detection for spatial data.

Density-Based Clustering Density-based clustering is a family of algorithms that started with DBSCAN [10]. This algorithm takes in two parameters ϵ and τ, and determines the clustering of a data object based on whether there are at least τ points within an ϵ distance from it. It may be noted that ϵ is a distance threshold applied on the spatial attributes only. The algorithm scans the database for unclustered points, and works by considering such points one by one. Consider the example in Fig. 6 where the search for a cluster starts at the point A. All circles in the figure are of radius ϵ and the example uses $\tau = 3$. Since there are at least 3 points within ϵ distance from A, all data points within ϵ distance are added to A's cluster. All of those points that have at least τ neighbors within an ϵ distance from them are used to further extend the search outward to include more points in the cluster. As shown in the figure, the cluster growth stops after including the points B and C, since neither of them satisfy the ϵ density constraint; in this case, however, there isn't even one new point within ϵ radius of either B and C. The point N in the example, gets labeled as a *noise* point since it does not satisfy the ϵ density constraint nor is it in the neighborhood of a cluster object that satisfies the ϵ density constraint. Since DBSCAN, there have been various extensions, notably OPTICS [1], that serves to identify clusters in data where objects are distributed in clusters of varying density.

8 Directions for Future Work

While the literature on anomaly detection has grown tremendously over the last fifteen years, there are still a lot of interesting problems yet to be addressed in the space of methods to detect anomalous regions. We will outline some potential research directions, with the hope of encouraging more work in this space.

- **Tracking Spatial Anomalies over Time:** Many of the techniques discussed in this chapter have been designed for identifying spatially coherent anomalous regions. A trivial extension of these to cover the time attribute would be to find spans in the time attribute for which these anomalies exist in space. For circular anomalies like those of the basic spatial scan statistic, these could give cylinders. However, such an approach does not allow to capture spatio-temporal anomalies that travel in space with time. For example, a weather anomaly caused by a tornado could travel in space with time, and thus, such a pattern would get split into multiple anomalies each of which have a small time span, under the trivial time extension method. Devising methods to extend spatial anomaly detection to include time in a more sophisticated manner would be an interesting direction for future work. Instead of just a timespan, the time model for an anomaly could be the direction and speed of the movement of the spatial anomaly.
- **Design of Neighborhood Region:** Local neighborhood is defined for local outliers using the set of adjacent cells in tessellated data, or as the k nearest neighbors in multi-dimensional point data. LHA uses a sheet of specified width as the neighborhood, which roughly extends the point neighborhood idea to regions. However in cases where the anomalies are not regularly shaped, this neighborhood may turn out to be non-intuitive. For example, see Fig. 7 where the middle black candidate hotspot is shown along with its sheet neighborhood in grey; note that the colors are used to demarcate the objects, and not to signify value attribute intensity as we have done previously. Consider two portions of the object and the neighborhood, those under *A* and *B*, respectively. Majority of the region covering the neighborhood falls under *A*, whereas majority of the region covering the candidate hotspot falls under *B*. Thus, the comparison approach ends up comparing mostly spatially distant regions while actually intending to

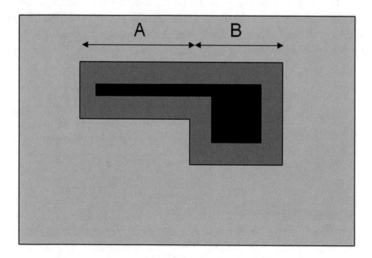

Fig. 7 Issues with local neighborhood definition

compare a region and its neighborhood. This is so since the long left extension of the candidate hotspot brings in a larger region to the neighborhood as compared to its fractional contribution to the anomaly itself. Addressing the neighborhood definition so that it remains intuitive even under arbitrarily shaped anomalies would lead to interesting directions.

- **Anomaly Detection with Varying Local Neighborhoods:** The local neighborhood definition would itself typically involve a parameter; this is typically k (in kNN) for outlier detection methods, and the sheet width ρ for LHA. Instead of fixing the value of such a parameter, it would be interesting to explore anomaly detection where the sheet width is variable. For example, we would then be asking for both an anomaly as well as the neighborhood sheet width under which its anomalous-ness is maximized, as the output.

9 Conclusions

In this chapter, we looked at various methods for anomaly detection. Table 1 summarizes the subset of techniques that were considered, in some detail. We started off by outlining a taxonomy of methods for anomaly detection, focusing on classifying them under features such as the neighborhood used for comparison, and size restrictions for the anomaly. This was followed by a discussion of point-anomaly (i.e., outlier) detection methods covering density based and spatial outliers. We then looked at global anomaly detection methods which is mostly due to the statistics community, where the literature around spatial scan statistics were developed. Our attention was then on discussing methods to find anomalies that contrast well with a local neighborhood. Under this, we saw the LHA method and discussed applicability of image segmentation methods. In the section on grouping-based anomaly detection, we outlined spatial data clustering methods, as well as anomaly detection methods developed for intrusion detection. We then discussed

Table 1 List of techniques for anomaly detection covered in this chapter

Category	Methods
General object anomalies	LOF [4]
Spatial object anomalies	SLOM [7]
Region anomalies—global	Spatial scan statistic [20]
	ULS scan statistic [25]
	Bump hunting [12]
Region anomalies—local	LHA [33]
Region anomalies—grouping	Clustering ensuring spatial convexity [8]
	CLAD [23]
Density-based clustering	DBSCAN [10]

how bottom-up clustering methods, such as those for density-based clustering, could be used to improve anomaly detection. This was followed by an outline of potential research direction in anomaly detection. We hope that this chapter provides the interested reader with a mixture of different perspectives towards the problem of anomaly detection. Apart from encouraging more research in this exciting area, we also hope that this would help accelerate cross-pollination of methods across communities towards extending the frontier in anomaly detection on spatial data.

References

1. Ankerst, M., Breunig, M.M., Kriegel, H.-P., Sander, J.: Optics: ordering points to identify the clustering structure. In: ACM Sigmod Record, vol. 28, pp. 49–60. ACM, New York (1999)
2. Balachandran, V., Deepak, P., Khemani, D.: Interpretable and reconfigurable clustering of document datasets by deriving word-based rules. Knowl. Inf. Syst. **32**(3), 475–503 (2012)
3. Bonnet, N., Cutrona, J., Herbin, M.: A 'no-threshold' histogram-based image segmentation method. Pattern Recogn. **35**(10), 2319–2322 (2002)
4. Breunig, M.M., Kriegel, H.-P., Ng, R.T., Sander, J.: Lof: identifying density-based local outliers. In: ACM Sigmod Record, vol. 29, pp. 93–104. ACM, New York (2000)
5. Celebi, M.E.: Partitional Clustering Algorithms. Springer, New York (2015)
6. Chaudhary, A., Szalay, A.S., Moore, A.W.: Very fast outlier detection in large multidimensional data sets. In: DMKD (2002)
7. Chawla, S., Sun, P.: Slom: a new measure for local spatial outliers. Knowl. Inf. Syst. **9**(4), 412–429 (2006)
8. Deepak, P., Deshpande, P., Visweswariah, K., Telang, A.: System and method for clustering ensuring convexity in subspaces. Prior Art Database (IP.COM) (2013)
9. Duczmal, L., Assuncao, R.: A simulated annealing strategy for the detection of arbitrarily shaped spatial clusters. Comput. Stat. Data Anal. **45**(2), 269–286 (2004)
10. Ester, M., Kriegel, H.-P., Sander, J., Xu, X.: A density-based algorithm for discovering clusters in large spatial databases with noise. In: Kdd, vol. 96, pp. 226–231 (1996)
11. Fan, J., Yau, D.K., Elmagarmid, A.K., Aref, W.G.: Automatic image segmentation by integrating color-edge extraction and seeded region growing. IEEE Trans. Image Process. **10**(10), 1454–1466 (2001)
12. Friedman, J.H., Fisher, N.I.: Bump hunting in high-dimensional data. Stat. Comput. **9**(2), 123–143 (1999)
13. Glaz, J., Pozdnyakov, V., Wallenstein, S.: Scan Statistics: Methods and Applications. Springer, Berlin (2009)
14. Grady, L., Schwartz, E.L.: Isoperimetric graph partitioning for image segmentation. IEEE Trans. Pattern Anal. Mach. Intell. **28**(3), 469–475 (2006)
15. Guha, S., Rastogi, R., Shim, K.: Rock: a robust clustering algorithm for categorical attributes. In: Proceedings of the 15th International Conference on Data Engineering, 1999, pp. 512–521. IEEE, New York (1999)
16. Hawkins, D.M.: Identification of Outliers, vol. 11. Springer, New York (1980)
17. He, Z., Xu, X., Deng, S.: Discovering cluster-based local outliers. Pattern Recogn. Lett. **24**(9), 1641–1650 (2003)
18. Jiang, M.-F., Tseng, S.-S., Su, C.-M.: Two-phase clustering process for outliers detection. Pattern Recogn. Lett. **22**(6), 691–700 (2001)
19. Knox, E.M., Ng, R.T.: Algorithms for mining distance based outliers in large datasets. In: Proceedings of the International Conference on Very Large Data Bases, pp. 392–403. Citeseer, New York (1998)

20. Kulldorff, M.: A spatial scan statistic. Commun. Stat. Theory Methods **26**(6), 1481–1496 (1997)
21. Kulldorff, M., Huang, L., Pickle, L., Duczmal, L.: An elliptic spatial scan statistic. Stat. Med **25**(22), 3929–3943 (2006)
22. MacQueen, J., et al.: Some methods for classification and analysis of multivariate observations. In: Proceedings of the Fifth Berkeley Symposium on Mathematical Statistics and Probability, California, vol. 1, pp. 281–297 (1967)
23. Mahoney, M.V., Chan, P.K., Arshad, M.H.: A machine learning approach to anomaly detection. Technical report, Tech. rep. CS-2003-06, Department of Computer Science, Florida Institute of Technology Melbourne (2003)
24. Neill, D.B., Moore, A.W., Cooper, G.F.: A bayesian spatial scan statistic. Adv. Neural Inf. Process. Syst. **18**, 1003 (2006)
25. Patil, G., Taillie, C.: Upper level set scan statistic for detecting arbitrarily shaped hotspots. Environ. Ecol. Stat. **11**(2), 183–197 (2004)
26. Portnoy L., Eskin E., Stolfo S.: Intrusion detection with unlabeled data using clustering. In Proceedings of ACM CSS Workshop on Data Mining Applied to Security (DMSA-2001). pp. 5–8 (2001)
27. Revol, C., Jourlin, M.: A new minimum variance region growing algorithm for image segmentation. Pattern Recogn. Lett. **18**(3), 249–258 (1997)
28. Schubert, E., Zimek, A., Kriegel, H.-P.: Local outlier detection reconsidered: a generalized view on locality with applications to spatial, video, and network outlier detection. Data Min. Knowl. Disc. **28**(1), 190–237 (2014)
29. Shaw, J.R.: Quickfill: an efficient flood fill algorithm. The Code Project (2004)
30. Shekhar, S., Lu, C.-T., Zhang, P.: Detecting graph-based spatial outliers: algorithms and applications (a summary of results). In: Proceedings of the Seventh ACM SIGKDD International Conference on Knowledge Discovery and Data Mining, pp. 371–376. ACM, New York (2001)
31. Shi, J., Malik, J.: Normalized cuts and image segmentation. IEEE Trans. Pattern Anal. Mach. Intell. **22**(8), 888–905 (2000)
32. Tango, T., Takahashi, K.: A flexibly shaped spatial scan statistic for detecting clusters. Int. J. Health Geogr. **4**(1), 11 (2005)
33. Telang, A., Deepak, P., Joshi, S., Deshpande, P., Rajendran, R.: Detecting localized homogeneous anomalies over spatio-temporal data. Data Min. Knowl. Disc. **28**(5–6), 1480–1502 (2014)
34. Turnbull, B.W., Iwano, E.J., Burnett, W.S., Howe, H.L., Clark, L.C.: Monitoring for clusters of disease: application to leukemia incidence in upstate new york. Am. J. Epidemiol. **132**(Suppl. 1), 136–143 (1990)
35. Yu, D., Sheikholeslami, G., Zhang, A.: Findout: finding outliers in very large datasets. Knowl. Inf. Syst. **4**(4), 387–412 (2002)
36. Zepeda-Mendoza, M.L., Resendis-Antonio, O.: Hierarchical agglomerative clustering. In: Encyclopedia of Systems Biology, pp. 886–887. Springer, Berlin (2013)

Anomaly Ranking in a High Dimensional Space: The UNSUPERVISED TREERANK Algorithm

S. Clémençon, N. Baskiotis, and N. Vayatis

Abstract Ranking unsupervised data in a multivariate feature space $\mathscr{X} \subset \mathbb{R}^d$, $d \geq 1$ by degree of abnormality is of crucial importance in many applications (e.g., fraud surveillance, monitoring of complex systems/infrastructures such as energy networks or aircraft engines, system management in data centers). However, the learning aspect of unsupervised ranking has only received attention in the machine-learning community in the past few years. The Mass-Volume (MV) curve has been recently introduced in order to evaluate the performance of any scoring function $s : \mathscr{X} \to \mathbb{R}$ with regard to its ability to rank unlabeled data. It is expected that relevant scoring functions will induce a preorder similar to that induced by the density function $f(x)$ of the (supposedly continuous) probability distribution of the statistical population under study. As far as we know, there is no efficient algorithm to build a scoring function from (unlabeled) training data with nearly optimal MV curve when the dimension d of the feature space is high. It is the major purpose of this chapter to introduce such an algorithm which we call the UNSUPERVISED TREERANK algorithm. Beyond its description and the statistical analysis of its performance, numerical experiments are exhibited in order to provide empirical evidence of its accuracy.

S. Clémençon (✉)
Institut Mines Telecom, LTCI UMR, Telecom ParisTech & CNRS No. 5141,
46 rue Barrault, 75013 Paris, France
e-mail: stephan.clemencon@telecom-paristech.fr

N. Baskiotis
Université Paris, 6 Pierre et Marie Curie, LIP6 UMR CNRS No. 7606,
Place Jussieu, 75005 Paris, France
e-mail: nicolas.baskiotis@lip6.fr

N. Vayatis
ENS Cachan, CMLA, UMR CNRS No. 8536, Cachan, France
e-mail: nicolas.vayatis@cmla.ens-cachan.fr

© Springer International Publishing Switzerland 2016
M.E. Celebi, K. Aydin (eds.), *Unsupervised Learning Algorithms*,
DOI 10.1007/978-3-319-24211-8_2

1   Introduction

The issue of ranking multivariate data by degree of abnormality, which shall
be referred to as *anomaly ranking* throughout the present chapter, is essential
for a wide variety of applications, ranging from fraud surveillance to distributed
fleet monitoring through system management in data centers (see, e.g., [1–3]).
Anomaly/novelty detection has been the subject of a good deal of attention in the
machine-learning literature these past ten years (see [4–8] or [9] among others). In
most of these papers, unsupervised learning methods are proposed to decide whether
an observation x in the feature space $\mathscr{X}$ lies among the $\alpha\%$ the most abnormal observations $\Omega_\alpha^* \subset \mathscr{X}$. However, the interest in applications is sometimes to learn how
to rank all possible observations $x_1, \dots, x_n$ by degree of abnormality. As pointed out
in [10], the anomaly ranking problem can be viewed as a *continuum* of "imbricated"
anomaly detection problems, requiring to recover from training data the collection
of nested sets $\{\Omega_\alpha^*; \ \alpha \in (0, 1)\}$. Indeed, in contrast to anomaly detection rules,
an anomaly ranking function would permit to prioritize the action check-list and
allows for a progressive examination of the situations predicted as most abnormal.
Although ad-hoc procedures (e.g., [11]) for ranking "outliers" among a statistical
population, either fully data-driven or based on prior expertise, have been proposed
in support of various problems (e.g., predictive maintenance, fraud surveillance),
the issue of measuring the performance of orderings induced by scoring functions
has only been considered in [10], where a Probability-Measure plot, termed Mass-
Volume curve (MV curve in abbreviated form), has been introduced for this specific
purpose. In [12], when the feature space $\mathscr{X}$ is supposed to be compact, the
connection between MV curve minimization and the (supervised) bipartite ranking
related to the discrimination between the distribution of the data observed and the
uniform distribution on $\mathscr{X}$ has been highlighted, offering the possibility, in theory
at least, to turn bipartite ranking algorithms into unsupervised versions based on
a pooled dataset formed by the original observations plus a sample of simulated
data uniformly distributed on $\mathscr{X}$. However, one faces computational difficulties
when trying to implement such an approach when the dimension of the feature
space increases: a simulated data sample of "reasonable" size would be unable
to fill a high dimensional space with enough points, consequently compromising
the supervised learning procedure. Building on the connection between anomaly
ranking and supervised bipartite ranking, the goal of this chapter is to introduce a
novel algorithm for anomaly ranking, referred to as UNSUPERVISED TREERANK.
It can be viewed as an extension of the TREERANK algorithm introduced in [13, 14]
when the LEAFRANK splitting procedure is a cost-sensitive version of the popular
CART algorithm with axis parallel splits, in the specific case where one of the two
distributions is known and coincides with the uniform distribution $\mathscr{U}_{\mathscr{X}}(dx)$ on $\mathscr{X}$.
In this case, the actual computation of the true cell volume (or $\mathscr{U}_{\mathscr{X}}$-measure) along
the recursive partitioning is straightforward and permits to avoid the simulation
stage.

The chapter is organized as follows. Key notions involved in the formulation of the *anomaly ranking* problem (including the form of the decision rules, the criterion to measure accuracy and the related optimal elements), as well as its connection with the (supervised) *bipartite ranking* problem, are recalled in Sect. 2. The UNSUPERVISED TREERANK algorithm to optimize the MV curve is described in Sect. 4. Numerical results empirically assessing the performance of the method promoted are displayed in Sect. 5. Finally, several concluding remarks are collected in Sect. 6.

2 Anomaly Ranking: Background and Preliminaries

We start off with introducing notations that shall be used throughout the chapter and recalling key notions related to the anomaly ranking problem. Throughout the chapter, the Lebesgue measure on $\mathbb{R}^d$ is denoted by λ, the indicator function of any event $\mathscr{E}$ by $\mathbb{I}\{\mathscr{E}\}$ and the generalized inverse of any cumulative distribution function $K(t)$ on $\mathbb{R}$ by $K^{-1}(u) = \inf\{t \in \mathbb{R} : K(t) \geq u\}$.

2.1 A Scoring Approach to Anomaly Ranking

We consider a generic random variable X taking its values in a measurable space $\mathscr{X}$, assumed to be a subset of the Euclidean space $\mathbb{R}^d$, $d \geq 1$, for simplicity. We denote by $F(dx)$ its supposedly continuous probability distribution and by $f(x)$ the related density function. The observations $X_1, \ldots, X_n$, with $n \geq 1$, are modeled as independent copies of the r.v. X. The most convenient way of defining a preorder on the feature space $\mathscr{X}$ is undoubtedly to transport the natural order on the real half-line $\mathbb{R}_+$ onto it by means of a *scoring function*, i.e., a Borel function $s : \mathscr{X} \to \mathbb{R}_+$: given two observations x and x' in $\mathscr{X}$, x will be said as more abnormal according to s than x' when $s(x) \leq s(x')$. We denote by $\mathscr{S}$ the set of all scoring functions on $\mathscr{X}$ that are integrable with respect to Lebesgue measure on $\mathscr{X}$. Notice incidentally that this condition is not restrictive insofar as the preorder induced by any scoring function is invariant by strictly increasing transformation (i.e., the scoring function s and its transform $T \circ s$ define the same preorder on the feature space $\mathscr{X}$ provided that the Borel transformation $T : \text{Im}(s) \to \mathbb{R}_+$ is strictly increasing on the image of the r.v. $s(X)$, which is denoted by $\text{Im}(s)$). An explanation for this integrability constraint is given in the next section.

Informally, the goal pursued is to build from the training dataset $\{X_1, \ldots, X_n\}$ a scoring function s such that, ideally, the smaller $s(X)$, the more abnormal the observation X. Stated this way, the set of optimal scoring rules in $\mathscr{S}$ is the set of strictly increasing transforms of the density function $f(x)$ that are integrable w.r.t. to Lebesgue measure

$$\mathscr{S}^* = \{T \circ f : T : \mathrm{Im}(f) \to \mathbb{R}_+ \text{strictly increasing}, \int_{\mathscr{X}} T \circ f(x) \lambda(dx) < +\infty\}.$$

The next section describes a criterion, whose optimal elements coincides with $\mathscr{S}^*$. It is of *functional* nature (i.e., it takes its values in a path space), which was expected, insofar as we previously pointed out that the anomaly scoring problem could be viewed as a continuum of (nested) anomaly detection problems.

The technical assumptions listed below are involved in the subsequent analysis.

$\mathbf{H_1}$ The r.v. $f(X)$ is continuous, i.e., $\forall c \in \mathbb{R}_+$, $\mathbb{P}\{f(X) = c\} = 0$.

$\mathbf{H_2}$ The density function $f(x)$ is bounded: $||f||_\infty \stackrel{def}{=} \sup_{x \in \mathscr{X}} |f(x)| < +\infty$.

2.2 Measuring Scoring Accuracy: The Mass-Volume Curve

Let $s \in \mathscr{S}$ be an arbitrary scoring function. Its level sets are denoted by $\Omega_{s,t} = \{x \in \mathscr{X} : s(x) \geq t\}$ for all $t \geq 0$. Observe that, since s is supposed to be λ-integrable, the measure $\lambda(\Omega_{s,t}) \leq (\int_{u \in \mathbb{R}_+} s(u)du)/t$ is finite for any $t > 0$. A natural measure of the anomaly ranking performance of a given scoring function $s \in \mathscr{S}$ has been introduced in [10]. It is the Probability-Measure plot, termed Mass-Volume curve (MV curve in short), given by:

$$t > 0 \mapsto (\mathbb{P}\{s(X) \geq t\}, \lambda(\{x \in \mathscr{X} : s(x) \geq t\})) = (F(\Omega_{s,t}), \lambda(\Omega_{s,t})) . \quad (1)$$

Connecting points corresponding to possible jumps, this parametric curve can be seen as the plot of a continuous mapping $\mathrm{MV}_s : \alpha \in (0, 1) \mapsto \mathrm{MV}_s(\alpha)$, starting at $(0, 0)$ and reaching $(1, \lambda(\mathrm{Supp}(F))$ in the case where the support $\mathrm{Supp}(F)$ of the distribution $F(dx)$ is compact, or having the vertical line "$\alpha = 1$" as an asymptote otherwise. See Fig. 1 for a typical MV curve.

Let $\alpha \in (0, 1)$. Denoting by $F_s(t)$ the cumulative distribution function of the r.v. $s(X)$, we have:

$$\mathrm{MV}_s(\alpha) = \lambda\left(\{x \in \mathscr{X} : s(x) \geq F_s^{-1}(1 - \alpha)\}\right), \quad (2)$$

when $F_s \circ F_s^{-1}(\alpha) = \alpha$. This functional criterion is invariant by increasing transform and induces a partial order over the set c. Let $(s_1, s_2) \in \mathscr{S}^2$, the ordering defined by s_1 is said to be more accurate than the one induced by s_2 when

$$\forall \alpha \in (0, 1), \quad \mathrm{MV}_{s_1}(\alpha) \leq \mathrm{MV}_{s_2}(\alpha).$$

As summarized by the result stated below, the MV curve criterion is adequate to measure the accuracy of scoring functions with respect to anomaly ranking. Indeed, we recall that, under the Assumptions $\mathbf{H_1}$–$\mathbf{H_2}$, the set

$$\Omega_\alpha^* \stackrel{def}{=} \{x \in \mathscr{X} : f(x) \geq Q^*(\alpha)\}$$

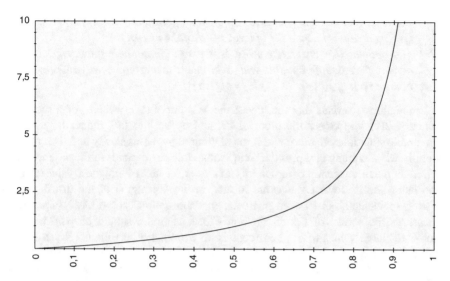

Fig. 1 A typical MV curve (x-axis: mass, y-axis: volume)

is the unique solution of the *minimum volume set* problem

$$\min_{\Omega \in \mathscr{B}(X)} \lambda(\Omega) \, \text{subject to} \, F(\Omega) \geq \alpha, \tag{3}$$

where $\mathscr{B}(\mathscr{X})$ denotes the ensemble made of all Borel subsets of $\mathscr{X}$ and $Q^*(\alpha) = F_f^{-1}(1 - \alpha)$ denotes the quantile of level $1 - \alpha$ of the r.v. $f(X)$. For small values of the mass level α, minimum volume sets are expected to contain the modes of the distribution, whereas their complementary sets correspond to "abnormal observations" when considering large values of α. Refer to [4, 15] for an account of minimum volume set theory and to [16] for related statistical learning results. This implies in particular that optimal scoring functions are those whose MV curves are minimum everywhere, as shown by the following summary of some crucial properties of the optimal MV curve, which plots the (minimum) volume $\lambda(\Omega_\alpha^*)$ against the mass $F(\Omega_\alpha^*) = \alpha$.

Proposition 1 ([10]). *Let the assumptions* $\mathbf{H_1}$–$\mathbf{H_2}$ *be fulfilled.*

(i) *The elements of the class* $\mathscr{S}^*$ *have the same* MV *curve and provide the best possible preorder on* $\mathscr{X}$ *in regard to the* MV *curve criterion:*

$$\forall (s, \alpha) \in \mathscr{S} \times (0, 1), \quad MV^*(\alpha) \leq MV_s(\alpha), \tag{4}$$

where $MV^*(\alpha) = MV_f(\alpha)$ *for all* $\alpha \in (0, 1)$.
(ii) *In addition, we have:* $\forall (s, \alpha) \in \mathscr{S} \times (0, 1)$,

$$0 \leq MV_s(\alpha) - MV^*(\alpha) \leq \lambda \left(\Omega_\alpha^* \Delta \Omega_{s, Q(s, \alpha)} \right), \tag{5}$$

where Δ denotes the symmetric difference operation between two sets, and $Q(s, \alpha)$ denotes the quantile of level $1 - \alpha$ of the r.v. $s(X)$.

(iii) *The optimal* MV *curve is convex. It is also differentiable provided that the density $f(x)$ is differentiable with a gradient taking non zero values on the boundary $\partial \Omega^*_\alpha = \{x \in \mathcal{X} : f(x) = Q^*(\alpha)\}$.*

Equation (5) reveals that the lowest the MV curve (everywhere) of a scoring function $s(x)$, the closer the preorder defined by $s(x)$ to that induced by $f(x)$. Favorable situations are those where the MV curve increases slowly and rises more rapidly when coming closer to the "one" value: this corresponds to the case where $F(dx)$ is much concentrated around its modes, $s(X)$ takes its highest values near the latter and its lowest values are located in the tail region of the distribution $F(dx)$. Incidentally, observe in particular that the optimal curve MV* somehow measures the spread of the distribution $F(dx)$: attention should be paid to the curve for large values of α when focus is on extremal observations (e.g., a light tail behavior corresponds to the situation where MV*(α) increases rapidly when approaching 1), whereas it should be examined for small values of α when modes of the underlying distributions are investigated (a flat curve near 0 indicates a high degree of concentration of $F(dx)$ near its modes).

Given this performance criterion, it becomes possible to develop a statistical theory for the anomaly scoring problem. In a statistical estimation perspective, the goal is to build from training data $X_1, \ldots, X_n$ a scoring function with MV curve as close as possible to MV*. Of course, closeness between (continuous) curves can be measured in many ways. The L_p-distances, $1 \leq p \leq +\infty$ can be used for this purpose for instance. Fix $\epsilon \in (0, 1)$ and consider in particular the losses related to the L_1-distance and to the sup-norm:

$$d_1(s,f) = \int_{\alpha=0}^{1-\epsilon} |\mathrm{MV}_s(\alpha) - \mathrm{MV}^*(\alpha)|\, d\alpha = \int_{\alpha=0}^{1-\epsilon} \mathrm{MV}_s(\alpha)d\alpha - \int_{\alpha=0}^{1-\epsilon} \mathrm{MV}^*(\alpha)d\alpha,$$

$$d_\infty(s,f) = \sup_{\alpha \in [0,1-\epsilon]} |\mathrm{MV}_s(\alpha) - \mathrm{MV}^*(\alpha)| = \sup_{\alpha \in [0,1-\epsilon]} \left(\mathrm{MV}_s(\alpha) - \mathrm{MV}^*(\alpha)\right).$$

We point out that $d_i(s,f)$, $i \in \{1, \infty\}$, is not a distance between the scoring functions s and f but measures the dissimilarity between the preorders they induce. In addition, notice that minimizing $d_1(s,f)$ boils down to minimizing the scalar quantity $\int_{\alpha=0}^{1-\epsilon} \mathrm{MV}_s(\alpha)d\alpha$, the area under the MV curve over $[0, 1-\epsilon]$.

In practice, the MV curve of a scoring function $s \in \mathcal{S}$ is generally unknown, just like the distribution $F(dx)$, and it must be estimated. A natural empirical counterpart can be obtained by plotting the stepwise graph of the mapping

$$\alpha \in (0,1) \mapsto \widehat{\mathrm{MV}}_s(\alpha) \overset{def}{=} \lambda\left(\left\{x \in \mathcal{X} : s(x) \geq \hat{F}_s^{-1}(1-\alpha)\right\}\right), \tag{6}$$

where $\hat{F}_s(t) = (1/n)\sum_{i=1}^n \mathbb{I}\{s(X_i) \leq t\}$ denotes the empirical cdf of the r.v. $s(X)$ and $\hat{F}_s^{-1}$ its generalized inverse. In [10], for a fixed $s \in \mathcal{S}$, consistency and asymptotic Gaussianity (in sup-norm) of the estimator (6) has been established, together with the asymptotic validity of a smoothed bootstrap procedure to build confidence regions in the MV space.

Remark 1 (Volume Estimation). We point out that the exact computation of (empirical) level sets of a scoring function can be numerically unfeasible and approximation schemes (*e.g., Monte Carlo procedures*) must be implemented in practice for this purpose. However, for scoring functions that are constant on the cells of a partition of the feature space $\mathscr{X}$ expressible as union of hypercubes (such as those produced by the UNSUPERVISED TREERANK algorithm, as shall be seen below), such approximation/estimation methods, which are extremely challenging in high dimension, can be avoided.

3 Turning Anomaly Ranking into Bipartite Ranking

We now explain in detail the connection between anomaly ranking and (supervised) bipartite ranking. Motivated by a wide variety of applications ranging from the design of diagnosis support tools in medicine to supervised anomaly detection in signal processing through credit-risk screening in finance, learning to rank data with a feedback provided by binary labels has recently received much attention in the machine-learning literature, see, e.g., [17–19]. Several performance criteria have been considered in order to formulate the issue of ranking observations in an order as close as possible to that induced by the ordinal output variable as a M-estimation problem: the (area under the) receiver operator characteristic curve (ROC curve in short) and the precision-recall curve, the normalized discounted cumulative gain criterion. Many practical ranking algorithms, supported by sound theoretical results extending the probabilistic theory of pattern recognition, have been introduced in the literature, see, e.g., [13, 20–22].

3.1 Bipartite Ranking and ROC Analysis

Let $G(dx)$ and $H(dx)$ be two probability distributions on the feature space $\mathscr{X}$, supposedly absolutely continuous with respect to each other. The ROC curve of any scoring function $s(x)$ is then defined as the Probability–Probability plot $t > 0 \mapsto (1 - H_s(t), 1 - G_s(t))$, where $H_s(dt)$ and $G_s(dt)$ respectively denote the images of the distributions H and G by the mapping $s : \mathscr{X} \to \mathbb{R}_+$. Connecting by convention possible jumps by line segments, the ROC curve of the scoring function $s(x)$ can always be viewed as the plot of a continuous mapping $\text{ROC}_s : \alpha \in (0, 1) \mapsto \text{ROC}_s(\alpha)$. It starts at $(0, 0)$ and ends at $(1, 1)$. At any point $\alpha \in (0, 1)$ such that $H_s \circ H_s^{-1}(\alpha) = \alpha$, we have: $\text{ROC}_s(\alpha) = 1 - G_s \circ H_s^{-1}(1 - \alpha)$. The curve ROC_s measures the capacity of s to discriminate between distributions H

and G. It coincides with the first diagonal when $H_s = G_s$. Observe also that the *stochastically larger* than H_s the distribution G_s, the closer to the left upper corner of the ROC space the curve ROC_s. One may refer to [23] for an account of ROC analysis and its applications.

The concept of ROC curve induces a partial order on $\mathscr{S}$. A scoring function s_1 is more accurate than s_2 iff: $\forall \alpha \in (0, 1)$, $\text{ROC}_{s_1}(\alpha) \geq \text{ROC}_{s_2}(\alpha)$. A Neyman-Pearson argument shows that the optimal ROC curve, denoted by ROC^*, is that of the likelihood ratio statistic $\phi(x) = dG/dH(x)$. It dominates any other ROC curve everywhere: $\forall (s, \alpha) \in \mathscr{S} \times (0, 1)$, $\text{ROC}_s(\alpha) \leq \text{ROC}^*(\alpha)$. The set $\mathscr{S}^*_{H,G} = \{T \circ \phi, \ T : \text{Im}\phi(X) \to \mathbb{R}_+ \text{strictly increasing}\}$ is the set of optimal scoring functions regarding the bipartite problem considered.

The goal of bipartite ranking is to build a scoring function with a ROC curve as high as possible, based on two independent *labeled* datasets: $(X_1^-, \ldots, X_m^-)$ and $(X_1^+, \ldots, X_q^+)$ made of independent realizations of H and G, respectively, with $m, q \geq 1$. Assigning the label $Y = +1$ to observations drawn from $G(dx)$ and label $Y = -1$ to those drawn from $H(dx)$, the objective can be also expressed as to rank/score any pooled set of observations (in absence of label information) so that, ideally, the higher the score of an observation X, the likelier its (hidden) label Y is positive.

The accuracy of any $s \in \mathscr{S}$ can be measured by:

$$D_p(s, s^*) = ||\text{ROC}_s - \text{ROC}^*||_p, \tag{7}$$

where $s^* \in \mathscr{S}^*_{H,G}$ and $p \in [1, +\infty]$. Observe that, in the case $p = 1$, one may write $D_1(s, s^*) = \text{AUC}^* - \text{AUC}(s)$, where $\text{AUC}(s) = \int_{\alpha=0}^1 \text{ROC}_s(\alpha) d\alpha$ is the *Area Under the ROC Curve* (AUC in short) and $\text{AUC}^* = \text{AUC}(\phi)$ is the maximum AUC. Hence, minimizing $D_1(s, s^*)$ boils down to maximizing the ROC summary $\text{AUC}(s)$. The popularity of this quantity arises from the fact it can be interpreted, in a probabilistic manner, as the *rate of concording pairs*

$$\text{AUC}(s) = \mathbb{P}\left\{s(X) < s(X')\right\} + \frac{1}{2}\mathbb{P}\left\{s(X) = s(X')\right\}, \tag{8}$$

where X and X' denote independent r.v.'s defined on the same probability space, drawn from H and G, respectively. An empirical counterpart of $\text{AUC}(s)$ can be straightforwardly derived from (8), paving the way for the implementation of "empirical risk minimization" strategies, see [18].

The algorithms proposed to optimize the AUC criterion or surrogate performance measures are too numerous to be listed in an exhaustive manner. Among methods well-documented in the literature, one may mention in particular the TREERANK method and its variants (see [13, 14, 24]), which relies on recursive AUC maximization, the RankBoost algorithm, which implements a boosting approach tailored for the ranking problem (see [20]), the SVMrank algorithm originally designed for ordinal regression (see [25]) and the RankRLS procedure proposed in [22].

3.2   A Bipartite View of Anomaly Ranking

With the notations of Sect. 3.1, we take $H(dx)$ as the uniform distribution $U(dx)$ on $[0, 1]^d$ and $G(dx)$ as $F(dx)$, the distribution of interest in the *anomaly ranking* problem. It follows immediately from the definitions and properties recalled in Sect. 2 that, for any scoring function $s \in \mathscr{S}$, the curves MV_s and ROC_s are symmetrical with respect to the first diagonal of the unit square $[0, 1]^2$. Hence, as stated in the next result, solving the anomaly ranking problem related to distribution $F(dx)$ is equivalent to solving the bipartite ranking problem related to the pair (U, F).

Theorem 1. *Suppose that assumptions* $\mathbf{H_1}$, $\mathbf{H_2}$ *hold true. Let* $U(dx)$ *be the uniform distribution on* $[0, 1]^d$. *For any* $(s, \alpha) \in \mathscr{S} \times (0, 1)$, *we have:* $ROC_s^{-1}(\alpha) = MV_s(\alpha)$. *We also have* $\mathscr{S}^* = \mathscr{S}_{U,F}^*$, *and*

$$\forall (s, s^*) \in \mathscr{S} \times \mathscr{S}^*, \quad D_p(s, s^*) = d_p(s, s^*),$$

for $1 \leq p \leq +\infty$. *In particular, we have:* $\forall s \in \mathscr{S}$,

$$1 - \int_{\alpha=0}^{1} MV_s(\alpha)d\alpha = \mathbb{P}\{s(W) < s(X)\} + \frac{1}{2}\mathbb{P}\{s(W) < s(X)\},$$

where W *and* X *are independent r.v.'s, drawn from* $U(dx)$ *and* $F(dx)$ *respectively.*

The proof is straightforward, it suffices to observe that $\phi = dG/dH = f$ in this context. Details are left to the reader. Incidentally, we point out that, under the assumptions listed above, the minimal area under the MV curve may be thus interpreted as a measure of dissimilarity between the distribution $F(dx)$ and the uniform distribution on $[0, 1]^d$. The closer $\int_0^1 MV^*(\alpha)d\alpha$ to $1/2$, the more similar to $U(dx)$ the distribution $F(dx)$.

Remark 2 (On the Support Assumption). In general, the support of $F(dx)$ is unknown, just like the distribution itself. However, the argument above remains valid in the case where $Supp(F(dx)) \subset [0, 1]^d$. The sole difference lies in the fact that the curve MV^* then ends at the point of mass-axis coordinate equal to one and volume-axis coordinate $\lambda(Supp(F)) \leq 1$, the corresponding curve ROC^* exhibiting a plateau: it reaches the value one from the false positive rate $\lambda(Supp(F))$. We point out that, when no information about the support is available, one may always carry out the analysis for the conditional distribution given $X \in \mathscr{C}$, $F \mid_{\mathscr{C}} (dx)$, where $\mathscr{C}$ denotes any compact set containing the observations $X_1, \ldots, X_n$. Observe in addition that, when $\mathscr{C} = \{x \in \mathscr{X} : f(x) > t\}$ for $t > 0$, the optimal MV curve related to $F \mid_{\mathscr{C}} (dx)$ coincides with that related to $F(dx)$ on $[0, F(\mathscr{C})]$, as can be immediately seen with the change of parameter $t \rightarrow t/ \int_{\mathscr{C}} f(x)dx$ in (1).

3.3 Extending Bipartite Methods via Uniform Sampling

Now that the connection between anomaly ranking and bipartite ranking has been highlighted, we show how to exploit it to extend efficient algorithms proposed in the supervised framework to MV curve minimization. Learning procedures are based on a training i.i.d. sample $X_1, \dots, X_n$, distributed according to the unknown probability measure $F(dx)$ with compact support, included in $[0, 1]^d$ say.

One may extend the use of any bipartite ranking algorithm $\mathscr{A}$ to the unsupervised context by simulating extra data, uniformly distributed on the unit hypercube, as follows.

ONE-CLASS SCORING VIA UNIFORM SAMPLING

Input: *unlabeled data sample* $\{X_1, \dots, X_n\}$, *bipartite ranking algorithm* $\mathscr{A}$, $m \geq 1$

1. *Sample additional data* $X_1^-, \dots, X_m^-$, *uniformly distributed over* $[0, 1]^d$.
2. *Assign a "negative" label to the sample* $\mathscr{D}_m^- = \{X_1^-, \dots, X_m^-\}$ *and a "positive" label to the original data* $\mathscr{D}_n^+ = \{X_1, \dots, X_n\}$.
3. *Run algorithm* $\mathscr{A}$ *based on the bipartite statistical population* $\mathscr{D}_m^- \cup \mathscr{D}_n^+$, *producing the anomaly scoring function* $s(x)$.

Except the choice of the algorithm $\mathscr{A}$ and the selection of its hyperparameters, the sole tuning parameter which must be set is the size m of the uniformly distributed sample. In practice, it should be chosen as large as possible, depending on the current computational constraints. From a practical perspective, it should be noticed that the computational cost of the sampling stage is reduced. Indeed, the d components of a r.v. uniformly distributed on the hypercube $[0, 1]^d$ being independent and uniformly distributed according to the uniform distribution on the unit interval, the "negative" sample can be thus generated by means of pseudo-random number generators (PRNG's), involving no complex simulation algorithm. Furthermore, uniform distributions on any (Borel) subset of $[0, 1]^d$ can be naturally simulated in a quite similar fashion, with an additional conditioning step.

We point out that, in the context of density estimation, a similar sampling technique for transforming this unsupervised problem into one of supervised function approximation is discussed in Sect. 14.2.4 in [26], where it is used in particular to build *generalized association rules*. This idea is also exploited in [7] for anomaly detection, see also [27]. In this respect, it should be mentioned that a variety of techniques, including that proposed in [6] where the SVM machinery has been extended to the unsupervised framework and now referred to as ONE CLASS SVM, have been proposed to recover the set Ω_α^* for a target mass level $\alpha \in (0, 1)$, fixed in advance. Therefore, even if the estimates produced of are of the form $\{x \in \mathscr{X} : \hat{f}(x) > t_\alpha\}$ and one could consider using the decision function $\hat{f}(x)$ as scoring function, one should keep in mind that there is no statistical guarantee that the ensembles $\{x \in \mathcal{X} : \hat{f}(x) > t\}$ are good estimates of density level sets for $t \neq t_\alpha$. This explains the poor performance of such a "plug-in" approach observed in practice.

4 The UNSUPERVISED TREERANK Algorithm

The major drawback of the simulation-based approach described above arises from the fact that sampling a "dense" dataset representative of the uniform distribution rapidly becomes unfeasible in practice as the dimension d of the space increases. In high dimension, uniformly distributed datasets of "reasonable" sizes will inevitably fill extremely sparsely the feature space, jeopardizing the supervised learning procedure. In contrast, the algorithm promoted in this section is not confronted with this difficulty, since it completely avoids simulation of uniform data. Like other recursive tree building methods, it is implemented in two steps: growing first a possibly overfitted decision tree model and selecting next a submodel with highest expected generalization ability by pruning the original tree.

4.1 Anomaly Ranking Trees

In supervised and unsupervised learning problems, decision trees are undoubtedly among the most popular techniques. One of the main reasons arises from the fact that they straightforwardly offer a visual model summary, taking the form of an easily interpretable binary tree graph, refer to [28, 29] or [30] for instance. Indeed, predictions can be generally described by means of a hierarchical combination of elementary rules of the type "$X^{(j)} \leq \kappa$" or "$X^{(j)} > \kappa$," comparing the value taken by a (quantitative) component of the input vector X (the *split variable*) to a certain threshold (the *split value*). In contrast to (supervised) learning problems such as classification or regression, which are of local nature, predictive rules for a global problem such as *anomaly ranking* cannot be described by a (tree-structured) partition of the feature space: cells (corresponding to the terminal leaves of the binary decision tree) must be ordered so as to define a scoring function. Hence, we define an *anomaly ranking trees* as a binary tree equipped with a "left-to-right" orientation, defining a tree-structured collection of anomaly scoring functions, as depicted by Fig. 2. The root node of an anomaly ranking tree $\mathcal{T}_J$ of depth $J \geq 0$ represents the whole feature space $\mathcal{X}$: $\mathcal{C}_{0,0} = \mathcal{X}$, while each internal node (j, k) with $j < J$ and $k \in \{0, \ldots, 2^j - 1\}$ corresponds to a subset $\mathcal{C}_{j,k} \subset \mathcal{X}$, whose left and right siblings respectively correspond to disjoint subsets $\mathcal{C}_{j+1,2k}$ and $\mathcal{C}_{j+1,2k+1}$ such that $\mathcal{C}_{j,k} = \mathcal{C}_{j+1,2k} \cup \mathcal{C}_{j+1,2k+1}$. Equipped with the left-to-right orientation, any subtree $\mathcal{T} \subset \mathcal{T}_J$ defines a preorder on $\mathcal{X}$: elements lying in the same terminal cell of $\mathcal{T}$ being equally ranked. The anomaly scoring function related to the oriented tree $\mathcal{T}$ can be written as:

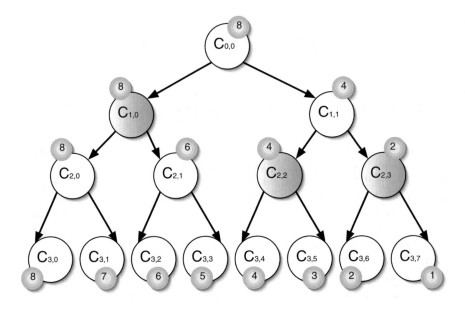

Fig. 2 An anomaly scoring function described by an oriented binary subtree $\mathcal{T}$. For any element $x \in \mathcal{X}$, the quantity $s_{\mathcal{T}}(x)$ can be computed very fast in a top-down manner using the heap structure: starting from the initial value 2^J at the root node, at each internal node $\mathcal{C}_{j,k}$, the score remains unchanged if x moves down to the *left* sibling and one subtracts $2^{J-(j+1)}$ from it if x moves down to the *right*

$$s_{\mathcal{T}}(x) = \sum_{\mathcal{C}_{j,k}:\ \text{terminal leaf of } \mathcal{T}} 2^J \left(1 - \frac{k}{2^j}\right) \cdot \mathbb{I}\{x \in \mathcal{C}_{j,k}\}. \tag{9}$$

Suppose that the feature space $\mathcal{X}$ is compact for simplicity. Then, observe that the MV curve of the anomaly scoring function $s_{\mathcal{T}}(x)$ is the piecewise linear curve connecting the knots:

$$(0, 0) \text{ and } \left(\sum_{l=0}^{k} F(\mathcal{C}_{j,l}), \sum_{l=0}^{k} \lambda(\mathcal{C}_{j,l})\right) \text{ for all terminal leaf } \mathcal{C}_{j,k} \text{ of } \mathcal{T}.$$

A statistical version can be computed by replacing the $F(\mathcal{C}_{m,l})$'s by their empirical counterpart. However, as pointed out in Remark 1, the evaluation of the volume $\lambda(\mathcal{C}_{m,l})$ can be problematic unless the geometry of the cells forming the partition is appropriate (e.g., union of hypercubes).

4.2 The Algorithm: Growing the Anomaly Ranking Tree

The TREERANK algorithm, a bipartite ranking technique optimizing the ROC curve in a recursive fashion, has been introduced in [13] and its properties have been investigated in [14] at length. Its output consists of a tree-structured scoring rule (9) with a ROC curve proved to be nearly optimal under mild assumptions. The growing stage is performed as follows. At the root, one starts with a constant scoring function $s_1(x) = \mathbb{I}\{x \in \mathscr{C}_{0,0}\} \equiv 1$ and after $m = 2^j + k$ iterations, $0 \le k < 2^j$, the current scoring function is

$$s_m(x) = \sum_{l=0}^{2k-1}(m - l) \cdot \mathbb{I}\{x \in \mathscr{C}_{j+1,l}\} + \sum_{l=k}^{2^j-1}(m - k - l) \cdot \mathbb{I}\{x \in \mathscr{C}_{j,l}\}$$

and the cell $\mathscr{C}_{j,k}$ is split in order to form a refined version of the scoring function,

$$s_{m+1}(x) = \sum_{l=0}^{2k}(m - l) \cdot \mathbb{I}\{x \in \mathscr{C}_{j+1,l}\} + \sum_{l=k+1}^{2^j-1}(m - k - l) \cdot \mathbb{I}\{x \in \mathscr{C}_{j,l}\}$$

namely, with maximum (empirical) AUC. Therefore, it happens that this problem boils down to solve a cost-sensitive binary classification problem on the set $\mathscr{C}_{j,k}$, see Sect. 3.3 in [14]. Indeed, one may write the AUC increment as

$$\text{AUC}(s_{m+1}) - \text{AUC}(s_m) = \frac{1}{2}H(\mathscr{C}_{j,k})G(\mathscr{C}_{j,k}) \times (1 - \Lambda(\mathscr{C}_{j+1,2k} \mid \mathscr{C}_{j,k})),$$

where

$$\Lambda(\mathscr{C}_{j+1,2k} \mid \mathscr{C}_{j,k}) \overset{def}{=} G(\mathscr{C}_{j,k} \setminus \mathscr{C}_{j+1,2k})/G(\mathscr{C}_{j,k}) + H(\mathscr{C}_{j+1,2k})/H(\mathscr{C}_{j,k}).$$

Setting $p = G(\mathscr{C}_{j,k})/(H(\mathscr{C}_{j,k}) + G(\mathscr{C}_{j,k}))$, the crucial point of the TREERANK approach is that the quantity $2p(1-p)\Lambda(\mathscr{C}_{j+1,2k} \mid \mathscr{C}_{j,k})$ can be interpreted as the cost-sensitive error of a classifier on $\mathscr{C}_{j,k}$ predicting positive label on $\mathscr{C}_{j+1,2k}$ and negative label on $\mathscr{C}_{j,k} \setminus \mathscr{C}_{j+1,2k}$ with cost p (respectively, $1-p$) assigned to the error consisting in predicting label $+1$ given $Y = -1$ (resp., label -1 given $Y = +1$), balancing thus the two types of error. Hence, at each iteration of the ranking tree growing stage, the TREERANK algorithm calls a *cost-sensitive* binary classification algorithm, termed LEAFRANK, in order to solve a statistical version of the problem above (replacing the theoretical probabilities involved by their empirical counterparts) and split $\mathscr{C}_{j,k}$ into $\mathscr{C}_{j+1,2k}$ and $\mathscr{C}_{j+1,2k+1}$. As described at length in [14], one may use cost-sensitive versions of celebrated binary classification algorithms such as CART or SVM for instance as LEAFRANK procedure, the performance depending on their ability to capture the geometry of the level sets of the likelihood ratio $dG/dH(x)$. The growing stage, which can be interpreted as a statistical version of an adaptive piecewise linear interpolation technique for approximating the optimal ROC curve, is generally followed by a pruning procedure, where children of a same parent node are recursively merged in order to produce a ranking subtree that maximizes an estimate of the AUC criterion, based on cross validation usually, see Sect. 4 in [14]. Under appropriate hypothese, consistency results and rate bounds for the TREERANK method (in the sup norm sense and for the AUC deficit both at the same time) are established in [13, 14]. A detailed experimental study can be found in [31].

As explained in Sect. 3.2, in the anomaly ranking context, the "negative distribution" is $H(dx) = U(dx)$ while $F(dx)$ plays the role of the "positive" distribution. Therefore, in the situation where LEAFRANK is chosen as a cost-sensitive version of the CART algorithm with axis parallel splits (see [28]), all the cells $\mathscr{C}_{j,k}$ can be expressed as union of hypercubes. The exact computation of the volume $U(\mathscr{C}_{j,k})$ is then elementarily feasible, as a function of the threshold values involved in the decision tree describing the split and of the volume of the parent node, as described below and illustrated by Fig. 3. We call this splitting rule UNSUPERVISED LEAFRANK. Suppose that the training data consist of i.i.d. observations $X_1, \ldots, X_n$, copies of the generic r.v. $X = (X^{(1)}, \ldots, X^{(d)})$ with probability distribution $F(dx)$ and taking its values in a compact feature space that can be expressed as union of hypercubes $\prod_{l=1}^{d} [a_{0,0}^{(l)}, b_{0,0}^{(l)}] \subset \mathbb{R}^d$ with $-\infty < a_{0,0} < b_{0,0} < +\infty$. Denoting by $\hat{F}_n = (1/n) \sum_{i=1}^{n} \delta_{X_i}$ the empirical distribution of the training observations, the UNSUPERVISED LEAFRANK algorithm is implemented in three steps as follows.

UNSUPERVISED LEAFRANK

- **Input.** Maximal depth $D \geq 1$ of the tree depicting the splitting rule.
- **Initialization.** Start from the root node $\mathscr{C}_{0,0}$, suppose to be the union of a finite number of disjoint hypercubes $[a_{0,0}^{(1)}, b_{0,0}^{(1)}] \times \cdots \times [a_{0,0}^{(d)}, b_{0,0}^{(d)}]$. The volume of the root cell $\lambda_{0,0} = \lambda(\mathscr{C}_{0,0})$ is obtained by summing the volumes $\prod_{i=1}^{d} (b_{0,0}^{(i)} - a_{0,0}^{(i)})$ of the hypercubes forming it.
- **Iterations** For $j = 1, \ldots, D$ and for $k = 0, \ldots, 2^j - 1$:

1. In a greedy fashion, compute

$$\mathscr{C}_{j+1,2k} = \arg\max_{\mathscr{C}} \left\{ \frac{\lambda(\mathscr{C})}{\lambda(\mathscr{C}_{j,k})} - \frac{\hat{F}_n(\mathscr{C})}{\hat{F}_n(\mathscr{C}_{j,k})} \right\},$$

over subsets $\mathscr{C}$ of $\mathscr{C}_{j,k} = \prod_{i=1}^{d} [a_{j,k}^{(i)}, b_{j,k}^{(i)}]$ of the form $\mathscr{C}_{j,k} \cap \{X^{(l)} \leq s_l\}$ or $\mathscr{C}_{j,k} \cap \{X^{(l)} > s_l\}$ for pairs "split variable—split value" $(X^{(l)}, s_l)$ with $l \in \{1, \ldots, d\}$ and $s_l \in \{X_i^{(l)} : i = 1, \ldots, n\}$. Stop if the maximum is not strictly positive and set $\mathscr{C}_{j+1,2k} = \mathscr{C}_{j,k}$, $\mathscr{C}_{j+1,2k+1} = \emptyset$.

2. If $\mathscr{C}_{j+1,2k} = \mathscr{C}_{j,k} \cap \{X^{(l)} \leq s_l\}$, then set

$$a_{j,k}^{(i)} = a_{j+1,2k}^{(i)} = a_{j+1,2k+1}^{(i)} \text{ and } b_{j,k}^{(i)} = b_{j+1,2k}^{(i)} = b_{j+1,2k+1}^{(i)} \text{ for } i \neq l$$

$$s_l = a_{j+1,2k+1}^{(l)} = b_{j+1,2k}^{(l)} \text{ and } b_{j,k}^{(l)} = b_{j+1,2k+1}^{(l)} \text{ and } a_{j,k}^{(l)} = a_{j+1,2k}^{(l)}$$

And if $\mathscr{C}_{j+1,2k} = \mathscr{C}_{j,k} \cap \{X^{(l)} > s_l\}$, then set

$$a_{j,k}^{(i)} = a_{j+1,2k}^{(i)} = a_{j+1,2k+1}^{(i)} \text{ and } b_{j,k}^{(i)} = b_{j+1,2k}^{(i)} = b_{j+1,2k+1}^{(i)} \text{ for } i \neq l$$

$$s_l = a_{j+1,2k}^{(l)} = b_{j+1,2k+1}^{(l)} \text{ and } b_{j,k}^{(l)} = b_{j+1,2k}^{(l)} \text{ and } a_{j,k}^{(l)} = a_{j+1,2k+1}^{(l)}$$

3. If $X^{(l)}$ denotes the split variable, compute $\lambda_{j+1,2k} \stackrel{def}{=} \lambda(\mathscr{C}_{j+1,2k})$ by summing over the disjoint hypercubes forming $\mathscr{C}_{j,k}$ the quantities

$$\lambda_{j,k} \times \frac{b_{j+1,2k}^{(l)} - a_{j+1,2k}^{(l)}}{b_{j,k}^{(l)} - a_{j,k}^{(l)}}.$$

Then, set $\lambda_{j+1,2k+1} \stackrel{def}{=} \lambda(\mathscr{C}_{j+1,2k+1}) = \lambda_{j,k} - \lambda_{j+1,2k}$.

- **Output** the left node $\mathscr{L} = \cup_k \mathscr{C}_{D,2k}$ and the right node $\mathscr{R} = C_{0,0} \setminus \mathscr{L}$. Compute $\lambda(\mathscr{L}) = \sum_k \lambda_{2^D,2k}$ as well as $\lambda(\mathscr{R}) = \lambda_{0,0} - \lambda(\mathscr{L})$.

Hence, only empirical counterparts of the quantities $F(\mathscr{C})$ for subset $\mathscr{C} \subset [0,1]^d$ candidates, $\hat{F}_n(\mathscr{C}) = (1/n) \sum_{i=1}^{n} \mathbb{I}\{X \in \mathscr{C}\}$, are required to estimate the cost-sensitive classification error and implement the splitting stage (AUC maximization). Hence, this approach does not require to sample any additional data, in contrast to that proposed in Sect. 3.3. This is a key advantage in practice, in contrast to "simulation-based" approaches: for high values of the dimension d, data are expected to lie very sparsely in $[0,1]^d$ and can be then very easily separated from those obtained by sampling a "reasonable" number of uniform observations, leading bipartite ranking algorithms to overfit. Similarly to the supervised case, the UNSUPERVISED TREERANK algorithm corresponds to a statistical version of an adaptive piecewise linear interpolation scheme of the optimal MV curve, see [13]. The growing stage is implemented by calling recursively the UNSUPERVISED LEAFRANK procedure, as follows. Again, rather than translating and rescaling the input vector, we assume that the compact feature space is $[0,1]^d$.

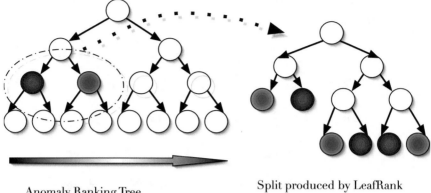

Anomaly Ranking Tree   Split produced by LeafRank

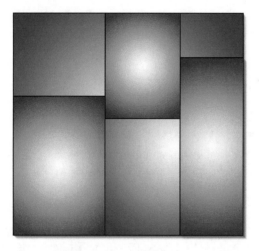

Fig. 3 Anomaly ranking tree produced by the UNSUPERVISED TREERANK algorithm. Ideally, the cells at the *bottom right* form the tail region, while those at the *bottom left* correspond to the modes of the underlying distribution

---

UNSUPERVISED TREERANK—GROWING STAGE

- **Input.** Maximal depth $D \geq 1$ of the anomaly ranking tree, maximal depth D_0 related to the implementation of UNSUPERVISED LEAFRANK. Training data $\mathcal{D}_n = \{X_1, \ldots, X_n\}$
- **Initialization.** Start from the root node $\mathcal{C}_{0,0} = [0, 1]^d$. Set $\lambda_{0,0} = \lambda(\mathcal{C}_{0,0}) = 1$.
- **Iterations** For $j = 1, \ldots, D$ and for $k = 0, \ldots, 2^j - 1$:

1. Considering the input space $\mathscr{C}_{j,k}$, its volume $\lambda_{j,k}$ and the sample composed of observations of the original sample $\mathscr{D}_n$, run the UNSUPERVISED LEAFRANK procedure to split the cell $\mathscr{C}_{j,k}$, yielding the left and right siblings $\mathscr{L}$ and $\mathscr{R}$ with volumes respectively given by $\lambda(\mathscr{L})$ and $\lambda(\mathscr{R})$.

2. Set $\mathscr{C}_{j+1,2k} = \mathscr{L}$, $\lambda_{j+1,2k} = \lambda(\mathscr{L})$, $\mathscr{C}_{j+1,2k+1} = \mathscr{R}$ and $\lambda_{j+1,2k+1} = \lambda(\mathscr{R})$

- **Output** the anomaly ranking tree $\mathscr{T}_D = \{\mathscr{C}_{j,k} = 0 \leq j \leq D, 0 \leq k < 2^j\}$.

4.3 Pruning the Anomaly Ranking Tree: Model Selection

Recursive partitioning techniques fragment the data, splitting rules becoming more and more unstable as the depth increases. The second stage of the UNSUPERVISED TREERANK algorithm consists in choosing an anomaly ranking subtree of that produced by the growing stage. In general, the growing stage is followed by a pruning procedure, where children of a same parent node are recursively merged, until the root of the original tree is reached. The goal pursued here is to select a ranking subtree that minimizes a certain estimator of the area under the MV curve criterion, the area under the empirical MV curve being of course a much too optimistic estimate.

Like for the CART algorithm in classification/regression or for the original TREERANK method (see Sect. 4 in [14]), a natural approach consists in using cross validation, as follows. The idea is to penalize the area under the empirical MV curve of a scoring function related to any anomaly ranking subtree candidate $\mathscr{T} \subset \mathscr{T}_D$, denoted by $\widehat{\mathrm{AMV}}(\mathscr{T})$, by the number of terminal leaves, denoted by $|\mathscr{T}|$, in a linear fashion, yielding the *complexity-penalized empirical area under the MV curve* criterion:

$$\widehat{\mathrm{CPAMV}}_v(\mathscr{T}) \stackrel{def}{=} \widehat{\mathrm{AMV}}(\mathscr{T}) + v \cdot |\mathscr{T}|, \qquad (10)$$

where $v > 0$ is a parameter tuning the trade-off between "goodness-of-fit" and "complexity" (as measured by $|\mathscr{T}|$). For each $v > 0$, define

$$\mathscr{T}_v^* = \arg\min_{\mathscr{T} \subset \mathscr{T}_D} \widehat{\mathrm{CPAMV}}_v(\mathscr{T}).$$

In practice, the $\mathscr{T}_v^*$'s are determined using *weakest link pruning*, i.e., by successively merging leaves of a same parent so as to produce the smallest increase in $\widehat{\mathrm{AMV}}(\mathscr{T})$ in a bottom-up manner, producing a decreasing sequence of anomaly ranking subtrees $\mathscr{T}_D = \mathscr{T}^{(2^D)} \subset \mathscr{T}^{(2^D-1)} \supset \cdots \supset \mathscr{T}^{(1)}$, denoting by $\mathscr{T}^{(1)}$ the anomaly ranking tree reduced to the root. The "best" model $\mathscr{T}_{v*}^*$ is then picked among this finite collection by cross validation.

Interpretation From a practical angle, a crucial advantage of the approach describes above lies in the interpretability of the anomaly ranking rules produced. In contrast to alternative techniques, they can be summarized by means of a left-to-right oriented binary tree graphic: observations are all the more considered as abnormal as they are located in terminal leaves at the right of the *anomaly ranking tree*. An arrow at the bottom of the tree indicates the direction in which the density decreases. Each splitting rule possibly involves the combination of elementary threshold rules of the type "$X^{(k)} > \kappa$" or "$X^{(k)} \leq \kappa$" with $\kappa \in \mathbb{R}$ in a hierarchical manner. In addition, it is also possible to rank the $X^{(k)}$'s depending on their *relative importance*, measured through the empirical *volume under the* MV *curve* decrease induced by splits involving $X^{(k)}$ as *split variable*, just like in the supervised setup, see Sect. 5.1 in [14] for further details. This permits to identify the variables which have most relevance to detect anomalies.

5   Numerical Experiments

We now illustrate the advantages of the UNSUPERVISED TREERANK algorithm by means of numerical experiments, based on unlabeled synthetic/real datasets.

Mixture of 2-d Gaussian Distributions We first display results based on a two-dimensional toy example to compare the performance of the algorithm described in the previous section with that of a simulation-based approach, namely the TREERANK algorithm fed with the original (positive) sample plus a (negative) sample of i.i.d. data uniformly distributed over a square containing the data to be ranked, as proposed in Sect. 3.3. For comparison purpose, the values of the complexity tuning parameters are the same in both cases (maximal depth in LEAFRANK/TREERANK), the positive and negative samples have the same size: $n = m = 2500$.

   Figure 4 respectively depicts the scoring function learnt with UNSUPERVISED TREERANK and Fig. 5 the related MV curves computed via fivefold cross validation. The latter must be compared with its analogue, Fig. 6, corresponding to the performance of the scoring function learnt by means of the simulated dataset. As the results summarized in Table 1 show, the UNSUPERVISED TREERANK algorithm outperforms its supervised version based on a simulated dataset, even in a small-dimension setting.

Real Dataset We have also implemented the UNSUPERVISED TREERANK algorithm to rank the input data of the breast cancer database ($n = 569$), lying in a feature space of dimension $d = 32$, see https://archive.ics.uci.edu/ml/datasets/. The MV curves produced by UNSUPERVISED TREERANK via fivefold cross validation are presented in Fig. 7: the mean of the area under the MV curve criterion is 0.037, while its standard deviation is 0.005. In contrast, in such a high dimensional space, simulation-based approaches completely fail: even if one increases drastically the

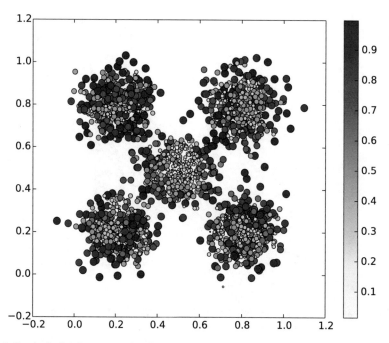

Fig. 4 Synthetic 2-d dataset: scoring function learnt by UNSUPERVISED TREERANK with 7 as ranking tree depth and 7 as LEAFRANK depth

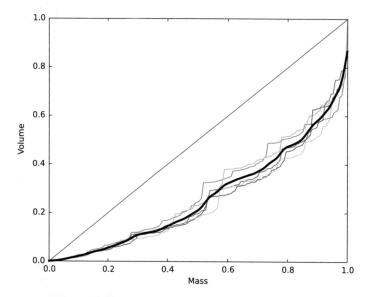

Fig. 5 Synthetic 2-d dataset: MV curves related to the scoring functions obtained via UNSUPER-VISED TREERANK through fivefold cross validation

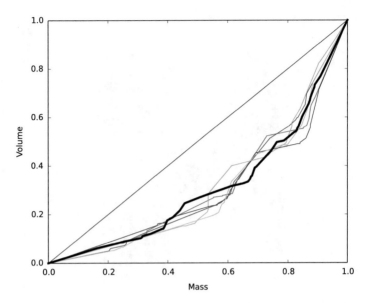

Fig. 6 Synthetic 2-d dataset: MV curves related to the scoring functions obtained via TREERANK and simulation through fivefold cross validation

Table 1 Area under the MV curve estimates: mean and standard error via fivefold cross validation

Method	Mean	Standard error
UNSUPERVISED TREERANK	0.265	0.016
TREERANK and simulation	0.306	0.0075

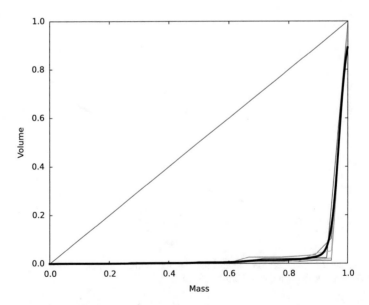

Fig. 7 Real dataset: MV curves related to the scoring functions obtained via UNSUPERVISED TREERANK in a 32-dimensional feature space through fivefold cross validation

number of simulated "negative" instances, supervised technique generally overfit and yield a null area under the MV curve for each replication of the cross validation scheme.

6 Conclusion

In this chapter, we have presented a novel algorithm for unsupervised anomaly ranking, cast as minimization of the Mass-Volume curve criterion, producing scoring rules described by oriented binary trees, referred to as *anomaly ranking trees*. It is called UNSUPERVISED TREERANK. Although it relies on the connection between anomaly ranking and supervised bipartite ranking, its major advantage consists in the fact that it does not involve any sampling stage: since the cells output by this recursive partitioning method can be expressed as unions of hypercubes, the exact computation of their volume is straightforward (avoiding the use of any Monte Carlo sampling scheme). In contrast to alternative techniques, this approach is thus quite tailored to large-scale and high dimensional (unlabeled) data, as empirically confirmed by experimental results. Though it is very promising, there is still room for improvement. Just like for supervised bipartite ranking (see [24]), combining aggregation and randomization, the main ingredients of *ensemble learning* techniques, could lead to dramatically improve stability and accuracy of anomaly ranking tree models both at the same time, while maintaining most of their advantages (e.g., scalability, interpretability). This will be the subject of future research.

References

1. Provost, F., Fawcett, T.: Adaptive fraud detection. Data Min. Knowl. Disc. **1**, 291–316 (1997)
2. Martin, R., Gorinevsky, D., Matthews, B.: Aircraft anomaly detection using performance models trained on fleet data. In: Proceedings of the 2012 Conference on Intelligent Data Understanding (2012)
3. Viswanathan, K., Choudur, L., Talwar, V., Wang, C., Macdonald, G., Satterfield, W.: Ranking anomalies in data centers. In: James, R.D. (ed.) Network Operations and System Management, pp. 79–87. IEEE, New York (2012)
4. Polonik, W.: Minimum volume sets and generalized quantile processes. Stoch. Process. Their Appl. **69**(1), 1–24 (1997)
5. Scott, C., Nowak, R.: Learning Minimum Volume Sets. J. Mach. Learn. Res. **7**, 665–704 (2006)
6. Schölkopf, B., Platt, J.C., Shawe-Taylor, J., Smola, A., Williamson, R.: Estimating the support of a high-dimensional distribution. Neural Comput. **13**(7), 1443–1471 (2001)
7. Steinwart, I., Hush, D., Scovel, C.: A classification framework for anomaly detection. J. Mach. Learn. Res. **6**, 211–232 (2005)
8. Vert, R., Vert, J.-P.: Consistency and convergence rates of one-class svms and related algorithms. J. Mach. Learn. Res. **7**, 817–854 (2006)
9. Park, C., Huang, J.Z., Ding, Y.: A computable plug-in estimator of minimum volume sets for novelty detection. Oper. Res. **58**(5), 1469–1480 (2010)

10. Clémençon, S., Jakubowicz, J.: Scoring anomalies: a M-estimation formulation. In: Proceedings of AISTATS, JMLR W&CP, vol. 31 (2013)
11. Han, J., Jin, W., Tung, A., Wang, W.: Ranking Outliers Using Symmetric Neighborhood Relationship. Lecture Notes in Computer Science, vol. 3918, pp. 148–188. Springer, Berlin (2006)
12. Clémençon, S., Robbiano, S.: Anomaly ranking as supervised bipartite ranking. In: Jebara, T., Xing, E.P. (eds.) Proceedings of the 31st International Conference on Machine Learning (ICML-14), pp. 343–351 (2014)
13. Clémençon, S., Vayatis, N.: Tree-based ranking methods. IEEE Trans. Inf. Theory 55(9), 4316–4336 (2009)
14. Clémençon, S., Depecker, M., Vayatis, N.: Adaptive partitioning schemes for bipartite ranking. Mach. Learn. 43(1), 31–69 (2011)
15. Einmahl, J.H.J., Mason, D.M.: Generalized quantile process. Ann. Stat. 20, 1062–1078 (1992)
16. Scott, C., Nowak, R.: Learning minimum volume sets. J. Mach. Learn. Res. 7, 665–704 (2006)
17. Duchi, J., Mackey, L., Jordan, M.: On the consistency of ranking algorithms. In: Proceedings of the 27th International Conference on Machine Learning (ICML-10) (2010)
18. Clémençon, S., Lugosi, G., Vayatis, N.: Ranking and empirical risk minimization of U-statistics. Ann. Stat. 36(2), 844–874 (2008)
19. Agarwal, S., Graepel, T., Herbrich, R., Har-Peled, S., Roth, D.: Generalization bounds for the area under the ROC curve. J. Mach. Learn. 6, 393–425 (2005)
20. Freund, Y., Iyer, R., Schapire, R., Singer, Y.: An efficient boosting algorithm for combining preferences. J. Mach. Learn. Res. 4, 933–969 (2003)
21. Rakotomamonjy, A.: Optimizing area under Roc curve with SVMs. In: Proceedings of the First Workshop on ROC Analysis in AI (2004)
22. Pahikkala, T., Tsivtsivadze, E., Airola, A., Boberg, J., Salakoski, T.: Learning to rank with pairwise regularized least-squares. In: Proceedings of SIGIR, pp. 27–33 (2007)
23. Fawcett, T.: An introduction to ROC analysis. Lett. Pattern Recogn. 27(8), 861–874 (2006)
24. Clémençon, S., Depecker, M., Vayatis, N.: Ranking forests. J. Mach. Learn. Res. 14, 39–73 (2013)
25. Herbrich, R., Graepel, T., Obermayer, K.: Large margin rank boundaries for ordinal regression. In: Advances in Large Margin Classifiers, pp. 115–132. MIT Press, Cambridge (2000)
26. Friedman, J., Hastie, T., Tibshirani, R.: The Elements of Statistical Learning. Springer, New York (2009)
27. Scott, C., Davenport, M.: Regression level set estimation via cost-sensitive classification. IEEE Trans. Signal Process. 55(6), 2752–2757 (2007)
28. Breiman, L., Friedman, J., Olshen, R., Stone, C.: Classification and Regression Trees. Wadsworth and Brooks, Belmont (1984)
29. Quinlan, J.R.: C4.5: Programs for Machine Learning. Morgan Kaufmann, Los Altos (1993)
30. Rokach, L., Maimon, O.: Data-Mining with Decision Trees: Theory and Applications, 2nd edn. Series in Machine Perception and Artificial Intelligence. World Scientific, Singapore (2014)
31. Clémençon, S., Depecker, M., Vayatis, N.: An empirical comparison of learning algorithms for nonparametric scoring: the treerank algorithm and other methods. Pattern Anal. Appl. 16(4), 475–496 (2013)

Genetic Algorithms for Subset Selection in Model-Based Clustering

Luca Scrucca

Abstract Model-based clustering assumes that the data observed can be represented by a finite mixture model, where each cluster is represented by a parametric distribution. The Gaussian distribution is often employed in the multivariate continuous case. The identification of the subset of relevant clustering variables enables a parsimonious number of unknown parameters to be achieved, thus yielding a more efficient estimate, a clearer interpretation and often improved clustering partitions. This paper discusses variable or feature selection for model-based clustering. Following the approach of Raftery and Dean (J Am Stat Assoc 101(473):168–178, 2006), the problem of subset selection is recast as a model comparison problem, and BIC is used to approximate Bayes factors. The criterion proposed is based on the BIC difference between a candidate clustering model for the given subset and a model which assumes no clustering for the same subset. Thus, the problem amounts to finding the feature subset which maximises such a criterion. A search over the potentially vast solution space is performed using genetic algorithms, which are stochastic search algorithms that use techniques and concepts inspired by evolutionary biology and natural selection. Numerical experiments using real data applications are presented and discussed.

1 Introduction

In the model-based approach to clustering, each cluster is represented by a parametric distribution, and then a finite mixture model is used to model the observed data. Parameters are estimated by optimising the fit, expressed by the likelihood (eventually penalised), between the data and the model. This approach, based on probabilistic models, includes a number of advantages, such as the choice of the number of clusters.

Recently, several authors have argued that the clustering structure of interest may be contained in a subset of available variables. Law et al. [22] proposed a method for obtaining feature saliency, based on the assumption that features are

L. Scrucca (✉)

Department of Economics, Università degli Studi di Perugia, Perugia, Italy

e-mail: luca@stat.unipg.it

© Springer International Publishing Switzerland 2016

M.E. Celebi, K. Aydin (eds.), *Unsupervised Learning Algorithms*,

DOI 10.1007/978-3-319-24211-8_3

conditionally independent given the clustering labels. Raftery and Dean [30] recast the problem of comparing two nested variable subsets as a model selection problem. They used the Bayesian information criterion (BIC) to compare two models, by assuming that the irrelevant variables are regressed on the entire set of relevant variables. Maugis et al. [23] improved on Raftery and Dean's approach by allowing the irrelevant variables to be explained by only a subset of relevant variables. Later, [24] introduced an algorithm leading to a general variable role modelling.

Regardless of the different approaches taken to solve the problem, all the authors agree that selecting the relevant variables enables parsimony of unknown parameters to be achieved, which results in more efficient estimates, a clearer interpretation of the parameters and often improved clustering partitions.

However, the problem of selecting the best subset of clustering variables is a non-trivial exercise, especially when numerous features are available. Therefore, search strategies for model selection are required to explore the vast solution space. The classical method to search for a feasible solution is to adopt a forward/backward stepwise selection strategy. The main criticism of this approach is that stepwise searching rarely finds the best overall model or even the best subset of any size.

This paper discusses a computationally feasible approach based on genetic algorithms (clustGAsel), which address the potentially daunting statistical and combinatorial problems presented by subset selection in model-based clustering via finite mixture modelling.

The reminder of this article is set out as follows. Section 2 reviews the model-based approach to clustering. Section 3 discusses subset selection as a model comparison problem, and introduces a BIC-type criterion to select the relevant clustering variables. A comparison with other existing approaches is also discussed. Section 4 introduces genetic algorithms to search the potentially vast solution space. Genetic operators for a global search over any subset size and for a pre-specified subset size are presented. Section 5 illustrates empirical results based on two real data examples. The final Section provides some concluding remarks and ideas for further improvements.

2   Model-Based Clustering

2.1   Finite Mixture Modelling

Model-based clustering assumes that the observed data are generated from a mixture of G components, each representing the probability distribution for a different group or cluster [14, 26]. The general form of a finite mixture model is $f(x) = \sum_{g=1}^{G} \pi_g f_g(x|\theta_g)$, where π_g represents the mixing probabilities, so that $\pi_g > 0$ and $\sum \pi_g = 1$, $f_g(\cdot)$ and θ_g are the density and the parameters of the g-th component $(g = 1, \ldots, G)$, respectively. With continuous data, we often take the density for each mixture component to be the multivariate Gaussian $\phi(x|\mu_g, \Sigma_g)$ with parameters $\boldsymbol{\theta}_g = (\boldsymbol{\mu}_g, \boldsymbol{\Sigma}_g)$. Thus, clusters are ellipsoidal, centred at the mean vector $\boldsymbol{\mu}_g$, with other geometric features, such as volume, shape and orientation, determined by $\boldsymbol{\Sigma}_g$. Parsimonious parameterisation of covariance matrices can be adopted through eigenvalue decomposition in the form $\boldsymbol{\Sigma}_g = \lambda_g \boldsymbol{D}_g \boldsymbol{A}_g \boldsymbol{D}_g^\top$, where λ_g is a scalar controlling the volume of the ellipsoid, $\boldsymbol{A}_g$ is a diagonal matrix specifying the shape of the density contours, and $\boldsymbol{D}_g$ is an orthogonal matrix which determines the orientation of the corresponding ellipsoid [2, 5]. Fraley et al. [15, Table 1] report some parameterisations of within-group covariance matrices available in the MCLUST software, and the corresponding geometric characteristics. Maximum likelihood estimates for this type of mixture models can be computed via the EM algorithm [11, 25]. A recent survey of finite mixture modelling for clustering is contained in [27].

2.2 BIC as a Criterion for Model Selection

One important aspect of finite mixture modelling is model choice, i.e. the selection of the number of mixture components and the parameterisation of covariance matrices. This is often pursued by using the Bayesian information criterion [BIC; 32]. For a recent review of BIC in model selection see [28].

From a Bayesian point of view, comparisons between models can be based on Bayes factors. Let us consider two candidate models, M_1 and M_2. The Bayes factor is defined as the ratio of the posterior odds to the prior odds:

$$B_{12} = \frac{p(M_1|X)/p(M_2|X)}{p(M_1)/p(M_2)} = \frac{p(X|M_1)}{p(X|M_2)}.$$

Model M_1 is favoured by the data if $B_{12} > 1$, but not if $B_{12} < 1$. When there are unknown parameters, B_{12} is equal to the ratio of the integrated likelihoods, with the integrated likelihood for model M_k (integrated over the model parameters) defined as

$$p(X|M_k) = \int p(X|\boldsymbol{\theta}_k, M_k) p(\boldsymbol{\theta}_k|M_k) d\boldsymbol{\theta}_k, \tag{1}$$

where $p(\boldsymbol{\theta}_k|M_k)$ is the prior distribution of the parameter vector $\boldsymbol{\theta}_k$ of model M_k. The integral in (1) is difficult to evaluate. However, assuming prior unit information, it can be approximated by

$$2\log p(X|M_k) \approx \mathrm{BIC}_k = 2\log p(X|\hat{\boldsymbol{\theta}}_k, M_k) - \nu_k \log(n),$$

where $p(X|\hat{\boldsymbol{\theta}}_k, M_k)$ is the maximised likelihood under model M_k, ν_k is the number of independent parameters to be estimated, and n is the number of observations available in the data. Kass and Raftery [20] showed that BIC provides an approximation to the Bayes factor for comparing two competing models, i.e.

$$2\log B_{12} = 2\log p(X|M_1) - 2\log p(X|M_2) \approx \mathrm{BIC}_1 - \mathrm{BIC}_2.$$

BIC has been widely used for mixture models, both for density estimation [31] and for clustering [13]. Keribin [21] showed that BIC is consistent for choosing the number of components in a mixture model, under the assumption that the likelihood is bounded. This may not be true in general of Gaussian mixture models (GMM), but it does hold if, for instance, the variance is bounded below, a constraint which is imposed in practice in the MCLUST software [15]. The use of BIC for model selection can also be seen as a way to penalise the likelihood based on model complexity. Finally, we note that other criteria have been proposed, such as the integrated complete-data likelihood (ICL) criterion [4].

3 Subset Selection in Model-Based Clustering

Raftery and Dean [30] discussed the problem of variable selection for model-based clustering by recasting the problem as a model selection problem. Their proposal is based on the use of BIC to approximate Bayes factors to compare mixture models fitted on nested subsets of variables. A generalisation of this approach was recently discussed by Maugis et al. [23, 24].

Let us suppose the set of available variables is partitioned into three disjoint parts: the set of previously selected variables, X_1; the set of variables under consideration for inclusion or exclusion from the active set, X_2; and the set of the remaining variables, X_3. Raftery and Dean [30] showed that the inclusion (or exclusion) of variables can be assessed by using the Bayes factor

$$B_{12} = \frac{p(X_2|X_1, M_1)p(X_1|M_1)}{p(X_1, X_2|M_2)}, \tag{2}$$

where $p(\cdot|M_k)$ is the integrated likelihood of model M_k ($k = 1, 2$). Model M_1 specifies that given X_1, X_2 is conditionally independent of the cluster membership, whereas model M_2 specifies that X_2 is relevant for clustering, once X_1 has been included in the model. An important aspect of this formulation is that set X_3 of remaining variables plays no role in (2). Minus twice the logarithm of the Bayes factor in Eq. (2) can be approximated by the following BIC difference:

$$\text{BIC}_{\text{diff}} = \text{BIC}_{\text{clust}}(X_1, X_2) - \text{BIC}_{\text{not clust}}(X_2|X_1), \tag{3}$$

where $\text{BIC}_{\text{clust}}(X_1, X_2)$ is the BIC value for the "best" clustering mixture model fitted using both X_1 and X_2 features, whereas $\text{BIC}_{\text{not clust}}(X_2|X_1)$ is the BIC value for no clustering for the same set of variables. Large, positive values of BIC_{diff} indicate that X_2 variables are relevant for clustering.

Raftery and Dean [30] adopted a stepwise greedy search algorithm to evaluate the inclusion or exclusion of a single feature from the already included set of variables. They show that the second term in the right-hand side of Eq. (3) can be written as

$$BIC_{\text{not clust}}(X_2|X_1) = BIC_{\text{clust}}(X_1) + BIC_{\text{reg}}(X_2|X_1), \tag{4}$$

i.e. the BIC value for the "best" clustering model fitted using X_1 plus the BIC value for the regression of the candidate X_2 variable on the X_1 variables. In all cases, the "best" clustering model is identified with respect to the number of mixture components (assuming $G \geq 2$) and to model parameterisations. In the original proposal, the variable X_2 in the linear regression model term was assumed to depend on all the variables in X_1. However, it may depend only on subset of them, or none (complete independence). To accommodate these situations, a subset selection step in the regression model has been proposed [23]. This approach is implemented in the R package `clustvarsel` [36].

The adopted stepwise algorithm is known to be suboptimal, as there is a risk of finding only a local optimum in the model space. To overcome this drawback and increase the chance of global optimisation, we propose the use of genetic algorithms to search the entire model space. However, moving on from the stepwise perspective, we need to reformulate the problem, as discussed in the next subsection.

3.1 The Proposed Approach

Let us suppose the set of variables previously included for clustering is empty and we want to evaluate the clustering model obtained from the candidate subset X_k of dimension $k \leq p$. The Bayes factor for comparing a candidate clustering model (M_k), fitted using the subset of k variables X_k, against the no clustering model (M_0) on the same set of variables, is given by

$$B_{k0} = \frac{p(X_k|M_k)}{p(X_k|M_0)}.$$

Using the same arguments discussed previously, we can approximate the above ratio by the following BIC difference:

$$BIC_k = BIC_{\text{clust}}(X_k) - BIC_{\text{not clust}}(X_k). \tag{5}$$

Thus, the goal is to maximise the difference between the maximum BIC value from the finite mixture model for clustering, i.e. assuming $G \geq 2$, and the maximum BIC value for no clustering, i.e. assuming $G = 1$, with both models estimated on the candidate subset of variables X_k.

By evaluating the criterion in Eq. (5) for a large number of candidate subsets of different sizes, we may choose the "best" subset as being the one which provides the largest BIC_k. However, the number of possible subsets of size k from a total of p variables is equal to $\binom{p}{k}$. Thus, the space of all possible subsets of size k, ranging from 1 to p, has a number of elements equal to $\sum_{k=1}^{p} \binom{p}{k} = 2^p - 1$. An exhaustive search would not be feasible, even for moderate values of p. Genetic algorithms are a natural candidate for searching this potentially very large model space. In Sect. 4, we discuss the application of genetic algorithms to address this problem. Two strategies are introduced to deal with (a) the general case of subset size not specified a priori, and (b) the case of a fixed subset size.

3.2 Models for No Clustering

When a no clustering model is fitted, an eigen decomposition of the marginal covariance matrix, $\Sigma = \lambda DAD^{\mathsf{T}}$, can be adopted. This yields three possible models: (1) spherical, λI; (2) diagonal, λA; and (3) full covariance. The numbers of estimated parameters are equal to 1, k, and $k(k - 1)/2$, respectively. Although these models include many interesting situations, there are some which are not included, but which may be useful in practical applications. For example, based on the above parameterisation, variables can be either all uncorrelated (case 1 and 2), or all correlated (case 3). However, a variable is often correlated with a subset of other variables, yet at the same time it can be nearly uncorrelated with the remainder. A simple strategy, although computationally demanding in high-dimensional cases, consists of calculating the BIC for each possible configuration of null covariance between pairs of variables. This would allow us to identify only a subset of correlated features for each variable, whereas the remaining features are left uncorrelated.

4 Genetic Algorithms

Genetic algorithms (GAs) are stochastic search algorithms, based on concepts of biological evolution and natural selection [18], which have been applied to find exact or approximate solutions to optimisation and search problems [16, 17].

A GA begins with a set of randomly generated individuals, or solutions, called the population. A fitness or performance is assigned to each individual in the population. The choice of the fitness function depends on the problem at hand. In optimisation problems, it is usually the objective function which needs to be minimised or maximised. A new population is formed by applying specific genetic operators. With the *selection* operator, individuals are selected to form new offspring according to their fitness value. Other genetic operators, such as *crossover* (by exchanging substrings of two individuals to obtain a new offspring) and *mutation* (randomly mutates individual bits), are applied probabilistically to the selected offspring to produce a new population of individuals. The new population is then used in the next iteration of the algorithm.

A GA is thus based on a sequence of cycles of evaluation and genetic operations, which are iterated for many generations. Generally, the overall fitness of the population improves, and better solutions are likely to be selected as the search continues. The algorithm usually terminates when any one of the following conditions is fulfilled: (1) a maximum number of generations has been produced, (2) a satisfactory fitness level has been reached, and (3) the algorithm has achieved a steady state.

Note that a GA is a stochastic search method. There is no guarantee, therefore, that the algorithm will find a solution in a given specific case. However, the algorithm can, on average, be reasonably expected to converge. Another popular stochastic algorithm is simulated annealing (SA), which resembles a GA in that it has one individual in the population, no crossover, and a decreasing mutation rate. SA may find local improvements more efficiently, but it is not efficient in exploring large solution spaces. On the contrary, GAs represent a very effective way of searching through vast, complex solution spaces and are, therefore, more suitable for subset selection [7, 38].

4.1 GAs for Subset Selection in Model-Based Clustering

In this section we describe specific points of our use of GAs for subset selection in model-based clustering.

4.1.1 Genetic Coding Scheme

Each subset of variables is encoded as a string, where each locus in the string is a binary code indicating the presence (1) or absence (0) of a given variable. For example, in a model-based clustering problem with $p = 5$ variables, the string 11001 represents a model where variables 1, 2, and 5 are included, whereas variables 3 and 4 are excluded from the model.

4.1.2 Generation of a Population of Models

The population size N (i.e. the number of models fitted at each generation) is an important parameter of the GA search. A sufficiently large set of models ensures that a large portion of the search space is explored at each step. On the other hand, if there are too many individuals (i.e. models) in the population, GA tends to slow down. After a certain limit, which depends on the encoding and the optimisation problem, increasing population size is not beneficial. Since there are $2^p - 1$ possible subsets for p variables, we set $N = \min(2^p - 1, 50)$ by default.

4.1.3   Fitness Function to Evaluate the Model Clustering

The BIC criterion in (5) is used to assign a fitness value to each model, which corresponds to an individual of the GA population.

4.1.4   Genetic Operators

They are at the core of any implementation of GAs, and typically include the following:

- *Selection*: this operator does not create any new solution. Instead, it selects models for mating, based on their fitness. The basic idea is that solutions with improved fitness have a higher probability of surviving. Thus, relatively good solutions from a population are likely to be selected, whereas it is highly probable that not-so-good solutions will be deleted. We adopted a *linear rank selection* scheme from among the many possible schemes [see 1, Chaps. 22–29] to assign a probability to each model. For a set of models forming the current GA population, let o_i $(i = 1, \ldots, N)$ be the rank obtained from their arrangement in non-increasing order, based on their fitness. Then, a new population is randomly extracted with probability of selection equal to $p_i = 2/N - 2/(N(N-1)) \times (o_i - 1)$. This ensures better models have a higher chance of being included in the next generation, but at the same time it does not impose too much pressure to avoid premature convergence.
- *Crossover*: this operator creates new offspring by means of the recombination of genetic material from their parents. Let p_c be the probability that two random strings (parents) are selected from the mating pool to generate a child. In *single point crossover*, one crossover point is selected, binary values from beginning of the string to the crossover point are copied from one parent, and the remainder are copied from the second parent. By default, we set $p_c = 0.8$.
- *Mutation*: this operator introduces random mutations in the population to ensure the search process can move to other areas of the search space. With probability p_m a randomly selected locus of a string can change from 0 to 1 or from 1 to 0. Thus, a randomly selected variable is either added to or removed from the model. The mutation probability is usually small, and we set $p_m = 0.1$ by default.
- *Elitism*: to improve the performance of GAs, a number of particularly fit individuals may survive through generations. By default the model with the best fitness is retained at each iteration.

4.2   Computational Issues

The computational effort needed by GAs mainly depends on the dimension of the search space, its complexity and the time required to calculate a fitness value for every individual of the population. In our case, this final step involves selecting the "best" clustering model for a given candidate subset of variables, which is a time-consuming process. However, since a portion of the search space has probably already been explored at each iteration, we can avoid re-evaluating the models already estimated, by saving new string solutions and the corresponding fitness values during the iterations. This may save a great deal of computing time. Another possibility would be to use parallel computing on a computer cluster (more on this point is included in the final Section).

4.3 Random-Key GAs to Select a Fixed Size Subset

Let us suppose we are interested in finding the "best" clustering model using a fixed size subset of the most relevant variables, e.g. $k = 3$. The genetic coding scheme discussed in Sect. 4.1 is likely to produce illegitimate solution strings since no constraints are imposed on the size of active variables. Therefore, we must employ a different genetic encoding which ensures GAs only search the space of feasible solutions.

Bean [3] introduced random-key GAs, which guarantee each string corresponds to a feasible encoding of a solution. In practice, random-key GAs indirectly explore the solution space, by searching the space of random keys and using the decoder to evaluate the fitness of the random key. Let $(u_1, \ldots, u_N)$ be a random sequence of N values from $\mathscr{U}(0, 1)$, and $(r_1, \ldots, r_N)$ be the corresponding ranks. Then, the solution string $(s_1, \ldots, s_N)$, with generic element $s_i = I(r_i > N - k)$, where $I(\cdot)$ is the indicator function, guarantees the resulting binary string has exactly k active variables (those for which $s_i = 1$) and $N - k$ non-active variables (those for which $s_i = 0$).

The adoption of the random-key encoding function always produces feasible solutions, but also mating will produces feasible offspring. As a result, the genetic operators for selection and crossover described in the previous section can be directly applied over the space of random-keys. Nevertheless, other genetic operators could also be adopted. In our implementation, for instance, we use *uniform constrained crossover*, which generates a child by selecting at random a value for each locus from the parents, i.e. each bit of the child string is given by uParent$_1 + (1 - u)$Parent$_2$, where $u \in \{0, 1\}$ is drawn at random. Finally, a small modification of the mutation operator is required. A randomly selected locus of a string can be substituted here with a random value extracted from $\mathscr{U}(0, 1)$.

5 Data Examples

In this section, we present some examples based on real data applications. The clustering partition obtained by the subset of selected variables is evaluated by calculating the classification error and the adjusted Rand index [ARI; 19]. The latter is a measure of agreement between two data partitions, whose expected value is zero when two random partitions are compared, and it achieves the maximum value of one when two partitions perfectly coincide.

5.1 Birds, Planes and Cars

Cook and Forzani [8] analysed a dataset from a pilot study to evaluate the possibility of distinguishing birds, planes, and cars on the basis of the sounds they emit. The sample consisted of 58 recordings identified as birds, 43 as cars, and 64 as planes. Each recording was processed and represented by 13 SDMFCCs (Scale-Dependent Mel-Frequency Cepstrum Coefficients). Further details can be found in the previously cited paper.

We focus here on the clustering of records based on the extracted standardised sound characteristics. A GMM fitted on all the available features selected a (VVV,3) model, i.e. a model with three components and different full covariance matrices within clusters. The accuracy was good with an ARI equal to 0.74.

Then, we investigated the possibility of improving the accuracy of the final partition by using a subset of sonic features. Figure 1 shows the GA search path using the BIC_k criterion in (5) as the fitness function. The green filled points represent the best fitness values at each generation, whereas the blue open circle points represent the corresponding fitness averages. Every generation consists of 50 binary strings, each of which represents a candidate feature subset. Other tuning parameters for genetic operators can be read from the output on the right-hand side of the graph. The algorithm soon achieves the optimal solution, and it terminates once there are no improvements in 200 consecutive iterations.

The subset selected by the GA approach has 11 sonic features. The corresponding GMM fitted on this subset has again the (VVV,3) structure. However, by removing two features from the complete set, we improve clustering accuracy, achieving an ARI of 0.84.

To show the clustering obtained with the selected subset, we may project the data using the methodology proposed by Scrucca [33], and recently extended to show the maximal separation of clusters in [35]. The resulting graph is reported in Fig. 2, where we can see that cars appear to be well separated from the other two groups (in fact, no cars are misclassified), but planes and birds overlap and they are clearly more difficult to separate.

5.2 Italian Wines

Forina et al. [12] reported data on 178 wines grown in the same region in Italy, yet obtained from three different cultivars (Barolo, Grignolino, and Barbera). A total of 13 measurements of chemical and physical properties were made for each wine,

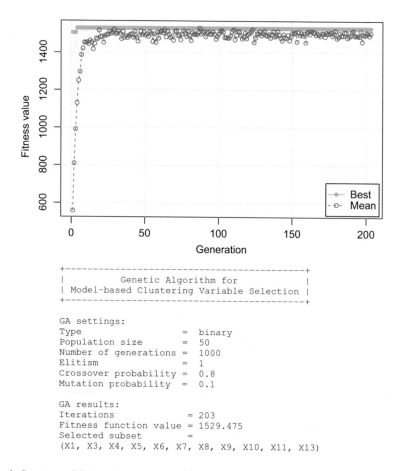

```
+---------------------------------------------------+
|                Genetic Algorithm for              |
| Model-based Clustering Variable Selection          |
+---------------------------------------------------+

GA settings:
Type                  =  binary
Population size       =  50
Number of generations =  1000
Elitism               =  1
Crossover probability =  0.8
Mutation probability  =  0.1

GA results:
Iterations              = 203
Fitness function value  = 1529.475
Selected subset         =
(X1, X3, X4, X5, X6, X7, X8, X9, X10, X11, X13)
```

Fig. 1 Summary of GA search applied to the birds–planes–cars data example

such as the level of alcohol, the level of magnesium, the colour intensity, etc. The dataset is available at the UCI Machine learning data repository http://archive. ics.uci.edu/ml/datasets/Wine.

Often, a preliminary step in cluster analysis is to perform a principal components (PCs) analysis, followed by a clustering algorithm on the features obtained. The first few PCs are usually retained, but, as discussed by Chang [6], these may not contain the most important information for clustering. Here we present an analysis based on random-key GAs, aimed at finding the best subset of PCs.

Table 1 reports the results from the application of random-key GAs to select subsets of PCs of various sizes, where the PCs are obtained from standardised variables. The subset with the largest BIC_k is the one with five PCs: the first two, the 5th, the 6th, and the last PCs. The Gaussian mixture model fitted using the best subset found enables us to correctly estimate the number of components which, in turn, yields a large adjusted Rand index and a small error rate. Another potentially

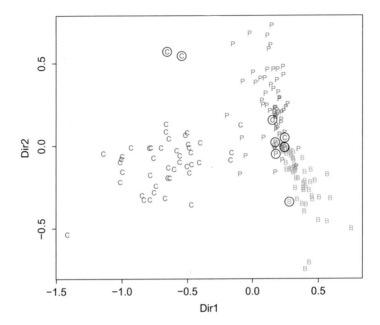

Fig. 2 Plot of data projected along the first two GMMDR directions for the birds–planes–cars data example. Points are marked according to the prevalent cluster membership: P = planes, C = cars, and B = birds. The misclassified points are identified by an *open circle*

Table 1 Results for random-key GAs for the selection of PCs of increasing subset size applied to Italian wine data (large values of BIC are preferable)

Size (k)	PCs subset	BIC_k	Model	G	Error (%)	ARI
1	5	45.04	V	2	60.11	0.0135
2	1 2	173.22	EEV	4	17.42	0.7099
3	1 2 6	200.77	EEV	3	15.73	0.5893
4	1 2 5 7	202.52	EII	7	31.46	0.5536
5	1 2 5 6 13	218.06	EEV	3	5.62	0.8300
6	1 2 3 5 6 13	213.38	VEV	3	1.12	0.9637
7	1 2 3 5 6 7 13	207.03	VEI	6	26.40	0.6722
8	1 2 3 5 6 7 11 13	193.78	VEI	6	26.40	0.6661
9	1 2 3 4 5 7 10 11 13	181.05	VEI	6	32.58	0.6040
10	1 2 3 4 5 6 7 8 10 13	175.27	VEI	5	17.42	0.7602
11	1 2 3 4 5 7 8 9 10 11 13	156.70	VEI	5	17.98	0.7394
12	1 2 3 4 5 6 7 8 10 11 12 13	128.60	VEI	4	16.85	0.7470
13	1 2 3 4 5 6 7 8 9 10 11 12 13	110.92	VEI	4	17.98	0.7372

interesting subset, with a BIC_k value very close to the maximum, is the subset which adds the third PC to the previous features. This enables us to achieve the largest adjusted Rand index and the smallest error rate. Note that, in general, the best

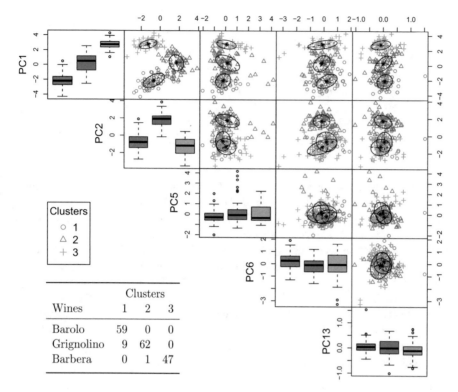

Fig. 3 Scatterplot matrix of PCs selected by random-key GAs with points marked according to the clustering obtained by the estimated GMM

subset of a given size k does not contain the first k PCs, and the selected subsets for increasing sizes do not necessarily form a hierarchy. Figure 3 shows the scatterplot matrix of the selected PCs with points marked in accordance with the predicted clusters partition. The estimated GMM allows us to correctly identify the Barolo wines, and all but one of the Barbera wines. The Grignolino appears to be the most difficult wine to classify, with nine cases assigned to the Barolo cluster.

A comparison with the clustGAsel approach discussed in this paper is reported in Table 2. We considered the original greedy search proposed by Raftery and Dean [30] as implemented in the R package clustvarsel version 1 [9], and the updated version 2 of the package [10] which includes a subset selection procedure also for the regressors [36]. Furthermore, we considered the methods of Maugis et al. [23, 24] as implemented, respectively, in the software SelvarClust and SelvarClustIndep. By looking at the results in Table 2, we can notice that clustvarsel ver. 1 selects all the PCs. This is due to the orthogonality of PCs that leads to overestimate the relevance of each component at the inclusion step. The improved ver. 2 of clustvarsel identifies a subset of five PCs, which are the same components selected by clustGAsel with the exception of the 7th PC that

Table 2 A comparison of subset selection procedures using different algorithms for the identification of relevant PCs for clustering using the Italian wine data

Algorithm	PCs subset	BIC_k	Model	G	Error (%)	ARI
clustvarsel ver. 1	1–13	110.92	VEI	4	17.98	0.7372
clustvarsel ver. 2	1 2 5 7 13	214.08	EEI	6	19.66	0.7279
SelvarClust	1 2 4 5 6 13	199.96	VEV	3	8.99	0.7410
SelvarClustIndep	1 2 5 6 13	207.11	VEV	3	7.30	0.7843
clustGAsel	1 2 5 6 13	218.06	EEV	3	5.62	0.8300

is included instead of the 6th PC. The BIC_k criterion is equal to 214.08, only slightly smaller than the highest value we found (218.06), but the final GMM has six mixture components. This yields an ARI = 0.7279, thus a less accurate clustering partition. The SelvarClust algorithm selected a six-components subset, with the 4th PC added to the best subset found by clustGAsel. The number of clusters is correctly identified, but the final partition is only slightly better than that provided by the model with all the PCs included (ARI = 0.7410). Furthermore, a better subset may be found for the size $k = 6$, as shown in Table 1. Finally, the SelvarClustIndep algorithm selected the same PCs identified by clustGAsel, the same number of clusters, but a different GMM model. The accuracy of the final partition is better than that obtained by previous methods (ARI = 0.7843), but worse than the accuracy provided by clustGAsel (ARI = 0.83).

6 Conclusions

In multivariate datasets, variables are often highly correlated with each other or do not carry much additional information about clustering. The performance of clustering algorithms can be severely affected by the presence of those variables which only serve to increase dimensionality, while adding redundant information. The elimination of such variables can potentially improve both the estimation and the clustering performance. This aspect of model selection has recently been discussed in literature.

In this paper, we have addressed this problem by means of genetic algorithms, using a BIC-type criterion as the fitness function for subset selection in model-based clustering. The criterion adopted generalises the method introduced by Raftery and Dean [30] by allowing a subset of clustering variables to be uncorrelated with the irrelevant features. The same problem was also addressed by Maugis et al. [23, 24] who proposed the use of a different criterion and a stepwise search.

The improvement in computational efficiency of GAs is clearly an area for further research. Considerable savings in elapsed time could be gained by evaluating each proposed model in parallel. A simple form of parallelism could be based on a master process which runs all the genetic operators, with the exception of the evaluation of the fitness function. The calculation of the BIC-type criterion in (5) for each individual of the current GA population would run in parallel, by worker processes operating on separate processors. Results from the worker processes could then be sent to the master, which would collect all the results before continuing with the next generation. There are some important issues in this standard master–slave parallelisation model, such as fault tolerance and reproducibility, which must be taken into account. See [37] for a discussion on these aspects.

The algorithms presented in this paper have been implemented using the GA package [34] for the open source statistical computing environment R [29].

References

1. Back, T., Fogel, D.B., Michalewicz, Z.: Evolutionary Computation 1: Basic Algorithms and Operators. IOP Publishing, Bristol and Philadelphia (2000)
2. Banfield, J., Raftery, A.E.: Model-based Gaussian and non-Gaussian clustering. Biometrics **49**, 803–821 (1993)
3. Bean, J.C.: Genetic algorithms and random keys for sequencing and optimization. ORSA J. Comput. **6**(2), 154–160 (1994)
4. Biernacki, C., Celeux, G., Govaert, G.: Assessing a mixture model for clustering with the integrated completed likelihood. IEEE Trans. Pattern Anal. Mach. Intell. **22**(7), 719–725 (2000)
5. Celeux, G., Govaert, G.: Gaussian parsimonious clustering models. Pattern Recogn. **28**, 781–793 (1995)
6. Chang, W.C.: On using principal components before separating a mixture of two multivariate normal distributions. Appl. Stat. **32**(3), 267–275 (1983)
7. Chatterjee, S., Laudato, M., Lynch, L.A.: Genetic algorithms and their statistical applications: an introduction. Comput. Stat. Data Anal. **22**, 633–651 (1996)
8. Cook, D.R., Forzani, L.: Likelihood-based sufficient dimension reduction. J. Am. Stat. Assoc. **104**(485), 197–208 (2009)
9. Dean, N., Raftery, A.E.: clustvarsel1: variable selection for model-based clustering. (2009). http://CRAN.R-project.org/package=clustvarsel, R package version 1.3
10. Dean, N., Raftery, A.E., Scrucca. L.: clustvarsel: variable selection for model-based clustering. (2014). http://CRAN.R-project.org/package=clustvarsel, R package version 2.1
11. Dempster, A.P., Laird, N.M., Rubin, D.B.: Maximum likelihood from incomplete data via the em algorithm (with discussion). J. R. Stat. Soc. Ser. B Stat. Methodol. **39**, 1–38 (1977)
12. Forina, M., Armanino, C., Castino, M., Ubigli, M.: Multivariate data analysis as a discriminating method of the origin of wines. Vitis **25**, 189–201 (1986). ftp://ftp.ics.uci.edu/pub/machine-learning-databases/wine, wine Recognition Database
13. Fraley, C., Raftery, A.E.: How many clusters? Which clustering method? Answers via model-based cluster analysis. Comput. J. **41**, 578–588 (1998)
14. Fraley, C., Raftery, A.E.: Model-based clustering, discriminant analysis, and density estimation. J. Am. Stat. Assoc. **97**(458), 611–631 (2002)
15. Fraley, C., Raftery, A.E., Murphy, T.B., Scrucca, L.: MCLUST version 4 for R: Normal mixture modeling for model-based clustering, classification, and density estimation. Technical Report 597, Department of Statistics, University of Washington (2012)
16. Goldberg, D.: Genetic Algorithms in Search, Optimization, and Machine Learning. Addison-Wesley Professional, Boston, MA (1989)
17. Haupt, R.L., Haupt, S.E.: Practical Genetic Algorithms, 2nd edn. Wiley, New York (2004)
18. Holland, J.H.: Genetic algorithms. Sci. Am. **267**(1), 66–72 (1992)

19. Hubert, L., Arabie, P.: Comparing partitions. J. Classif. **2**, 193–218 (1985)
20. Kass, R.E., Raftery, A.E.: Bayes factors. J. Am. Stat. Assoc. **90**, 773–795 (1995)
21. Keribin, C.: Consistent estimation of the order of mixture models. Sankhya Ser. A **62**(1), 49–66 (2000)
22. Law, M.H.C., Figueiredo, M.A.T., Jain, A.K.: Simultaneous feature selection and clustering using mixture models. IEEE Trans. Pattern Anal. Mach. Intell. **26**(9), 1154–1166 (2004)
23. Maugis, C., Celeux, G., Martin-Magniette, M.L.: Variable selection for clustering with gaussian mixture models. Biometrics **65**(3), 701–709 (2009)
24. Maugis, C., Celeux, G., Martin-Magniette, M.L.: Variable selection in model-based clustering: a general variable role modeling. Comput. Stat. Data Anal. **53**(11), 3872–3882 (2009)
25. McLachlan, G.J, Krishnan, T.: The EM Algorithm and Extensions, 2nd edn. Wiley, Hoboken, NJ (2008)
26. McLachlan, G.J., Peel, D.: Finite Mixture Models. Wiley, New York (2000)
27. Melnykov, V., Maitra, R.: Finite mixture models and model-based clustering. Stat. Surv. **4**, 80–116 (2010)
28. Neath, A.A., Cavanaugh, J.E.: The Bayesian information criterion: background, derivation, and applications. Wiley Interdiscip. Rev. Comput. Stat. **4**(2),199–203 (2012). doi:10.1002/wics.199
29. R Core Team (2014) R: A Language and Environment for Statistical Computing. R Foundation for Statistical Computing, Vienna, Austria. http://www.R-project.org/
30. Raftery, A.E., Dean, N.: Variable selection for model-based clustering. J. Am. Stat. Assoc. **101**(473), 168–178 (2006)
31. Roeder, K., Wasserman, L.: Practical bayesian density estimation using mixtures of normals. J. Am. Stat. Assoc. **92**(439), 894–902 (1997)
32. Schwartz, G.: Estimating the dimension of a model. Ann. Stat. **6**, 31–38 (1978)
33. Scrucca, L.: Dimension reduction for model-based clustering. Stat. Comput. **20**(4), 471–484 (2010). doi:10.1007/s11222-009-9138-7
34. Scrucca, L.: GA: A package for genetic algorithms in R. J. Stat. Softw. **53**(4), 1–37 (2013). http://www.jstatsoft.org/v53/i04/
35. Scrucca, L.: Graphical tools for model-based mixture discriminant analysis. Adv. Data Anal. Classif. **8**(2), 147–165 (2014)
36. Scrucca, L., Raftery, A.E.: Clustvarsel: A package implementing variable selection for model-based clustering in R http://arxiv.org/abs/1411.0606. J. Stat. Soft. Available at http://arxiv.org/abs/1411.0606 (2014, submitted)
37. Ševčíková, H.: Statistical simulations on parallel computers. J. Comput. Graph. Stat. **13**(4), 886–906 (2004)
38. Winker, P., Gilli, M.: Applications of optimization heuristics to estimation and modelling problems. Comput. Stat. Data Anal. **47**(2), 211–223 (2004)

Clustering Evaluation in High-Dimensional Data

Nenad Tomašev and Miloš Radovanović

Abstract Clustering evaluation plays an important role in unsupervised learning systems, as it is often necessary to automatically quantify the quality of generated cluster configurations. This is especially useful for comparing the performance of different clustering algorithms as well as determining the optimal number of clusters in clustering algorithms that do not estimate it internally. Many clustering quality indexes have been proposed over the years and different indexes are used in different contexts. There is no unifying protocol for clustering evaluation, so it is often unclear which quality index to use in which case. In this chapter, we review the existing clustering quality measures and evaluate them in the challenging context of high-dimensional data clustering. High-dimensional data is sparse and distances tend to concentrate, possibly affecting the applicability of various clustering quality indexes. We analyze the stability and discriminative power of a set of standard clustering quality measures with increasing data dimensionality. Our evaluation shows that the curse of dimensionality affects different clustering quality indexes in different ways and that some are to be preferred when determining clustering quality in many dimensions.

1 Introduction

Unsupervised learning arises frequently in practical machine learning applications and clustering is considered to be one of the most important unsupervised learning tasks. Dividing the data up into clusters helps with uncovering the hidden structure in the data and is a useful step in many data processing pipelines.

N. Tomašev (✉)
Artificial Intelligence Laboratory and Jožef Stefan International Postgraduate School,
Jožef Stefan Institute, Jamova 39, 1000 Ljubljana, Slovenia
e-mail: nenad.tomasev@gmail.com

M. Radovanović
Faculty of Sciences, Department of Mathematics and Informatics,
University of Novi Sad, Trg Dositeja Obradovića 4, 21000 Novi Sad, Serbia
e-mail: radacha@dmi.uns.ac.rs

© Springer International Publishing Switzerland 2016
M.E. Celebi, K. Aydin (eds.), *Unsupervised Learning Algorithms*,
DOI 10.1007/978-3-319-24211-8_4

Clustering is a difficult problem and it is often not clear how to reach an optimal cluster configuration for non-trivial datasets. This problem extends not only to the choice of the clustering algorithm itself, but also to choosing among the generated cluster configurations in case of non-deterministic clustering methods.

In order to be able to compare different cluster configurations and establish preference among them, various clustering quality indexes have been introduced. Each quality index quantifies how well a configuration conforms to some desirable properties.

Internal clustering quality indexes usually measure compactness and separation between the clusters, while the *external* indexes are based on the ground truth about the optimal partition of the data. *Relative* quality indexes are primarily used for choosing the best and most stable output among the repetition of a single clustering method with varying parameters.

The main problem with clustering evaluation is that there is no single preferred clustering quality criterion and it is often not clear which criterion to use for a given application. Different quality indexes should be preferred in different contexts.

High intrinsic data dimensionality is known to be challenging for many standard machine learning methods, including the unsupervised learning approaches. Surprisingly, it has been given comparatively little attention in the context of clustering evaluation.

This chapter gives an overview of the existing approaches for clustering quality evaluation and discusses the implications of high intrinsic data dimensionality for their selection and applicability.

2 Basic Notation

This section introduces the basic notation that will be used throughout the text.

Even though clustering is usually applied to unlabeled data, ground truth information might be available, so the presented definitions include this possibility. Alternatively, these labels might also represent categories in the data that do not correspond to the underlying clusters.

Let $T = \{(x_1, y_1), (x_2, y_2) \ldots (x_N, y_N)\}$ represent the training data given over $X \times Y$, where X is the feature space and Y the finite label space.

A significant portion of the analysis presented in Sect. 7.3 has to do with assessing the influence of prominent neighbor points on clustering quality estimation. Therefore, we introduce the basic notation for k-nearest neighbor sets as follows.

The k-neighborhood of x_i will be denoted by $D_k(x_i) = \{(x_{i1}, y_{i1}), (x_{i2}, y_{i2}) \ldots (x_{ik}, y_{ik})\}$. Any $x \in D_k(x_i)$ is a neighbor of x_i and x_i is a reverse neighbor of any $x \in D_k(x_i)$. An occurrence of an element in some $D_k(x_i)$ is referred to as k-occurrence. $N_k(x)$ denotes the k-occurrence frequency of x. In the supervised case, the occurrences can be further partitioned based on the labels of the reverse neighbor points.

In the context of clustering, a hard partition of the data is defined as a collection $\{C_1, C_2 \ldots C_K\}$ of non-overlapping data subsets where K is the number of clusters. In general, the clusters are required to cover the data, so $\bigcup_{i=1}^{K} C_i = T$. In some cases when outlier removal is done along with the clustering, this condition can be relaxed.

Alternative definitions exist for fuzzy or soft partitioning of the data, as well as overlapping partitioning. The overlap usually arises in subspace clustering methods that allow for the same point to belong to different clusters in different projections [1].

Since each data point is assigned to exactly one cluster under the examined framework, denote the cluster label of x_i by $\hat{y}_i$. Given the ground truth and a clustering of the data, it is possible to define the following quantities:

TP: True positive count, as $\text{TP} = |x_i, x_j : y_i = y_j \wedge \hat{y}_i = \hat{y}_j|$.
FP: False positive count, as $\text{FP} = |x_i, x_j : y_i \neq y_j \wedge \hat{y}_i = \hat{y}_j|$.
TN: True negative count, as $\text{TN} = |x_i, x_j : y_i \neq y_j \wedge \hat{y}_i \neq \hat{y}_j|$.
FN: False negative count, as $\text{FN} = |x_i, x_j : y_i = y_j \wedge \hat{y}_i \neq \hat{y}_j|$.

Many clustering methods and evaluation techniques take into account the *centroid* of each cluster which is obtained by averaging across all dimensions. The centroid of cluster i will be denoted by $\bar{x}_i$. Similarly, the centroid (mean) of the entire sample will be denoted by $\bar{x}$.

3 Problems in Analyzing High-Dimensional Data

The term "Curse of Dimensionality" was first coined by Bellman [8] to denote difficulties that arise when working with high-dimensional data. An increase in data dimensionality induces an increase of the containing volume, which in turn leads to sparsity. Sparse data is difficult to handle, since it becomes very hard to obtain reliable density estimates. The amount of data required to derive statistically sound estimates rises exponentially with the number of dimensions. Even the big datasets for large-scale industrial applications do not usually contain enough data to entirely overcome these issues.

The curse of dimensionality poses new challenges for many types of machine learning and pattern recognition methods, including similarity search [20, 21, 90], classification [73], kernel methods [9, 27], privacy-preserving data preprocessing [4, 5], artificial neural networks [87], and clustering [63].

Some types of learning methods are affected by high embedding data dimensionality, while others only in case of high intrinsic data dimensionality. There are certain data domains, like time-series data, where a high embedding data dimensionality does not necessarily imply a high intrinsic dimensionality, due to correlation between the measurements.

The manifold assumption is often used in practical machine learning approaches, where data is assumed to lie on a lower-dimensional manifold in a high-dimensional space. Restricting the analysis to such manifolds via projections can improve the performance of many methods [74, 92]. The intrinsic dimensionality of the data can be estimated in various ways [17, 29, 37, 62].

There are two phenomena in particular that might be highly relevant for analyzing the behavior of clustering evaluation in intrinsically high-dimensional data: distance concentration and hubness. They are discussed in Sects. 3.1 and 3.2, respectively.

3.1 Distance Concentration

Many clustering evaluation measures take into account either the average or the extremal distances between points that belong to certain clusters in the partitioning.

The concentration of distances in intrinsically high-dimensional data is a highly negative aspect of the curse of dimensionality and a very counterintuitive property of the data [34, 61]. In many dimensions, the relative contrast between the distances calculated on pairs of examples drawn from the same distribution is known to decrease. This makes it hard to distinguish between close and distant points, which is essential in many practical applications.

Let d_M be the maximal observed distance from a fixed query point to other points in the sample and d_m the minimal observed distance. The relative contrast (RC) is then defined as $\rho_d^n = \frac{d_M - d_m}{d_m}$. This definition can be extended to cover all pairs of points. With increasing data dimensionality, relative contrast vanishes, $\lim_{d \to \infty} \rho_d^n = 0$. Distance concentration usually stems from the fact that the expected value for the distance increases, while the variance remains constant, or shrinks.

Distance concentration has an impact on instance-based learning and k-nearest neighbor methods in particular. The difference between nearest and farthest neighbors can appear to vanish in intrinsically high-dimensional data, therefore the very notion of nearest neighbors in many dimensions has been questioned [11, 26, 39]. However, data that follows multiple distributions is less prone to the negative effects of distance concentration, so the nearest neighbor methods remain potentially useful in high dimensions, and can even benefit from it [10, 94].

Establishing the exact conditions for stability of distance functions in high-dimensional data is non-trivial [43, 49]. Many standard metrics are known to be affected by severe distance concentration. Fractional distances are somewhat less susceptible to concentration, though it still occurs [34]. Redesigning metrics for high-dimensional data analysis might lead to certain improvements [3]. Secondary distances have shown promising results in practical applications, including the shared-neighbor distances [42, 46, 91], local scaling, NICDM, and global scaling (mutual proximity) [71].

Not all learning methods are equally susceptible to the distance concentration phenomenon [48]. As for clustering in particular, many quality indexes that are used for clustering evaluation rely on explicit or implicit contrasts over radii or distance sums. While it might be non-trivial to express their dependency on data dimensionality explicitly in closed-form rules, it is reasonable to assume that they would be at least somewhat affected by distance concentration when it occurs.

3.2 Hubness: The Long Tail of Relevance and the Central Tendencies of Hubs

Hubness is a common property of intrinsically high-dimensional data that has recently been shown to play an important role in clustering.

It denotes a tendency of the data to give rise to hubs in the k-nearest neighbor graph as exceedingly frequent and influential nearest neighbors. With increasing dimensionality, the resulting distribution of the neighbor occurrence frequency becomes increasingly skewed. An illustrative example is shown in Fig. 1.

The neighbor occurrence frequency N_k is by convention used to quantify the hubness degree of individual data points in the sample. The hubness of the entire sample is then defined as the skewness of the neighbor occurrence frequency distribution, as per the following formula:

$$SN_k(x) = \frac{m_3(N_k(x))}{m_2^{3/2}(N_k(x))} = \frac{1/N \sum_{i=1}^{N} (N_k(x_i) - k)^3}{(1/N \sum_{i=1}^{N} (N_k(x_i) - k)^2)^{3/2}} \tag{1}$$

Substantial hubness has been shown to arise in most data domains of interest for practical information systems, including text, images, audio signals, and time series.

Hubness is often detrimental to data analysis and can interfere with many standard types of machine learning methods [64–66]. However, it turns out that it is possible to design hubness-aware learning methods that improve on the performance of standard algorithms simply by taking the neighbor occurrence

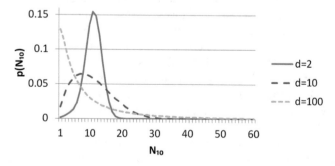

Fig. 1 The change in the distribution shape of 10-occurrence frequency (N_{10}) in i.i.d. Gaussian data with increasing dimensionality when using the Euclidean distance, averaged over 50 randomly generated data sets [78]

Fig. 2 Probability density
function of observing a point
at distance r from the mean of
a multivariate d-dimensional
normal distribution, for
$d = 1, 3, 20, 100$ [65]

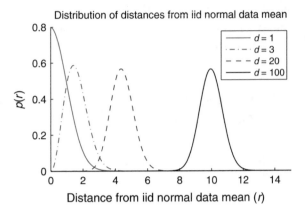

frequency distribution explicitly into account by learning training data neighbor
occurrence models. Hubness-aware methods have been proposed for data reduction [15], ranking [80], representation learning [82], metric learning [71, 72, 76, 79],
classification [77, 78, 81], outlier detection [67], and clustering [83].

Points in certain regions of the data space are more likely to become hubs
than others, under standard distance measures. This can be related to the distance
concentration. The concentration causes points to lie approximately on hyperspheres around cluster means. Non-negligible variance ensures that some points
remain closer to the means than others [34, 35] and these points tend to be closer
on average to the remaining points in the sample as well. While this might hold
for any particular dimensionality, the phenomenon is amplified in high-dimensional
data and the central points have a much higher probability of being included in k-nearest neighbor sets of other points in the data. This line of reasoning suggests that
hubness might be closely linked to centrality in such cases.

As an illustration, consider Fig. 2. It depicts the distribution of Euclidean
distances of all points to the true data mean, for i.i.d. Gaussian data, for different
d values. The distribution of distances is the Chi distribution with d degrees
of freedom [65]. As previously mentioned, the variance of distance distributions
is asymptotically constant with respect to increasing d, unlike the means that
asymptotically behave like $\sqrt{d}$ [34, 65]. Their ratio tends to 0 as $d \to \infty$, which
reaffirms the hyper-sphere view of high-dimensional data.

Figure 3 further illustrates why hubs are relevant for clustering and why hubness
can be used as a measure of local cluster centrality in high-dimensional data.
It shows the interaction between norm, hubness, and density estimates in synthetic
Gaussian data.

Density estimates represent a reasonable way to estimate cluster centrality and
interior regions close to compact cluster centers tend to have a higher density
than other regions of the data space, as shown in Fig. 3a. However, this natural
interpretation of density no longer holds in high-dimensional data (Fig. 3b) and it
is no longer possible to rely on density as the primary indicator of centrality.

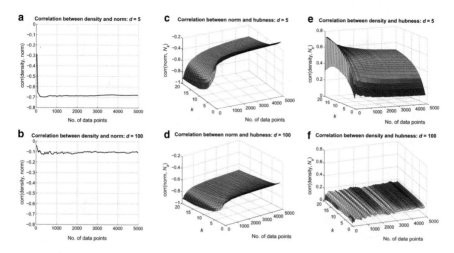

Fig. 3 Interaction between norm, hubness, and density in the simulated setting, in low- and high-dimensional scenarios. (**a**) Correlation between density and norm for $d = 5$. (**b**) Correlation between density and norm for $d = 100$. (**c**) Correlation between norm and hubness for $d = 5$. (**d**) Correlation between norm and hubness for $d = 100$. (**e**) Correlation between density and hubness for $d = 5$. (**f**) Correlation between density and hubness for $d = 100$ [83]

Unlike density, hubness exhibits the opposite trend. It is not highly correlated with centrality in low-dimensional data (Fig. 3c), but an increase in dimensionality gives rise to a very strong correlation between the two (Fig. 3d), so it becomes possible to use hubness as a centrality indicator.

Figure 4 compares the major cluster hubs to medoids and other points in the same simulated setting, based on their pairwise distances and distances to the distribution mean. It can be seen in Fig. 4a, b that in high dimensions the hub is equally informative about the location of the cluster center as the medoid, while in low dimensions the hub and medoid are unrelated. However, this does not imply that hubs and medoids then correspond to the same data points, as Fig. 4c, d shows that the distances from hubs to medoids remain non-negligible. This is also indicated in Fig. 4e, f that shows the ratio between hub to medoid distance and average pairwise distance. In addition, Fig. 4f suggests that in high dimensions hubs and medoids become relatively closer to each other.

Due to its properties, hubness can be exploited for high-dimensional data clustering [84].

In this chapter we have demonstrated that it can also be used as a tool to analyze the behavior of certain clustering quality indexes.

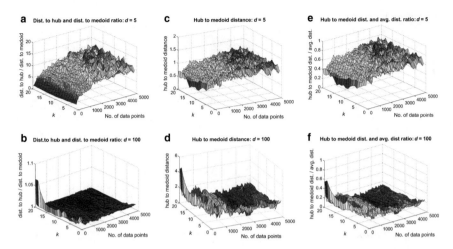

Fig. 4 Interaction between hubs, medoids, and other points in the simulated setting, expressed through distances, in low- and high-dimensional scenarios. (**a**) Ratio between distance to hub and distance to medoid for $d = 5$. (**b**) Ratio between distance to hub and distance to medoid for $d = 100$. (**c**) Hub to medoid distance for $d = 5$. (**d**) Hub to medoid distance for $d = 100$. (**e**) Ratio between hub to medoid distance and average pairwise distance for $d = 5$. (**f**) Ratio between hub to medoid distance and average pairwise distance for $d = 100$ [83]

4   Clustering Techniques for High-Dimensional Data

Clustering high-dimensional data can be challenging and many different techniques have been proposed to this end.

The manifold assumption is often used for subspace clustering of text documents [47, 50], images, and data streams [2, 59]. It is not always necessary to explicitly form a mapping to a feature subspace. The process can be simulated by iterative or non-iterative feature weighting that affects the influence of different features on the proximity measures [6, 19, 53]. Increasing the weights of low-variance dimensions is a common approach [12].

Standard clustering methods designed for low-dimensional data do not perform well in many dimensions and model-based approaches are often over-parametrized [14]. However, they can be applied in conjunction with subspace methods, in order to perform partitioning in lower-dimensional feature subspaces [47]. Different types of hybrid approaches can be used in subspace clustering methods, including density-based techniques, K-means and its extensions and decision trees [51].

Despite that, subspace clustering methods should not be viewed simply as a variation of standard methods with a different notion of similarity that is based on a particular subspace, as different clusters might lie in different subspaces of the data.

Common alternatives to subspace clustering include approximate expectation maximization (EM) [23], spectral clustering [89], shared-neighbor methods [46, 91, 93] and relevant set correlation [41, 88], and clustering ensembles [31, 32].

Hubness-based clustering has recently been proposed for high-dimensional clustering problems [83, 84] and has been successfully applied in some domains like document clustering [40].

Careful data preprocessing and feature selection in particular are often an integral part of the clustering process [13, 28].

5 Clustering Quality Indexes: An Overview

Clustering evaluation is a complex task and many different approaches have been proposed over the years. Most approaches incorporate measures of compactness and separation between the proposed clusters. In this chapter we examine 18 different clustering quality indexes.

Silhouette index is a very well-known clustering evaluation approach that introduces clustering quality scores for each individual point and calculates the final quality index as an average of the point-wise quality estimates [69]. Each point-wise estimate for a point $x_p \in C_i$ is derived from two quantities: $a_{i,p}$ and $b_{i,p}$ which correspond to the average distance to other points within the same cluster and the minimal average distance to points from a different cluster, respectively. Formally, $a_{i,p} = \frac{1}{|C_i|-1} \sum_{x_q \in C_i, q \neq p} \| x_q - x_p \|$ and $b_{i,p} = \min_{j \in \{1...K\}, i \neq j} \frac{1}{|C_j|} \sum_{x_q \in C_j} \| x_q - x_p \|$.

$$ \text{SIL}(x_p) = \frac{a_{i,p} - b_{i,p}}{\max a_{i,p}, b_{i,p}} \qquad (2) $$

$$ \text{SIL} = \frac{1}{N} \sum_{p=1}^{N} \text{SIL}(x_p) \qquad (3) $$

The time complexity of the standard Silhouette criterion is $O(dN^2)$, making it difficult to scale to large datasets.

Simplified Silhouette index is an approximation of the original Silhouette coefficient that computes the intra-cluster and inter-cluster distances as distances to the respective cluster centroids [85], thus achieving a significant speed-up and an overall time complexity of $O(dNK)$.

Dunn index is a simple index that has been frequently used in the past. It is defined as a ratio between the minimal between-cluster distance and the maximal cluster diameter [25].

$$DN = \min_{i,j \in \{1...K\}, i \neq j} \left(\frac{\min_{x_p \in C_i} \min_{x_q \in C_j} \| x_p - x_q \|}{\max_{l \in \{1...K\}} \max_{x_p, x_q \in C_l} \| x_p - x_q \|} \right) \tag{4}$$

The value of the index increases both when the separation between the clusters is improved or when the within-cluster dispersion decreases. The cluster diameter and the between-cluster distance can be defined in more than one way, leading to possible generalizations of the original Dunn index.

Davies–Bouldin index is also based on a ratio between the intra-cluster and inter-cluster distances [22]. It is defined as follows:

$$DB = \frac{1}{K} \sum_{i=1}^{K} \max_{j \neq i} \left(\frac{\frac{1}{|C_i|} \sum_{x_p \in C_i} \| x_p - \bar{x}_i \| + \frac{1}{|C_j|} \sum_{x_q \in C_j} \| x_q - \bar{x}_j \|}{\| \bar{x}_i - \bar{x}_j \|} \right) \tag{5}$$

The time complexity of computing the Davies–Bouldin index is $O(d(K^2 + N))$. Smaller values of Davies–Bouldin index correspond to better cluster configurations.

Isolation index is defined simply as an average proportion of neighbors in the data that agree with the query point in terms of their cluster label [60]. Assume that $\delta_{p,k} = \frac{|x_q \in D_k(x_p):(\nexists C_i : x_p, x_q \in C_i)|}{k}$ is the local neighborhood disagreement ratio. The isolation index is defined as follows:

$$IS = \frac{1}{N} \sum_{p=1}^{N} (1 - \delta_{p,k}) \tag{6}$$

A weighted version of the isolation index was also later proposed [36].

C-index is derived from the intra-cluster distances and their extremal values [45]. Let $I_{p,q}$ be the indicator function that equals 1 when p and q belong to the same cluster and 0 otherwise. The factor θ that the index is based on is then defined as $\theta = \sum_{p,q \in \{1...N\}, p \neq q} I_{p,q} \| x_p - x_q \|$. The final C-index is merely the normalized version of the θ factor, taking into account the maximal and minimal values that θ could possibly take on the data.

$$CInd = \frac{\theta - \min\theta}{\max\theta - \min\theta} \tag{7}$$

$C\sqrt{K}$ **index** takes individual features into account and measures their corresponding contributions to the inter-cluster and intra-cluster distances [68]. Let $SST_l = \sum_{p=1}^{N} \| x_p^l - \bar{x}^l \|^2$ be the contribution of the feature l to the average total divergence from the overall data mean $\bar{x}$. The contribution of l to the between-cluster distances is then given as $SSB_l = SST_l - \sum_{i=1}^{K} \sum_{x_p \in C_i} (x_p^l - \bar{x}_i^l)^2$. The $C\sqrt{K}$ index is calculated from SST_l and SSB_l as follows:

$$C\sqrt{K}\text{Ind} = \frac{1}{d \cdot \sqrt{K}} \sum_{l=1}^{d} \sqrt{\frac{SSB_l}{SST_l}} \qquad (8)$$

The division by $\sqrt{K}$ is used to counterbalance the expected increase of the $\sqrt{\frac{SSB_l}{SST_l}}$ ratio with increasing K. This is necessary in order to prevent the index from being biased toward preferring configurations that contain more clusters by design.

Calinski–Harabasz index is defined as a variance ratio between inter-cluster and intra-cluster dispersions [16]. Let V_B and V_W be the inter-cluster and intra-cluster dispersion matrices.

$$V_B = \sum_{i=1}^{K} |C_i|(\bar{x}_i - \bar{x})(\bar{x}_i - \bar{x})^T \qquad (9)$$

$$V_W = \sum_{i=1}^{K} \sum_{x_p \in C_i} (x_p - \bar{x}_i)(x_p - \bar{x}_i)^T \qquad (10)$$

The Calinski–Harabasz index is then defined as:

$$CH = \frac{\text{trace}(V_B)}{\text{trace}(V_W)} \cdot \frac{N - K}{K - 1} \qquad (11)$$

Compact and well-separated cluster configurations are expected to have high inter-cluster variance and a comparatively low intra-cluster variance, leading to the high values of the Calinski–Harabasz index. The traces of the inter-cluster and intra-cluster dispersion matrices can be calculated efficiently in $O(dN)$ time.

Fowlkes–Mallows index is defined as the square root of the product of precision and recall given the ground truth of the data [33]. In particular, if $\text{prec} = \frac{TP}{TP+FP}$ and $\text{recall} = \frac{TP}{TP+FN}$ then the Fowlkes–Mallows index can be calculated as $FM = \sqrt{\text{prec} \cdot \text{recall}}$. A higher Fowlkes–Mallows index corresponds to a better match between the two compared partitions of the data. It can also be used to compare two different generated clustering outputs and has been used for this in case of hierarchical clustering.

Goodman–Kruskal index is derived from the concepts of concordant and discordant distance pairs [7]. A pair of distances is said to be concordant in case the distance between the objects from the same cluster is lower than the distance between the objects belonging to different clusters. Denote the total number of concordant pairs in the data w.r.t. the partitioning induced by the clustering under evaluation as S_+ and the number of discordant distance pairs as S_-. Goodman–Kruskal is then defined simply as a ratio involving these two quantities.

$$GK = \frac{S_+ - S_-}{S_+ + S_-} \qquad (12)$$

G_+ **index** is another simple index derived from the concepts of concordance and discordance among pairs of distances in the data. Unlike the Goodman–Kruskal index, it only takes into account the discordant counts. Let $t = \frac{N(N-1)}{2}$ be the number of distances defined by N data points. The G_+ index is given as the count of discordant distance pairs normalized by the total number of distance comparisons, namely $G_+ = \frac{2S_-}{t(t-1)}$. The higher values of the G_+ index correspond to a lower clustering quality, unlike in most other indexes. This is why we have used $\bar{G}_+ = 1 - G_+$ in our experiments instead.

Hubert's Γ statistic can be used to calculate the correlation between two data matrices. It is often used for calculating the correlation between the original and projected data, though it is also possible to apply it for quantifying the correlation between the two distance matrices, which makes it possible to apply for evaluating clustering quality. Namely, let Ψ_T be the distance matrix on the training data. We define the post-clustering centered distance matrix $\bar{\Psi}_T$ to be the distance matrix where $d(x_p, x_q)$ has been replaced by $d(\bar{x}_{y_p}, \bar{x}_{y_q})$ which is the distance between the corresponding centroids. Hubert's statistic is defined as follows:

$$\Gamma_N = \frac{\sum_{p,q=1}^{N} \Psi_T(p,q) \cdot \bar{\Psi}_T(p,q)}{\sqrt{\sum_{p,q=1}^{N}(\Psi_T(p,q) - \mu_{\Psi_T})^2 \cdot \sum_{p,q=1}^{N}(\bar{\Psi}_T(p,q) - \mu_{\bar{\Psi}_T})^2}} = \frac{\sigma_{\Psi_T, \bar{\Psi}_T}}{\sqrt{\sigma_{\Psi_T}^2 \cdot \sigma_{\bar{\Psi}_T}^2}}$$

(13)

This is the normalized version of Γ and it is used in most practical cases. It is also possible to use the non-normalized alternative that is defined simply as the average cross-matrix element product, namely $\Gamma = \frac{1}{N} \sum_{p,q=1}^{N} \Psi_T(p,q) \cdot \bar{\Psi}_T(p,q)$.

McClain–Rao index represents the quotient between the mean intra-cluster and inter-cluster distances. Again, let us denote for simplicity by $I_{p,q}$ the indicator function that equals 1 when p and q belong to the same cluster and 0 otherwise. Let b_d and w_d represent the number of inter-cluster and intra-cluster pairs, so that $b_d = \sum_{i=1}^{K} |C_i|(N - |C_i|)$ and $w_d = \sum_{i=1}^{K} \frac{|C_i|(|C_i|-1)}{2}$.

$$\text{MCR} = \frac{\sum_{p,q \in \{1...N\}, p \neq q} I_{p,q} \parallel x_p - x_q \parallel}{\sum_{p,q \in \{1...N\}, p \neq q} (1 - I_{p,q}) \parallel x_p - x_q \parallel} \cdot \frac{b_d}{w_d}$$

(14)

PBM index is given as a normalized squared ratio between inter-cluster and intra-cluster distances, where they are calculated w.r.t. cluster centroids.

$$\text{PBM} = \left(\frac{\sum_{p=1}^{N} \parallel x_p - \bar{x} \parallel}{\sum_{x_q \in C_j} \parallel x_q - \bar{x}_j \parallel} \cdot \frac{\max_{i,j \in \{1...K\}, i \neq j} \parallel \bar{x}_i - \bar{x}_j \parallel}{K} \right)^2$$

(15)

Point-biserial index is a maximization criterion based on inter-cluster and intra-cluster distances [55]. Assume that $t = \frac{N(N-1)}{2}$ corresponds to the number of pairs of points in the data. Let b_d and w_d represent the number of inter-cluster and intra-cluster pairs. Furthermore, let d_w represent the average intra-cluster distance and d_b represent the average inter-cluster distance. Denote by σ_d the standard distance deviation.

$$\text{PBS} = \frac{(d_b - d_w) \cdot \frac{\sqrt{w_d \cdot b_d}}{t}}{\sigma_d} \tag{16}$$

RS index denotes the ratio between the sum of squared distances between the clusters and the total squared distance sum, where distances are calculated between the points and the respective centroids.

$$\text{RS} = \frac{\sum_{p=1}^{N} \| x_p - \bar{x} \|^2 - \sum_{i=1}^{K} \sum_{x_p \in C_i} \| x_p - \bar{x}_i \|^2}{\sum_{p=1}^{N} \| x_p - \bar{x} \|^2} \tag{17}$$

Rand index is an index that is based on simple pairwise label comparisons with either the ground truth or alternative cluster configurations. Let a be the number of pairs of points in the same cluster with the same label, b the number of pairs of points in same cluster with different labels, c the number of pairs of points in different clusters with the same label, and d the number of pairs of points in different clusters with different labels. Rand index of clustering quality is then defined as the following ratio:

$$\text{RAND} = \frac{a + d}{a + b + c + d} \tag{18}$$

However, the Rand index has known weaknesses, such as bias toward partitions with larger numbers of clusters. An improved version of the Rand index was proposed by [44], referred to as the **adjusted Rand index (ARI)**, and is considered to be one of the most successful cluster validation indices [70]. ARI can be computed as:

$$\text{ARI} = \frac{\binom{N}{2}(a + d) - [(a + b)(a + c) + (c + d)(b + d)]}{\binom{N}{2}^2 - [(a + b)(a + c) + (c + d)(b + d)]} \tag{19}$$

SD index is a combination of two factors: scatter and separation. They will be denoted by F_S and F_D, respectively. The total SD index is defined as $\text{SD} = \alpha F_S + F_D$. The scatter component of the SD index is derived from the global and within-cluster variance vectors over different data features. Denote by V_T and V_{C_i} such d-dimensional feature variance vectors, for $i \in \{1 \ldots K\}$. The scatter component is defined as the ratio between the average within-cluster variance vector norm and the global variance vector norm, $F_S = \frac{\sum_{i=1}^{K} \| V_{C_i} \|}{K \cdot \| V_T \|}$. On the other hand, separation is defined via between-centroid distances, $F_D = \frac{\max_{i \neq j} \| \bar{x}_i - \bar{x}_j \|}{\min_{i \neq j} \| \bar{x}_i - \bar{x}_j \|} \cdot \sum_{i=1}^{K} \frac{1}{\sum_{j=1, j \neq i}^{K} \| \bar{x}_i - \bar{x}_j \|}$.

τ **index** represents the correlation between the distance matrix of the data and a
binary matrix corresponding to whether pairs of points belong to the same cluster
or not. It can be written down in terms of concordance and discordance rates.
Again, let $t = \frac{N(N-1)}{2}$ be the number of distances defined by N data points.
Similarly, let b_d and w_d represent the number of inter-cluster and intra-cluster
pairs, respectively. Let $t_{bw} = \binom{b_d}{2} + \binom{w_d}{2}$ be the total number of distance pairs
that can not be considered discordant since they belong to the same distance type.

$$\tau = \frac{S_+ - S_-}{\left(\frac{t(t-1)}{2} - t_{bw}\right)\frac{t(t-1)}{2}} \tag{20}$$

This is by no means an exhaustive list of clustering quality measures. Many more
exist and continue to be developed.

6 Clustering Quality Indexes: Existing Surveys

Several studies were conducted in order to elucidate the overall applicability of
various clustering quality indexes. Different studies have focused on different
aspects of the clustering problem and different properties that good quality measures
are expected to have. Detailed surveys of the employed methods discuss the
motivation behind the different approaches that can be taken [38].

In one such study [52], the authors have identified five primary criteria to
evaluate the clustering quality indexes by, as follows: non-monotonicity, robustness
to noise, proper handling of cluster configurations with highly varying cluster
density, handling of subclusters, and performance under skewed distributions. The
authors have compared 11 different internal clustering quality measures according
to the proposed criteria on a series of carefully crafted examples and have ranked
the examined indexes accordingly.

Standard clustering evaluation approaches that were tailored for partitional
clustering methods [18] are not universal and not well suited for other types of
tasks like the evaluation of nested clustering structures, for which they need to be
extended and adapted [24].

An inherent problem in many existing studies lies in the fact that one has to
know the ground truth of all the clustering problems used to evaluate the indexes
on the level of detail required for properly estimating the degree to which they
satisfy each of the proposed benchmarking criteria. This is why most examples
in such studies are taken to be 2-dimensional [54]. Indeed, it is easy to construct
2-dimensional cluster configurations of arbitrary shape, compactness, and density.
More importantly, it is possible to visually inspect the data and confirm the
correctness of the assumptions.

In contrast, high-dimensional data is rarely used for such testing and we believe
that this was a major flaw in the experimental design of many approaches that
aim to establish which clustering quality indexes are to be preferred in practical

applications. This is not to say that the conclusions reached in those studies are any less valuable nor that they cannot be generalized. However, high-dimensional data is often counterintuitive and is known to exhibit different properties than low-dimensional data.

Since most practical applications involve clustering intrinsically high-dimensional data, it is necessary to carefully examine the behavior of clustering quality indexes with increasing data dimensionality in order to better understand their role in high-dimensional clustering evaluation.

Existing studies on the feasibility of clustering evaluation in high-dimensional clustering tasks have mostly been focused on subspace clustering [56, 57] as it is a popular clustering approach in such cases.

This study does not aim to replicate the types of experiments or results that are already available in the literature. Instead, it aims to complement them in order to provide insight into the overall behavior of clustering quality indexes in many dimensions.

It should be noted that there are experimental approaches that can be applied both in the low-dimensional and the high-dimensional case. Most notably, the clustering quality indexes can be compared based on their ability to select the optimal cluster configuration among a series of generated cluster configurations for each dataset, while varying the number of clusters [85]. The main problem with this approach lies in determining a priori what constitutes the optimal clustering configuration for a given dataset. This is often taken to correspond to the number of classes in the data, which is problematic. Even in cases when the ground truth information is available, preferring a clustering configuration produced by a clustering algorithm that has the same number of clusters as the number of clusters in the data is not necessarily optimal. Namely, those proposed clusters need not match the ground truth clusters exactly. In case of mismatch, another configuration might provide a more natural partitioning of the data and this is usually difficult to estimate unless it is possible to visualize the clustering outcomes and inspect the results.

Supervised evaluation can be used as an alternative [30], when labels are available, though many real-world datasets violate the cluster assumption in such way that the classes do not correspond to clusters nor does the data cluster well. In such cases it is not advisable to use the average cluster homogeneity of the selected configurations as the primary evaluation measure.

7 Clustering Evaluation in Many Dimensions

In order to understand the potential problems that might arise when applying different clustering quality criteria to many-dimensional datasets, it is necessary to consider the basic underlying question of what it means to cluster well in terms of index scores.

Most clustering quality indexes are relative and can be used either as an objective function to optimize or as a criterion to make comparisons between different cluster configurations on the same feature representation of the same dataset, under a chosen distance measure. Ideally, it would be very useful if an index could be applied to perform meaningful comparisons across different feature representations and metric selections as well, or across datasets.

Unlike supervised measures like accuracy or the F_1 score, clustering quality indexes are sometimes non-trivial to interpret. For instance, it is not immediately clear what a Silhouette index of 0.62 means. It is also unclear whether it means the same thing in 2 dimensions as it does in 100 dimensions, assuming the same underlying data distribution.

Different clustering quality indexes capture similar but slightly different aspects of clustering quality. This is why ensembles of clustering quality indexes are sometimes used instead of single quality measures [58, 86]. The implicit assumption of the ensemble approach is that the ensemble constituents exhibit equal sensitivity to the varying conditions in the data. If this assumption is violated, a different index combination would be required for each dataset, which would be highly impractical.

Additionally, if an index is to be used for cluster configuration selection over different sampled feature subspaces on a dataset, then the stability w.r.t. dimensionality and representation is a strict requirement for the comparisons to make sense.

The experiments presented here are aimed at clarifying the issues regarding the sensitivity of clustering quality indexes to data dimensionality, in order to determine whether it is possible to safely use certain indexes for cross-dimensional comparisons and evaluation.

7.1 Experimental Protocol

In order to evaluate the performance and robustness of different cluster configuration quality estimation indexes, a series of intrinsically high-dimensional multi-cluster datasets was generated. Each synthetic dataset consisted of about 10,000 points of i.i.d. Gaussian data based on a diagonal covariance matrix. In turn, ten datasets were generated for each dimensionality, experimental context, and tested cluster number. Tests were run for data with $C \in \{2, 3, 5, 10, 20\}$ in order to see how the number of clusters correlates with index performance.

Two different experimental contexts were examined: well-defined and separated clusters and clusters with significant overlap between them. It was assumed that different indexes might exhibit different behavior in the two cases.

The degree of overlap between the clusters in the overlapping case was controlled via a constrained iterative procedure for generating feature-specific distribution means. Since the diagonal covariance matrices were used, the diagonal entry $\sigma_{C_i}^l$ corresponding to the dispersion of l-th feature in cluster C_i and the mean $\mu_{C_i}^l$ were used to determine where $\mu_{C_j}^l$ should be placed for $j \neq i$ and vice versa. The procedure was executed as follows, for any given dimension (feature).

A permutation $\{i_1, i_2 \ldots i_C\}$ of cluster indexes was randomly selected. The mean and standard deviation of the first Gaussian, $\mu^l_{C_{i_0}}$ and $\sigma^l_{C_{i_0}}$ were selected randomly from a fixed range. For each $p \in \{2 \ldots C\}$, cluster C_{i_p} got randomly paired with a cluster C_{i_q}, for $1 \leq q < p$. The mean and standard deviation for feature l in cluster q were then generated by the following rules.

$$\mu^l_{C_{i_q}} = \mu^l_{C_{i_p}} \pm \alpha \sigma^l_{C_{i_p}} \tag{21}$$

$$\sigma^l_{C_{i_q}} = \beta(1 + N_{0,1})\sigma^l_{C_{i_p}} \tag{22}$$

In the experiments presented in this chapter, the parameter values of $\alpha = 0.5$ and $\beta = 0.75$ were used.

On each dataset in both experimental contexts, the indexes were run both on ground truth cluster configurations as well as a series of cluster configurations produced by repeated runs of K-means clustering. The clustering was repeated 10 times for each dataset. These results will be presented separately. Euclidean distance was used in all cases.

The experiments were based on the clustering quality index implementations in Hub Miner (https://github.com/datapoet/hubminer) [75], within the learning. unsupervised.evaluation.quality package. The experimental framework for tests with increasing dimensionality on synthetic Gaussian data was implemented in the QualityIndexEvalInHighDim class in the experimental sub-package.

All clustering quality indexes in Hub Miner are implemented so as to correspond to a maximization problem. Most existing indexes follow this approach. For those indexes that naturally correspond to minimization problems like Davies Bouldin or G_+, an inverse or a complement was computed and reported instead, in order to present the results in a more consistent way.

7.2 Sensitivity to Increasing Dimensionality

Since the synthetic datasets in all dimensionalities were generated from the same distribution type, differing only in the number of dimensions, the underlying assumption was that the robust clustering quality indexes would yield similar quality estimates in all cases, on average. In contrast, the indexes that are sensitive to data dimensionality would either display different average estimates or would become more or less stable in terms of the variance of quality predictions.

7.2.1 Sensitivity of the Average Quality Assessment

The average clustering quality estimation by different indexes was calculated for each generated dataset and experimental context. These have been subsequently averaged over collections of datasets in order to calculate the average estimates

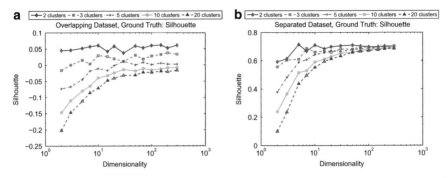

Fig. 5 Average clustering quality index values for the Silhouette index with increasing dimensionality, when evaluated on ground truth cluster labels. (**a**) Overlapping clusters. (**b**) Well-separated clusters

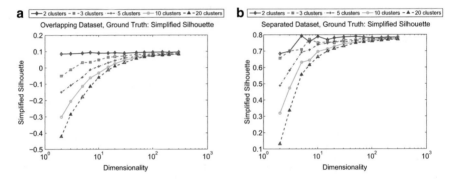

Fig. 6 Average clustering quality index values for the simplified Silhouette index with increasing dimensionality, when evaluated on ground truth cluster labels. (**a**) Overlapping clusters. (**b**) Well-separated clusters

associated with the used Gaussian mixtures for each tested data dimensionality. The results are shown in Figs. 5, 6, 7, 8, 9, 10, 11, 12, 13, 14, 15, 16, 17, 18, 19, 20, and 21.

Some indexes seem to be robust to increasing dimensionality w.r.t. the average cluster configuration quality estimates and these include: isolation index, C-index, $C\sqrt{K}$-index, Calinski–Harabasz, $\bar{G}_+$ index, Goodman–Kruskal, RS index, and τ-index. In these cases, the cluster configuration quality estimates remain similar when the dimensionality is increased. It should be noted that, though there are no emerging trends in these indexes, some of them do exhibit substantial variance in their output.

In other cases, it is possible to observe changes in the clustering quality estimates and there can be substantial differences between the output of a same index for different dimensionalities of the data, both in case of clearly separated clusters as well as overlapping clusters with the same controlled degree of overlap.

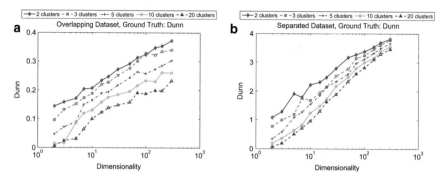

Fig. 7 Average clustering quality index values for the Dunn index with increasing dimensionality, when evaluated on ground truth cluster labels. (**a**) Overlapping clusters. (**b**) Well-separated clusters

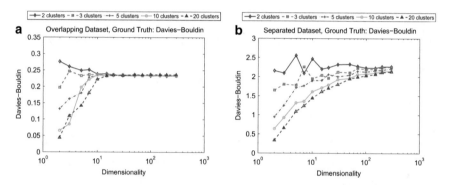

Fig. 8 Average clustering quality index values for the inverted Davies–Bouldin index with increasing dimensionality, when evaluated on ground truth cluster labels. (**a**) Overlapping clusters. (**b**) Well-separated clusters

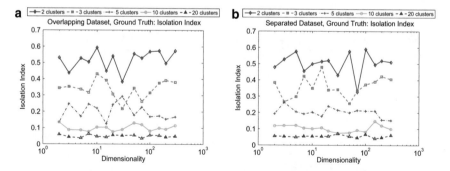

Fig. 9 Average clustering quality index values for the isolation index with increasing dimensionality, when evaluated on ground truth cluster labels. (**a**) Overlapping clusters. (**b**) Well-separated clusters

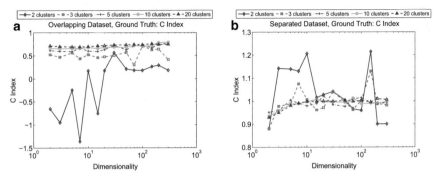

Fig. 10 Average clustering quality index values for the C index with increasing dimensionality, when evaluated on ground truth cluster labels. (**a**) Overlapping clusters. (**b**) Well-separated clusters

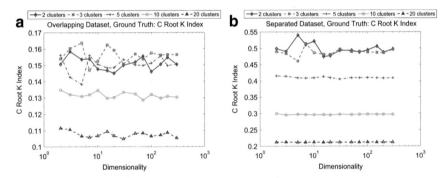

Fig. 11 Average clustering quality index values for the $C\sqrt{K}$ index with increasing dimensionality, when evaluated on ground truth cluster labels. (**a**) Overlapping clusters. (**b**) Well-separated clusters

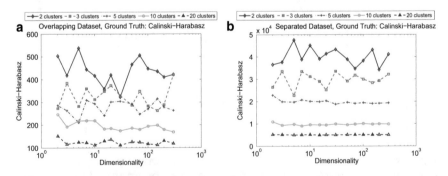

Fig. 12 Average clustering quality index values for the Calinski–Harabasz index with increasing dimensionality, when evaluated on ground truth cluster labels. (**a**) Overlapping clusters. (**b**) Well-separated clusters

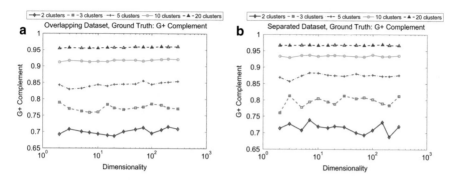

Fig. 13 Average clustering quality index values for the $\bar{G}_+$ index with increasing dimensionality, when evaluated on ground truth cluster labels. (**a**) Overlapping clusters. (**b**) Well-separated clusters

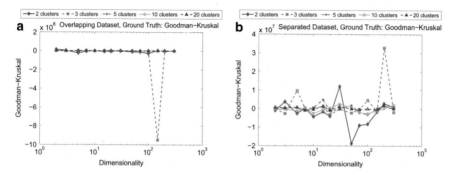

Fig. 14 Average clustering quality index values for the Goodman–Kruskal index with increasing dimensionality, when evaluated on ground truth cluster labels. (**a**) Overlapping clusters. (**b**) Well-separated clusters

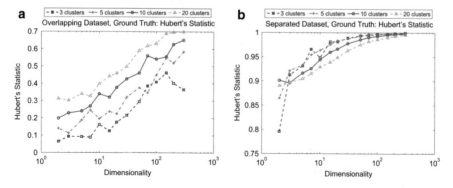

Fig. 15 Average clustering quality index values for the Hubert's statistic with increasing dimensionality, when evaluated on ground truth cluster labels. (**a**) Overlapping clusters. (**b**) Well-separated clusters

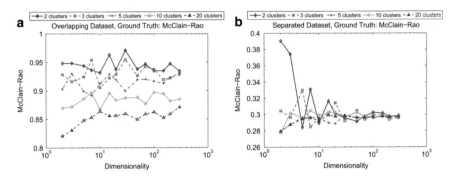

Fig. 16 Average clustering quality index values for the McClain–Rao index with increasing dimensionality, when evaluated on ground truth cluster labels. (**a**) Overlapping clusters. (**b**) Well-separated clusters

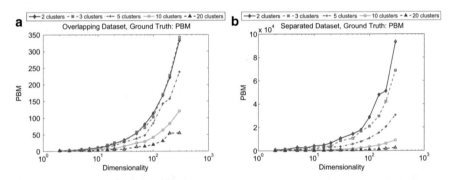

Fig. 17 Average clustering quality index values for the PBM index with increasing dimensionality, when evaluated on ground truth cluster labels. (**a**) Overlapping clusters. (**b**) Well-separated clusters

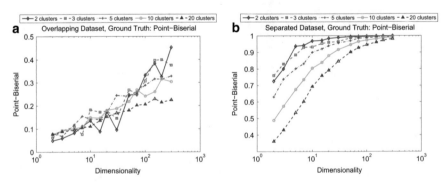

Fig. 18 Average clustering quality index values for the point-biserial index with increasing dimensionality, when evaluated on ground truth cluster labels. (**a**) Overlapping clusters. (**b**) Well-separated clusters

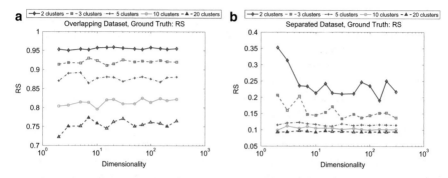

Fig. 19 Average clustering quality index values for the RS index with increasing dimensionality, when evaluated on ground truth cluster labels. (**a**) Overlapping clusters. (**b**) Well-separated clusters

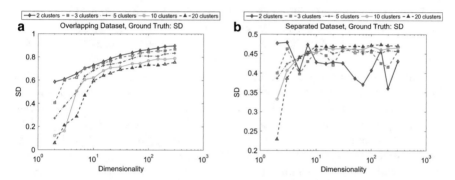

Fig. 20 Average clustering quality index values for the SD index with increasing dimensionality, when evaluated on ground truth cluster labels. (**a**) Overlapping clusters. (**b**) Well-separated clusters

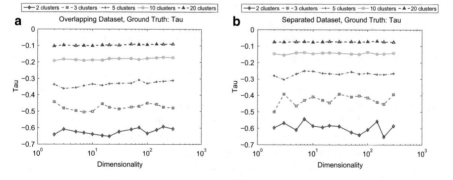

Fig. 21 Average clustering quality index values for the τ-index with increasing dimensionality, when evaluated on ground truth cluster labels. (**a**) Overlapping clusters. (**b**) Well-separated clusters

The estimated clustering quality increases with increasing dimensionality for the Silhouette index, the simplified Silhouette index, Dunn index, inverted Davies–Bouldin index, Hubert's statistic, PBM, and point-biserial index. These indexes seem to be prone to preferring high-dimensional configurations and interpreting them as better. While it might be the case that introducing more dimensions produces additional separation in commonly used distance measures, this makes it hard to compare the quality of clustering across dimensions and across different feature subsets of the data.

Being able to reliably compare the clustering quality across dimensions is especially important in certain subspace clustering applications where the dimensionality of the observed subspaces varies. It is also of importance in applications where multiple feature representations and feature subsets of the same underlying data are being used in the experiments. In contrast, robustness to changes in dimensionality has no impact on the simple clustering use cases when the dimensionality is fixed and no subspaces are observed. Such use cases are more common in low-dimensional clustering tasks.

There are also notable differences in terms of the preferred number of clusters. The following indexes consistently prefer fewer clusters in the data, regardless of the dimensionality: Silhouette index, simplified Silhouette index, Dunn, Davies Bouldin, isolation index, Calinski–Harabasz index, $\bar{G}_+$ index, PBM index, and RS index. This tendency is also present in $C\sqrt{K}$-index, though to a somewhat lower degree, at least when it comes to lower cluster numbers. It is interesting to see that there are cases where this tendency is present only in case of evaluating clearly separated cluster configurations (point-biserial index, Fig. 18), as well as cases where the preference is present only for overlapping cluster configurations (McClain–Rao index, Fig. 16). Additionally, there is the surprising case of Hubert's statistic, which seems to exhibit a preference for fewer clusters in clearly separated cases and a preference for more clusters in the case of overlapping distributions.

The tendencies of the clustering quality estimates to either increase or remain approximately constant w.r.t. dimensionality do not depend on the number of clusters in the data, which shows that they are an intrinsic property of some of the compared clustering quality indexes.

Furthermore, dependencies on representational dimensionality do not merely affect the cluster configuration quality estimation for ground truth clustering configurations, but for the output of the K-means algorithm as well, which is important in practical terms. Figures 22 and 23 show the behavior of the Fowlkes–Mallows and adjusted Rand indexes with increasing data dimensionality. Both indexes prefer high-dimensional cluster configurations generated by K-means on the synthetic Gaussian data, meaning that K-means was actually more successful in finding the designated true clustering in high dimensions.

Surprisingly, it seems that the tendencies that were observed on ground truth cluster configurations do not necessarily generalize to the tendencies of the same indexes when evaluating the cluster configurations produced by the repeated K-means clustering. While in some cases the dependency curves might take the same shape, it is not always the case. Many indexes that seem not to depend on data dimensionality when it comes to ground truth estimates exhibit a strong dependency on data dimensionality in terms of the K-means quality estimates.

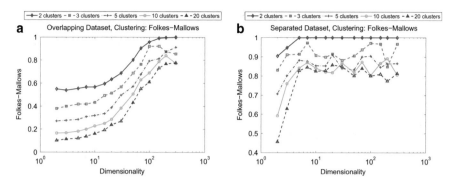

Fig. 22 Average clustering quality index values for the Fowlkes–Mallows index with increasing dimensionality, when evaluated on *K*-means clustering results. (**a**) Overlapping clusters. (**b**) Well-separated clusters

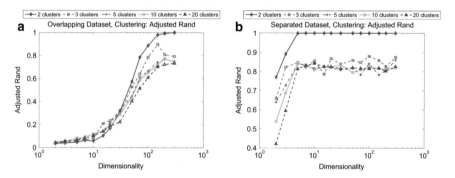

Fig. 23 Average clustering quality index values for the adjusted Rand index with increasing dimensionality, when evaluated on *K*-means clustering results. (**a**) Overlapping clusters. (**b**) Well-separated clusters

For instance, Fig. 24 gives an example of the behavior of the Calinski–Harabasz index and the $C\sqrt{K}$-index when evaluating the configurations produced by K-means clustering on data of increasing dimensionality. In both cases, the average index values clearly decrease with data dimensionality, unlike what was previously observed for the ground truth configurations. Apparently, changing properties of distance distributions as dimensionality increases (i.e., increasing expected values, stable variances) cause these indexes to produce values that are not comparable across dimensionalities. For the $C\sqrt{K}$-index it seems that normalization by $d\sqrt{K}$ may not be the most appropriate choice to counter the effects of dimensionality on computed contributions of features to the average total divergence from the overall data mean, and to between-cluster distances (concretely, division by d may be too strong, since increasing expected values already cancel out, at least partially, in the ratio of *SSB* and *SST*).

Yet, not all indexes exhibit trends that are either constant or monotonically increasing/decreasing. The behavior of several indexes with increasing data dimen-

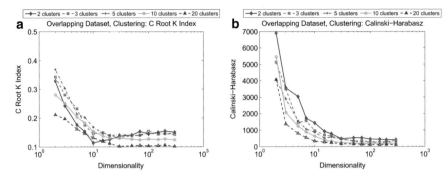

Fig. 24 Average clustering quality index values for the $C\sqrt{K}$ and Calinski–Harabasz index with increasing dimensionality, when evaluated on K-means clustering results. These trends clearly differ from the trends that the same indexes exhibit on ground truth configurations. (**a**) $C\sqrt{K}$, Overlapping clusters. (**b**) Calinski–Harabasz, Overlapping clusters

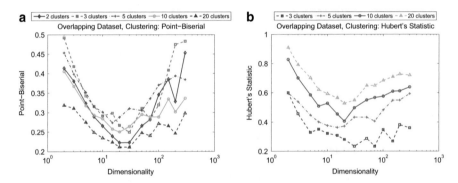

Fig. 25 Average clustering quality index values for the point-biserial index and Hubert's statistic with increasing dimensionality, when evaluated on K-means clustering results. (**a**) Point-biserial, Overlapping clusters. (**b**) Hubert's-Statistic, Overlapping clusters

sionality can be captured by U-shaped curves, as is the case with Hubert's statistic and the point-biserial index. This is shown in Fig. 25. In those cases the average estimates first decrease with increasing dimensionality and then start increasing after reaching a local minimum. This sort of behavior is highly surprising and also suggests that some indexes might have either ideal operating points in terms of the underlying data dimensionality or ranges of dimensionality values that should be avoided.

Overall, for indexes which implicitly or explicitly rely on some combination of notions of within- and between-cluster distances it can be seen that changes in distance distributions which result from increase of dimensionality affect the produced trends in scores in various ways, implying that some indexes adjust to these changes better than others.

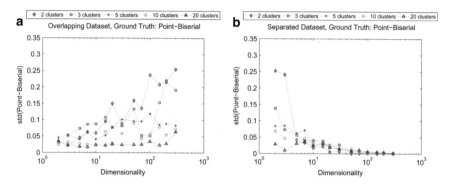

Fig. 26 Standard deviation of the clustering quality index values for the point-biserial index with increasing dimensionality, when evaluated on ground truth cluster labels. (**a**) Overlapping clusters. (**b**) Well-separated clusters

7.2.2 Stability in Quality Assessment

Increasing data dimensionality also impacts the performance of the clustering quality indexes in terms of their stability and associated predictive variance. Similarly to the average index performance, different indexes are influenced in different ways.

Consider, for example, the point-biserial index. The standard deviation of the index values with increasing data dimensionality is given in Fig. 26, both for the overlapping and the clearly separated context. In case of overlapping clusters, the variance of the point-biserial index increases for configurations with a low number of clusters as the dimensionality is increased. In case of separated clusters, the variance decreases with increasing dimensionality.

This is not the case for other quality measures. In case of PBM (Fig. 27), the variance increases both for separated and overlapping cluster configurations. For $\bar{G}_+$ index, the variance does not increase or decrease with dimensionality, but there is a consistent difference between the stability of the estimation when applied to configurations with different numbers of clusters. The stability of the $\bar{G}_+$ index (Fig. 28) decreases for more complex cluster configurations and the standard deviation is the lowest in case of 2 clusters.

It is also interesting to note that the stability of different quality indexes also depends on whether it is determined based on the ground truth configuration quality estimates or the clustering configuration estimates. These trends need to be analyzed separately. The differences are shown in Figs. 29, 30, and 31, for the simplified Silhouette index, Dunn index, and isolation index.

Stability is important in clustering evaluation in order to allow for meaningful interpretations of the results. If the estimates for different repeated samples drawn from the same underlying distribution vary greatly, interpretation of the estimated quality becomes difficult. Stable indexes are to be preferred when used for external comparisons, especially when the number of data samples is low.

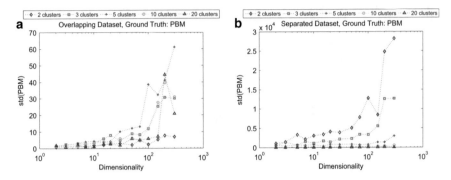

Fig. 27 Standard deviation of the clustering quality index values for the PBM index with increasing dimensionality, when evaluated on ground truth cluster labels. (**a**) Overlapping clusters. (**b**) Well-separated clusters

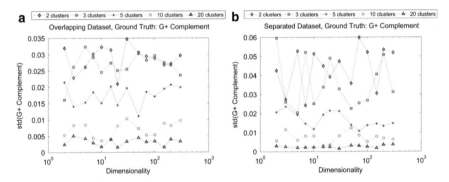

Fig. 28 Standard deviation of the clustering quality index values for the $\bar{G}_+$ index with increasing dimensionality, when evaluated on ground truth cluster labels. (**a**) Overlapping clusters. (**b**) Well-separated clusters

7.3 Quantifying the Influence of Hubs

It turns out that the behavior of many of the evaluated clustering quality indexes can be partially captured and quantified by hubness in the data. Previous studies have shown hubs to cluster poorly in some cases [63] and that proper handling of hub points is required in order to improve clustering quality. Further studies have found that the contribution of hub points to the Silhouette index depends on their hubness and that hubs contribute more to the final Silhouette index estimate [83].

In our experiments we have been able to confirm that it is not only Silhouette index that is susceptible to being influenced by the hubness in the data and the positioning of hub points in the cluster configuration. Many different indexes exhibit similar types of dependencies on hub points, though some also exhibit the reverse trends.

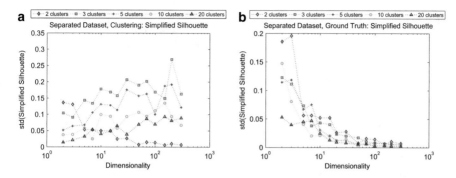

Fig. 29 Different trends in stability of the simplified Silhouette index with increasing dimensionality, depending on whether it was applied to the ground truth cluster labels or the K-means clustering results. These comparisons were performed on synthetic data with well-separated clusters. (**a**) Ground truth evaluation. (**b**) Evaluating K-means results

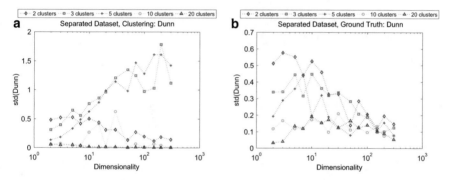

Fig. 30 Different trends in stability of the Dunn index with increasing dimensionality, depending on whether it was applied to the ground truth cluster labels or the K-means clustering results. These comparisons were performed on synthetic data with well-separated clusters. (**a**) Ground truth evaluation. (**b**) Evaluating K-means results

Formally, we denote **hubs** are points $x_h \in D$ such that $N_k(x_h)$ belongs to the top one third of the N_k distribution. We will denote the set of all hubs in T by H_k^T. In contrast, anti-hubs are the rarely occurring points in the data and their number is by convention taken to match the number of hubs on a dataset. The remaining points are referred to as being regular.

Figure 32 shows the contributions of hubs, regular points, and anti-hubs to the final point-biserial index score with increasing dimensionality, for cluster configurations produced by K-means. The contribution of hubs is higher for this index for well-separated configurations while significantly lower for overlapping configurations where anti-hubs dominate the scores. This is unlike the Silhouette index and simplified Silhouette index, where hubs contribute substantially more in both cases. The contributions to the Silhouette index and simplified Silhouette index for overlapping configurations are shown in Fig. 33.

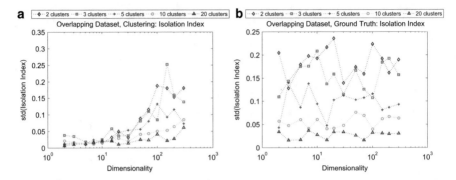

Fig. 31 Different trends in stability of the isolation index with increasing dimensionality, depending on whether it was applied to the ground truth cluster labels or the K-means clustering results. These comparisons were performed on synthetic data with overlapping clusters. (**a**) Ground truth evaluation. (**b**) Evaluating K-means results

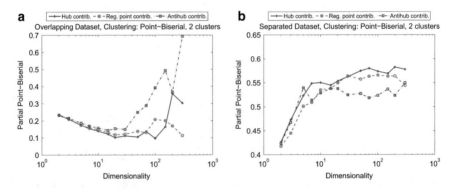

Fig. 32 The contributions of different point types to final index values for the point-biserial index in case of 2 clusters, for cluster configurations produced by K-means. (**a**) Overlapping clusters. (**b**) Well-separated clusters

Hubs seem to be the most influential points in determining the $\bar{G}_+$ and Davies–Bouldin scores as well, as seen in Fig. 34.

In contrast, anti-hubs seem to be contributing more to the partial sums in McClain–Rao index and the difference becomes more apparent with increasing the number of clusters in the data, as shown in Fig. 35.

Whether hub points contribute substantially more or less than regular points for any given index might affect the robustness of the index and its sensitivity to increasing data dimensionality.

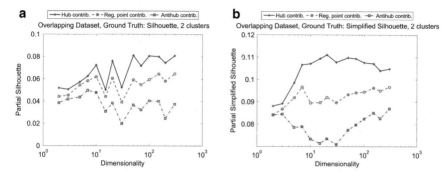

Fig. 33 The contributions of different point types to final index values for the Silhouette and simplified Silhouette index in case of 2 overlapping clusters, for ground truth cluster configurations. (**a**) Silhouette index. (**b**) Simplified Silhouette index

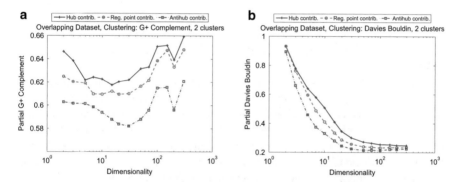

Fig. 34 The contributions of different point types to final index values for the $\bar{G}_+$ and Davies–Bouldin index in case of 2 overlapping clusters, for cluster configurations produced by K-means. (**a**) $\bar{G}_+$ index, Overlapping clusters. (**b**) Davies–Bouldin, Well-separated clusters

8 Perspectives and Future Directions

Robust and effective clustering evaluation plays an essential role in selecting the optimal clustering methods and cluster configurations for a given task. It is therefore important to understand the behavior of clustering quality measures in challenging contexts, like high data dimensionality.

We have examined a series of frequently used clustering quality measures under increasing data dimensionality in synthetic Gaussian data. The experiments have demonstrated that different quality metrics are influenced in different ways by rising dimensionality in the data. Dimensionality of the data can affect both the mean quality value assigned by an index as well as the stability of the quality estimation. Furthermore, the same index can behave differently depending on the presence or absence of overlap between different clusters as well as whether the comparisons are being made to the ground truth or an output of a clustering algorithm.

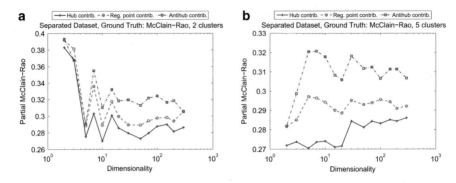

Fig. 35 The contributions of different point types to final index values for the McClain–Rao index in case of clearly separated clusters, for ground truth cluster configurations. (**a**) McClain–Rao index, 2 clusters. (**b**) McClain–Rao index, 5 clusters

These initial results show that selecting the appropriate clustering quality index for high-dimensional data clustering is non-trivial and should be approached carefully.

Robustness to increasing data dimensionality would be highly beneficial in practical clustering quality estimation. Our results indicate that the dimensionality of the test data needs to be factored in any meaningful cross-index comparison, as it is otherwise possible for the reported results to be an artifact of the dimensionality of the test data. This is another reason why many surveys of clustering evaluation that were conducted in the past on low-dimensional data might not be relevant for practical clustering tasks, where data is usually high-dimensional.

This study was performed on synthetic data, as such an evaluation gives more control over the experimental protocol and the parameters of the context. In order for the findings to be truly relevant, a similar detailed study should be performed on real-world data as well, by using repeated sub-sampling of larger high-dimensional datasets. Of course, the availability of high-dimensional clustering benchmarks with ground truth and low violation of the cluster assumption is not ideal.

Additionally, since it was demonstrated on several occasions that a better handling of hub points may result in better overall clustering quality in many-dimensional problems [40, 83, 84], we intend to consider either extending the existing clustering quality indexes or proposing new ones that would incorporate this finding into account.

References

1. Achtert, E.: Hierarchical Subspace Clustering. Ludwig Maximilians Universitat, Munich (2007)
2. Aggarwal, C.: On high dimensional projected clustering of uncertain data streams. In: Proceedings of the 25th IEEE International Conference on Data Engineering (ICDE), pp. 1152–1154 (2009)

3. Aggarwal, C.C.: Re-designing distance functions and distance-based applications for high dimensional data. ACM Sigmod Rec. **30**(1), 13–18 (2001)
4. Aggarwal, C.C.: On k-anonymity and the curse of dimensionality. In: Proceedings of the 31st International Conference on Very Large Data Bases (VLDB), pp. 901–909 (2005)
5. Aggarwal, C.C.: On randomization, public information and the curse of dimensionality. In: Proceedings of the 23rd IEEE International Conference on Data Engineering (ICDE), pp. 136–145 (2007)
6. Bai, L., Liang, J., Dang, C., Cao, F.: A novel attribute weighting algorithm for clustering high-dimensional categorical data. Pattern Recogn. **44**(12), 2843–2861 (2011)
7. Baker, F.B., Hubert, L.J.: Measuring the power of hierarchical cluster analysis. J. Am. Stat. Assoc. **70**(349), 31–38 (1975)
8. Bellman, R.E.: Adaptive Control Processes – A Guided Tour. Princeton University Press, Princeton (1961)
9. Bengio, Y., Delalleau, O., Le Roux, N.: The curse of dimensionality for local kernel machines. Technical Report 1258, Departement d'Informatique et Recherche Operationnelle, Université de Montréal, Montreal, Canada (2005)
10. Bennett, K.P., Fayyad, U.M., Geiger, D.: Density-based indexing for approximate nearest-neighbor queries. In: Proceedings of the 5th ACM SIGKDD International Conference on Knowledge Discovery and Data Mining, pp. 233–243 (1999)
11. Beyer, K., Goldstein, J., Ramakrishnan, R., Shaft, U.: When is "nearest neighbor" meaningful? In: Proceedings of the International Conference on Database Theory (ICDT), ACM, New York, NY, pp. 217–235 (1999)
12. Bohm, C., Kailing, K., Kriegel, H.P., Kroger, P.: Density connected clustering with local subspace preferences. In: Proceedings of the Fourth IEEE International Conference on Data Mining (ICDM), pp. 27–34 (2004)
13. Bouguila, N., Almakadmeh, K., Boutemedjet, S.: A finite mixture model for simultaneous high-dimensional clustering, localized feature selection and outlier rejection. Expert Syst. Appl. **39**(7), 6641–6656 (2012)
14. Bouveyron, C., Brunet-Saumard, C.: Model-based clustering of high-dimensional data: a review. Comput. Stat. Data Anal. **71**, 52–78 (2014)
15. Buza, K., Nanopoulos, A., Schmidt-Thieme, L.: INSIGHT: efficient and effective instance selection for time-series classification. In: Proceedings of the 15th Pacific-Asia Conference on Knowledge Discovery and Data Mining (PAKDD), Part II, pp. 149–160 (2011)
16. Caliński, T., Harabasz, J.: A dendrite method for cluster analysis. Commun. Stat. Simul. Comput. **3**(1), 1–27 (1974)
17. Carter, K., Raich, R., Hero, A.: On local intrinsic dimension estimation and its applications. IEEE Trans. Signal Process. **58**(2), 650–663 (2010)
18. Celebi, M.E. (ed.): Partitional Clustering Algorithms. Springer, Berlin (2014)
19. Chen, X., Ye, Y., Xu, X., Huang, J.Z.: A feature group weighting method for subspace clustering of high-dimensional data. Pattern Recogn. **45**(1), 434–446 (2012)
20. Chávez, E., Navarro, G.: A probabilistic spell for the curse of dimensionality. In: Algorithm Engineering and Experimentation, pp. 147–160. Springer, Berlin (2001)
21. Chávez, E., Navarro, G.: Probabilistic proximity search: fighting the curse of dimensionality in metric spaces. Inf. Process. Lett. **85**(1), 39–46 (2003)
22. Davies, D.L., Bouldin, D.W.: A cluster separation measure. IEEE Trans. Pattern Anal. Mach. Intell. **1**(2), 224–227 (1979)
23. Draper, B., Elliott, D., Hayes, J., Baek, K.: EM in high-dimensional spaces. IEEE Trans. Syst. Man Cybern. **35**(3), 571–577 (2005)
24. Draszawka, K., Szymanski, J.: External validation measures for nested clustering of text documents. In: Ryżko, D., Rybinski, H, Gawrysiak, P., Kryszkiewicz, M. (eds.) Emerging Intelligent Technologies in Industry. Studies in Computational Intelligence, vol. 369, pp. 207–225. Springer, Berlin (2011)
25. Dunn, J.C.: Well-separated clusters and optimal fuzzy partitions. J. Cybern. **4**(1), 95–104 (1974)

26. Durrant, R.J., Kabán, A.: When is 'nearest neighbour' meaningful: a converse theorem and implications. J. Complex. **25**(4), 385–397 (2009)
27. Evangelista, P.F., Embrechts, M.J., Szymanski, B.K.: Taming the curse of dimensionality in kernels and novelty detection. In: Applied Soft Computing Technologies: The Challenge of Complexity, pp. 425–438. Springer, Berlin (2006)
28. Fan, W., Bouguila, N., Ziou, D.: Unsupervised hybrid feature extraction selection for high-dimensional non-Gaussian data clustering with variational inference. IEEE Trans. Knowl. Data Eng. **25**(7), 1670–1685 (2013)
29. Farahmand, A.M., Szepesvári, C.: Manifold-adaptive dimension estimation. In: Proceedings of the 24th International Conference on Machine Learning (ICML), ACM, New York, NY, pp. 265–272 (2007)
30. Färber, I., Günnemann, S., Kriegel, H.P., Kröger, P., Müller, E., Schubert, E., Seidl, T., Zimek, A.: On using class-labels in evaluation of clusterings. In: MultiClust: 1st International Workshop on Discovering, Summarizing and Using Multiple Clusterings Held in Conjunction with KDD (2010)
31. Fern, X.Z., Brodley, C.E.: Random projection for high dimensional data clustering: a cluster ensemble approach. In: Proceedings of 20th International Conference on Machine learning (ICML), pp. 186–193 (2003)
32. Fern, X.Z., Brodley, C.E.: Cluster ensembles for high dimensional clustering: an empirical study. Technical Report CS06-30-02, Oregon State University (2004)
33. Fowlkes, E.B., Mallows, C.L.: A method for comparing two hierarchical clusterings. J. Am. Stat. Assoc. **78**(383), 553–569 (1983)
34. François, D., Wertz, V., Verleysen, M.: The concentration of fractional distances. IEEE Trans. Knowl. Data Eng. **19**(7), 873–886 (2007)
35. France, S., Carroll, D.: Is the distance compression effect overstated? Some theory and experimentation. In: Proceedings of the 6th International Conference on Machine Learning and Data Mining in Pattern Recognition (MLDM), pp. 280–294 (2009)
36. Frederix, G., Pauwels, E.J.: Shape-invariant cluster validity indices. In: Proceedings of the 4th Industrial Conference on Data Mining (ICDM), pp. 96–105 (2004)
37. Gupta, M.D., Huang, T.S.: Regularized maximum likelihood for intrinsic dimension estimation. Comput. Res. Rep. (2012). CoRR abs/1203.3483
38. Halkidi, M., Batistakis, Y., Vazirgiannis, M.: On clustering validation techniques. J. Intell. Inf. Syst. **17**, 107–145 (2001)
39. Hinneburg, A., Aggarwal, C., Keim, D.A.: What is the nearest neighbor in high dimensional spaces? In: Proceedings of the 26th International Conference on Very Large Data Bases (VLDB), pp. 506–515. Morgan Kaufmann, New York, NY (2000)
40. Hou, J., Nayak, R.: The heterogeneous cluster ensemble method using hubness for clustering text documents. In: Lin, X., Manolopoulos, Y., Srivastava, D., Huang, G. (eds.) Proceedings of the 14th International Conference on Web Information Systems Engineering (WISE). Lecture Notes in Computer Science, vol. 8180, pp. 102–110. Springer, Berlin (2013)
41. Houle, M.E.: The relevant-set correlation model for data clustering. J. Stat. Anal. Data Min. **1**(3), 157–176 (2008)
42. Houle, M.E., Kriegel, H.P., Kröger, P., Schubert, E., Zimek, A.: Can shared-neighbor distances defeat the curse of dimensionality? In: Proceedings of the 22nd International Conference on Scientific and Statistical Database Management (SSDBM), pp. 482–500 (2010)
43. Hsu, C.M., Chen, M.S.: On the design and applicability of distance functions in high-dimensional data space. IEEE Trans. Knowl. Data Eng. **21**(4), 523–536 (2009)
44. Hubert, L., Arabie, P.: Comparing partitions. J. Classif. **2**(1), 193–218 (1985)
45. Hubert, L.J., Levin, J.R.: A general statistical framework for assessing categorical clustering in free recall. Psychol. Bull. **83**(6), 1072 (1976)
46. Jarvis, R.A., Patrick, E.A.: Clustering using a similarity measure based on shared near neighbors. IEEE Trans. Comput. **22**, 1025–1034 (1973)
47. Jing, L., Ng, M., Huang, J.: An entropy weighting k-means algorithm for subspace clustering of high-dimensional sparse data. IEEE Trans. Knowl. Data Eng. **19**(8), 1026–1041 (2007)

48. Kabán, A.: On the distance concentration awareness of certain data reduction techniques. Pattern Recogn. **44**(2), 265–277 (2011)
49. Kaban, A.: Non-parametric detection of meaningless distances in high dimensional data. Stat. Comput. **22**(2), 375–385 (2012)
50. Li, T., Ma, S., Ogihara, M.: Document clustering via adaptive subspace iteration. In: Proceedings of the 27th Annual International ACM SIGIR Conference on Research and Development in Information Retrieval, pp. 218–225 (2004)
51. Liu, B., Xia, Y., Yu, P.S.: Clustering through decision tree construction. In: Proceedings of the 26th ACM SIGMOD International Conference on Management of Data, pp. 20–29 (2000)
52. Liu, Y., Li, Z., Xiong, H., Gao, X., Wu, J.: Understanding of internal clustering validation measures. In: Proceedings of the 10th IEEE International Conference on Data Mining (ICDM), pp. 911–916 (2010)
53. Lu, Y., Wang, S., Li, S., Zhou, C.: Particle swarm optimizer for variable weighting in clustering high-dimensional data. Mach. Learn. **82**(1), 43–70 (2011)
54. Maulik, U., Bandyopadhyay, S.: Performance evaluation of some clustering algorithms and validity indices. IEEE Trans. Pattern Anal. Mach. Intell. **24**(12), 1650–1654 (2002)
55. Milligan, G.W.: A monte carlo study of thirty internal criterion measures for cluster analysis. Psychometrika **46**(2), 187–199 (1981)
56. Moise, G., Zimek, A., Kröger, P., Kriegel, H.P., Sander, J.: Subspace and projected clustering: experimental evaluation and analysis. Knowl. Inf. Syst. **21**(3), 299–326 (2009)
57. Müller, E., Günnemann, S., Assent, I., Seidl, T.: Evaluating clustering in subspace projections of high dimensional data. Proc. VLDB Endowment **2**(1), 1270–1281 (2009)
58. Naldi, M.C., Carvalho, A., Campello, R.J.: Cluster ensemble selection based on relative validity indexes. Data Min. Knowl. Disc. **27**(2), 259–289 (2013)
59. Ntoutsi, I., Zimek, A., Palpanas, T., Kröger, P., Kriegel, H.P.: Density-based projected clustering over high dimensional data streams. In: Proceedings of the 12th SIAM International Conference on Data Mining (SDM), pp. 987–998 (2012)
60. Pauwels, E.J., Frederix, G.: Cluster-based segmentation of natural scenes. In: Proceedings of the 7th IEEE International Conference on Computer Vision (ICCV), vol. 2, pp. 997–1002 (1999)
61. Pestov, V.: On the geometry of similarity search: Dimensionality curse and concentration of measure. Inf. Process. Lett. **73**(1–2), 47–51 (2000)
62. Pettis, K.W., Bailey, T.A., Jain, A.K., Dubes, R.C.: An intrinsic dimensionality estimator from near-neighbor information. IEEE Trans. Pattern Anal. Mach. Intell. **1**(1), 25–37 (1979)
63. Radovanović, M.: Representations and Metrics in High-Dimensional Data Mining. Izdavačka knjižarnica Zorana Stojanovića, Novi Sad, Serbia (2011)
64. Radovanović, M., Nanopoulos, A., Ivanović, M.: Nearest neighbors in high-dimensional data: The emergence and influence of hubs. In: Proceedings of the 26th International Conference on Machine Learning (ICML), pp. 865–872 (2009)
65. Radovanović, M., Nanopoulos, A., Ivanović, M.: Hubs in space: popular nearest neighbors in high-dimensional data. J. Mach. Learn. Res. **11**, 2487–2531 (2010)
66. Radovanović, M., Nanopoulos, A., Ivanović, M.: Time-series classification in many intrinsic dimensions. In: Proceedings of the 10th SIAM International Conference on Data Mining (SDM), pp. 677–688 (2010)
67. Radovanović, M., Nanopoulos, A., Ivanović, M.: Reverse nearest neighbors in unsupervised distance-based outlier detection. IEEE Trans. Knowl. Data Eng. **27**(5), 1369–1382 (2015)
68. Ratkowsky, D., Lance, G.: A criterion for determining the number of groups in a classification. Aust. Comput. J. **10**(3), 115–117 (1978)
69. Rousseeuw, P.J.: Silhouettes: a graphical aid to the interpretation and validation of cluster analysis. J. Comput. Appl. Math. **20**, 53–65 (1987)
70. Santos, J.M., Embrechts, M.: On the use of the adjusted rand index as a metric for evaluating supervised classification. In: Proceedings of the 19th International Conference on Artificial Neural Networks (ICANN), Part II. Lecture Notes in Computer Science, vol. 5769, pp. 175–184. Springer, Berlin (2009)

71. Schnitzer, D., Flexer, A., Schedl, M., Widmer, G.: Using mutual proximity to improve content-based audio similarity. In: Proceedings of the 12th International Society for Music Information Retrieval Conference (ISMIR), pp. 79–84 (2011)
72. Schnitzer, D., Flexer, A., Schedl, M., Widmer, G.: Local and global scaling reduce hubs in space. J. Mach. Learn. Res. **13**(1), 2871–2902 (2012)
73. Serpen, G., Pathical, S.: Classification in high-dimensional feature spaces: Random subsample ensemble. In: Proceedings of the International Conference on Machine Learning and Applications (ICMLA), pp. 740–745 (2009)
74. Talwalkar, A., Kumar, S., Rowley, H.A.: Large-scale manifold learning. In: Proceedings of the IEEE Conference on Computer Vision and Pattern Recognition (CVPR), pp. 1–8 (2008)
75. Tomašev, N.: Hub miner: A hubness-aware machine learning library. (2014). http://dx.doi.org/10.5281/zenodo.12599
76. Tomašev, N.: Taming the empirical hubness risk in many dimensions. In: Proceedings of the 15th SIAM International Conference on Data Mining (SDM), pp. 1–9 (2015)
77. Tomašev, N., Mladenić, D.: Nearest neighbor voting in high dimensional data: learning from past occurrences. Comput. Sci. Inf. Syst. **9**(2), 691–712 (2012)
78. Tomašev, N., Mladenić, D.: Hub co-occurrence modeling for robust high-dimensional kNN classification. In: Proceedings of the European Conference on Machine Learning (ECML), pp. 643–659. Springer, Berlin (2013)
79. Tomašev, N., Mladenić, D.: Hubness-aware shared neighbor distances for high-dimensional k-nearest neighbor classification. Knowl. Inf. Syst. **39**(1), 89–122 (2013)
80. Tomašev, N., Leban, G., Mladenić, D.: Exploiting hubs for self-adaptive secondary re-ranking in bug report duplicate detection. In: Proceedings of the Conference on Information Technology Interfaces (ITI) (2013)
81. Tomašev, N., Radovanović, M., Mladenić, D., Ivanović, M.: Hubness-based fuzzy measures for high-dimensional k-nearest neighbor classification. Int. J. Mach. Learn. Cybern. **5**(3), 445–458 (2014)
82. Tomašev, N., Rupnik, J., Mladenić, D.: The role of hubs in cross-lingual supervised document retrieval. In: Proceedings of the Pacific-Asia Conference on Knowledge Discovery and Data Mining (PAKDD), pp. 185–196. Springer, Berlin (2013)
83. Tomašev, N., Radovanović, M., Mladenić, D., Ivanović, M.: The role of hubness in clustering high-dimensional data. IEEE Trans. Knowl. Data Eng. **26**(3), 739–751 (2014)
84. Tomašev, N., Radovanović, M., Mladenić, D., Ivanović, M.: Hubness-based clustering of high-dimensional data. In: Celebi, M.E. (ed.) Partitional Clustering Algorithms, pp. 353–386. Springer, Berlin (2015)
85. Vendramin, L., Campello, R.J.G.B., Hruschka, E.R.: Relative clustering validity criteria: a comparative overview. Stat. Anal. Data Min. **3**(4), 209–235 (2010)
86. Vendramin, L., Jaskowiak, P.A., Campello, R.J.: On the combination of relative clustering validity criteria. In: Proceedings of the 25th International Conference on Scientific and Statistical Database Management, p. 4 (2013)
87. Verleysen, M., Francois, D., Simon, G., Wertz, V.: On the effects of dimensionality on data analysis with neural networks. In: Proceedings of the 7th International Work-Conference on Artificial and Natural Neural Networks, Part II: Artificial Neural Nets Problem Solving Methods, pp. 105–112. Springer, Berlin (2003)
88. Vinh, N.X., Houle, M.E.: A set correlation model for partitional clustering. In: Zaki, M.J., Yu, J.X., Ravindran, B., Pudi, V. (eds.) Advances in Knowledge Discovery and Data Mining. Lecture Notes in Computer Science, vol. 6118, pp. 4–15. Springer, Berlin (2010)
89. Wu, S., Feng, X., Zhou, W.: Spectral clustering of high-dimensional data exploiting sparse representation vectors. Neurocomputing **135**, 229–239 (2014)
90. Yianilos, P.N.: Locally lifting the curse of dimensionality for nearest neighbor search. In: Proceedings of the Eleventh Annual ACM-SIAM Symposium on Discrete Algorithms (SODA), pp. 361–370 (2000)

91. Yin, J., Fan, X., Chen, Y., Ren, J.: High-dimensional shared nearest neighbor clustering algorithm. In: Fuzzy Systems and Knowledge Discovery. Lecture Notes in Computer Science, vol. 3614, pp. 484–484. Springer, Berlin (2005)
92. Zhang, Z., Wang, J., Zha, H.: Adaptive manifold learning. IEEE Trans. Pattern Anal. Mach. Intell. **34**(2), 253–265 (2012)
93. Zheng, L., Huang, D.: Outlier detection and semi-supervised clustering algorithm based on shared nearest neighbors. Comput. Syst. Appl. **29**, 117–121 (2012)
94. Zimek, A., Schubert, E., Kriegel, H.P.: A survey on unsupervised outlier detection in high-dimensional numerical data. Stat. Anal. Data Min. **5**(5), 363–387 (2012)

Combinatorial Optimization Approaches for Data Clustering

Paola Festa

Abstract Target of cluster analysis is to group data represented as a vector of measurements or a point in a multidimensional space such that the most similar objects belong to the same group or cluster. The greater the similarity within a cluster and the greater the dissimilarity between clusters, the better the clustering task has been performed. Starting from the 1990s, cluster analysis has emerged as an important interdisciplinary field, applied to several heterogeneous domains with numerous applications, including among many others social sciences, information retrieval, natural language processing, galaxy formation, image segmentation, and biological data.

Scope of this paper is to provide an overview of the main types of criteria adopted to classify and partition the data and to discuss properties and state-of-the-art solution approaches, with special emphasis to the combinatorial optimization and operational research perspective.

1 Introduction

Given a finite set of *objects*, cluster analysis aims to group them such that the most similar ones belong to the same group or *cluster*, and dissimilar objects are assigned to different clusters. In the scientific literature, the objects are also called *entities* or *patterns* and are usually represented as a vector of measurements or a point in a multidimensional space. Clearly, it can be easily guessed that the greater the similarity within a cluster and the greater the dissimilarity between clusters, the better the clustering task has been performed.

Starting from the 1990s, cluster analysis has emerged as an important inter-disciplinary field, involving different scientific research communities, including mathematics, theoretical and applied statistics, genetics, biology, biochemistry, computer science, and engineering. It has been applied to several domains with numerous applications, ranging from social sciences to biology. Starting from one

P. Festa (✉)

Department of Mathematics and Applications "R. Caccioppoli",
University of Napoli FEDERICO II, Compl. MSA, Via Cintia, 80126 Napoli, Italy
e-mail: paola.festa@unina.it

© Springer International Publishing Switzerland 2016   109
M.E. Celebi, K. Aydin (eds.), *Unsupervised Learning Algorithms*,
DOI 10.1007/978-3-319-24211-8_5

of the pioneering paper of Rao, which appeared in 1971 [60], more recent surveys on clustering algorithms and their applications can be found in [8–11, 21, 22, 40, 41, 49, 53, 53, 77], and very recently in [29, 45, 71]. Nice books and edited books are [12, 56, 57, 76].

In cluster analysis, the criterion for evaluating the quality of a clustering strongly depends upon the specific application in which it is to be used. The cluster task can be mathematically formulated as a constrained fractional nonlinear 0–1 programming problem, and there are no computationally efficient procedures for solving such a problem, which becomes computationally tractable only under restrictive hypotheses.

This paper surveys the main types of clustering and criteria for homogeneity or separation, with special emphasis to the optimization and operational research perspective. In fact, first a few mathematical models of the problem are reported under several different criterion adopted to classify the data and some classical state-of-the-art exact methods are described that use the mathematical model of the problem. Then, the most famous and applied state-of-the-art clustering algorithms are reviewed, underlying and comparing their main ingredients.

The remainder of this paper is organized as follows. The next section lists some among the most useful applications of the problem. In Sect. 3, the cluster analysis task is formally stated and the most used distance measures between the various entities are described. State-of-the-art mathematical formulations of the problem along with some classical state-of-the-art exact methods are described in Sect. 4. In Sect. 5, properties and state-of-the-art solution approaches are discussed. Concluding remarks are given in the last section.

2   Applications

Cluster analysis applies in several heterogeneous domains with numerous applications, whose number grows increasingly in recent years. As the increasingly sophisticated technology allows the storage of increasingly large amounts of data, the availability of efficient techniques for generating new information by examining the resulting large databases becomes ever more urgent.

Some of cluster analysis applications are listed in the following.

- **Social sciences**: in this context, clustering helps in understanding how people analyze and catalogue life experiences.
- **Information retrieval**: in this context, clustering is used to create groups of documents with the goal of improving the efficiency and effectiveness of the retrieval, such as in the case of thousands of Web pages retrieved as a result of a query to a search engine (see, e.g., [16–18, 74]).
- **Natural language processing**: typically, large vocabularies of words of a given natural language must be clustered w.r.t. corpora of very high size [70].

- **Business**: in this application, one can be interested in analyzing information about potential customers, in order to cluster them for some sort of marketing activities [58].
- **Galaxy formation**: in this context, a study has been conducted on the formation of galaxies by gas condensation with massive dark halos [75].
- **Image segmentation**: in this case, the segmentation is achieved by searching for closed contours of the elements in the image [73].
- **Biological data**: one among the most prominent interests of the biologists is the analysis of huge data containing genetic information, such as to find groups of genes having similar functions (see among others [3, 20, 28, 30, 50, 54, 55]).

3 Problem Definition and Distance Measures Definition

In cluster analysis, we are given

◇ a set of N objects (entities, patterns) $O = \{o_1, \ldots, o_N\}$;
◇ a set of M of pre-assigned clusters $S = \{S_1, \ldots, S_M\}$;
◇ a function $d : O \times O \mapsto R$ that assigns to each pair $o_i, o_j \in O$ a "metric distance" or "similarity" $d_{ij} \in R$ (usually, $d_{ij} \geq 0$, $d_{ii} = 0$, $d_{ij} = d_{ji}$, for $i, j = 1, \ldots, N$)

and the objective is to assigning the objects in O to some cluster in S while optimizing some distance criteria in such a way that the greater the similarity (or proximity, homogeneity) within a cluster and the greater the difference between clusters, the better or more distinct the clustering [38, 40].

According to [38] and [40], the problem involves the following five steps:

1. pattern representation (optionally including feature extraction and/or selection);
2. definition of a pattern proximity (or similarity) measure appropriate to the data domain;
3. clustering or grouping;
4. data abstraction (if needed), and
5. assessment of output (if needed).

In more detail, the first step, usually referred to as pattern representation, refers to the number of classes, the number of available patterns, and the number, type, and scale of the features available to the clustering algorithm. Typical of this first step are the process of *feature selection*, i.e., the identification of the most effective subset of the original features to use in clustering, and the process of *feature extraction*, i.e., the transformations of the input features to produce new salient features.

Pattern proximity (similarity) is usually measured by a distance function defined on pairs of patterns. In the scientific literature, the patterns or data objects are usually represented as a vector of measurements or a point in a multidimensional space.

Formally, a data object o_i, $i = 1, \ldots, N$ can be represented as the following numerical vector

$$\overrightarrow{A}_i = \{a_{ij} \mid 1 \leq j \leq L\},$$

where

- a_{ij} is the value of the jth feature for the ith data object and
- L is the number of features.

Then, the proximity d_{ij} between two objects o_i and o_j is measured by a proximity function d of corresponding vectors $\overrightarrow{A}_i$ and $\overrightarrow{A}_j$.

Several different scientific communities have used and discussed a variety of distance measures (see, for example, [4, 37, 38, 41]). Some of them are listed in the following.

3.1 Euclidean Distance

Given two objects o_i and $o_j \in O$, their Euclidean distance in L-dimensional space is defined as follows:

$$d_{ij} = \sqrt{\sum_{k=1}^{L} (a_{ik} - a_{jk})^2} = \|\mathbf{a}_i - \mathbf{a}_j\|_2. \tag{1}$$

Euclidean distance has an intuitive meaning and it is usually used to evaluate the proximity of objects in two- or three-dimensional space. In general, it works well when the data set has "compact" or "isolated" clusters [51].

3.2 Pearson's Correlation Coefficient

The Pearson's correlation coefficient measures the similarity between the *shapes of two patterns*, also known as *profiles*.

Given two objects o_i and $o_j \in O$, their Pearson's correlation coefficient is defined as follows:

$$d_{ij} = \frac{\sum_{k=1}^{L} [(a_{ik} - \mu_{o_i}) \cdot ((a_{jk} - \mu_{o_j}))]}{\sqrt{\sum_{k=1}^{L} a_{ik} - \mu_{o_i})^2} \cdot \sqrt{\sum_{k=1}^{L} a_{jk} - \mu_{o_j})^2}}, \tag{2}$$

where μ_{o_i} and μ_{o_j} are the mean value for $\overrightarrow{A}_i$ and $\overrightarrow{A}_j$, respectively.

This correlation coefficient views each object as a random variable with L observations and measures the similarity between two objects by calculating the linear relationship between the distributions of the two corresponding random variables. One drawback of the Pearson's correlation coefficient is that it assumes an approximate Gaussian distribution of the patterns and may not be robust for non-Gaussian distributions, as experimentally shown by Bickel in 2001 [7].

3.3 City-Block or Manhattan

City-block or Manhattan distance simulates the distance between points in a city road grid. It measures the absolute differences between two object attributes.

Given two objects o_i and $o_j \in O$, their City-block or Manhattan distance is defined as follows:

$$d_{ij} = \sum_{k=1}^{L} |a_{ik} - a_{jk}|. \tag{3}$$

3.4 Cosine or Uncentered Correlation

Cosine or uncentered correlation is a geometric correlation defined by the angle between two objects.

Given two objects o_i and $o_j \in O$, their cosine or uncentered correlation is defined as follows:

$$D_{ij} = \frac{\sum_{k=1}^{L} a_{ik} \cdot a_{jk}}{\sum_{k=1}^{L} a_{ik}^2 \sum_{k=1}^{L} a_{jk}^2}. \tag{4}$$

Note that,

- the larger the value of D_{ij}, the lower the angle between the objects;
- $D_{ij} \in [-1, 1]$: $D_{ij} = -1$ implies that the angle between vectors representing o_i and o_j is a right angle; while $D_{ij} = 1$ implies that the angle between o_i and o_j is 0;
- $d_{ij} = 1 - |D_{ij}|$.

4 Mathematical Formulations of the Problem

In 1971, Rao [60] presented several mathematical formulations of the clustering problem, depending on several different criteria adopted to classify the data. Some of them are reported in the following.

4.1 Minimize (Maximize) the Within (Between)-Clusters Sum of Squares

Defining a set of $N \times M$ Boolean decision variables $x_{ik} \in \{0, 1\}$, $i = 1, \ldots, N$, $k = 1, \ldots, M$, such that in a solution

$$x_{ik} = \begin{cases} 1, & \text{if } o_i \in O \text{ is in cluster } S_k; \\ 0, & \text{otherwise,} \end{cases}$$

the problem admits the following fractional nonlinear mathematical formulation:

$$(DC-1) \quad \min \sum_{k=1}^{M} \left\{ \frac{\sum_{i=1}^{N-1} \sum_{j=i+1}^{N} d_{ij}^2 x_{ik} x_{jk}}{\sum_{i=1}^{N} x_{ik}} \right\}$$

s.t.

$$(DC-1.1) \quad \sum_{k=1}^{M} x_{ik} = 1, \qquad\qquad i = 1, \ldots, N$$

$$(DC-1.2) \quad x_{ik} \geq 0 \text{ and integer}, \qquad i = 1, \ldots, N, \ k = 1, \ldots, M.$$

Constraints (DC-1.1) assure that each o_i, $i = 1, \ldots, N$, belongs to only one cluster.

Since (DC-1) is a fractional nonlinear 0–1 programming problem, it is difficult to solve. Exceptions are the two cases described in the following two paragraphs.

4.1.1 Cardinality of Each Cluster A Priori Known

A special case of problem (DC-1) occurs when the cardinality of each cluster is a priori known, i.e., when

$$|S_k| = n_k, \text{ s.t. } \sum_{k=1}^{M} n_k = N.$$

In this case, mathematical formulation (DC-1) can be slightly modified to take into account this further information as follows:

$$(DC-2) \quad \min \sum_{k=1}^{M} \frac{1}{n_k} \left\{ \sum_{i=1}^{N-1} \sum_{j=i+1}^{N} d_{ij}^2 x_{ik} x_{jk} \right\}$$

s.t.

$$(DC-2.1) \quad \sum_{k=1}^{M} x_{ik} = 1, \qquad\qquad i = 1, \ldots, N$$

$$(DC - 2.2) \quad \sum_{i=1}^{N} x_{ik} = n_k, \qquad k = 1,\ldots,M$$

$$(DC - 2.3) \quad x_{ik} \geq 0 \text{ and integer,} \quad i = 1,\ldots,N, \; k = 1,\ldots,M.$$

In formulation (DC-2), the objective remains to minimize (maximize) the within (between)-clusters sum of squares. As in (DC-1), constraints (DC-2.1) guarantee that each object o_i, $i = 1,\ldots,N$, belongs to only one cluster. The additional set of constraints (DC-2.2) impose that for each cluster S_k, $k = 1,\ldots,M$, the number of objects in S_k is equal to n_k.

(DC-2) is still a nonlinear 0–1 formulation, but its objective function has lost the fractional characteristics, since here $\{n_k\}_{k=1}^{M}$ are known in advance. Among the first approaches to solve this problem are to be counted the Boolean methods proposed by Hammer and Rudeanu in [35] to be applied once the clustering problem is interpreted as a constrained nonlinear Boolean programming problem. Another pioneering approach has been described by Rao in [60]. It first linearizes the objective function by adding a further set of constraints. Then, it solves the resulting problem by applying any known method for linear integer problems. The resulting linearized 0–1 problem admits the following formulation:

$$(DC' - 2) \quad \min \sum_{k=1}^{M} \frac{1}{n_k} \left\{ \sum_{i=1}^{N-1} \sum_{j=i+1}^{N} d_{ij}^2 \, y_{ij}^k \right\}$$

s.t.

$$(DC' - 2.1) \quad x_{ik} + x_{jk} - y_{ij}^k \leq 1, \qquad i = 1,\ldots,N-1, \; j = i+1,\ldots,N,$$

$$k = 1,\ldots,M$$

$$(DC' - 2.2) \quad \sum_{k=1}^{M} x_{ik} = 1, \qquad i = 1,\ldots,N$$

$$(DC' - 2.3) \quad \sum_{i=1}^{N} x_{ik} = n_k, \qquad k = 1,\ldots,M$$

$$(DC' - 2.4) \quad x_{ik}, \; y_{ij}^k \geq 0 \text{ and integer,} \quad i,j = 1,\ldots,N, \; k = 1,\ldots,M.$$

It must be underlined that, although easier to solve formulation (DC'-2) presents a number of constraints that grows rapidly with N and M and therefore can only be used for small instances of the problem.

4.1.2 Bipartition of the Patterns

Another special case occurs when $M = 2$, i.e., when there are only two pre-assigned clusters S_1 and S_2. In this particular scenario, introducing N Boolean decision variables $x_i \in \{0, 1\}$, $i = 1, \ldots, N$, such that in a solution

$$x_i = \begin{cases} 1, & \text{if } o_i \in O \text{ is in cluster } S_1; \\ 0, & \text{if } o_i \in O \text{ is in cluster } S_2, \end{cases}$$

the clustering task can be modeled as the following fractional nonlinear 0–1 programming problem with no additional constraints:

$$(DC-3) \quad \min \left\{ \frac{\sum_{i=1}^{N-1} \sum_{j=i+1}^{N} d_{ij}^2 x_i x_j}{\sum_{i=1}^{N} x_i} + \frac{\sum_{i=1}^{N-1} \sum_{j=i+1}^{N} d_{ij}^2 (1 - x_i)(1 - x_j)}{N - \sum_{i=1}^{N} x_i} \right\}$$

s.t.

$$(DC-3.1) \quad x_i \in \{0, 1\}, \ i = 1, \ldots, N.$$

Based on the above mathematical formulation, Rao [60] designed an exact method that he presented as a Branch and Bound algorithm, but that is quite a method of exploring the feasible region making use of appropriate lower bounds obtained by means of any relaxation. The root of the decision tree corresponds to the empty solution in which any object has been assigned yet. An object o_i is selected and a branch is emanated: for convenience, the assignment of o_i to cluster S_1 corresponds to the right branch and that to cluster S_2 to the left branch. At a generic iteration, each current leaf of the tree corresponds to a partial solution characterized by some objects assigned to S_1, some assigned to S_2, while others are yet to be assigned.

Generally speaking, in order to design any Branch and Bound approach, it is necessary to specify the following key issues:

- the branching rule;
- the bounding strategy;
- the fathoming criterion.

About the branching rule, let us suppose that at a generic iteration t the algorithm is visiting node t of the tree that corresponds to a partial solution, where we have a set (possibly empty) of objects in S_1 and a set (possibly empty) of objects in S_2. Let $\overline{O} = O \setminus \{S_1 \cup S_2\}$ be the set of objects that are still to be assigned and let x_{best} be the incumbent solution, i.e., the best feasible solution found so far (at the beginning, x_{best} either is a known feasible solution or it is empty). If $\overline{O} = \emptyset$, then a complete solution has been found and x_{best} is eventually updated. Otherwise, a branching operation should be performed. Before performing a further branching, the so-called bounding criterion is tested, i.e., a lower bound for the objective function is computed (see hereafter how to compute a valid lower bound) and if the lower bound is greater than or equal to the objective function value corresponding to x_{best}, then no branching is performed since any feasible solution that can be generated from node t will be no better than the incumbent solution itself. Conversely, if the bounding criterion fails, a branching operation is performed by assigning to cluster S_1 an object $k \in \overline{O}$ such that

$$o_k = \arg\min_{j \in \overline{O}} \sum_{i \in S_1} d_{ij},$$

i.e., o_k is a not yet assigned object that corresponds to the minimum sum of the distances to all other objects already assigned to cluster S_1.

Generally speaking, the strategy for selecting the next sub-problem to be investigated determines how the Branch and Bound algorithm should proceed through the search tree and can have a significant effect on the behavior of the algorithm (fathoming rule). In [60], the proposed approach adopts a depth first search (DFS, for short) strategy, where the node with the largest level in the search tree is chosen for exploration. The algorithm continues branching until the end of the tree is reached or the bounding criterion indicates any branching is further needed. When this happens, the algorithm back-tracks along the tree until a node is reached with an unexplored branch to the left. A graphical representation of the branching rule is depicted in Fig. 1.

To obtain a lower bound, Rao [60] proposed several different strategies, based on the following further definitions and notation.

At a given node t of the branching tree, let x be a Boolean partial solution defined as in formulation (DC-3) and $\overline{O} = O \setminus \{S_1 \cup S_2\}$. Moreover, let

Fig. 1 Graphical representation of the branching rule

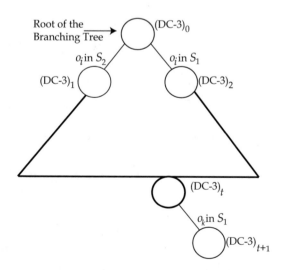

$\diamond$ $K_1 = \sum\limits_{i,j \in S_1} d_{ij}$;

$\diamond$ $K_2 = \sum\limits_{i,j \in S_2} d_{ij}$;

$\diamond$ D be the $n \times n$ symmetric distance matrix;

$\diamond$ F be a $|\overline{O}| \times |S_1|$ sub-matrix of D containing the distance between each currently not yet assigned object $k \in \overline{O}$ and an object in S_1;

$\diamond$ $F_i, i = 1, \ldots, |\overline{O}|$, be the sum of the distances in row i of matrix F;

$\diamond$ H be a $|S_2| \times |S_1|$ sub-matrix of D containing the distance between each object in S_2 and an object in S_1;

$\diamond$ $H_i, i = 1, \ldots, |S_1|$, be the sum of the distances in row i of matrix H;

$\diamond$ C be a $|\overline{O}| \times |\overline{O}|$ sub-matrix of D containing the distance between each pair of unassigned objects, whose diagonal elements are assigned a very high value $(+\infty)$.

Without loss of generality, matrices F and H are assumed arranged in such a way that

$$F_i < F_j, \text{ if } i < j;$$
$$H_i < H_j, \text{ if } i < j.$$

The objective function can be rewritten as follows:

$$Z = \min \left\{ \frac{K_1 + \sum\limits_{i \in S_1, j \in \overline{O}} d_{ij} x_i x_j + \sum\limits_{i,j \in \overline{O}} d_{ij} x_i x_j}{|S_1| + \sum\limits_{i \in \overline{O}} x_i} \right.$$

$$\left. + \frac{K_2 + \sum\limits_{i \in S_2, j \in \overline{O}} d_{ij}(1 - x_i)(1 - x_j) + \sum\limits_{i,j \in \overline{O}} d_{ij}(1 - x_i)(1 - x_j)}{|S_2| + \left(|\overline{O}| - \sum\limits_{i \in \overline{O}} x_i \right)} \right\}, \quad (5)$$

where $x_i \in \{0, 1\}$.

For any fixed value for $\bar{n} = \sum\limits_{i \in \overline{O}} x_i$, by setting $n' = |\overline{O}| - \bar{n}$, $p = |S_1| + \bar{n}$, and $t = |S_2| + n'$, Z in (5) can be rewritten as follows:

$$Z' = \min t \cdot \left[\sum\limits_{i \in S_1, j \in \overline{O}} d_{ij} x_i x_j + \sum\limits_{i,j \in \overline{O}} d_{ij} x_i x_j \right]$$

$$+ p \cdot \left[\sum\limits_{i \in S_2, j \in \overline{O}} d_{ij}(1 - x_i)(1 - x_j) + \sum\limits_{i,j \in \overline{O}} d_{ij}(1 - x_i)(1 - x_j) \right], \quad (6)$$

given that the denominator (equal to pt), $K_1 t$, and $K_2 p$ are constant.

By varying $\bar{n}$ in its range from 0 to $|\overline{O}|$, $|\overline{O}| + 1$ lower bounds for Z' (6) are computed and the minimum among them is kept as lower bound for Z (5). One way to obtain such a lower bound is the following:

$$Z' \geq u_1 + u_2 + u_3,$$

where

$$u_1 = t \cdot \min \sum_{i \in S_1, j \in \overline{O}} d_{ij} x_i x_j = t \cdot \sum_{i=1}^{\bar{n}} F_i;$$

$$u_2 = p \cdot \min \sum_{i \in S_2, j \in \overline{O}} d_{ij}(1 - x_i)(1 - x_j) = p \cdot \sum_{i=1}^{n'} H_i;$$

$$u_3 = \left[t \cdot \min \sum_{i, j \in \overline{O}} d_{ij} x_i x_j \right] + \left[p \cdot \min \sum_{i, j \in \overline{O}} d_{ij}(1 - x_i)(1 - x_j) \right].$$

Note that, the number of addenda in the summation to compute u_3 is $v = \frac{\bar{n} \cdot (\bar{n}-1) + n'(n'-1)}{2}$, lying either in the upper or in the lower triangular part of the symmetric matrix C. Consequently, a lower bound for u_3 can be obtained by multiplying $\min\{t, p\}$ by the sum of the v smallest elements of one of the triangle of matrix C. The reader interested in learning further techniques to compute a lower bound can refer to Rao's paper [60].

4.2 Optimizing the Within Clusters Distance

If one is interested in minimizing the total within clusters distance, the clustering task objective can be formulated as follows:

$$(DC - 4) \quad \min \sum_{k=1}^{M} \left\{ \sum_{i=1}^{N-1} \sum_{j=i+1}^{N} d_{ij} \, x_{ik} \, x_{jk} \right\}$$

s.t.

$$(DC - 4.1) \quad \sum_{k=1}^{M} x_{ik} = 1, \qquad\qquad i = 1, \ldots, N$$

$$(DC - 4.2) \quad x_{ik} \geq 0 \text{ and integer}, \qquad i = 1, \ldots, N, \ k = 1, \ldots, M.$$

Note that, the objective function of (DC-4) is similar to the objective function of (DC-2), with the only differences that here the factors $\frac{1}{n_k}$, $k = 1, \ldots, M$, must not appear and coherently constraints (DC-2.2) do not require to be imposed.

A further useful possible target of a clustering task is to minimize the maximum within cluster distance. In this case, one is basically interested in minimizing the maximum distance within clusters and the problem can be modeled as a linear integer programming problem as follows:

(DC − 5)   min Z

s.t.

(DC − 5.1) $d_{ij} x_{ij} + d_{ij} x_{jk} - Z \leq d_{ij}, \ i = 1, \ldots, N - 1$
$$j = i + 1, \ldots, N$$
$$k = 1, \ldots, M$$

(DC − 5.2) $\displaystyle\sum_{k=1}^{M} x_{ik} = 1, \qquad\qquad i = 1, \ldots, N, \ k = 1, \ldots, M$

(DC − 5.3) $x_{ik} \geq 0$ and integer, $i = 1, \ldots, N, \ k = 1, \ldots, M$

(DC − 5.4) $Z \geq 0$ and integer.

In the case of minimizing the distance between the objects inside the same cluster, in 2010 Nascimento et al. [55] slightly modified Rao's model (DC-4) as follows:

(DC − 6) $\displaystyle\min \sum_{i=1}^{N-1} \sum_{j=i+1}^{N} d_{ij} \sum_{k=1}^{M} x_{ik} \cdot x_{jk}$

s.t.

(DC − 6.1) $\displaystyle\sum_{k=1}^{M} x_{ik} = 1, \qquad\qquad i = 1, \ldots, N$

(DC − 6.2) $\displaystyle\sum_{i=1}^{N} x_{ik} \geq 1, \qquad\qquad k = 1, \ldots, M$

(DC − 6.3) $x_{ik} \in \{0, 1\}, \qquad\qquad i = 1, \ldots, N, \ k = 1, \ldots, M.$

Note that, a set of additional constraints (DC-6.2) are imposed to guarantee that each cluster S_k, $k = 1, \ldots, M$, contains at least one object. In the attempt of remedying to the nonlinearity of the objective function, Nascimento et al. proposed a linearization of the model (DC-6). In more detail, by defining a further decision vector $y \in \mathscr{R}^{N \times N}$, the linearized version of formulation (DC-6) is the following:

$$(LDC - 6) \quad \min \sum_{i=1}^{N-1} \sum_{j=i+1}^{N} d_{ij} \cdot y_{ij}$$

s.t.

$$(LDC - 6.1) \quad \sum_{k=1}^{M} x_{ik} = 1, \qquad i = 1, \ldots, N$$

$$(LDC - 6.2) \quad \sum_{i=1}^{N} x_{ik} \geq 1, \qquad k = 1, \ldots, M$$

$$(LDC - 6.3) \quad x_{ik} \in \{0, 1\}, \qquad i = 1, \ldots, N, \ k = 1, \ldots, M$$

$$(LDC - 6.4) \quad y_{ij} \geq x_{ik} + x_{jk} - 1, \ i = 1, \ldots, N, \ j = i+1, \ldots, N, \ k = 1, \ldots, M$$

$$(LDC - 6.5) \quad y_{ij} \geq 0, \qquad i = 1, \ldots, N, \ j = i+1, \ldots, N.$$

Constraints (LCD-6.4) and (LCD-6.5) guarantee that $y_{ij} = 1$ if $x_{ik} = x_{jk} = 1$, i.e., if objects $o_i, o_j \in O$ are in the same cluster. Therefore, it can be easily seen that the objective function of model (LCD-6) aims at minimizing the distance between objects in the same cluster. Note that, model (LCD-6) has $\frac{N \cdot (N-1) \cdot (M+1)}{2}$ more constraints than (DC-6) but it is linear and therefore "easier" to be solved.

Exact methods [35, 60] and the above-described mathematical formulations of the problem can be used only for small- sized instances, since the number of constraints characterizing the models increases very rapidly with both the number of objects N and the number of pre-assigned clusters M.

5 A Review of the Most Popular Clustering Techniques

According to Jain et al. [40] (see the taxonometric representation of clustering methods in Fig. 2), state-of-the-art clustering algorithms can be mainly divided into two families: *partitioning* and *hierarchical algorithms*.

A partitioning method partitions the set of data objects into non-overlapping clusters such that each data object belongs to exactly one cluster. Instead, in a hierarchical approach a cluster is permitted to have subclusters and the result of the clustering task is a set of nested clusters that can be organized in a tree. Each node of the tree corresponds to a cluster and it is the union of its children (subclusters). Clearly, the leaves have no subclusters and the root node represents the cluster containing all the objects.

Besides *exclusive/non-overlapping* versus *overlapping* clustering, in the literature several different further types of clusterings can be found, such as *fuzzy clustering* and *probabilistic clustering*.

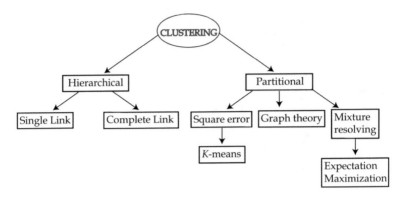

Fig. 2 A taxonomy of clustering methods

In a fuzzy clustering, clusters are viewed as fuzzy sets and a *membership weight function* $W : O \times S \mapsto [0, 1]$ is defined such that for each object $o_i \in O, i = 1, \ldots, N$, and for each cluster $S_k, k = 1, \ldots, M$, W_{ik} measures a "level" of membership of o_i to cluster S_k. Clearly, $W_{ik} = 0$ corresponds to "absolutely $o_i \notin S_k$, while $W_{ik} = 1$ corresponds to "absolutely $o_i \in S_k$.

Similarly to fuzzy clustering, a probabilistic clustering requires the computation of the probability p_{ik} with which each object $o_i \in O, i = 1, \ldots, N$, belongs to each cluster $S_k, k = 1, \ldots, M$. Clearly, it must be imposed that these probabilities must sum to 1. In general, probabilistic clustering are converted to an exclusive/non-overlapping clustering, where each object is assigned to the cluster in which its probability is highest.

5.1 Hierarchical Clustering Algorithms

The most popular hierarchical clustering algorithms are the single-link algorithm [65], complete-link [46], and minimum-variance algorithms [72]. The single-link and the complete-link approaches differ in how they define the similarity between a pair of clusters: in the single-link approach, this distance is the minimum of the distances between all pairs of patterns drawn from the two clusters (one pattern from the first cluster, the other from the second); in a complete-link algorithm, this distance is the maximum of all pairwise distances between patterns in the two clusters. In either case, two clusters are merged to form a larger cluster based on minimum distance criteria.

5.2 Partitioning Clustering Algorithms

The most popular partitioning clustering algorithms are the squared error algorithms (among them the most famous are the k-means method [52] and the k-medoid method [43, 44]), graph-theoretic algorithms [78], and mixture-resolving and mode-seeking algorithms [38].

5.2.1 Squared Error Algorithms and the k-Means/k-Medoid Algorithms

The squared error criterion is the most intuitive and used criterion among partitioning clustering techniques. As the Euclidean distance, it tends to work well with "isolated" and "compact" clusters.

Given a clustering S of a set of patterns O containing M clusters, the squared error for S, also known as *scatter*, is defined as

$$E^2(O, S) = \sum_{j=1}^{M} \sum_{i=1}^{n_j} \|\mathbf{a}_i^{(j)} - \mathbf{c}_j\|^2,$$

where

◇ n_j is the number of objects in cluster $j \in \{1, \dots, M\}$;
◇ $\mathbf{a}_i^{(j)}$ the ith pattern belonging to cluster $j \in \{1, \dots, M\}$;
◇ $\mathbf{c}_j$ is the centroid of cluster $j \in \{1, \dots, M\}$.

The framework of a generic squared error algorithm is described in Fig. 3.

The k-means algorithm [52] and the k-medoid algorithm [43] are the simplest and maybe most famous approaches that use a squared error criterion. k-means relies on the definition of a centroid usually as the mean of a group of objects and it is typically applied to objects in a continuous n-dimensional space. Conversely, k-medoid algorithm relies on the definition of a *medoid* as the most representative object for a group of objects and in principle can be used in a wider range of contexts compared to k-means, since it only requires a suitable definition of a proximity

---

algorithm squared-error ($O = \{o_1, \dots, o_N\}$, M)
1 Select an initial partition $S = \{S_1, \dots, S_M\}$ of the patterns in O with a fixed number M of clusters and cluster centers.
2 Assign each pattern o_i, $i = 1, \dots, N$, to its closest cluster center and compute the new cluster centers as the centroids of the clusters.
   Repeat this step until convergence is achieved, i.e., until the cluster membership is stable.
3 Merge and split clusters based on some heuristic information, optionally repeating step 2.
end squared-error

Fig. 3 Framework of a general squared error algorithm

```
algorithm k-means (O = {o₁,...,oₙ}, M)
1 Select from O a set of M patterns as initial centroids.
2 Repeat
        Form M clusters by assigning each pattern to its closest centroid.
        Re-compute the centroid of each cluster.
3 until a stopping criterion is met.
end k-means
```

Fig. 4 Framework of the k-means algorithm

```
algorithm bisection k-means (O = {o₁,...,oₙ}, M, n)
1 S := {S₁} := {O}
2 Repeat
        Select and remove a cluster Sₜ from the set S
        For n trials, bisect Sₜ in S¹ₜ and S²ₜ using k-means and retaining the best
        bisection Sᵇ¹ₜ and Sᵇ²ₜ
        S := S ∪ {Sᵇ¹ₜ, Sᵇ²ₜ}
3 until |S| = M
end bisection k-means
```

Fig. 5 Framework of the bisection k-means algorithm

measure among objects. It has to be furthermore underlined that in a k-medoid algorithm, the medoid is by definition an object of the given set of objects O, while in a k-means approach this property is not necessarily satisfied by a centroid.

A typical k-means algorithm is described in Fig. 4. It starts with a random initial partition of the data objects and keeps reassigning objects to "close" clusters until a convergence criterion is met. The most used stopping criteria are no (or minimal) reassignment of patterns to new cluster centers or minimal decrease in squared error. In this latter case, when the stopping criterion is related to the squared error, to evaluate a clustering it is generally used as objective function a function that takes into account the squared error: among different clustering of the same set of objects it is preferred the one corresponding to the minimum squared error, since this means that its centroids better represent the objects in their own cluster.

The main drawback of the k-means algorithm is that the quality of the solution it returns strongly depends upon the selection of the initial partition. If the initial partition is not properly chosen, the approach may converge to a local minimum of the criterion function value. In the attempt to overcome this important drawback of the k-means approach, a variant of the method has been proposed called *bisection* k-means [67, 68]. The *bisection* k-means initially splits the objects into two clusters, then further splits one of the just created clusters, and so on, iteratively, until M clusters are individuated (obtaining, as side effect, hierarchical clusters).

A typical bisection k-means approach is described in Fig. 5. In line 1, the algorithm initializes the set of clusters to contain only one cluster S_1 containing all objects. Then, in the loop at lines 2–3, until a set S of M clusters is attained, the algorithm iteratively selects and removes from the current set S a cluster S_t and for a given in input number n of trials it bisects S_t into two clusters, retaining the best bisection S_t^{b1} and S_t^{b2} (corresponding, for example, to the lowest squared error) and adding it to the set S under construction.

The study and the design of efficient initialization methods for the k-means technique is still a research topic that attracts the efforts by various scientific communities. A recent survey and comparative study of the existing methods has been conducted by Celebi et al. [15], who cited as particularly interesting two hierarchical initialization methods named Var-Part and PCA-Part proposed in 2007 by Su and Dy [69]. These two methods are not only linear, deterministic, and order-invariant. Besides these nice characteristics, in a recent paper by Celebi and Kingravi [13], a discriminant analysis-based approach has been proposed that addresses a common deficiency of these two methods. A deep experimental analysis showed that the two methods are highly competitive with state-of-the-art best random initialization methods to date and that the proposed approach significantly improves the performance of both hierarchical methods. Finally, in [14], Celebi and Kingravi presented an in-depth comparison of six linear, deterministic, and order-invariant initialization methods.

5.2.2   Graph-Theoretic Algorithms

Most of the graph-theoretic algorithms are divisive approaches, i.e., they start with one cluster that contains all the objects and then at each step they split a cluster. Specularly, agglomerative algorithms start with each pattern in a distinct (singleton) cluster, and successively merge clusters together until a stopping criterion is satisfied.

The graph-theoretic algorithms use a graph representation of the Data sets. Formally, given the set of objects $O = \{o_1, \ldots, o_N\}$ and the distance function $d : O \times O \mapsto R$, a weighted undirected graph $G = (V, E, w)$ can be defined such that

- $V = O$;
- edges in E indicate the relationship between objects;
- $w_{ij} = d_{ij}, \forall\, i, j \in V$ (i.e., $o_i, o_j \in O$).

The graph G is usually called *proximity graph*. It is very easy to see that each cluster can correspond to a *connected component* of G, i.e., a subset of nodes/objects that are connected to one another [63]. A further stronger possible graph-theoretic approaches family looks for *cliques* in G, i.e., sets of nodes/objects in the graph that are completely connected to each other [6, 55].

The easiest and maybe most famous graph-theoretic divisive clustering algorithm is based on the construction of a minimal spanning tree (MST) of the graph G [78]. Once obtained a MST, it generates clusters by deleting from the MST the edges with the largest weights.

5.2.3 Mixture-Resolving Algorithms

The idea behind this family of clustering algorithms is that the objects in O are drawn from one of several distributions (usually, Gaussian) and the goal is to identify the parameters of each distribution (e.g., a maximum likelihood estimate). Traditional approaches to this problem involve obtaining (iteratively) a maximum likelihood estimate of the parameter vectors of the component densities.

Among these techniques, it has to be cited the Expectation Maximization (EM) algorithm [19], a general purpose maximum likelihood algorithm for missing-data problems. The EM algorithm has been applied to the problem of parameter estimation.

5.3 Efficient Metaheuristic Approaches

Metaheuristics for data clustering have been designed only in the last 20 years. They include artificial neural networks [39, 64], evolutionary approaches such as genetic algorithms [42, 59], simulated annealing [47], and tabu search [2, 31].

Starting from 2010, GRASP (Greedy Randomized Adaptive Search Procedure) algorithms [24, 30, 55] have also been proposed that model the problem of pattern clustering as a combinatorial optimization problem defined on the weighted graph $G = (V, E, w)$ representing the data sets, as also used by graph-theoretic algorithms. GRASP, originally proposed by Feo and Resende [23] for set covering, has been applied in a wide range of problem areas [25–27, 61, 62]. It is a multi-start process, where at each iteration, a greedy randomized solution is constructed to be used as a starting solution for local search. Local search repeatedly substitutes the current solution by a better solution in the neighborhood of the current solution. If there is no better solution in the neighborhood, the current solution is returned as a local minimum and the search stops. The best local minimum found over all GRASP iterations is output as the final solution.

In 2010, Nascimento et al. [55] proposed a GRASP algorithm to cluster biological data sets. The greedy randomized solution is iteratively built, starting from a complete graph $\left(|E| = \frac{N \cdot (N-1)}{2}\right)$, indicating that the data set forms a unique cluster. Then, at each iteration of the construction procedure edges are gradually eliminated from the graph, creating unconnected full subgraphs (cliques), each representing a cluster. The edge elimination follows a greedy randomized criterion that selects at random one edge in a subset (RCL—Restricted Candidate List) of higher weighted edges. Once a greedy randomized clustering $S = \{S_1, \dots, S_M\}$ has been obtained, a local search procedure is applied starting from S to attempt to improve it. The local search works in an iterative fashion, until no better solution is found in the neighborhood. At each iteration, it replaces the current solution S by a better solution in the neighborhood $N(S)$ obtained by transferring in S an object from a cluster to another one.

In 2011, Frinhani et al. [30] described several hybrid GRASP with path-relinking heuristics [32, 48, 62] for data clustering. In a GRASP with path-relinking [1, 48, 62], at each GRASP iteration an elite set of good-quality solutions is stored and eventually updated. In fact, the local optimal current solution is combined with a randomly selected solution from the elite set using the path-relinking operator. The best of the combined solutions is a candidate for inclusion in the elite set and is added to the elite set if it meets quality and diversity criteria. The path-relinking procedure proposed in [30] applies to a pair of solutions S_s (starting solution) and S_t (target solution). Initially, the procedure computes the symmetric difference $\Delta(S_s, S_t)$, i.e., the set of moves needed to reach S_t (target solution) from S_s (initial solution). A path of solutions in the solution space is generated linking S_s and S_t and the best solution S^* in this path is returned by the algorithm. At each step, the procedure examines all moves $m \in \Delta(S, S_t)$ from the current solution S and selects the one corresponding to the least cost solution, i.e., the one which minimizes the objective function evaluates in $S \oplus m$, the solution resulting from applying move m to solution S. The best move m^* is made, producing solution $S \oplus m^*$. The set of available moves is updated. If necessary, the best solution S^* is updated. The procedure terminates when S_t is reached, i.e., when $\Delta(S, S_t) = \emptyset$.

In 2013 [28], a Biased Random-Key Genetic Algorithm (BRKGA) has been proposed for data clustering. It is well known that in the attempt of finding good quality solutions for a combinatorial optimization problem, Genetic Algorithms (GAs) [33, 36] implement the concept of *survival of the fittest*, making an analogy between a solution and an *individual* in a *population*. In particular, each individual of the current population represents a feasible solution, that is encoded by a corresponding *chromosome* that consists of a string of *genes*. Each gene can take on a value, called an *allele*, from some alphabet. For each chromosome it is possible to evaluate its *fitness level*, which is clearly correlated to the corresponding objective function value of the solution it encodes. GAs keep proceeding over a number of iterations, called *generations*, evolving a population of chromosomes. This *evolution* is implemented by simulating the process of natural selection through mating and mutation.

Genetic algorithms with random keys, or *random-key genetic algorithms* (RKGA), were introduced by Bean [5]. In a RKGA, chromosomes are represented as vectors of randomly generated real numbers in the interval $(0, 1]$. A deterministic algorithm, called a *decoder*, takes as input a solution vector and associates with it a solution of the combinatorial optimization problem for which an objective value or fitness can be computed.

A RKGA evolves a population of random-key vectors over a number of generations. The initial population is made up of $p > 0$ vectors of random-keys. Each component of the solution vector is generated independently at random in the real interval $(0, 1]$. After the fitness of each individual is computed by the decoder in generation t, the population is partitioned into two groups of individuals: a small group of p_e *elite* individuals, i.e., those with the best fitness values, and the remaining set of $p - p_e$ *non-elite* individuals. To evolve the population, a new generation of individuals must be produced. All elite individuals of the population

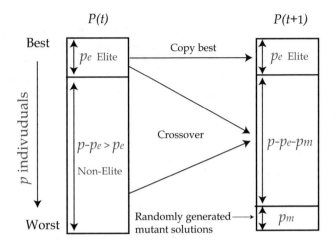

Fig. 6 BRKGA: generation $t + 1$ from generation t

of generation t are copied without modification to the population of generation $t+1$. RKGAs implement mutation by introducing *mutants* into the population. A mutant is simply a vector of random keys generated in the same way that an element of the initial population is generated. At each generation (see Fig. 6), a small number (p_m) of mutants are introduced into the population. With the p_e elite individuals and the p_m mutants accounted for in population $k + 1$, $p - p_e - p_m$ additional individuals need to be produced to complete the p individuals that make up the new population. This is done by producing $p - p_e - p_m$ offspring through the process of mating or crossover.

Bean [5] selects two parents at random from the entire population to implement mating in a RKGA. A *biased random-key genetic algorithm*, or BRKGA [34], differs from a RKGA in the way parents are selected for mating. In a BRKGA, each element is generated combining one element selected at random from the elite partition in the current population and one from the non-elite partition. Repetition in the selection of a mate is allowed and therefore an individual can produce more than one offspring in the same generation. As in RKGA, *parametrized uniform crossover* [66] is used to implement mating in BRKGAs. Let $\rho_e > 0.5$ be the probability that an offspring inherits the vector component of its elite parent. Let m denote the number of components in the solution vector of an individual. For $i = 1, \ldots, m$, the ith component C_i of the offspring vector C takes on the value of the ith component E_i of the elite parent E with probability ρ_e and the value of the ith component $\bar{E}_i$ of the non-elite parent $\bar{E}$ with probability $1 - \rho_e$.

When the next population is complete, i.e., when it has p individuals, fitness values are computed by the decoder for all of the newly created random-key vectors and the population is partitioned into elite and non-elite individuals to start a new generation.

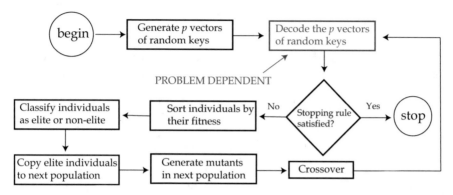

Fig. 7 BRKGA: flow chart

A BRKGA searches the solution space of the combinatorial optimization problem indirectly by searching the continuous m-dimensional hypercube, using the decoder to map solutions in the hypercube to solutions in the solution space of the combinatorial optimization problem where the fitness is evaluated. As underlined in [34], it is evident then that any BRKGA heuristic is based on a general-purpose metaheuristic framework (see Fig. 7), in which there is a portion totally independent from the specific problem to be solved. The only connection to the combinatorial optimization problem being solved is the problem-dependent portion of the algorithm, where the decoder produces solutions from the vectors of random-keys and computes the fitness of these solutions. Therefore, to describe a BRKGA for a specific combinatorial optimization problem, one needs only to show how solutions are encoded as vectors of random keys and how these vectors are decoded to feasible solutions of the optimization problem. We report in the following encoding and decoding schemes proposed in [28] for data clustering problems.

A solution is denoted as $x = (x_1, \ldots, x_N)$, such that for all $i = 1, \ldots, N$, $x_i \in \{1, \ldots, M\}$, i.e.,

$$\forall\, i = 1, \ldots, N, \quad x_i = k, \text{ iff } o_i \in S_k,\ k \in \{1, \ldots, M\}.$$

5.4 Encoding

A solution is encoded as a random-key vector $X = (X_1, X_2, \ldots, X_N)$, where for $i = 1, \ldots, N$, component X_i is a real number in the interval $(0, 1]$.

5.5 Decoding

Given an encoded solution $X = (X_1, X_2, \ldots, X_N)$ and representing the set S as an array of $|S| = M$ elements, decoding consists of three steps and produces a string $x = (x_1, x_2, \ldots, x_N)$.

In the first step, the string x is computed such that, for each $i = 1, \ldots, N$,

$$x_i = \left\lceil X_i \cdot \frac{1}{\Delta} \right\rceil, \qquad \Delta = \frac{1}{M}.$$

In the second step, starting from x, a locally optimal solution $\hat{x} = (\hat{x}_1, \hat{x}_2, \ldots, \hat{x}_N)$ with respect to the 2-swap neighborhood is computed.

Finally, in the third step, a chromosome adjustment is made to the encoded solution such that the resulting chromosome $\hat{X}$ decodes directly to $\hat{x}$ using only the first step of the decoding process.

For $j = 1, \ldots, |S| = M$, let

$$l_j = (j-1) \cdot \Delta;$$
$$u_j = j \cdot \Delta.$$

Then, to complete the adjustment, for each $i = 1, \ldots, N$, it is simply computed

$$\hat{X}_i = l_j + X_i \cdot \Delta,$$

where j is the index of the subset in S to which object o_i belongs according to solution $\hat{x}$.

6 Concluding Remarks

The scope of this paper is to provide an overview of the main types of clustering and criteria for homogeneity or separation, with special emphasis to the optimization and operational research perspective, providing a few mathematical models of the problem under several different criterion adopted to classify the data.

Besides mathematical models that can be efficiently used to find exact solutions only in case of small-sized instances, there are hundreds of approximate techniques, proposed by researchers from several heterogenous communities, and more or less efficient, depending on the input data and the type of information they contain.

The conclusion is that unfortunately there is no single best approach that wins in every aspect. Given the intrinsic difficulties to be faced when clustering data, a person interested in performing this task should select a group of algorithms that seem to be the most appropriate, taking into account also the specific data set under analysis.

References

1. Aiex, R.M., Binato, S., Resende, M.G.C.: Parallel GRASP with path-relinking for job shop scheduling. Parallel Comput. **29**, 393–430 (2003)
2. Al-Sultan, K.S.: A tabu search approach to clustering problems. Pattern Recognit. **28**, 1443–1451 (1995)
3. Alon, U., Barkai, N., Notterman, D.A., Gish, K., Ybarra, S., Mack, D., et al.: Broad patterns of gene expression revealed by clustering analysis of tumor and normal colon tissues probed by oligonucleotide arrays. In: Proceedings of the National Academy of Sciences, pp. 6745–6750 (1999)
4. Anderberg, M.R.: Cluster Analysis for Applications. Academic, New York (1973)
5. Bean, J.C.: Genetic algorithms and random keys for sequencing and optimization. ORSA J. Comput. **6**, 154–160 (1994)
6. Ben-Dor, A., Shamir, R., Yakhini, Z.: Clustering gene expression patterns. J. Comput. Biol. **6**(3/4), 281–297 (1999)
7. Bickel, D.R.: Robust cluster analysis of DNA microarray data: an application of nonparametric correlation dissimilarity. In: Proceedings of the American Statistical Association (2001)
8. Boginski, V., Butenko, S., Pardalos, P.M.: Network models of massive datasets. Comput. Sci. Inf. Syst. **1**(1), 75–89 (2004)
9. Boginski, V., Butenko, S., Pardalos, P.M.: Mining market data: a network approach. Comput. Oper. Res. **33**(11), 3171–3184 (2006)
10. Busygin, S., Prokopyev, O.A., Pardalos, P.M.: Feature selection for consistent biclustering via fractional 0–1 programming. J. Comb. Optim. **10**(1), 7–21 (2005)
11. Busygin, S., Prokopyev, O.A., Pardalos, P.M.: Biclustering in data mining. Comput. Oper. Res. **39**(9), 2964–2987 (2008)
12. Butenko, S., Chaovalitwongse, W.A., Pardalos, P.M. (eds.): Clustering Challenges in Biological Networks. World Scientific, Singapore (2009)
13. Celebi, M.E., Kingravi, H.A.: Deterministic initialization of the k-means algorithm using hierarchical clustering. Int. J. Pattern Recogn. Artif. Intell. **26**(7), 1250018 (2012)
14. Celebi, M.E., Kingravi, H.: Linear, deterministic, and order-invariant initialization methods for the k-means clustering algorithm. In: Celebi, M.E. (ed.) Partitional Clustering Algorithms, pp. 79–98. Springer, Cham (2014)
15. Celebi, M.E., Kingravi, H.A., Vela, P.A.: A comparative study of efficient initialization methods for the k-means clustering algorithm. Expert Syst. Appl. **40**(1), 200–210 (2013)
16. Charikar, M., Chekuri, C., Feder, T., Motwani, R.: Incremental clustering and dynamic information retrieval. In: Proceedings of the 29th Annual ACM Symposium on Theory of Computing, pp. 626–634 (1997)
17. Cutting, D.R., Karger, D.R., Pedersen, J.O., Tukey, J.W.: Scatter/gather: a cluster-based approach to browsing large document collections. In: Fifteenth Annual International ACM SIGIR Conference on Research and Development in Information Retrieval, pp. 318–329 (1992)
18. Cutting, D.R., Karger, D.R., Pedersen, J.O.: Constant interaction-time scatter/gather browsing of very large document collections. In: Proceedings of the Sixteenth Annual International ACM SIGIR Conference on Research and Development in Information Retrieval, pp. 126–134 (1993)
19. Dempster, A.P., Laird, N.M., Rubin, D.B.: Maximum likelihood from incomplete data via the EM algorithm. J. R. Stat. Soc. **B39**, 1–38 (1977)
20. Ding, C.H.Q., Dubchak, I.: Multi-class protein fold recognition using support vector machines and neural networks. Bioinformatics **17**(4), 349–358 (2001)
21. Fan, N., Pardalos, P.M.: Multi-way clustering and biclustering by the Ratio cut and Normalized cut in graphs. J. Comb. Optim. **23**(2), 224–251 (2012)
22. Fan, Y.-J., Chaovalitwongse, W.A., Liu, C.-C., Sachdeo, R.C., Iasemidis, L.D., Pardalos, P.M.: Optimisation and data mining techniques for the screening of epileptic patients. Int. J. Bioinform. Res. Appl. **5**(2), 187–196 (2009)

23. Feo, T.A., Resende, M.G.C.: A probabilistic heuristic for a computationally difficult set covering problem. Oper. Res. Lett. **8**, 67–71 (1989)
24. Feo, T.A., Resende, M.G.C.: Greedy randomized adaptive search procedures. J. Glob. Optim. **6**, 109–133 (1995)
25. Festa, P., Resende, M.G.C.: GRASP: an annotated bibliography. In: Ribeiro, C.C., Hansen, P. (eds.) Essays and Surveys on Metaheuristics, pp. 325–367. Kluwer, Norwell (2002)
26. Festa, P., Resende, M.G.C.: An annotated bibliography of GRASP – part I: algorithms. Int. Trans. Oper. Res. **16**(1), 1–24 (2009)
27. Festa, P., Resende, M.G.C.: An annotated bibliography of GRASP – part II: applications. Int. Trans. Oper. Res. **16**(2), 131–172 (2009)
28. Festa, P.: A biased random-key genetic algorithm for data clustering. Math. Biosci. **245**(1), 76–85 (2013)
29. Festa, P.: On data clustering: exact and approximate solutions. In: Butenko, S., et al. (eds.) Examining Robustness and Vulnerability of Networked Systems, pp. 65–82. IOS Press, Fairfax (2014)
30. Frinhani, R.M.D., Silva, R.M.A., Mateus, G.R., Festa, P., Resende, M.G.C.: Grasp with path-relinking for data clustering: a case study for biological data. In: Proceedings of the 10th International Symposium on Experimental Algorithms, SEA'11, pp. 410–420. Springer, Berlin/Heidelberg (2011)
31. Glover, F.: Tabu search and adaptive memory programing – advances, applications and challenges. In: Barr, R.S., Helgason, R.V., Kennington, J.L. (eds.) Interfaces in Computer Science and Operations Research, pp. 1–75. Kluwer, Boston (1996)
32. Glover, F., Laguna, M., Martí, R.: Fundamentals of scatter search and path relinking. Control Cybern. **39**, 653–684 (2000)
33. Goldberg, D.E.: Genetic Algorithms in Search, Optimization, and Machine Learning. Addison-Wesley, Boston (1989)
34. Gonçalves, J.F., Resende, M.G.C.: Biased random-key genetic algorithms for combinatorial optimization. J. Heuristics **17**(5), 487–525 (2011)
35. Hammer, P.L., Rudeanu, S.: Boolean Methods in Operations Research and Related Areas. Springer, Heidelberg (1968)
36. Holland, J.H.: Adaptation in Natural and Artificial Systems. MIT Press, Cambridge (1975)
37. Hubert, L., Arabie, P.: Comparing partitions. J. Classif. **2**, 193–218 (1985)
38. Jain, A.K., Dubes, R.C.: Algorithms for Clustering Data. Prentice-Hall, Englewood Cliffs (1988)
39. Jain, A.K., Mao, J.: Neural networks and pattern recognition. In: Zurada, J.M., Marks, R.J., Robinson, C.J. (eds.) Computational Intelligence: Imitating Life, pp. 194–212. IEEE Press, New York (1994)
40. Jain, A.K., Murty, M.N., Flynn, P.J.: Data clustering: a review. ACM Comput. Surv. **31**(3), 264–323 (1999)
41. Jiang, D., Tang, C., Zhang, A.: Cluster analysis for gene expression data: a survey. IEEE Trans. Knowl. Data Eng. **16**(11), 1370–1386 (2004)
42. Jones, D., Beltramo, M.A.: Solving partitioning problems with genetic algorithms. In: Proceedings of the Fourth International Conference on Genetic Algorithms, pp. 442–449 (1991)
43. Kaufman, L., Rousseeuw, P.J.: Statistical Data Analysis Based on the L1-Norm and Related Methods. North-Holland, Amsterdam (1987)
44. Kaufman, L., Rousseeuw, P.J.: Finding Groups in Data: An Introduction to Cluster Analysis. Wiley, New York (2005)
45. Kocheturov, A., Batsyn, M., Pardalos, P.M.: Dynamics of cluster structures in a financial market network. Phys. A Stat. Mech. Appl. **413**, 523–533 (2014)
46. King, B.: Step-wise clustering procedures. J. Am. Stat. Assoc. **69**, 86–101 (1967)
47. Klein, R.W., Dubes, R.C.: Experiments in projection and clustering by simulated annealing. Pattern Recogn. **22**, 213–220 (1989)

48. Laguna, M., Martí, R.: GRASP and path relinking for 2-layer straight line crossing minimization. INFORMS J. Comput. **11**, 44–52 (1999)
49. Liu, C.-C., Chaovalitwongse, W.A., Pardalos, P.M., Uthman, B.M.: Dynamical feature extraction from brain activity time series. In: Encyclopedia of Data Warehousing and Mining, pp. 729–735. IDEA Group, Hershey (2009)
50. Ma, P.C.H., Chan, K.C.C., Yao, X., Chiu, D.K.Y.: An evolutionary clustering algorithm for gene expression microarray data analysis. IEEE Trans. Evol. Comput. **10**(3), 296–314 (2006)
51. Mao, J., Jain, A.K.: A self-organizing network for hyperellipsoidal clustering (hec). IEEE Trans. Neural Netw. **7**, 16–29 (1996)
52. McQueen, J.: Some methods for classification and analysis of multivariate observations. In: Proceedings of the 5.th Berkeley Symposium on Mathematical Statistics and Probability, pp. 281–297 (1967)
53. Mucherino, A., Papajorgji, P., Pardalos, P.M.: A survey of data mining techniques applied to agriculture. Oper. Res. **9**(2), 121–140 (2009)
54. Nascimento, M.C.V.: Uma heurística GRASP para o problema de dimensionamento de lotes com múltiplas plantas. PhD thesis, USP (2007)
55. Nascimento, M.C.V., Toledo, F.M.B., de Carvalho, A.C.P.L.F.: Investigation of a new GRASP-based clustering algorithm applied to biological data. Comput. Oper. Res. **37**(8), 1381–1388 (2010)
56. Pardalos, P.M., Hansen, P. (eds.) : Data Mining and Mathematical Programming. American Mathematical Society, Providence (2008)
57. Pardalos, P.M., Coleman, T.F., Xanthopoulos, P. (eds.): Optimization and Data Analysis in Biomedical Informatics. Springer Series: Fields Institute Communications, vol. 63, 150 p. Springer, New York (2012). ISBN 978-1-4614-4132-8
58. Porter, M.: Location, competition, and economic development: local clusters in a global economy. Econ. Dev. Q. **14**(1), 15–34 (2000)
59. Raghavan, V.V., Birchand, K.: A clustering strategy based on a formalism of the reproductive process in a natural system. In: Second International Conference on Information Storage and Retrieval, pp. 10–22 (1979)
60. Rao, M.R.: Cluster analysis and mathematical programming. J. Am. Stat. Assoc. **66**(335), 622–626 (1971)
61. Resende, M.G.C., Ribeiro, C.C.: Greedy randomized adaptive search procedures. In: Glover, F., Kochenberger, G. (eds.) Handbook of Metaheuristics, pp. 219–249. Kluwer, New York (2002)
62. Resende, M.G.C., Ribeiro, C.C.: GRASP with path-relinking: recent advances and applications. In: Ibaraki, T., Nonobe, K., Yagiura, M. (eds.) Metaheuristics: Progress as Real Problem Solvers, pp. 29–63. Springer, New York (2005)
63. Shamir, R., Sharan, R.: CLICK: a clustering algorithm for gene expression analysis. In: Proc. Eighth Int'l Conf. Intelligent Systems for Molecular Biology (ISMB '00) (2000)
64. Sethi, I., Jain, A.K. (eds.): Artificial Neural Networks and Pattern Recognition: Old and New Connections. Elsevier, New York (1991)
65. Sneath, P.H.A., Sokal, R.R.: Numerical Taxonomy. Freeman, San Francisco (1973)
66. Spears, W.M., DeJong, K.A.: On the virtues of parameterized uniform crossover. In: Proceedings of the Fourth International Conference on Genetic Algorithms, pp. 230–236 (1991)
67. Steinbach, M., Karypis, G., Kumar, V.: On the virtues of parameterized uniform crossover. In: Proceedings of World Text Mining Conference, KDD2000 (2000)
68. Steinbach, M., Karypis, G., Kumar, V.: A comparison of document clustering techniques. In: Proceedings of KDD Workshop on Text Mining Conference, KDD2000 (2000)
69. Su, T., Dy, J.G.: In search of deterministic methods for initializing k-means and Gaussian mixture clustering. Intell. Data Anal. **11**(4), 319–338 (2007)
70. Ushioda, A., Kawasaki, J.: Hierarchical clustering of words and application to NLP tasks. In: Ejerhed, E., Dagan, I. (eds.) Fourth Workshop on Very Large Corpora, pp. 28–41. Association for Computational Linguistics, Cambridge (1996)

71. Valery, K.A., Koldanov, A.P., Pardalos, P.M.: A general approach to network analysis of statistical data sets In: Pardalos, P.M., Resende, M.G.C., Vogiatzis, C., Walteros, J.L. (eds.) Learning and Intelligent Optimization. 8th International Conference, Lion 8, Gainesville, FL, 16–21 February 2014. Revised Selected Papers. Lecture Notes in Computer Science, vol. 8426, pp. 88–97. Springer, Berlin/Heidelberg (2014)
72. Ward Jr., J.H.: Hierarchical grouping to optimize an objective function. J. Am. Stat. Assoc. **58**, 236–244 (1963)
73. White, S.D.M., Frenk, C.S.: An optimal graph theoretic approach to data clustering: theory and its application to image segmentation. Astrophys. J. **379**(Part 1), 52–72 (1991)
74. Willet, P.: Recent trends in hierarchic document clustering: a critical review. Inf. Process. Manag. **24**(5), 577–597 (1988)
75. Wu, Z., Leahy, R.: Galaxy formation through hierarchical clustering. IEEE Trans. Pattern Anal. Mach. Intell. **15**(11), 1101–1013 (1993)
76. Xanthopoulos, P., Pardalos, P.M., Trafalis, T.B.: Robust Data Mining. Springer Briefs in Optimization, vol. XII, 59 p. Springer, New York (2013)
77. Xu, R., Wunsch, D.: Survey of clustering algorithms. IEEE Trans. Neural Netw. **16**(3), 645–678 (2005)
78. Zahn, C.T.: Graph-theoretical methods for detecting and describing gestalt clusters. IEEE Trans. Comput. **C-20**, 68–86 (1971)

Kernel Spectral Clustering and Applications

Rocco Langone, Raghvendra Mall, Carlos Alzate, and Johan A. K. Suykens

Abstract In this chapter we review the main literature related to kernel spectral clustering (KSC), an approach to clustering cast within a kernel-based optimization setting. KSC represents a least-squares support vector machine-based formulation of spectral clustering described by a weighted kernel PCA objective. Just as in the classifier case, the binary clustering model is expressed by a hyperplane in a high dimensional space induced by a kernel. In addition, the multi-way clustering can be obtained by combining a set of binary decision functions via an Error Correcting Output Codes (ECOC) encoding scheme. Because of its model-based nature, the KSC method encompasses three main steps: training, validation, testing. In the validation stage model selection is performed to obtain tuning parameters, like the number of clusters present in the data. This is a major advantage compared to classical spectral clustering where the determination of the clustering parameters is unclear and relies on heuristics. Once a KSC model is trained on a small subset of the entire data, it is able to generalize well to unseen test points. Beyond the basic formulation, sparse KSC algorithms based on the Incomplete Cholesky Decomposition (ICD) and $L_0, L_1, L_0 + L_1$, Group Lasso regularization are reviewed. In that respect, we show how it is possible to handle large-scale data. Also, two possible ways to perform hierarchical clustering and a soft clustering method are presented. Finally, real-world applications such as image segmentation, power load time-series clustering, document clustering, and big data learning are considered.

1 Introduction

Spectral clustering (SC) represents the most popular class of algorithms based on graph theory [11]. It makes use of the Laplacian's spectrum to partition a graph into weakly connected subgraphs. Moreover, if the graph is constructed based on

R. Langone (✉) • R. Mall • J.A.K. Suykens
KU Leuven ESAT-STADIUS, Kasteelpark Arenberg 10 B-3001 Leuven (Belgium)
e-mail: rocco.langone@esat.kuleuven.be

C. Alzate
IBM's Smarter Cities Technology Center Dublin (Ireland)

© Springer International Publishing Switzerland 2016
M.E. Celebi, K. Aydin (eds.), *Unsupervised Learning Algorithms*,
DOI 10.1007/978-3-319-24211-8_6

any kind of data (vector, images etc.), data clustering can be performed.[1] SC began to be popularized when Shi and Malik introduced the Normalized Cut criterion to handle image segmentation [59]. Afterwards, Ng and Jordan [51] in a theoretical work based on matrix perturbation theory have shown conditions under which a good performance of the algorithm is expected. Finally, in the tutorial by Von Luxburg the main literature related to SC has been exhaustively summarized [63]. Although very successful in a number of applications, SC has some limitations. For instance, it cannot handle big data without using approximation methods like the Nyström algorithm [19, 64], the power iteration method [37], or linear algebra-based methods [15, 20, 52]. Furthermore, the generalization to out-of-sample data is only approximate.

These issues have been recently tackled by means of a spectral clustering algorithm formulated as weighted kernel PCA [2]. The technique, named kernel spectral clustering (KSC), is based on solving a constrained optimization problem in a primal-dual setting. In other words, KSC is a Least-Squares Support Vector Machine (LS-SVM [61]) model used for clustering instead of classification.[2] By casting SC in a learning framework, KSC allows to rigorously select tuning parameters such as the natural number of clusters which are present in the data. Also, an accurate prediction of the cluster memberships for unseen points can be easily done by projecting test data in the embedding eigenspace learned during training. Furthermore, the algorithm can be tailored to a given application by using the most appropriate kernel function. Beyond that, by using sparse formulations and a fixed-size [12, 61] approach, it is possible to readily handle big data. Finally, by means of adequate adaptations of the core algorithm, hierarchical clustering and a soft clustering approach have been proposed.

The idea behind KSC is similar to the earlier works introduced in [16, 17]. In these papers the authors showed that a general weighted kernel k-means objective is mathematically equivalent to a weighted graph partitioning objective such as ratio cut, normalized cut and ratio association. This equivalence allows, for instance, to use the weighted kernel k-means algorithm to directly optimize the graph partitioning objectives, which eliminates the need for any eigenvector computation when this is prohibitive. Although quite appealing and mathematically sound, the algorithm presents some drawbacks. The main issues concern the sensitivity of the final clustering results to different types of initialization techniques, the choice of the shift parameter, and the model selection (i.e., how to choose the number of clusters). Furthermore, the out-of-sample extension problem is not discussed. On the other hand, as we will see later, these issues are not present in the KSC algorithm.

The remainder of the Chapter is organized as follows. After presenting the basic KSC method, the soft KSC algorithm will be summarized. Next, two possible

---

[1]In this case the given data points represent the node of the graph and their similarity the corresponding edges.

[2]This is a considerable novelty, since SVMs are typically known as classifiers or function approximation models rather than clustering techniques.

ways to accomplish hierarchical clustering will be explained. Afterwards, some sparse formulations based on the Incomplete Cholesky Decomposition (ICD) and L_0, L_1, $L_0 + L_1$, Group Lasso regularization will be described. Lastly, various interesting applications in different domains such as computer vision, power-load consumer profiling, information retrieval, and big data clustering will be illustrated. All these examples assume a static setting. Concerning other applications in a dynamic scenario the interested reader can refer to [29, 33] for fault detection, to [32] for incremental time-series clustering, to [25, 28, 31] in case of community detection in evolving networks and [54] in relation to human motion tracking.

2 Notation

x^T	Transpose of the vector x
A^T	Transpose of the matrix A
I_N	$N \times N$ Identity matrix
1_N	$N \times 1$ Vector of ones
$\mathscr{D}_{tr} = \{x_i\}_{i=1}^{N_{tr}}$	Training sample of N_{tr} data points
$\varphi(\cdot)$	Feature map
$\mathscr{F}$	Feature space of dimension d_h
$\{\mathscr{A}_p\}_{p=1}^k$	Partitioning composed of k clusters
$\mathscr{G} = (\mathscr{V}, \mathscr{E})$	Set of N vertices $\mathscr{V} = \{v_i\}_{i=1}^N$ and m edges $\mathscr{E}$ of a graph
$\lvert \cdot \rvert$	Cardinality of a set

3 Kernel Spectral Clustering (KSC)

3.1 Mathematical Formulation

3.1.1 Training Problem

The KSC formulation for k clusters is stated as a combination of $k - 1$ binary problems [2]. In particular, given a set of training data $\mathscr{D}_{tr} = \{x_i\}_{i=1}^{N_{tr}}$, the primal problem is:

$$\min_{w^{(l)}, e^{(l)}, b_l} \frac{1}{2} \sum_{l=1}^{k-1} w^{(l)^T} w^{(l)} - \frac{1}{2} \sum_{l=1}^{k-1} \gamma_l e^{(l)^T} V e^{(l)} \tag{1}$$

$$\text{subject to} \quad e^{(l)} = \Phi w^{(l)} + b_l 1_{N_{tr}}, l = 1, \ldots, k-1.$$

The $e^{(l)} = [e_1^{(l)}, \ldots, e_i^{(l)}, \ldots, e_{N_{tr}}^{(l)}]^T$ are the projections of the training data mapped in the feature space along the direction $w^{(l)}$. For a given point x_i, the model in the primal form is:

$$e_i^{(l)} = w^{(l)^T} \varphi(x_i) + b_l. \tag{2}$$

The primal problem (1) expresses the maximization of the weighted variances of the data given by $e^{(l)^T} V e^{(l)}$ and the contextual minimization of the squared norm of the vector $w^{(l)}$, $\forall l$. The regularization constants $\gamma_l \in \mathbb{R}^+$ mediate the model complexity expressed by $w^{(l)}$ with the correct representation of the training data. $V \in \mathbb{R}^{N_{tr} \times N_{tr}}$ is the weighting matrix and Φ is the $N_{tr} \times d_h$ feature matrix $\Phi = [\varphi(x_1)^T; \ldots; \varphi(x_{N_{tr}})^T]$, where $\varphi : \mathbb{R}^d \to \mathbb{R}^{d_h}$ denotes the mapping to a high-dimensional feature space, b_l are bias terms.

The dual problem corresponding to the primal formulation (1), by setting $V = D^{-1}$ becomes[3]:

$$D^{-1} M_D \Omega \alpha^{(l)} = \lambda_l \alpha^{(l)} \tag{3}$$

where Ω is the kernel matrix with ijth entry $\Omega_{ij} = K(x_i, x_j) = \varphi(x_i)^T \varphi(x_j)$. $K : \mathbb{R}^d \times \mathbb{R}^d \to \mathbb{R}$ means the kernel function. The type of kernel function to utilize is application-dependent, as it is outlined in Table 1. The matrix D is the graph degree matrix which is diagonal with positive elements $D_{ii} = \sum_j \Omega_{ij}$, M_D is a centering matrix defined as $M_D = I_{N_{tr}} - \frac{1}{1_{N_{tr}}^T D^{-1} 1_{N_{tr}}} 1_{N_{tr}} 1_{N_{tr}}^T D^{-1}$, the $\alpha^{(l)}$ are vectors of dual variables, $\lambda_l = \frac{N_{tr}}{\gamma_l}$, $K : \mathbb{R}^d \times \mathbb{R}^d \to \mathbb{R}$ is the kernel function. The dual clustering model for the ith point can be expressed as follows:

$$e_i^{(l)} = \sum_{j=1}^{N_{tr}} \alpha_j^{(l)} K(x_j, x_i) + b_l, j = 1, \ldots, N_{tr}, l = 1, \ldots, k - 1. \tag{4}$$

The cluster prototypes can be obtained by binarizing the projections $e_i^{(l)}$ as $\text{sign}(e_i^{(l)})$. This step is straightforward because, thanks to the presence of the bias term b_l, both the $e^{(l)}$ and the $\alpha^{(l)}$ variables get automatically centered around zero. The set of the most frequent binary indicators form a code-book $\mathscr{CB} = \{c_p\}_{p=1}^k$, where each codeword of length $k - 1$ represents a cluster.

Interestingly, problem (3) has a close connection with SC based on a random walk Laplacian. In this respect, the kernel matrix can be considered as a weighted graph $\mathscr{G} = (\mathscr{V}, \mathscr{E})$ with the nodes $v_i \in \mathscr{V}$ represented by the data points x_i. This graph has a corresponding random walk in which the probability of leaving a vertex is distributed among the outgoing edges according to their weight: $p_{t+1} = P p_t$, where $P = D^{-1} \Omega$ indicates the transition matrix with the ijth entry denoting

---

[3]By choosing $V = I$, problem (3) is identical to kernel PCA [48, 58, 62].

Table 1 Types of kernel functions for different applications

Application	Kernel name	Mathematical expression
Vector data	RBF	$K(x_i, x_j) = \exp(-\|x_i - x_j\|_2^2/\sigma^2)$
Images	RBF_{χ^2}	$K(h^{(i)}, h^{(j)}) = \exp(-\dfrac{\chi_{ij}^2}{\sigma_\chi^2})$
Text	Cosine	$K(x_i, x_j) = \dfrac{x_i^T x_j}{\|x_i\|\|x_j\|}$
Time-series	RBF_{cd}	$K(x_i, x_j) = \exp(-\|x_i - x_j\|_{cd}^2/\sigma_{cd}^2)$

In this table RBF means Radial Basis Function, and σ denotes the bandwidth of the kernel. The symbol $h^{(i)}$ indicates a color histogram representing the ith pixel of an image, and to compare two histograms $h^{(i)}$ and $h^{(j)}$ the χ^2 statistical test is used [55]. Regarding time-series data, the symbol cd means correlation distance [36], and $\|x_i - x_j\|_{cd} = \sqrt{\frac{1}{2}(1 - R_{ij})}$, where R_{ij} can indicate the Pearson or Spearman's rank correlation coefficient between time-series x_i and x_j

the probability of moving from node i to node j in one time-step. Moreover, the stationary distribution of the Markov Chain describes the scenario where the random walker stays mostly in the same cluster and seldom moves to the other clusters [14, 46, 47, 47].

3.1.2 Generalization

Given the dual model parameters $\alpha^{(l)}$ and b_l, it is possible to assign a membership to unseen points by calculating their projections onto the eigenvectors computed in the training phase:

$$e_{\text{test}}^{(l)} = \Omega_{\text{test}}\alpha^{(l)} + b_l 1_{N_{\text{test}}} \tag{5}$$

where Ω_{test} is the $N_{\text{test}} \times N$ kernel matrix evaluated using the test points with entries $\Omega_{\text{test},ri} = K(x_r^{\text{test}}, x_i)$, $r = 1, \ldots, N_{\text{test}}$, $i = 1, \ldots, N_{\text{tr}}$. The cluster indicator for a given test point can be obtained by using an Error Correcting Output Codes (ECOC) decoding procedure:

- the score variable is binarized
- the indicator is compared with the training code-book $\mathcal{CB}$ (see previous Section), and the point is assigned to the nearest prototype in terms of Hamming distance.

The KSC method, comprising training and test stage, is summarized in Algorithm 1, and the related Matlab package is freely available on the Web.[4]

---

[4]http://www.esat.kuleuven.be/stadius/ADB/alzate/softwareKSClab.php.

Algorithm 1: KSC algorithm [2]

Data: Training set $\mathscr{D}_{\text{tr}} = \{x_i\}_{i=1}^{N_{\text{tr}}}$, test set $\mathscr{D}_{\text{test}} = \{x_m^{\text{test}}\}_{m=1}^{N_{\text{test}}}$ kernel function
  $K : \mathbb{R}^d \times \mathbb{R}^d \rightarrow \mathbb{R}$ positive definite and localized ($K(x_i, x_j) \rightarrow 0$ if x_i and x_j belong to
  different clusters), kernel parameters (if any), number of clusters k.
Result: Clusters $\{\mathscr{A}_1, \ldots, \mathscr{A}_k\}$, codebook $\mathscr{C}\mathscr{B} = \{c_p\}_{p=1}^k$ with $\{c_p\} \in \{-1, 1\}^{k-1}$.

1 compute the training eigenvectors $\alpha^{(l)}$, $l = 1, \ldots, k-1$, corresponding to the $k-1$ largest
  eigenvalues of problem (3)
2 let $A \in \mathbb{R}^{N_{\text{tr}} \times (k-1)}$ be the matrix containing the vectors $\alpha^{(1)}, \ldots, \alpha^{(k-1)}$ as columns
3 binarize A and let the code-book $\mathscr{C}\mathscr{B} = \{c_p\}_{p=1}^k$ be composed by the k encodings of
  $Q = \text{sign}(A)$ with the most occurrences
4 $\forall i, i = 1, \ldots, N_{\text{tr}}$, assign x_i to A_{p^*} where $p^* = \text{argmin}_p d_H(\text{sign}(\alpha_i), c_p)$ and $d_H(.,.)$ is the
  Hamming distance
5 binarize the test data projections $\text{sign}(e_m^{(l)})$, $m = 1, \ldots, N_{\text{test}}$, and let $\text{sign}(e_m) \in \{-1, 1\}^{k-1}$
  be the encoding vector of x_m^{test}
6 $\forall m$, assign x_m^{test} to A_{p^*}, where $p^* = \text{argmin}_p d_H(\text{sign}(e_m), c_p)$.

3.1.3 Model Selection

In order to select tuning parameters like the number of clusters k and eventually
the kernel parameters, a model selection procedure based on grid search is adopted.
First, a validation set $\mathscr{D}_{\text{val}} = \{x_i\}_{i=1}^{N_{\text{val}}}$ is sampled from the whole dataset. Then, a grid
of possible values of the tuning parameters is constructed. Afterwards, a KSC model
is trained for each combination of parameters and the chosen criterion is evaluated
on the partitioning predicted for the validation data. Finally, the parameters yielding
the maximum value of the criterion are selected. Depending on the kind of data, a
variety of model selection criteria have been proposed:

- *Balanced Line Fit (BLF)*. It indicates the amount of collinearity between
  validation points belonging to the same cluster, in the space of the projections.
  It reaches its maximum value 1 in case of well-separated clusters, represented as
  lines in the space of the $e_{\text{val}}^{(l)}$ (see, for instance, the bottom left side of Fig. 1)
- *Balanced Angular Fit or BAF* [39]. For every cluster, the sum of the cosine
  similarity between the validation points and the cluster prototype, divided by the
  cardinality of that cluster, is computed. These similarity values are then summed
  up and divided by the total number of clusters.
- *Average Membership Strength abbr. AMS* [30]. The mean membership per cluster
  denoting the mean degree of belonging of the validation points to the cluster is
  computed. These mean cluster memberships are then averaged over the number
  of clusters.
- *Modularity* [49]. This quality function is well suited for network data. In the
  model selection scheme, the Modularity of the validation subgraph corresponding to a given partitioning is computed, and the parameters related to the highest
  Modularity are selected [26, 27].
- *Fisher Criterion*. The classical Fisher criterion [8] used in classification has been
  adapted to select the number of clusters k and the kernel parameters in the KSC

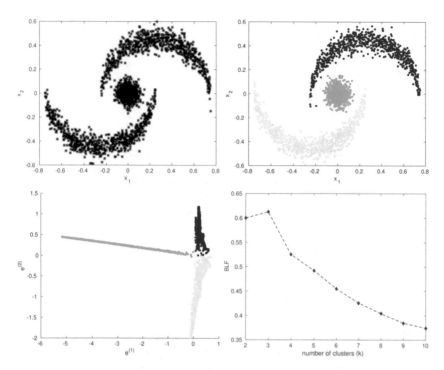

Fig. 1 KSC partitioning on a toy dataset. *(Top)* Original dataset consisting of three clusters *(left)* and obtained clustering results *(right)*. *(Bottom)* Points represented in the space of the projections $[e^{(1)}, e^{(2)}]$ *(left)*, for an optimal choice of k (and $\sigma^2 = 4.36 \cdot 10^{-3}$) suggested by the BLF criterion *(right)*. We can notice how the points belonging to one cluster tend to lie on the same line. A perfect line structure is not attained due to a certain amount of overlap between the clusters

framework [4]. The criterion maximizes the distance between the means of the two clusters while minimizing the variance within each cluster, in the space of the projections $e_{\text{val}}^{(l)}$.

In Fig. 1 an example of clustering obtained by KSC on a synthetic dataset is shown. The BLF model selection criterion has been used to tune the bandwidth of the RBF kernel and the number of clusters. It can be noticed how the results are quite accurate, despite the fact that the clustering boundaries are highly nonlinear.

3.2 Soft Kernel Spectral Clustering

Soft kernel spectral clustering (SKSC) makes use of Algorithm 1 in order to compute a first hard partitioning of the training data. Next, soft cluster assignments are performed by computing the cosine distance between each point and some cluster prototypes in the space of the projections $e^{(l)}$. In particular, given the projections for the training points $e_i = [e_i^{(1)}, \ldots, e_i^{(k-1)}]$, $i = 1, \ldots, N_{tr}$ and the corresponding hard assignments q_i^p we can calculate for each cluster the cluster prototypes $s_1, \ldots, s_p, \ldots, s_k$, $s_p \in \mathbb{R}^{k-1}$ as:

$$s_p = \frac{1}{n_p} \sum_{i=1}^{n_p} e_i \tag{6}$$

where n_p is the number of points assigned to cluster p during the initialization step by KSC. Then the cosine distance between the ith point in the projections space and a prototype s_p is calculated by means of the following formula:

$$d_{ip}^{cos} = 1 - e_i^T s_p / (||e_i||_2 ||s_p||_2). \tag{7}$$

The soft membership of point i to cluster q can be finally expressed as:

$$sm_i^{(q)} = \frac{\prod_{j \neq q} d_{ij}^{cos}}{\sum_{p=1}^k \prod_{j \neq p} d_{ij}^{cos}} \tag{8}$$

with $\sum_{p=1}^k sm_i^{(p)} = 1$. As pointed out in [7], this membership represents a subjective probability expressing the belief in the clustering assignment.

The out-of-sample extension on unseen data consists simply of calculating Eq. (5) and assigning the test projections to the closest centroid.

An example of soft clustering performed by SKSC on a synthetic dataset is depicted in Fig. 2. The AMS model selection criterion has been used to select the bandwidth of the RBF kernel and the optimal number of clusters. The reader can appreciate how SKSC provides more interpretable outcomes compared to KSC.

The SKSC method is summarized in Algorithm 2 and a Matlab implementation is freely downloadable.[5]

Algorithm 2: SKSC algorithm [30]

Data: Training set $\mathscr{D}_{tr} = \{x_i\}_{i=1}^{N_{tr}}$ and test set $\mathscr{D}_{test} = \{x_m^{test}\}_{m=1}^{N_{test}}$, kernel function
     $K : \mathbb{R}^d \times \mathbb{R}^d \to \mathbb{R}$ positive definite and localized ($K(x_i, x_j) \to 0$ if x_i and x_j belong to
     different clusters), kernel parameters (if any), number of clusters k.
Result: Clusters $\{\mathscr{A}_1, \ldots, \mathscr{A}_p, \ldots, \mathscr{A}_k\}$, soft cluster memberships $sm^{(p)}$, $p = 1, \ldots, k$,
     cluster prototypes $\mathscr{S}\mathscr{P} = \{s_p\}_{p=1}^k$, $s_p \in \mathbb{R}^{k-1}$.

1 Initialization by solving Eq. (4).
2 Compute the new prototypes $s_1, \ldots, s_k$ [Eq. (6)].
3 Calculate the test data projections $e_m^{(l)}$, $m = 1, \ldots, N_{test}$, $l = 1, \ldots, k - 1$.
4 Find the cosine distance between each projection and all the prototypes [Eq. (7)] $\forall m$, assign
   x_m^{test} to cluster A_p with membership $sm^{(p)}$ according to Eq. (8).

---

[5] http://www.esat.kuleuven.be/stadius/ADB/langone/softwareSKSClab.php.

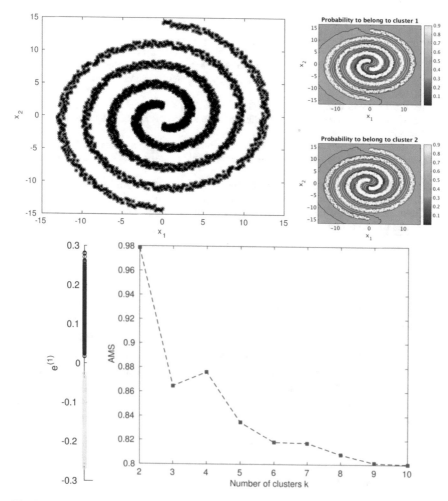

Fig. 2 SKSC partitioning on a synthetic dataset. *(Top)* Original dataset consisting of two clusters *(left)* and obtained soft clustering results *(right)*. *(Bottom)* Points represented in the space of the projection $e^{(1)}$ *(left)*, for an optimal choice of k (and $\sigma^2 = 1.53 \cdot 10^{-3}$) as detected by the AMS criterion *(right)*

3.3 Hierarchical Clustering

In many cases, clusters are formed by sub-clusters which in turn might have substructures. As a consequence, an algorithm able to discover a hierarchical organization of the clusters provides a more informative result, incorporating several scales in the analysis. The flat KSC algorithm has been extended in two ways in order to deal with hierarchical clustering.

Algorithm 3: HKSC algorithm [4]

Data: Training set $\mathscr{D}_{\text{tr}} = \{x_i\}_{i=1}^{N_{\text{tr}}}$, Validation set $\mathscr{D}_{\text{val}} = \{x_i\}_{i=1}^{N_{\text{val}}}$ and test set
$\mathscr{D}_{\text{test}} = \{x_m^{\text{test}}\}_{m=1}^{N_{\text{test}}}$, RBF kernel function with parameter σ^2, maximum number of
clusters $k_{\max}$, set of $R\sigma^2$ values $\{\sigma_1^2, \ldots, \sigma_R^2\}$, Fisher threshold θ.

Result: Linkage matrix Z

1 For every combination of parameter pairs (k, σ^2) train a KSC model using Algorithm 1,
  predict the cluster memberships for validation points and calculate the related Fisher
  criterion

2 $\forall k$, find the maximum value of the Fisher criterion across the given range of σ^2 values. If
  the maximum value is greater than the Fisher threshold θ, create a set of these optimal
  (k_*, σ_*^2) pairs.

3 Using the previously found (k_*, σ_*^2) pairs train a clustering model and compute the cluster
  memberships for the test set using the out-of-sample extension.

4 Create the linkage matrix Z by identifying which clusters merge starting from the bottom of
  the tree which contains max k_* clusters.

3.3.1 Approach 1

This approach, named hierarchical kernel spectral clustering (HKSC), was proposed
in [4] and exploits the information of a multi-scale structure present in the data given
by the Fisher criterion (see end of Sect. 3.1.3). A grid search over different values of
k and σ^2 is performed to find tuning parameter pairs such that the criterion is greater
than a specified threshold value. The KSC model is then trained for each pair and
evaluated at the test set using the out-of-sample extension. A specialized linkage
criterion determines which clusters are merging based on the evolution of the cluster
memberships as the hierarchy goes up. The whole procedure is summarized in
Algorithm 3.

3.3.2 Approach 2

In [42] and [41] an alternative hierarchical extension of the basic KSC algorithm
was introduced, for network and vector data, respectively. In this method, called
agglomerative hierarchical kernel spectral clustering (AH-KSC), the structure of the
projections in the eigenspace is used to automatically determine a set of increasing
distance thresholds. At the beginning, the validation point with maximum number
of similar points within the first threshold value is selected. The indices of all
these points represent the first cluster at level 0 of hierarchy. These points are then
removed from the validation data matrix, and the process is repeated iteratively
until the matrix becomes empty. Thus, the first level of hierarchy corresponding
to the first distance threshold is obtained. To obtain the clusters at the next level of
hierarchy the clusters at the previous levels are treated as data points, and the whole
procedure is repeated again with other threshold values. This step takes inspiration
from [9]. The algorithm stops when only one cluster remains. The same procedure
is applied in the test stage, where the distance thresholds computed in the validation

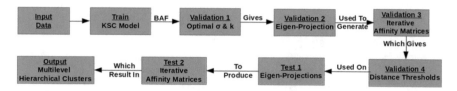

Fig. 3 AH-KSC algorithm. Steps of AH-KSC method as described in [42] with addition of the step where the optimal σ and k are estimated

phase are used. An overview of all the steps involved in the algorithm is depicted in Fig. 3. In Fig. 4 an example of hierarchical clustering performed by this algorithm on a toy dataset is shown.

3.4 Sparse Clustering Models

The computational complexity of the KSC algorithm depends on solving the eigenvalue problem (3) related to the training stage and computing Eq. (5) which gives the cluster memberships of the remaining points. Assuming that we have N_{tot} data and we use N_{tr} points for training and $N_{test} = N_{tot} - N_{tr}$ as test set, the runtime of Algorithm 1 is $O(N_{tr}^2) + O(N_{tr}N_{test})$. In order to reduce the computational complexity, it is then necessary to find a reduced set of training points, without loosing accuracy. In the next sections two different methods to obtain a sparse KSC model, based on the Incomplete Cholesky Decomposition (ICD) and L_1 and L_0 penalties, respectively, are discussed. In particular, thanks to the ICD, the KSC computational complexity for the training problem is decreased to $O(R^2 N_{tr})$ [53], where R indicates the reduced set size.

3.4.1 Incomplete Cholesky Decomposition

One of the KKT optimality conditions characterizing the Lagrangian of problem (1) is:

$$w^{(l)} = \Phi^T \alpha^{(l)} = \sum_{i=1}^{N_{tr}} \alpha_i^{(l)} \varphi(x_i). \tag{9}$$

From Eq. (9) it is evident that each training data point contributes to the primal variable $w^{(l)}$, resulting in a non-sparse model. In order to obtain a parsimonious model a reduced set method based on the Incomplete Cholesky Decomposition (ICD) was proposed in [3, 53]. The technique is based on finding a small number $R \ll N_{tr}$ of points $\mathscr{R} = \{\hat{x}_r\}_{r=1}^{R}$ and related coefficients $\zeta^{(l)}$ with the aim of approximating $w^{(l)}$ as:

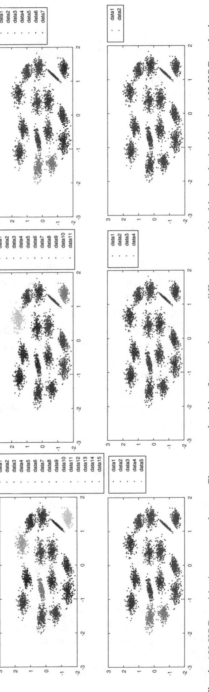

Fig. 4 AH-KSC partitioning on a toy dataset. Cluster memberships for a toy dataset at different hierarchical levels obtained by the AH-KSC method

$$w^{(l)} \approx \hat{w}^{(l)} = \sum_{r=1}^{R} \zeta_r^{(l)} \varphi(\hat{x}_r). \tag{10}$$

As a consequence, the projection of an arbitrary data point x into the training embedding is given by:

$$e^{(l)} \approx \hat{e}^{(l)} = \sum_{r=1}^{R} \zeta_r^{(l)} K(x, \hat{x}_r) + \hat{b}_l. \tag{11}$$

The set $\mathcal{R}$ of points can be obtained by considering the pivots of the ICD performed on the kernel matrix Ω. In particular, by assuming that Ω has a small numerical rank, the kernel matrix can be approximated by $\Omega \approx \hat{\Omega} = GG^T$, with $G \in \mathbb{R}^{N_{tr} \times R}$. If we plug in this approximated kernel matrix in problem (3), the KSC eigenvalue problem can be written as:

$$\hat{D}^{-1} M_{\hat{D}} U \Psi^2 U^T \hat{\alpha}^{(l)} = \hat{\lambda}_l \hat{\alpha}^{(l)}, l = 1, \dots, k \tag{12}$$

where $U \in \mathbb{R}^{N_{tr} \times R}$ and $V \in \mathbb{R}^{N_{tr} \times R}$ denotes the left and right singular vectors deriving from the singular value decomposition (SVD) of G, and $\Psi \in \mathbb{R}^{N_{tr} \times N_{tr}}$ is the matrix of the singular values. If now we pre-multiply both sides of Eq. (12) by U^T and replace $\hat{\delta}^{(l)} = U^T \hat{\alpha}^{(l)}$, only the following eigenvalue problem of size $R \times R$ must be solved:

$$U^T \hat{D}^{-1} M_{\hat{D}} U \Psi^2 \hat{\delta}^{(l)} = \hat{\lambda}_l \hat{\delta}^{(l)}, l = 1, \dots, k. \tag{13}$$

The approximated eigenvectors of the original problem (3) can be computed as $\hat{\alpha}^{(l)} = U \hat{\delta}^{(l)}$, and the sparse parameter vector can be found by solving the following optimization problem:

$$\min_{\zeta^{(l)}} \| w^{(l)} - \hat{w}^{(l)} \|_2^2 = \min_{\zeta^{(l)}} \| \Phi^T \alpha^{(l)} - \chi^T \zeta^{(l)} \|_2^2 . \tag{14}$$

The corresponding dual problem can be written as follows:

$$\Omega^{\chi\chi} \delta^{(l)} = \Omega^{\chi\phi} \alpha^{(l)}, \tag{15}$$

where $\Omega_{rs}^{\chi\chi} = K(\tilde{x}_r, \tilde{x}_s)$, $\Omega_{ri}^{\chi\phi} = K(\tilde{x}_r, x_i)$, $r, s = 1, \dots, R, i = 1, \dots, N_{tr}$ and $l = 1, \dots, k - 1$. Since the size R of problem (13) can be much smaller than the size N_{tr} of the starting problem, the sparse KSC method[6] is suitable for big data analytics.

---

[6]A C implementation of the algorithm can be downloaded at: http://www.esat.kuleuven.be/stadius/ADB/novak/softwareKSCICD.php.

3.4.2 Using Additional Penalty Terms

In this part we explore sparsity in the KSC technique by using an additional penalty term in the objective function (14). In [3], the authors used an L_1 penalization term in combination with the reconstruction error term to introduce sparsity. It is well known that the L_1 regularization introduces sparsity as shown in [66]. However, the resulting reduced set is neither the sparsest nor the most optimal w.r.t. the quality of clustering for the entire dataset. In [43], we introduced alternative penalization techniques like Group Lasso [65] and [21], L_0 and $L_1 + L_0$ penalizations. The Group Lasso penalty is ideal for clusters as it results in groups of relevant data points. The L_0 regularization calculates the number of non-zero terms in the vector. The L_0-norm results in a non-convex and NP-hard optimization problem. We modify the convex relaxation of L_0-norm based on an iterative re-weighted L_1 formulation introduced in [10, 22]. We apply it to obtain the optimal reduced sets for sparse kernel spectral clustering. Below we provide the formulation for Group Lasso penalized objective (16) and re-weighted L_1-norm penalized objectives (17).

The Group Lasso [65] based formulation for our optimization problem is:

$$\min_{\beta \in \mathbb{R}^{N_{tr} \times (k-1)}} \|\Phi^\mathsf{T}\alpha - \Phi^\mathsf{T}\beta\|_2^2 + \lambda \sum_{l=1}^{N_{tr}} \sqrt{\rho_l}\|\beta_l\|_2, \tag{16}$$

where $\Phi = [\phi(x_1), \ldots, \phi(x_{N_{tr}})]$, $\alpha = [\alpha^{(1)}, \ldots, \alpha^{(k-1)}]$, $\alpha \in \mathbb{R}^{N_{tr} \times (k-1)}$ and $\beta = [\beta_1, \ldots, \beta_{N_{tr}}]$, $\beta \in \mathbb{R}^{N_{tr} \times (k-1)}$. Here $\alpha^{(i)} \in \mathbb{R}^{N_{tr}}$ while $\beta_j \in \mathbb{R}^{k-1}$ and we set $\sqrt{\rho_l}$ as the fraction of training points belonging to the cluster to which the lth training point belongs. By varying the value of λ we control the amount of sparsity introduced in the model as it acts as a regularization parameter. In [21], the authors show that if the initial solutions are $\hat{\beta}_1, \hat{\beta}_2, \ldots, \hat{\beta}_{N_{tr}}$ then if $\|X_l^\mathsf{T}(y - \sum_{i \neq l} X_i \hat{\beta}_i)\| < \lambda$, then $\hat{\beta}_l$ is zero otherwise it satisfies: $\hat{\beta}_l = (X_l^\mathsf{T} X_l + \lambda/\|\hat{\beta}_l\|)^{-1} X_l^\mathsf{T} r_l$ where $r_l = y - \sum_{i \neq l} X_i \hat{\beta}_i$.

Analogous to this, the solution to the Group Lasso penalization for our problem can be defined as: $\|\phi(x_l)(\Phi^\mathsf{T}\alpha - \sum_{i \neq l} \phi(x_i)\hat{\beta}_i)\| < \lambda$ then $\hat{\beta}_l$ is zero otherwise it satisfies: $\hat{\beta}_l = (\Phi^\mathsf{T}\Phi + \lambda/\|\hat{\beta}_l\|)^{-1} \phi(x_l) r_l$ where $r_l = \Phi^\mathsf{T}\alpha - \sum_{i \neq l} \phi(x_i)\hat{\beta}_i$. The Group Lasso penalization technique can be solved by a blockwise co-ordinate descent procedure as shown in [65]. The time complexity of the approach is $O(\text{maxiter} * k^2 N_{tr}^2)$ where maxiter is the maximum number of iterations specified for the co-ordinate descent procedure and k is the number of clusters obtained via KSC. From our experiments we observed that on an average ten iterations suffice for convergence.

Concerning the re-weighted L_1 procedure, we modify the algorithm related to classification as shown in [22] and use it for obtaining the reduced set in our clustering setting:

$$\min_{\beta \in \mathbb{R}^{N_{tr} \times (k-1)}} \|\Phi^{\mathsf{T}}\alpha - \Phi^{\mathsf{T}}\beta\|_2^2 + \rho \sum_{i=1}^{N_{tr}} \epsilon_i + \|\Lambda\beta\|_2^2$$

$$\text{such that} \quad \|\beta_i\|_2^2 \leq \epsilon_i, i = 1, \ldots, N_{tr} \tag{17}$$

$$\epsilon_i \geq 0,$$

where Λ is matrix of the same size as the β matrix i.e. $\Lambda \in \mathbb{R}^{N_{tr} \times (k-1)}$. The term $\|\Lambda\beta\|_2^2$ along with the constraint $\|\beta_i\|_2^2 \leq \epsilon_i$ corresponds to the L_0-norm penalty on β matrix. Λ matrix is initially defined as a matrix of ones so that it gives equal chance to each element of β matrix to reduce to zero. The constraints on the optimization problem forces each element of $\beta_i \in \mathbb{R}^{(k-1)}$ to reduce to zero. This helps to overcome the problem of sparsity per component which is explained in [3]. The ρ variable is a regularizer which controls the amount of sparsity that is introduced by solving this optimization problem.

In Fig. 5 an example of clustering obtained using the Group Lasso formulation (16) on a toy dataset is depicted. We can notice how the sparse KSC model is able to obtain high quality generalization using only four points in the training set.

4 Applications

The KSC algorithm has been successfully used in a variety of applications in different domains. In the next sections we will illustrate various results obtained in different fields such as computer vision, information retrieval and power load consumer segmentation.

4.1 Image Segmentation

Image segmentation relates to partitioning a digital image into multiple regions, such that pixels in the same group share a certain visual content. In the experiments performed using KSC only the color information is exploited in order to segment the given images.[7] More precisely, a local color histogram with a 5×5 pixels window around each pixel is computed using minimum variance color quantization of 8 levels. Then, in order to compare the similarity between two histograms $h^{(i)}$ and $h^{(j)}$, the positive definite χ^2 kernel $K(h^{(i)}, h^{(j)}) = \exp(-\frac{\chi_{ij}^2}{\sigma_\chi^2})$ has been adopted

---

[7]The images have been extracted from the Berkeley image database [45].

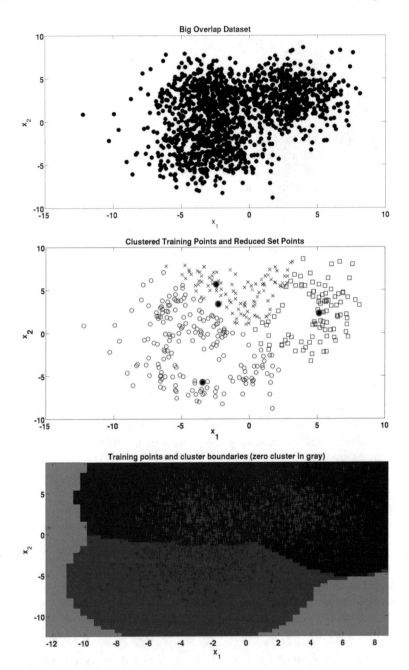

Fig. 5 Sparse KSC on toy dataset. *(Top)* Gaussian mixture with three highly overlapping components. *(Center)* clustering results, where the reduced set points are indicated with *red circles*. *(Bottom)* generalization boundaries

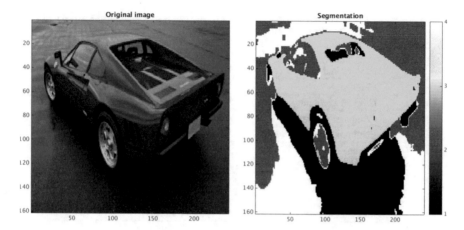

Fig. 6 Image segmentation. *(Left)* original image. *(Right)* KSC segmentation

[19]. The symbol χ_{ij}^2 denotes the χ_{ij}^2 statistical test used to compare two probability distributions [55], σ_χ as usual indicates the bandwidth of the kernel. In Fig. 6 an example of segmentation obtained using the basic KSC algorithm is given.

4.2 Scientific Journal Clustering

We present here an integrated approach for clustering scientific journals using KSC. Textual information is combined with cross-citation information in order to obtain a coherent grouping of the scientific journals and to improve over existing journal categorizations. The number of clusters k in this scenario is fixed to 22 since we want to compare the results with respect to the 22 essential science indicators (ESI) shown in Table 2.

The data correspond to more than six million scientific papers indexed by the Web of Science (WoS) in the period 2002–2006. The type of manuscripts considered is article, letter, note, and review. Textual information has been extracted from titles, abstracts and keywords of each paper together with citation information. From these data, the resulting number of journals under consideration is 8305.

The two resulting datasets contain textual and cross-citation information and are described as follows:

- **Term/Concept by Journal dataset:** The textual information was processed using the term frequency—inverse document frequency (TF-IDF) weighting procedure [6]. Terms which occur only in one document and stop words were not considered into the analysis. The Porter stemmer was applied to the remaining terms in the abstract, title, and keyword fields. This processing leads to a term-by-document matrix of around six million papers and 669, 860 term dimensionality.

Table 2 The 22 science fields according to the essential science indicators (ESI)

Field	Name	Field	Name
1	Agricultural sciences	12	Mathematics
2	Biology and biochemistry	13	Microbiology
3	Chemistry	14	Molecular biology and genetics
4	Clinical medicine	15	Multidisciplinary
5	Computer science	16	Neuroscience and behavior
6	Economics and business	17	Pharmacology and toxicology
7	Engineering	18	Physics
8	Environment/Ecology	19	Plant and animal science
9	Geosciences	20	Psychology/Psychiatry
10	Immunology	21	Social sciences
11	Materials sciences	22	Space science

The final journal-by-term dataset is a $8305 \times 669, 860$ matrix. Additionally, latent semantic indexing (LSI) [13] was performed on this dataset to reduce the term dimensionality to 200 factors.

- **Journal cross-citation dataset:** A different form of analyzing cluster information at the journal level is through a cross-citation graph. This graph contains aggregated citations between papers forming a journal-by-journal cross-citation matrix. The direction of the citations is not taken into account which leads to an undirected graph and a symmetric cross-citation matrix.

The cross-citation and the text/concept datasets are integrated at the kernel level by considering the following linear combination of kernel matrices[8]:

$$\Omega^{\text{integr}} = \rho \Omega^{\text{cross} - \text{cit}} + (1 - \rho)\Omega^{\text{text}}$$

where $0 \le \rho \le 1$ is a user-defined integration weight which value can be obtained from internal validation measures for cluster distortion,[9] $\Omega^{\text{cross} - \text{cit}}$ is the cross-citation kernel matrix with ijth entry $\Omega_{ij}^{\text{cross} - \text{cit}} = K(x_i^{\text{cross} - \text{cit}}, x_j^{\text{cross} - \text{cit}})$, $x_i^{\text{cross} - \text{cit}}$ is the ith journal represented in terms of cross-citation variables, Ω^{text} is the textual kernel matrix with ijth entry $\Omega_{ij}^{\text{text}} = K(x_i^{\text{text}}, x_j^{\text{text}})$, x_i^{text} is the ith journal represented in terms of textual variables and $i, j = 1, \ldots, N$.

The KSC outcomes are depicted in Tables 3 and 4. In particular, Table 3 shows the results in terms of internal validation of cluster quality, namely mean silhouette value (MSV) [57] and Modularity [49, 50], and in terms of agreement with existing categorizations (adjusted rand index or ARI [23] and normalized mutual information (NMI [60]). Finally, Table 4 shows the top 20 terms per cluster, which indicate a coherent structure and illustrate that KSC is able to detect the text categories present in the corpus.

---

[8]Here we use the cosine kernel described in Table 1.

[9]In our experiments we used the mean silhouette value (MSV) as an internal cluster validation criterion to select the value of ρ which gives more coherent clusters.

Table 3 Text clustering quality

	Internal validation				External validation		
	MSV textual	MSV cross-cit.	MSV integrated	Modularity cross-cit.	Modularity ISI 254	ARI 22 ESI	NMI 22 ESI
22 ESI fields	0.057	0.016	0.063	0.475	0.526*	1.000	1.000
Cross-citations	0.093	0.057	0.189	**0.547**	0.442	0.278	0.516
Textual (LSI)	0.118	0.035	0.130	0.505	0.451	0.273	0.516
Hierarch. Ward's method $\rho = 0.5$	0.121	0.055	0.190	**0.547**	**0.488**	**0.285**	**0.540**
Integr. Terms+Cross-citations $\rho = 0.5$	0.138	**0.064**	**0.201**	0.533	0.465	0.294	0.557
Integr. LSI+Cross-citations $\rho = 0.5$	0.145	0.062	0.197	0.527	0.465	0.308	**0.560**

Spectral clustering results of several integration methods in terms of mean Silhouette value (MSV), Modularity, adjusted Rand index (ARI), and normalized mutual information (NMI). The first four rows correspond to existing clustering results used for comparison. The last two rows correspond to the proposed spectral clustering algorithms. For external validation, the clustering results are compared with respect to the 22 ESI fields and the ISI 254 subject categories. The highest value per column is indicated in bold while the second highest value appears in italic. For MSV, a standard t-test for the difference in means revealed that differences between highest and second highest values are statistically significant at the 1 % significance level (p-value $< 10^8$). The selected method for further comparisons is the integrated LSI+Cross-citations approach since it wins in external validation with one highest value (NMI) and one second highest value (Modularity)

Table 4 Text clustering results

	Best 20 terms		Best 20 terms
Cluster 1	Diabet therapi hospit arteri coronari physician renal hypertens mortal syndrom cardiac nurs chronic infect pain cardiovascular symptom serum cancer pulmonari	Cluster 12	Algebra theorem manifold let finit infin polynomi invari omega singular inequ compact lambda graph conjectur convex proof asymptot bar phi
Cluster 2	Polit war court reform parti legal gender urban democraci democrat civil capit feder discours economi justic privat liber union welfar	Cluster 13	Pain surgeri injuri lesion muscl bone brain ey surgic nerv mri ct syndrom fractur motor implant arteri knee spinal stroke
Cluster 3	Diet milk fat intak cow dietari fed meat nutrit fatti chees vitamin ferment fish dry fruit antioxid breed pig egg	Cluster 14	Rock basin fault sediment miner ma tecton isotop mantl volcan metamorph seismic sea magma faci earthquak ocean cretac crust sedimentari
Cluster 4	Alloi steel crack coat corros fiber concret microstructur thermal weld film deform ceram fatigu shear powder specimen grain fractur glass	Cluster 15	Web graph fuzzi logic queri schedul semant robot machin video wireless neural node internet traffic processor retriev execut fault packet
Cluster 5	Infect hiv vaccin viru immun dog antibodi antigen pathogen il pcr parasit viral bacteri dna therapi mice bacteria cat assai	Cluster 16	Student school teacher teach classroom instruct skill academ curriculum literaci learner colleg write profession disabl faculti english cognit peer gender
Cluster 6	Psycholog cognit mental adolesc emot symptom child anxieti student sexual interview school abus psychiatr gender attitud mother alcohol item disabl	Cluster 17	Habitat genu fish sp forest predat egg nest larva reproduct taxa bird season prei nov ecolog island breed mate genera
Cluster 7	Text music polit literari philosophi narr english moral book essai write discours philosoph fiction ethic poetri linguist german christian religi	Cluster 18	Star galaxi solar quantum neutrino orbit quark gravit cosmolog decai nucleon emiss radio nuclei relativist neutron cosmic gaug telescop hole
Cluster 8	Firm price busi trade economi invest capit tax wage financi compani incom custom sector bank organiz corpor stock employ strateg	Cluster 19	Film laser crystal quantum atom ion beam si nm dope thermal spin silicon glass scatter dielectr voltag excit diffract spectra
Cluster 9	Nonlinear finit asymptot veloc motion stochast elast nois turbul ltd vibrat iter crack vehicl infin singular shear polynomi mesh fuzzi	Cluster 20	Polym catalyst ion bond crystal solvent ligand hydrogen nmr molecul atom polymer poli aqueou adsorpt methyl film spectroscopi electrod bi
Cluster 10	Soil seed forest crop leaf cultivar seedl ha shoot fruit wheat fertil veget germin rice flower season irrig dry weed	Cluster 21	Receptor rat dna neuron mice enzym genom transcript brain mutat peptid kinas inhibitor metabol cancer mrna muscl ca2 vitro chromosom
Cluster 11	Soil sediment river sea climat land lake pollut wast fuel wind ocean atmospher ic emiss reactor season forest urban basin	Cluster 22	Cancer tumor carcinoma breast therapi prostat malign chemotherapi tumour surgeri lesion lymphoma pancreat recurr resect surgic liver lung gastric node

Best 20 terms per cluster according to the integrated results (LSI+cross-citation) with $\rho = 0.5$. The terms found display a coherent structure in the clusters

4.3 Power Load Clustering

Accurate power load forecasts are essential in electrical grids and markets particularly for planning and control operations [5]. In this scenario, we apply KSC for finding power load smart meter data that are similar in order to aggregate them and improve the forecasting accuracy of the global consumption signal. The idea is to fit a forecasting model on the aggregated load of each cluster (aggregator). The k predictions are summed to form the final disaggregated prediction. The number of clusters and the time series used for each aggregator are determined via KSC [1]. The forecasting model used is a periodic autoregressive model with exogenous variables (PARX) [18]. Table 5 (taken from [1]) shows the model selection and disaggregation results. Several kernels appropriate for time series were tried including a Vector Autoregressive (VAR) kernel [Add: Cuturi, Autoregressive kernels for time series, arXiv], Triangular Global Alignment (TGA) kernel [Add: Cuturi, Fast Global Alignment Kernels, ICML 2011] and an RBF kernel with Spearman's distance. The results show an improvement of 20.55 % with the similarity based on Spearman's correlation in the forecasting accuracy compared to not using clustering at all (i.e., aggregating all smart meters). The BLF was also able to detect the number of clusters that maximize the improvement (six clusters in this case), as shown in Fig. 7.

4.4 Big Data

KSC has been shown to be effective in handling big data at a desktop PC scale. In particular, in [39], we focused on community detection in big networks containing millions of nodes and several million edges, and we explained how to scale our

Table 5 Kernel comparisons for power load clustering data

Kernel	Cluster number (BLF)	Cluster number (MAPE)	MAPE (%)
VAR	7	13	2.85
TGA	5	8	2.61
Spearman	**6**	**6**	**2.59**
RBF-DB6-11	4	5	3.02
kmeans-DB6-11	−	16	2.9
Random	−	3	2.93 ± 0.03

Model selection and forecasting results in terms of the mean absolute percentage error (MAPE). RBF-DB6-11 refers to using the RBF kernel on the detail coefficients using wavelets (DB6, 11 levels). The winner is the Spearman-based kernel with a improvement of 20.55 % compared to the baseline MAPE of the disaggregated forecast equal to 3.26 %. For this kernel, the number of clusters k found by the BLF also coincides with the number of aggregators needed to maximize the improvement

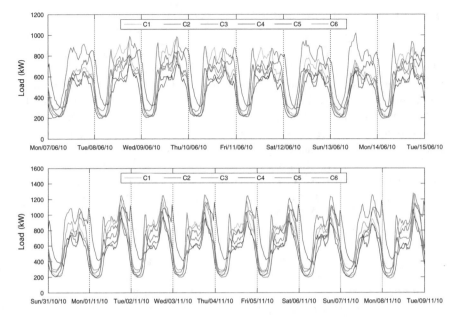

Fig. 7 Power load clustering results. Visualization of the six clusters obtained by KSC. *(Top)* aggregated load in summer. *(Bottom)* aggregated load in winter. The daily cycles are clearly visible and the clusters capture different characteristics of the consumption pattern. This clustering result improves the forecasting accuracy by 20.55 %

method by means of three steps.[10] First, we select a smaller subgraph that preserves the overall community structure by using the FURS algorithm [38], where hubs in dense regions of the original graph are selected via a greedy activation–deactivation procedure. In this way the kernel matrix related to the extracted subgraph fits the main memory and the KSC model can be quickly trained by solving a smaller eigenvalue problem. Then the BAF criterion described in Sect. 3.1.3, which is memory and computationally efficient, is used for model selection.[11] Finally, the out-of-sample extension is used to infer the cluster memberships for the remaining nodes forming the test set (which is divided into chunks due to memory constraints).

In [42] the hierarchical clustering technique summarized in Sect. 3.3.2 has been used to perform community detection in real-life networks at different resolutions. In the experiment conducted on seven networks from the Stanford SNAP datasets (http://snap.stanford.edu/data/index.html), the method has been shown to be able to detect complex structures at various hierarchical levels, by not suffering of

---

[10]A *Matlab* implementation of the algorithm can be downloaded at: http://www.esat.kuleuven.be/stadius/ADB/mall/softwareKSCnet.php.

[11]In [40] this model selection step has been eliminated by proposing a self-tuned method where the structure of the projections in the eigenspace is exploited to automatically identify an optimal cluster structure.

any resolution limit. This is not the case for other state-of-the-art algorithms like Infomap [56], Louvain [9], and OSLOM [24]. In particular, we have observed that Louvain method is often not able to detect high quality clusters at finer levels of granularity ($<$ 1000 clusters). On the other hand, OSLOM cannot identify good quality coarser clusters (i.e. number of clusters detected are always $>$ 1000), and Infomap method produces only two levels of hierarchy. Moreover, in general Louvain method works best in terms of the Modularity criterion, and it always performs worse than hierarchical KSC w.r.t. cut-conductance [35]. Regarding Infomap, in most of the cases the clusters at one level of hierarchy perform good w.r.t. only one quality metric. Concerning OSLOM, this algorithm in the majority of the datasets has poorer performances than KSC in terms of both Modularity and cut-conductance.

An illustration of the community structure obtained on the *Cond-mat* network of collaborations between authors of papers submitted to Condense Matter category in *Arxiv* [34] is shown in Fig. 8. This network is formed by 23, 133 nodes and 186, 936 edges. For the analysis and visualization of bigger networks, and the detailed comparison of KSC with other community detection methods in terms of computational efficiency and quality of detected communities, the reader can refer to [42].

Finally, in [44], we propose a deterministic method to obtain subsets from big vector data which are a good representative of the inherent clustering structure. We first convert the large-scale dataset into a sparse undirected k-NN graph using a Map-Reduce framework. Then, the FURS method is used to select a few representative nodes from this graph, corresponding to certain data points in the original dataset. These points are then used to quickly train the KSC model,

Fig. 8 Large-scale community detection. Community structure detected at one particular hierarchical level by the AH-KSC method summarized in Sect. 3.3.2, related to the *Cond-Mat* collaboration network

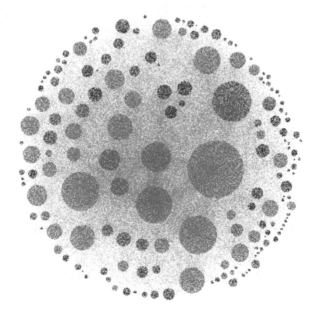

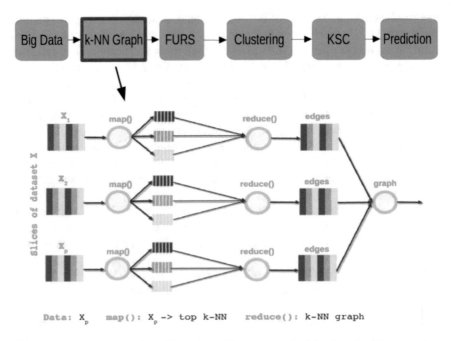

Fig. 9 Big data clustering. *(Top)* illustration of the steps involved in clustering big vector data using KSC. *(Bottom)* map-reduce procedure used to obtain a representative training subset by constructing a k-NN graph

while the generalization property of the method is exploited to compute the cluster memberships for the remainder of the dataset. In Fig. 9 a summary of all these steps is sketched.

5 Conclusions

In this chapter we have discussed the kernel spectral clustering (KSC) method, which is cast in an LS-SVM learning framework. We have explained that, like in the classifier case, the clustering model can be trained on a subset of the data with optimal tuning parameters, found during the validation stage. The model is then able to generalize to unseen test data thanks to its out-of-sample extension property. Beyond the core algorithm, some extensions of KSC allowing to produce probabilistic and hierarchical outputs have been illustrated. Furthermore, two different approaches to sparsify the model based on the Incomplete Cholesky Decomposition (ICD) and L_1 and L_0 penalties have been described. This allows to handle large-scale data at a desktop scale. Finally, a number of applications in various fields ranging from computer vision to text mining have been examined.

Acknowledgements EU: The research leading to these results has received funding from the European Research Council under the European Union's Seventh Framework Programme (FP7/2007–2013)/ERC AdG A-DATADRIVE-B (290923). This chapter reflects only the authors' views, the Union is not liable for any use that may be made of the contained information. Research Council KUL: GOA/10/09 MaNet, CoE PFV/10/002 (OPTEC), BIL12/11T; PhD/Postdoc grants. Flemish Government: FWO: projects: G.0377.12 (Structured systems), G.088114N (Tensor-based data similarity); PhD/Postdoc grants. IWT: projects: SBO POM (100031); PhD/Postdoc grants. iMinds Medical Information Technologies SBO 2014. Belgian Federal Science Policy Office: IUAP P7/19 (DYSCO, Dynamical systems, control and optimization, 2012–2017.)

References

1. Alzate, C., Sinn, M.: Improved electricity load forecasting via kernel spectral clustering of smart meters. In: ICDM, pp. 943–948 (2013)
2. Alzate, C., Suykens, J.A.K.: Multiway spectral clustering with out-of-sample extensions through weighted kernel PCA. IEEE Trans. Pattern Anal. Mach. Intell. **32**(2), 335–347 (2010)
3. Alzate, C., Suykens, J.A.K.: Sparse kernel spectral clustering models for large-scale data analysis. Neurocomputing **74**(9), 1382–1390 (2011)
4. Alzate, C., Suykens, J.A.K.: Hierarchical kernel spectral clustering. Neural Networks **35**, 21–30 (2012)
5. Alzate, C., Espinoza, M., De Moor, B., Suykens, J.A.K.: Identifying customer profiles in power load time series using spectral clustering. In: Proceedings of the 19th International Conference on Neural Networks (ICANN 2009), pp. 315–324 (2009)
6. Baeza-Yates, R., Ribeiro-Neto, B.: Modern Information Retrieval. Addison-Wesley, Boston (1999)
7. Ben-Israel, A., Iyigun, C.: Probabilistic d-clustering. J. Classif. **25**(1), 5–26 (2008)
8. Bishop, C.M.: Pattern Recognition and Machine Learning (Information Science and Statistics). Springer, New York (2006)
9. Blondel, V.D., Guillaume, J.L., Lambiotte, R., Lefebvre, E.: Fast unfolding of communities in large networks. J. Stat. Mech. Theory Exp. **2008**(10), P10,008 (2008)
10. Candes, E.J., Wakin, M.B., Boyd, S.: Enhancing sparsity by reweighted l1 minimization. J. Fourier Anal. Appl. (Special Issue on Sparsity) **14**(5), 877–905 (2008)
11. Chung, F.R.K.: Spectral Graph Theory. American Mathematical Society, Providence (1997)
12. De Brabanter, K., De Brabanter, J., Suykens, J.A.K., De Moor, B.: Optimized fixed-size kernel models for large data sets. Comput. Stat. Data Anal. **54**(6), 1484–1504 (2010)
13. Deerwester, S.C., Dumais, S.T., Landauer, T.K., Furnas, G.W., Harshman, R.A.: Indexing by latent semantic analysis. J. Am. Soc. Inf. Sci. **41**(6), 391–407 (1990)
14. Delvenne, J.C., Yaliraki, S.N., Barahona, M.: Stability of graph communities across time scales. Proc. Natl. Acad. Sci. **107**(29), 12755–12760 (2010)
15. Dhanjal, C., Gaudel, R., Clemenccon, S.: Efficient eigen-updating for spectral graph clustering (2013) [arXiv/1301.1318]
16. Dhillon, I., Guan, Y., Kulis, B.: Kernel k-means, spectral clustering and normalized cuts. In: 10th ACM Knowledge Discovery and Data Mining Conf., pp. 551–556 (2004)
17. Dhillon, I., Guan, Y., Kulis, B.: Weighted graph cuts without eigenvectors a multilevel approach. IEEE Trans. Pattern Anal. Mach. Intell. **29**(11), 1944–1957 (2007)
18. Espinoza, M., Joye, C., Belmans, R., De Moor, B.: Short-term load forecasting, profile identification and customer segmentation: a methodology based on periodic time series. IEEE Trans. Power Syst. **20**(3), 1622–1630 (2005)
19. Fowlkes, C., Belongie, S., Chung, F., Malik, J.: Spectral grouping using the Nyström method. IEEE Trans. Pattern Anal. Mach. Intell. **26**(2), 214–225 (2004)

20. Frederix, K., Van Barel, M.: Sparse spectral clustering method based on the incomplete cholesky decomposition. J. Comput. Appl. Math. **237**(1), 145–161 (2013)
21. Friedman, J., Hastie, T., Tibshirani, R.: A note on the group lasso and a sparse group lasso (2010) [arXiv:1001.0736]
22. Huang, K., Zheng, D., Sun, J., Hotta, Y., Fujimoto, K., Naoi, S.: Sparse learning for support vector classification. Pattern Recogn. Lett. **31**(13), 1944–1951 (2010)
23. Hubert, L., Arabie, P.: Comparing partitions. J. Classif. **1**(2), 193–218 (1985)
24. Lancichinetti, A., Radicchi, F., Ramasco, J.J., Fortunato, S.: Finding statistically significant communities in networks. PLoS ONE **6**(4), e18961 (2011)
25. Langone, R., Suykens, J.A.K.: Community detection using kernel spectral clustering with memory. J. Phys. Conf. Ser. **410**(1), 012100 (2013)
26. Langone, R., Alzate, C., Suykens, J.A.K.: Modularity-based model selection for kernel spectral clustering. In: Proc. of the International Joint Conference on Neural Networks (IJCNN 2011), pp. 1849–1856 (2011)
27. Langone, R., Alzate, C., Suykens, J.A.K.: Kernel spectral clustering for community detection in complex networks. In: Proc. of the International Joint Conference on Neural Networks (IJCNN 2012), pp. 2596–2603 (2012)
28. Langone, R., Alzate, C., Suykens, J.A.K.: Kernel spectral clustering with memory effect. Phys. A Stat. Mech. Appl. **392**(10), 2588–2606 (2013)
29. Langone, R., Alzate, C., De Ketelaere, B., Suykens, J.A.K.: Kernel spectral clustering for predicting maintenance of industrial machines. In: IEEE Symposium Series on Computational Intelligence and data mining SSCI (CIDM) 2013, pp. 39–45 (2013)
30. Langone, R., Mall, R., Suykens, J.A.K.: Soft kernel spectral clustering. In: Proc. of the International Joint Conference on Neural Networks (IJCNN 2013), pp. 1–8 (2013)
31. Langone, R., Mall, R., Suykens, J.A.K.: Clustering data over time using kernel spectral clustering with memory. In: SSCI (CIDM) 2014, pp. 1–8 (2014)
32. Langone, R., Agudelo, O.M., De Moor, B., Suykens, J.A.K.: Incremental kernel spectral clustering for online learning of non-stationary data. Neurocomputing **139**, 246–260 (2014)
33. Langone, R., Alzate, C., De Ketelaere, B., Vlasselaer, J., Meert, W., Suykens, J.A.K.: Ls-svm based spectral clustering and regression for predicting maintenance of industrial machines. Eng. Appl. Artif. Intell. **37**, 268–278 (2015)
34. Leskovec, J., Kleinberg, J., Faloutsos, C.: Graph evolution: densification and shrinking diameters. ACM Trans. Knowl. Discov. Data **1**(1), 2 (2007)
35. Leskovec, J., Lang, K.J., Mahoney, M.: Empirical comparison of algorithms for network community detection. In: Proceedings of the 19th International Conference on World Wide Web, WWW '10, pp. 631–640. ACM, New York (2010)
36. Liao, T.W.: Clustering of time series data - a survey. Pattern Recogn. **38**(11), 1857–1874 (2005)
37. Lin, F., Cohen, W.W.: Power iteration clustering. In: ICML, pp. 655–662 (2010)
38. Mall, R., Langone, R., Suykens, J.: FURS: fast and unique representative subset selection retaining large scale community structure. Soc. Netw. Anal. Min. **3**(4), 1–21 (2013)
39. Mall, R., Langone, R., Suykens, J.A.K.: Kernel spectral clustering for big data networks. Entropy (Special Issue on Big Data) **15**(5), 1567–1586 (2013)
40. Mall, R., Langone, R., Suykens, J.A.K.: Self-tuned kernel spectral clustering for large scale networks. In: IEEE International Conference on Big Data (2013)
41. Mall, R., Langone, R., Suykens, J.A.K.: Agglomerative hierarchical kernel spectral data clustering. In: Symposium Series on Computational Intelligence (SSCI-CIDM), pp. 1–8 (2014)
42. Mall, R., Langone, R., Suykens, J.A.K.: Multilevel hierarchical kernel spectral clustering for real-life large scale complex networks. PLoS ONE **9**(6), e99966 (2014)
43. Mall, R., Mehrkanoon, S., Langone, R., Suykens, J.A.K.: Optimal reduced sets for sparse kernel spectral clustering. In: Proc. of the International Joint Conference on Neural Networks (IJCNN 2014), pp. 2436–2443 (2014)
44. Mall, R., Jumutc, V., Langone, R., Suykens, J.A.K.: Representative subsets for big data learning using kNN graphs. In: IEEE International Conference on Big Data, pp. 37–42 (2014)

45. Martin, D., Fowlkes, C., Tal, D., Malik, J.: A database of human segmented natural images and its application to evaluating segmentation algorithms and measuring ecological statistics. In: Proc. 8th Int'l Conf. Computer Vision, vol. 2, pp. 416–423 (2001)
46. Meila, M., Shi, J.: Learning segmentation by random walks. In: T.K. Leen, T.G. Dieterich, V. Tresp (eds.) Advances in Neural Information Processing Systems, vol. 13. MIT Press, Cambridge (2001)
47. Meila, M., Shi, J.: A random walks view of spectral segmentation. In: Artificial Intelligence and Statistics AISTATS (2001)
48. Mika, S., Schölkopf, B., Smola, A.J., Müller, K.R., Scholz, M., Rätsch, G.: Kernel PCA and de-noising in feature spaces. In: Kearns, M.S., Solla, S.A., Cohn, D.A. (eds.) Advances in Neural Information Processing Systems, vol. 11. MIT Press, Cambridge (1999)
49. Newman, M.E.J.: Modularity and community structure in networks. Proc. Natl. Acad. Sci. U. S. A. **103**(23), 8577–8582 (2006)
50. Newman, M.E.J., Girvan, M.: Finding and evaluating community structure in networks. Phys. Rev. E **69**(2), 026113 (2004)
51. Ng, A.Y., Jordan, M.I., Weiss, Y.: On spectral clustering: analysis and an algorithm. In: Dietterich, T.G., Becker, S., Ghahramani, Z. (eds.) Advances in Neural Information Processing Systems 14, pp. 849–856. MIT Press, Cambridge (2002)
52. Ning, H., Xu, W., Chi, Y., Gong, Y., Huang, T.S.: Incremental spectral clustering by efficiently updating the eigen-system. Pattern Recogn. **43**(1), 113–127 (2010)
53. Novak, M., Alzate, C., langone, R., Suykens, J.A.K.: Fast kernel spectral clustering based on incomplete cholesky factorization for large scale data analysis. Internal Report 14–119, ESAT-SISTA, KU Leuven (Leuven, Belgium) (2015)
54. Peluffo, D., Garcia, S., Langone, R., Suykens, J.A.K., Castellanos, G.: Kernel spectral clustering for dynamic data using multiple kernel learning. In: Proc. of the International Joint Conference on Neural Networks (IJCNN 2013), pp. 1085–1090 (2013)
55. Puzicha, J., Hofmann, T., Buhmann, J.: Non-parametric similarity measures for unsupervised texture segmentation and image retrieval. In: Computer Vision and Pattern Recognition, pp. 267–272 (1997)
56. Rosvall, M., Bergstrom, C.T.: Maps of random walks on complex networks reveal community structure. Proc. Natl. Acad. Sci. **105**(4), 1118–1123 (2008)
57. Rousseeuw, P.J.: Silhouettes: a graphical aid to the interpretation and validation of cluster analysis. J. Comput. Appl. Math. **20**(1), 53–65 (1987)
58. Schölkopf, B., Smola, A.J., Müller, K.R.: Nonlinear component analysis as a kernel eigenvalue problem. Neural Comput. **10**, 1299–1319 (1998)
59. Shi, J., Malik, J.: Normalized cuts and image segmentation. IEEE Trans. Pattern Anal. Mach. Intell. **22**(8), 888–905 (2000)
60. Strehl, A., Ghosh, J.: Cluster ensembles - a knowledge reuse framework for combining multiple partitions. J. Mach. Learn. Res. **3**, 583–617 (2002)
61. Suykens, J.A.K., Van Gestel, T., De Brabanter, J., De Moor, B., Vandewalle, J.: Least Squares Support Vector Machines. World Scientific, Singapore (2002)
62. Suykens, J.A.K., Van Gestel, T., Vandewalle, J., De Moor, B.: A support vector machine formulation to PCA analysis and its kernel version. IEEE Trans. Neural Netw. **14**(2), 447–450 (2003)
63. von Luxburg, U.: A tutorial on spectral clustering. Stat. Comput. **17**(4), 395–416 (2007)
64. Williams, C.K.I., Seeger, M.: Using the Nyström method to speed up kernel machines. In: Advances in Neural Information Processing Systems, vol. 13. MIT Press, Cambridge (2001)
65. Yuan, M., Lin, Y.: Model selection and estimation in regression with grouped variables. J. R. Stat. Soc. **68**(1), 49–67 (2006)
66. Zhu, J., Rosset, S., Hastie, T., Tibshirani, R.: 1-norm svms. In: Neural Information Processing Systems, vol. 16 (2003)

Uni- and Multi-Dimensional Clustering Via Bayesian Networks

Omid Keivani and Jose M. Peña

Abstract This chapter discusses model based clustering via Bayesian networks. Both uni-dimensional and multi-dimensional clustering methods are discussed. The main idea for uni-dimensional clustering via Bayesian networks is to use the Bayesian structural clustering algorithm, which is a greedy algorithm that makes use of the EM algorithm. On the other hand, for multi-dimensional clustering we investigate latent tree models which according to our knowledge, are the only model based approach to multi-dimensional clustering. There are generally two approaches for learning latent tree models: Greedy search and feature selection. The former is able to cover a wider range of models, but the latter is more time efficient. However, latent tree models are unable to capture dependency between partitions through attributes. So we propose two approaches to overcome this shortcoming. Our first approach extends the idea of Bayesian structural clustering for uni-dimensional clustering, while the second one is a combination of feature selection methods and the main idea of multi-dimensional classification with Bayesian networks. We test our second approach on both real and synthetic data. The results show the goodness of our approach in finding meaningful and novel partitions.

1 Introduction

Clustering, also known as unsupervised learning, is a task aiming to group objects together based on their similarities. Different notions of similarity will lead to significantly different clustering algorithms. There are generally two popular notions for cluster similarity: distances among objects and cluster distributions. The first approach tries to find homogeneous clusters such that each cluster is as different as possible from others. These methods are called distance based. The second approach tries to model the probability distribution that gave rise to each cluster. This approach is called model based.

O. Keivani (✉) • J.M. Peña
ADIT, IDA, Linköping University, Linköping, Sweden
e-mail: omid.keivani@liu.se; jose.m.pena@liu.se

© Springer International Publishing Switzerland 2016
M.E. Celebi, K. Aydin (eds.), *Unsupervised Learning Algorithms*,
DOI 10.1007/978-3-319-24211-8_7

Each approach has its own merits and drawbacks. For example, distance based methods are simpler and more time efficient. On the other hand, model based approaches are more flexible and will find the structure of the mechanism that produced the data instead of just cluster assignments or representatives for clusters (cluster centers). For further discussion about model based approaches and their advantages, see [1]. In this chapter, we will only consider model based approaches via Bayesian networks (BNs).

Datasets are usually represented by a $n \times m$ matrix, where n is the number of objects in dataset and m is the number of attributes. Therefore, the aim of a model based clustering method would be to find a model, which describes the existing structures within the dataset best. Model based clustering typically consists in modeling the probability distribution of the cluster variable, i.e., $P(C)$, and the probability distribution of the attributes given the cluster variable, i.e., $P(A_1, A_2, \ldots, A_m | C)$. The probability distribution of the attributes can be obtained by combining the two previous distributions into a so-called finite mixture model, i.e.

$$P(A_1, A_2, \ldots, A_m) = \sum_c P(C) P(A_1, A_2, \ldots, A_m | C). \tag{1}$$

In clustering, we are however more interested in the probability distribution of the cluster variable given the attributes, i.e.

$$P(C | A_1, A_2, \ldots, A_m) \propto P(C) P(A_1, A_2, \ldots, A_m | C). \tag{2}$$

However, when the number of attributes is large, it may be cumbersome to work with $P(A_1, \ldots, A_m | C)$. Instead, probabilistic graphical models (PGMs) such as BNs can be used to factorize this distribution, making it possible to work with it efficiently. Specifically,

$$P(C | A_1, A_2, \ldots, A_m) \propto P(C) \prod_i P(A_i | \mathrm{Pa}_G(A_i), C), \tag{3}$$

where $\mathrm{Pa}_G(A_i)$ are the attributes that are parents of A_i in the directed and acyclic graph G (DAG), which is known as the structure of the BN. Figure 1 shows a BN with four variables where, $\mathrm{Pa}(C) = \{A, B\}$, $\mathrm{Pa}(D) = \{C, B\}$, and $\mathrm{Pa}(A) = \mathrm{Pa}(B) = \emptyset$. The second component of the BN is a set of parameters [the conditional probability distributions in the r.h.s of (3)] typically estimated through maximum likelihood (ML) from data. Clearly, knowing the correct structure of the BN is crucial for finding the correct clustering. There are generally two approaches to this task: The structure is provided by an expert beforehand, or the structure is learnt from some data at hand. In this chapter, we will review the latter approach, which includes works such as [2–5].

Most of the model based approaches to clustering assume that there exists only one cluster variable, i.e., uni-dimensional clustering. However, data may be

Fig. 1 A simple BN with
four random variables

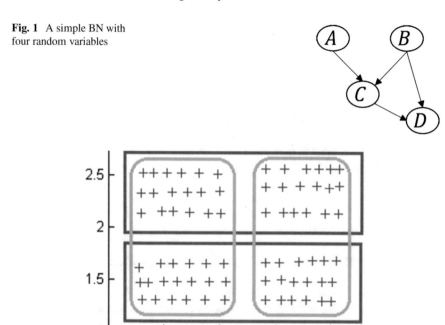

Fig. 2 A toy example for multi-dimensional clustering

multifaceted or, in other words, the objects may be grouped differently according to different criteria. As a toy example, consider Fig. 2. We can group this data into two clusters, both horizontally and vertically. The situation where data is clustered simultaneously according to different criteria is known as multi-dimensional clustering. While there are many works on model based multi-dimensional classification [6–10], there exist just a few works on model based multi-dimensional clustering [11–14]. However, there exist other methods in the BN literature that find and introduce latent variables in the network [15, 16]. However, these approaches have not been applied to clustering domains and we are not aware of their capabilities under clustering assumptions, i.e., cluster variables should not have any parents. Hence, we are not going to discuss them. This chapter will review model based multi-dimensional clustering works and contribute with two new algorithms for this task.

The rest of the chapter is organized as follows. Section 2 discusses model based uni-dimensional approaches, while Sect. 3 will discuss multi-dimensional clustering. Section 4 contains our proposed model based method for multi-dimensional clustering and some preliminary results are shown in Sect. 5. Finally, we will draw some conclusion and summarize the discussed methods in Sect. 6.

2 Uni-Dimensional Clustering

Uni-dimensional clustering or simply clustering means that only one cluster variable
is available and we are looking for a model that best describes the data. Large
number of methods are available for uni-dimensional clustering [2–5]. In model
based clustering with BNs, we normally assume that we have one cluster variable C,
which is hidden and is the parent of all attributes, and for each object, we are looking
for the value of variable C that maximizes the posterior probability of the cluster
variable given that object:

$$c = \text{argmax}_C P(C|A_1, A_2, \ldots, A_m). \qquad (4)$$

According to Bayes formula:

$$P(C|A_1, A_2, \ldots, A_m) \propto P(C)P(A_1, A_2, \ldots, A_m|C). \qquad (5)$$

However, as we mentioned before, working with $P(A_1, A_2, \ldots, A_m|C)$ may be
inconvenient. With the help of a BN such as G, this term would be factorize as
follow:

$$P(C|A_1, A_2, \ldots, A_m) \propto P(C) \prod_i P(A_i|\text{Pa}_G(A_i), C). \qquad (6)$$

But, since the cluster variable C is hidden, computing the ML parameters for G is
not an easy task. So, we have to estimate the ML parameters by iterating two steps:
Fractional completion of the database and using ML estimation on the completed
data. This process is known as expectation maximization (EM) [17]. However, EM
merely learns the parameters. In order to learn both the structure and its parameters
simultaneously, Peña et al. [2] proposed a method, which has been used by many
others [3–5] afterwards. The main idea is to consider the expected value of C given
the values of the attributes in that object as its real value. In this way we will have a
complete data at hand. Having a complete data, any parameter estimation algorithm,
such as ML or maximum a posteriori (MAP) can be used to estimate parameters
in the r.h.s of (6). Figure 3 shows the general algorithm for learning BNs from
incomplete data. However, as a special case, where one variable is hidden (cluster
variable), we can use it for clustering so we call it Bayesian structural clustering
(BSC). It is also known as Bayesian structural EM if we use EM algorithm as
our parameter search step [18]. Although there exist other methods for parameter
search step such as matrix decomposition methods [19], which is used for learning
parameters of a hidden markov model, but we do not know if it can be applied
to clustering as well. Furthermore, there exist other variants of EM (we will point
them later on), but usually EM algorithm will be used for parameter learning step
and different methods merely differ in their structure search step. Some methods
do both the structure learning and parameter learning steps. However, since the

1. Choose initial structure and initial set of parameter values

2. Parameter search step

3. Compute $P(C|A_1 \ldots A_m)$ for all objects to complete the dataset

4. Structure search step

5. Re-estimate parameter values for the new structure

6. If no change in the structure has been done

   then stop

   else go to 2.

Fig. 3 General algorithm for clustering via BNs (Bayesian structural clustering)

structure learning step is time consuming, there exist another group of methods which ignore this step and only focus on the parameter learning step. We will discuss both approaches in the following.

2.1 Known Structure

In some problems, the structure of the BN may be pre-defined. This could be the case when an expert has some knowledge about the problem and defines a specific structure for it or we may define a structure according to some assumptions. The former situation requires a strong and reliable expert, while the latter is very common and is widely used. The most well-known case for the latter is naive Bayes (NB) structure.

2.1.1 Naive Bayes

The simplest structure for a BN is to assume that all attributes are independent of each other given the cluster variable. This structure is called NB, also known as latent class model (LCM). NB structure is fixed and does not require any learning procedure. In this structure, the cluster variable is the sole parent of all attributes and no other edges are allowed. Figure 4 shows an NB structure. According to this assumption, we can rewrite (3) as follows:

$$P(C|A_1, A_2, \ldots, A_m) \propto P(C) \prod_{i=1}^{m} P(A_i|C). \tag{7}$$

Having the known structure, we only need the parameters ($P(C)$ and $P(A_i|C)$) to be able to compute (7). Since the cluster variable is hidden, we are not able to use MLE or MAP to estimate the parameters. So, learning the parameters would

Fig. 4 Naive Bayes structure

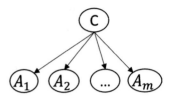

Input

    G: A BN structure

    θ^0: Random parameters

    D: Dataset

Output

    θ^k: Estimated parameters

Main

    1. while {stopping criterion}

    2.   Compute the posterior probability of cluster variable given attributes for each object

    3.   Compute MLE/MAP parameters θ^{k+1}(according to now complete data) (θ^{k+1})

    4. End while

Fig. 5 EM method for a known BN structure

not be as straightforward as it is in classification problems (where class nodes are observed). However, we can consider this problem as a parameter learning problem with missing data. Usually, in the literature, the EM algorithm will be used to estimate the parameter from missing data [20]. Figure 5 shows the procedure of EM when the BN structure is known. The algorithm requires an incomplete dataset D and since it assumes that the structure is known, it should also take a BN structure (G) along with its initial (random) parameters (θ^0) as its input. At the second line, for each object in D the algorithm will compute the posterior probability of the cluster variable (7) and complete the dataset. Line 3 computes either MLE or MAP parameters according to the completed dataset. These two steps should be done until a stopping criterion is met. Usually the difference between the log-likelihood of two consecutive steps ($\mathrm{LL}(\theta_{k+1}|D, G) - \mathrm{LL}(\theta_k|D, G)$) will be used as a stopping criterion. Having both structure and parameters, we can compute the posterior probability of cluster variable given each object and since we did not learn the structure, this method would have low time complexity. Although the NB assumption may not be realistic and we expect it to have detrimental effects, this model has proved to be very powerful and produce acceptable results. However, [21, 22] used expectation model averaging (EMA) instead of EM as the parameter search step to incorporate information about conditional (in)dependencies between attributes in parameters and compensate for the lack of conditional (in)dependencies that exist in the NB model.

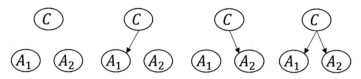

Fig. 6 All possible selective naive Bayes structures with two attributes

2.1.2 Expectation Model Averaging

The main idea of [21, 22] for learning a BN is the same as for learning NB. However, they differ in parameter search step. Santafé et al. [21] propose and use EMA algorithm instead of EM for the parameter search step. Like the EM algorithm, EMA has two steps. First, it will compute expected sufficient statistics and then it will average over the parameters of all possible selective naive Bayes (SNB) structures. SNB is an NB structure, where each attribute may or may not depend on cluster variable. Figure 6 shows all possible SNBs with two attributes. E step calculates the posterior probabilities of the cluster variable given each object to complete the dataset and will compute all expected sufficient statistics. Expected sufficient statistics should be computed for all SNB structures, so three different types of variables exist. Cluster variable (which has no parents) and an attribute with and without parent.

Having all sufficient statistics at hand, we can run model averaging (MA) algorithm to reestimate parameters for NB structure (this step mimics the M step in EM algorithm). In order to do so, MA algorithm computes $P(\theta|D)$ by averaging over the MAP parameters of all possible SNB structures. Computing these values for a single structure S is an easy task. We know that computing MAP parameters requires both prior knowledge $(P(\theta))$ and actual sufficient statistics computed from the complete data $(P(D|\theta, S))$:

$$P(\theta|S, D) \propto P(D|\theta, S) \times P(\theta). \tag{8}$$

We can use the expected sufficient statistics, computed in the E step, as an approximation for real ones. Also, knowing that the Dirichlet distribution is in the class of conjugate priors for multinomial distributions, assuming Dirichlet distribution as our prior knowledge helps us do our computation in closed form [23]. MAP parameters can be easily computed from the expected sufficient statistics and hyper parameters of the Dirichlet distribution [23].

However, the aim of EMA algorithm is to average over parameters of all possible SNBs. Hence, the MA calculations need a summation over all possible SNBs (2^m). Note that, since we need to average over all possible structures, there would be a lot of repetitions in the computations that need to be done only once. Hence, the algorithm would be time efficient. For more details on how to average parameters over all possible structures efficiently, see [21].

In this way, although the structure that we will have is an NB, its parameters will take into account some information about the conditional (in)dependencies between variables represented by all SNBs. In another attempt, to further improve the parameter search step, Santafé et al. [22] used the same idea as EMA, but this time they replaced SNB structures with tree augmented naive Bayes (TAN). TAN structure provides more realistic conditional (in)dependencies, so we would expect to have better results.

2.1.3 Expectation Model Averaging: Tree Augmented Naive Bayes

To improve the quality of the model which will be learnt by EMA, Santafé et al. [22] used a TAN model instead of SNB to capture more conditional (in)dependencies and incorporating them into parameter search step. The classical TAN model [24] is a structure in which the attributes form a tree and the hidden variable (cluster) is the parent of all attributes. Santafé et al. [22] removes these two constraints to widen the range of representable conditional (in)dependencies. First, they allow the attributes to form a forest instead of a tree and second, the cluster variable may or may not be the parent of an attribute. Also, they consider a fixed ordering for their structure. L^{π}_{TAN} refers to such a TAN with a fixed ordering of π. This new TAN is a superset of NB, SBN, and TAN models. For example, for a given ordering of $\pi = \{A_1, A_2, A_3\}$, the possible parent sets for A_2 will be:

$$\text{Pa}_{A_2} = \{\emptyset\}, \{C\}, \{A_1\}, \{A_1, C\}. \tag{9}$$

Santafé et al. [22] considers $\pi = C, A_1, A_2, \ldots, A_m$ for all cases. So, the number of parents for ith attribute will be 2^i. Like EMA algorithm, EMA-TAN has two steps. In the first step, just like EMA, we have to compute the expected sufficient statistics. The second step is to run MA algorithm according to the completed dataset. The MA step of EMA-TAN is the same as the MA step of EMA algorithm, which we discussed before. The only difference is that, now the structure S is a TAN instead of SNB. Once again, MAP parameters can be easily computed from the expected sufficient statistics and hyper parameters of the Dirichlet distribution for a single structure. So, we can average over all MAP parameters of all members of L^{π}_{TAN}. The averaging process is just like the EMA algorithm [22].

Since L^{π}_{TAN} is a superset of SNB, EMA-TAN will incorporate more information regarding the conditional (in)dependencies between variables in its parameter learning step in comparison to EMA. However, both EMA and EMA-TAN suffer from the NB assumption in their structure. NB structure is a very basic and simple minded model which is not true in many real-world problems. This is why many researchers tried to extend this model by removing the conditional independence assumptions. We will focus on them next.

2.2 Unknown Structure

There are many cases in which only raw data is available and no information regarding the structure of the BN is at hand. In these cases, the structure should be learnt from data itself. Several methods for learning BNs from complete data exist [20, 23]. However, in a clustering problem, since the cluster variable is hidden, we have to learn the structure from incomplete data. In order to do so, we have to use BSC (Fig. 3). There are two main steps in BSC: Parameter search and structure learning steps. As we saw earlier, some methods assume that the structure is known and only focus on the parameter search step. In this section we are going to mention those which learn the structure and try to remove the NB assumption. However, these extensions will happen at the cost of time complexity. The key point is to propose an algorithm that balances well between time complexity of the algorithm and quality of structure. Peña et al. [2] tried to extend NB in a way that both improved its model quality and keep its simplicity and they called it constructive induction learning.

2.2.1 Extended Naive Bayes

Peña et al. [2] tried to introduce a model, which maintains the simplicity of NB and still improves its model quality. This model is the same as NB with only one difference. The model allows some attributes to group together under the same node as fully correlated attributes (supernodes), so the number of nodes in the structure can be smaller than the original number of attributes. Hence, the posterior probability of a class given all attributes (7) can be rewritten as:

$$P(C|A_1, A_2, \ldots, A_m) \propto P(C) \prod_{i=1}^{e} P(A_i|C), \tag{10}$$

where e is the number of variables in the structure (supernodes and common nodes) and $e \leq m$. In this way a better performance than NB will be achieved at low cost. In order to find supernodes, we have to choose a score first. A good and efficient score is one which is both factorable and has a closed form computation. Both, log-likelihood and log marginal likelihood scores have these properties [2, used the latter]. Under standard assumptions, log marginal likelihood is:

$$\mathrm{LML}(D|G) = \log p(D|G) \propto \sum_{i=1}^{m} \log \prod_{j=1}^{q_i} \frac{(r_i - 1)!}{(N_{ij} + r_i - 1)!} \prod_{k=1}^{r_i} N_{ijk}!, \tag{11}$$

where m is the number of attributes, r_i the number of states that the ith attribute has, q_i the number of states that the parent set of the ith attribute has, N_{ijk} the number of cases in the dataset where A_i has its kth value and its parent set takes its jth value and $N_{ij} = \sum_{k=1}^{r_i} N_{ijk}$.

Fig. 7 Initial structure for
backward structure search

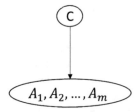

For the search algorithm, Peña et al. [2] used constructive induction by using
forward and backward search. In a forward structure search (FSS), we have to
consider NB as the initial point. Then all pairs will be joined (creating a new variable
which is the Cartesian product of two variables) and the one which increases the
score most, will be chosen as supernode. This process will be iterated until no join
action can improve the score of the current structure. On the other hand, backward
structure search (BSS) use a fully correlated model as its starting point (Fig. 7).
We have to consider all possible splitting actions (breaking one variable into two
separate and conditionally independence variables). Then we will choose the action
which leads to the highest increase in score. This process should be iterated until
no splitting action can improve the score of the current structure. Using either FSS
or BSS as the structure search and EM as the parameter search step of the general
algorithm for clustering via BNs (Fig. 3) we are able to learn a locally optimal BN
and its parameters and use it for clustering.

Peña et al. [3] enhances parameter search step and called it Bayesian structural
BC+EM. The main idea of BC+EM algorithm is to alternate between bound and
collapse (BC) [25] method and the EM algorithm. BC method bounds the set of
possible estimates and then collapse them into a unique value according to a convex
combination. The estimation of BC algorithm will be used as an initial point for
EM algorithm. This way we will achieve a faster convergence rate and more robust
algorithm in comparison to traditional EM. In their second effort to further improve
the structure search step, Peña et al. [4] proposed recursive Bayesian multinets
(RBMNs).

2.2.2 Recursive Bayesian Multinets

Peña et al. [4] in their effort to improve the structure for clustering, proposed
RBMNs. Bayesian multinets (BMNs) consist of a distinguished variable Z and a
set of component BNs for $\{A_1, \ldots, A_m, C\} \setminus Z$. Generally, the distinguished variable
can be either the cluster variable or one of the attributes [26] but, in RBMNs, the
distinguished variable is not allowed to be the cluster variable. Figure 8 shows
an example of a BMN. As we can see, since BMNs only have one distinguished
variable it will always be a depth 1 decision tree (consider the distinguished variable
as root and the component BNs as leaves). RBMNs are extensions of BMNs, which
can have more than one distinguished variable. Lets say a RBMN is a BMN, which

Fig. 8 An example of a BMN for data clustering with five attributes $(A_1, A_2, \ldots, A_5)$, one cluster variable (C) and two component BNs $(g_1$ and $g_2)$. $Z = A_3$ is the distinguished variable

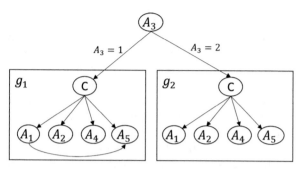

Fig. 9 An example of a RBMN for data clustering with five attributes $(A_1, A_2, \ldots, A_5)$ and one cluster variable (C). $Z = \{A_1, A_2, A_4\}$ is the set of distinguished variables

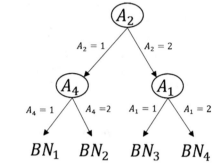

each of its components (leaves) can either be a BMN or just a BN. This way we can have a decision tree with a depth more than one. Peña et al. [4] referred to RBMNs as *distinguished decision tree* and defined it as follows:

- The cluster variable cannot be a distinguished variable.
- Every internal node is a distinguished variable.
- Every internal node has as many children or branches coming out from it as states for the variable represented by the node.
- No path to the leaf should contain a repeated variable.

So, RBMNs are more general than both BMNs and BNs. A BMN is a RBMN with only one internal node, while a BN is a RBMN with no internal nodes. Figure 9 is an example of RBMN structure. According to the fourth condition of RBMNs definition, in Fig. 9, BN_1 and BN_2 should only contain $\{A_1, A_3, A_5, C\}$ variables and BN_3 and BN_4 should include $\{A_3, A_4, A_5, C\}$ and no more.

The most compelling advantage of BMNs, or generally RBMNs, is that they can encode context-specific (in)dependencies [27], while the conditional (in)dependencies that are encoded by a BN are context-non-specific. A conditional (in)dependency is context specific if two sets of variables are conditionally independent of each other given a third set with a specific configuration and dependent given the same third set but with another configuration for its variables. This means that RBMNs are more flexible than simple BNs and they can support a wider range of conditional (in)dependencies. Also, RBMNs are very intuitive. Many examples can be found that have different models according to different values of one or more variables (distinguished variables). For example, the body types of male and female are different so they should have different models so, in this case, the Sex variable would be the distinguished variable. Age, Birth place, Height, Weight, and etc., are few examples of many possible distinguished variables in an example. So, now that the preliminaries of RBMNs have been introduced, it is time to provide a learning algorithm for the clustering purpose. We can use the general algorithm for clustering via BNs (Fig. 3). As for the parameter search step, both EM and BC + EM are eligible. Peña et al. [4] used an ENB for component BNs, so the process of learning component BNs will be the same as Sect. 2.2.1. Recall the fact that, in order to be able to search efficiently for an optimal BN among the vast number of possible BNs, the chosen score has to be both decomposable and in closed form. For learning ENB, log marginal likelihood (11) has been used, Thiesson et al. [26] extends this score for BMNs and [4] extends it further for RBMNs. Under standard assumptions, the log marginal likelihood for BMNs will be:

$$\log p(D|G_{\text{BMN}}) = \log p(D^Z) + \sum_{g=1}^{|Z|} \log p(D^{X,z}|bn_g), \tag{12}$$

where D^Z is the data restricted to distinguished variable Z, $|Z|$ is the number of component BNs (number of values of the distinguished variable), bn_g is the gth component BN, X is the set of all attributes and the cluster variable ($\{A_1, A_2, \ldots, A_m, C\}$) and $D^{X,z}$ is the data limited to $X \setminus \{Z\}$ and those objects in which $Z = z$. The first term on the r.h.s of (12) is the log marginal likelihood of a BN with only one variable (distinguished variable) and the second term is the sum of log marginal likelihood for all component BNs separately.

Moreover, under the same assumption as for BMNs, log marginal likelihood for RBMNs is:

$$\log p(D|G_{\text{RBMN}}) = \sum_{l=1}^{L} [\log p(D^{t(\text{root},l)}) + \log p(D^{X,t(\text{root},l)}|bn_l)], \tag{13}$$

where L is the number of leaves, $t(\text{root}, l)$ are the variables which are in the path between root and the lth leaf, $D^{t(\text{root},l)}$ is the data restricted to those variables in $t(\text{root}, l)$, bn_l is the lth component BN, and $D^{X,t(\text{root},l)}$ is the data restricted to variables which are not in $t(\text{root}, l)$ and those cases which are consistent with $t(\text{root}, l)$ set values. The first term in the summation of (13) is the log marginal likelihood of a trivial BN with a single node with as many states as leaves in the distinguished decision tree, which is easy to compute (11). Also, the second term is

Table 1 Different choices for clustering via BNs

Algorithm	Parameter search	Score	Structure	Structure search
Peña et al. [2]	BC + EM	LML	ENB	FSS/BSS
Peña et al. [3]	BC + EM	LML	Any BN	HC
Peña et al. [4]	BC + EM	LML	RBMN	FSS/BSS
Pham and Ruz [5]	EM	MI	CL multinet	MWST
Pham and Ruz [5]	EM	CMI	TAN	MWST
Pham and Ruz [5]	EM	CMI	SBN	MWST
Santafe et al. 2006 [21]	EMA	LL	NB	–
Santafe et al. 2006 [22]	EMA-TAN	LL	NB	–

the log marginal likelihood for each leaf (component BN). So, under the mentioned assumptions, both BMNs (12) and RBMNs (13) scores are factorable and have closed form calculation.

Now that we have a factorable and closed form score for RBMNs we can proceed with the learning algorithm for clustering purpose. Peña et al. [4] has used constructive induction as their learning algorithm. It starts with an empty distinguished structure and at each iteration, it will increase the depth of the structure by one. In each iteration, each leaf should be replaced with the best BMN that has the highest log marginal likelihood (12) for variables in $X \setminus t(\text{root}, l)$. This task should be iterated until either the depth of the structure reaches a specific number or there exist no more BMNs which can be replace a component BN, such that it increases the log marginal likelihood (13).

Generally, to do clustering with BNs, one can use BSC (Fig. 3) and use different methods for its parameter learning (EM, BC+EM, EMA, or EMA-TAN) and structure learning (HC, FSS, BSS, or MWST) steps. The vital point is that, no matter which structure learning algorithm we choose, we have to select a scoring function that is both factorable and in closed form. For example, [5] select log-likelihood as the scoring function, EM as the parameter search, TAN and maximum-weighted spanning tree (MWST) for structure learning step to propose a new clustering algorithm via BNs. Table 1 shows different choices that lead to different methods. Each of these methods has their own pros and cons. For example, BC + EM is more robust and has a faster convergence rate in comparison to EM or the tree structure which will be learnt by Pham and Ruz [5] according to (conditional) mutual information (CMI) is a global optimal tree while other structures in other methods are locally optimal. Also, the methods by Santafe et al. have the advantage of incorporating some information about conditional independencies into parameters (which obviates the need to do structure search step) and gives it the upper hand to do the calculations faster. The one thing that all of these methods have in common is the fact that they want to balance between time complexity and model quality.

3 Multi-Dimensional Clustering

As we said before, there are cases where data is multifaceted. This means that different experts may cluster data differently (Fig. 2). This disparity may emanate from choosing different subsets of attributes. However, two experts may consider all attributes and still do the clustering task differently. In Fig. 2, considering X or Y dimension, will result in different clustering outputs. As another example, consider that we want to partition customers of a supermarket. Considering "pet food" attribute, we can partition them into those who have pets and those who have not. However, if we consider another attribute, for example "cigar," then we may partition differently (smokers and non-smokers). Generally, there exist two types of methods for this purpose: Multi uni-dimensional and multi-dimensional. We refer to those that assume that all partitions are independent of each other as multi uni-dimensional and call the ones which consider potential relationship between partitions (this may result in either a connected or disconnected model), multi-dimensional. Galimberti and Soffritti [28] and Guan et al. [29] are examples of multi uni-dimensional clustering. One can just select an attribute subset and assign a latent variable to it to achieve a clustering model. And since it is assumed that all partitions are totally independent of each other, this problem could be reduced to a simple feature selection problem. Herman et al. [30] introduced and discussed many measures for this purpose. In this chapter we won't discuss multi uni-dimensional clustering. On the other hand, for multi-dimensional clustering, Zhang [14] introduced a model called latent tree models (LTMs) in 2004. In the next section we will focus on this model and its extensions.

3.1 Latent Tree Models

LTMs have two types of variables: Observed variables which are the attributes and latent variables. Each latent variable stands for a partition and can only be an internal node while each attribute should be a leaf. Figure 10 shows a possible LTM structure, which has three partitions which are correlated ($C_2 \leftarrow C_1 \rightarrow C_3$) and three conditionally independent attribute subsets $\{A_1, A_2, A_3\}, \{A_4\}, \{A_5, A_6\}$. From now on, we use the words "latent variable," "partition," and "cluster variable" interchangeably. Table 2 shows the general steps of LTM to perform multi-dimensional clustering. In the first step, we can either assume a pre-defined number for partitions or figure it out in the learning process. Like in the first step, one can set a pre-defined number for clusters in each partition for the second step. As for the third step, any structure learning algorithm can be use to learn the required structures. Also, one can use EM as its parameter estimation step. Since we can have a single partition, uni-dimensional clustering is a special case of LTMs. In the next section we will discuss both cases where the structure is known a priori or it is unknown.

Fig. 10 LTM structure with six observed and three latent variables

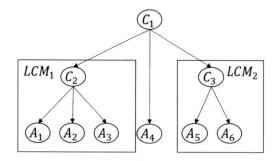

Table 2 Latent tree model steps for multi-dimensional clustering

1. Determine the number of cluster variables (partitions)
2. Determine the number of clusters for each partition (values for each cluster variable)
3. Learning the LTM structure
   (a) Finding the relation between attributes and cluster variables (bridge model)
   (b) Finding the relations only between attributes
   (c) Finding the relations between cluster variables
4. Estimate model parameters

3.1.1 Known Structure

Knowing LTM structure obviates the need for steps 1–3 in Table 2. This means that we know the number of partitions and their cardinalities along with the relations between all variables (both latent and observed) so we only need to learn the parameters. As we mentioned before, when hidden variables (missing data) exist the first choice to learn parameters is the EM algorithm. However, due to its local optimality and its high time complexity, many researchers tried to improve it or propose a new algorithm. For LTMs, Mourad et al. [13] introduced LCM-based EM.

The main idea is to first detect all NB structures (they called it LCM) in a LTM and learn their parameters locally by running EM and then predict the latent variables values according to the estimated parameters. These two steps will iterate until we have an estimation for all parameters of all LCMs. Ultimately, a parameter refinement will be done by running EM algorithm globally. Due to using local EMs, this method will reduce computation time. But, especially in multi-dimensional clustering, knowing the structure is hardly the case. So we will focus on the case where the structure is unknown.

3.1.2 Unknown Structure

Lack of prior knowledge about the structure, leaves us with no choice but to learn from data all the components of a LTM. These are number of partitions, their cardinalities, relations between all variables and parameters. One possible option to do so is to use BSC algorithm (Fig. 3). Zhang [14] used such an algorithm by making some assumptions. They assumed that attributes cannot have any relations with each other and the initial structure is an NB structure. Also, two greedy search algorithms have been used for structure search step. One to determine the structure and number of partitions and the other for deciding the cardinalities of latent variables. The first greedy search has three operators: Node introduction, node elimination, and neighbor relocation. And the second one only has one operator which increases cardinality by 1. Due to running two hill climbing algorithms during its process, this method is also called double hill climbing (DHC). It is obvious that such an algorithm is very inefficient because of its double use of hill climbing search. In order to do the search once and reduce the search space, Chen et al. [11] introduced EAST algorithm ([13] called it advanced greedy search (AGS)). This algorithm has five operators, which are divided into three groups. The first group contains node introduction and cardinality increase operators. These two operators will expand the current structure so we call it expansion group. Second group only has the node relocation operator. This operator decides if any edge changes are needed for the current structure. And the last group includes node elimination and cardinality decrease operators. Since these two operators lead to simpler models, this group is called thin group [11]. These three groups will be applied to the current structure step-by-step and after each step, the current model may or may not change. At the end of these three steps, if the structure does not changed, then the algorithm will be terminated.

BIC is a common scoring criterion to decide between current structure and its neighbors. However, Zhang and Kocka [31] mentioned that, initiating the search with a naive structure and using BIC, we will always select cardinality increase over the node introduction operator in expansion step. So, they introduced a so-called improvement ratio criterion and use it instead of BIC in this step:

$$\mathrm{IR}^{\mathrm{BIC}}(T', T|D) = \frac{\mathrm{BIC}(T', D) - \mathrm{BIC}(T, D)}{\dim(T') - \dim(T)}. \tag{14}$$

The numerator of (14) is the difference of BIC scores between the candidate structure (T') and current structure (T) while denominator is their dimension difference.

Pouch latent tree model (PLTM) is another example of learning LTM structure based on score-search methods [32, 33]. It assumes that all attributes are continuous and they can be merged into one node called pouch node (pouch node is the same as supernodes in RBMNs, except that in a pouch, all attributes are continuous). Also, it introduces two new search operators (in addition to the five operators of EAST algorithm) for its hill climbing (pouching (PO) and unpouching (UP) for merging and splitting attributes, respectively). Yet, score-search based methods still suffer from tedious searches and the problem will be doubled in presence of latent variables (need for running EM before the first step).

So, there is another class of methods which uses a feature selection algorithm to group attributes and then try to construct the structure with inductive learning.

Liu et al. [12] proposed a new method called bridged islands (BI) and then [34] extend it. In the next section we will try to extend this algorithm by overcoming some of its deficiencies, so we will have a close scrutiny of this method.

BI do the clustering task by introducing a new concept called sibling clusters. Sibling clusters are sets of attributes which are grouped under the same latent variable. BI learns a LTM in four steps:

1. Form sibling clusters
2. Assign a latent variable to each sibling cluster and decide its cardinality
3. Learn a Chow-Liu tree for latent variables
4. Refine the model

To form sibling clusters we have an active subset of attributes. At first, all attributes are in the active set. Initially, the pair with the highest mutual information (MI) will be chosen and then each time an attribute which has the highest MI with the previously selected attributes will be added to them. The MI between a single attribute (A) and a set of attributes (W) is estimated as follows:

$$I(A; W) = \max_{Z \in W} I(A; Z). \tag{15}$$

With each attribute addition we will have a new naive model containing the selected attributes and one latent variable (g_1) and use EAST algorithm to find the best possible structure with exactly two latent variables (g_2) (for the same subset of attributes). The process of adding attributes continues until uni-dimensionality (UD) test fails. UD test runs on two LTMs (g_1 and g_2) to see which structure has larger score. If the model with two latent variables (g_2) has the larger score, then we say that UD test has failed and we stop adding more attributes. Two possible structures which can be formed according to the mentioned process is shown in Fig. 11. At this point we have to choose one of the attribute sets in g_2 ($\{A_1, A_2, A_3\}$ or $\{A_4, A_5\}$) as sibling cluster. We choose the one that contains the attributes of the pair with the highest MI (if the attributes of such a pair were on different sets, then we choose randomly). Finally, the attributes in the cluster sibling set have to be removed from the active set and the process restarts with the new active set until |Active Set| ≤ 2. Now, we have all sibling clusters and thus we know the number of partitions (we assign a latent variable to each sibling cluster), so we have to decide the cardinality

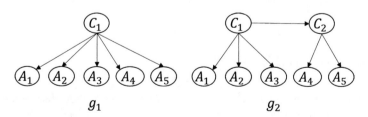

Fig. 11 Two possible structures for the UD test during forming the sibling clusters

of each latent variable. This step can be done easily using a greedy search. Third step is to find the relations between latent variables such that they form a tree. This step can be easily done by using the well-known Chow-Liu algorithm [24]. The last step tries to correct the probable mistakes in previous steps. It will investigate if any node relocation action or changing any latent variable cardinality will lead to a greater score, if it does that action should be taken and the model will change for one last time. Ultimately, EM algorithm runs globally to estimate the parameters. Note that, at the end of each step (after the model changes), EM algorithm will be run locally to refine parameters. However, Mourad et al. [13] discussed that when the number of attributes is large (e.g., 1000), it is better to use a forest structure instead of a tree. They justify it with the fact that, when the number of attributes is large, there exist many cluster variables which are not correlated to each other. So, they suggest to first learn a tree structure and then use any independence test to remove edges between cluster variables. It is obvious that BI is much faster than EAST algorithm [12, 34].

Another method which uses feature selection as its first step is binary trees (BT) [13]. Like BI, at the first step BT finds a pair with the highest MI and set a latent variable as their parents. Then it will remove the selected attributes from the observed variables set and complete the data for the introduced latent variable, next it will add (now observed) latent variable to the observed variables set. These steps will be repeated until only two variables left in the observed variables set. This algorithm is named as LCM-based LTM learning, because at each step a LCM structure is learnt.

In all of the mentioned methods (DHC, EAST, BI, and BT) there exists a step for finding the cluster variable cardinality. The simple way is to check the score of the structure for all possible cardinalities of all cluster variables. However, when the number of latent variables is large this approach would be intractable. So, generally a greedy search will be done to find a local optimum. But, this method still remains inefficient for large number of latent variables. There exist several approaches which tackle this issue [16, 35, 36], but we won't discuss them here.

With all this being said, there is a vital point in multi-dimensional clustering which we didn't heed to. In any multi-dimensional learning algorithm, each partition should express different concept from other partitions. This means that the learning algorithm should ensure the novelty of each partition. In the next section we will describe how LTM methods guarantee this property and how we can interpret the meaning of each partition.

3.2 Cluster Variables Novelty

This section describes how LTMs guarantee the novelty of their discovered partitions. If different partitions express different concepts, then we say that they are novel. So, let us begin by showing how we can infer the concept of a partition in LTMs. Information curve is widely used for this purpose [11–13]. Information curve is a measure which detects the most influential attributes for a latent variable in LTM structure. Both pairwise mutual information (PMI) and cumulative mutual information (CMI) are part of an information curve. PMI is the MI between a single latent variable (C) and an attribute (A_i) $(\text{MI}(A_i, C))$, while CMI is the MI between a single latent variable (C) and a set of attributes $\{A_1, A_2, \ldots, A_i\}$, where $i = 1, 2, \ldots, m$ $(\text{MI}(C, \{A_1, \ldots, A_i\}))$. The following value is based on CMI and it is called the information coverage:

$$\text{IC} = \frac{\text{MI}(C, A_1, \ldots, A_i)}{\text{MI}(C, A_1, \ldots, A_m)}. \tag{16}$$

When IC equals to 1, we can say that the first i attributes describe C perfectly and C is independent of all other attributes given the first i attributes.

So, to choose a set of attributes which best describe a latent variable C (influential set) we can select an attribute which has the highest PMI with C and then, at each step, add the attribute which increases the IC value most. A threshold for IC can be used as a stopping criterion. For example, we can stop adding attributes whenever IC reaches 90%. One can use conditional entropy instead of IC in the same manner. At each step the attribute which decreases the conditional entropy of the latent variable C most, has to be added to the influential set. Whenever the conditional entropy of C given its influential set reaches zero (close to zero), we can stop adding attributes. Moreover, Herman et al. [30] discussed and introduced alternative MI-based methods for selecting informative attribute sets, which can be used to determine influential sets.

Having a set of influential attributes for each partition, we can infer the concept of each partition in the LTM structure. For instance, let "sex" and "age" be the only two attributes in the influential set of partition C, then we can infer that C is primarily based on these attributes and C thus gives us information regarding the age and sex of a person. Thus, in a LTM structure, if the influential sets for two partitions C_1 and C_2 are exactly the same (or nearly the same) we can claim that one of them is not novel.

LTMs guarantee the novelty of its partitions by not allowing them to share any attributes. This fact will ensure that the intersection of any two influential sets will be a null set. So, the novelty condition is always satisfied. However, not allowing partitions to share attributes result in two severe shortcomings. The first issue is straightforward, if an attribute belongs to a partition but does not belong to its influential set, then it is not allowed to be selected in any other influential set. Thus, many attributes will be useless, while this is not true in reality. The second problem is more subtle and we explain it with an example. Consider a case with two attributes "wet grass" and "slippery road" and two partitions "rain" and "sprinkler." The most intuitive model would be the one in Fig. 12. We can see that "wet grass" is shared between two partitions and removing it from "rain" will increase the uncertainty of this partition and removing it from "sprinkle" will leave this partition with no attribute, which force us to remove the partition itself. In either

Fig. 12 A multi-dimensional clustering with two partitions and two attributes

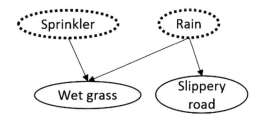

case, not sharing "wet grass" between two partitions will be harmful. This motivates us to propose a new method for multi-dimensional clustering by overcoming the mentioned shortcomings of LTMs.

4 Our Approach

In this section we aim to propose a new method to learn a BN for multi-dimensional clustering by overcoming the aforementioned disadvantages of the previous methods. The most compelling drawbacks of LTMs were their conditional independence assumption between attributes and limiting each attribute to belong to only one partition. However, it is obvious that there exist many models which violate these assumptions. Also, there exist lots of multi-dimensional classification algorithms without such restrictions [8, 37–39], which have been applied to different problems successfully [40–43]. We try to borrow some ideas from these algorithms and apply them to the clustering field. The most intuitive model for multi-dimensional classification via BNs is called multi-dimensional Bayesian network classification (MBC). In such a model, there exist three structures for two subset of variables (class variables and attributes) (Fig. 13). The main algorithm for learning MBC is the same as BI for clustering without some of its restrictions. Unlike LTMs, there exist edges between attributes ($A_5 \rightarrow A_3$) and also two or more classes may share the same attribute ($C_2 \rightarrow A_3 \leftarrow C_3$). Van Der Gaag and De Waal [6] and de Waal and van der Gaag [44] used tree and polytree structures respectively for both class and attributes and learn the bridge structure with a greedy algorithm with accuracy as its stopping criterion. Bielza et al. [7] used a fixed bridge structure and a greedy algorithm for learning class and attribute structures. However, in clustering, the problem is that no knowledge about the value or number of class variables exist. In other words not only all classes are latent, but also we don't know their numbers.

A straightforward method would be to start from a naive model with one partition and then use BSC algorithm with seven operators (adding an edge, removing an edge, reversing an edge, partition introduction (PI), partition removal (PR), cardinality increment (CI) and cardinality decrement (CD)) for its structure learning step, to improve the structure. Generally, the only restriction for the structure is that there cannot exist an edge from an attribute toward a partition. The first three operators are standard operators, thus there is no need to discuss them. PI is the act

Fig. 13 A possible MBC model with four classes and five attributes

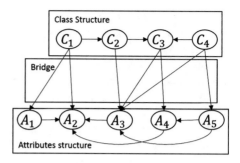

Input

D: Data matrix with n objects and m attributes

$G_{current}$: Initial BN structure

$\theta_{current}$: Initial random parameters for $G_{current}$

Output

$G_{current}$: BN structure in the kth iteration

$\theta_{current}$: Parameters of $G_{current}$

Main

1. Run EM to estimate the parameters
2. Compute the posterior probability of cluster variable given attributes (for each object) and complete the dataset
3. $BIC_{old} \leftarrow$ Compute the BIC score of the current model ($G_{current}$)
4. Find all neighbors of $G_{current}$ with the following operators: add, remove, reverse, PI, PR, CI, CD
5. Compute BIC score for all neighbors and select the one with the highest BIC score (G_{new}, BIC_{new})
6. If $BIC_{new} > BIC_{old} + \varepsilon$
    $G_{current} \leftarrow G_{new}$
    Compute ML parameters for $G_{current}$
    Go back to 1.
7. Else
    Return $G_{current}$

Fig. 14 The general algorithm for learning multi-dimensional clustering

of adding a binary latent variable to the model and consider it as the parent of all attributes and PR will remove a latent variable. The last two operators are for adjusting the cardinality of latent variables. Figure 14 shows the general schema of such an algorithm. The algorithm starts with an initial structure and its random parameters. At the first step, we run EM to estimate the parameters and complete the dataset according to posterior probabilities. Third step will compute the BIC score of the current DAG according to the completed data. Next, we have to find all neighbors of the current structure (with the help of mentioned operators) and compute the BIC score for each one of them. If the current model has the highest score in comparison to all its neighbors then we terminate the algorithm and return the current model as the best one, otherwise we will reestimate the best neighbor parameters and replace the current best model with it and run the same procedure again. Note that using BIC score will ensure the novelty of the partitions. If a partition is not novel then the log-likelihood of the new structure (resulting from using PI operator) will not increase and only the penalty term (structure complexity) will grow. This will lead the new structure to have lower BIC score in comparison to current one.

However, searching in such a large space, would be practically intractable for large number of attributes. In order to reduce the search space, we can either group operators and apply them sequentially (like EAST algorithm) or group random variables and making different sections in the original structure (like MBC) and learn one section at a time. Furthermore, like BSC for uni-dimensional case, any parameter learning and structure learning algorithm can be used for parameter and structure search steps, respectively.

Also, we can follow the idea of BI [12] and form cluster siblings to avoid the time-consuming greedy search required by our first proposed method (Fig. 14). Forming sibling clusters was the first step of BI. This step is done via an approximation for MI (15) and the stopping criterion was the result of a UD test. However, (15) will not consider the MI between the attributes and the cluster variable, which results in detecting attributes of a sibling cluster as redundant in the interpretation of each partition. This is due to the fact that BI or generally all LTM methods, merely use MI between attributes (15) to learn the structure and then measure the MI between the attributes and the cluster variable to interpret the meaning of each cluster. So, we propose a method that consider the dependence between attributes and partitions in the learning procedure. Figure 15 shows the steps of our approach.

Just like MBC, we have three different steps for learning. The first step tries to learn the bridge model, while the second one will learn partition structure and the third step aims to learn the attributes structure. Again, one can limit the attribute and cluster structures to an empty or tree structure. Learning the bridge model resembles the BI algorithm. At first, all attributes belong to the active set ({Active}). The pair with the highest MI in the active set will be chosen. If the pair includes two attributes

1. While there is an attribute left in the active set
    1. Find the pair with the highest MI in the active set (at least one of the variables should be an attribute)
    2. If the pair consist of two attributes (A_i, A_j)
        Set a latent variable (C_k) as their parents
        Remove them from the active set
        Add C_k to the active set
        $k = k + 1$
        Run EM locally to determine the values of C_k for each object 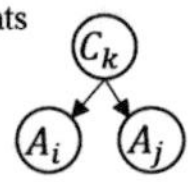
    3. If the pair consist of one attribute and one latent variable (A_i, C_j)
        Add A_i to the child set of C_j
        Remove A_i from the active set
        Run EM locally to update the values of C_j for each object
2. Use a search algorithm to find the best structure for partitions
3. Use a search algorithm to find the best structure for attributes

Fig. 15 Our second proposed method for multi-dimensional clustering via BN

(A_i and A_j), then we have to introduce a new latent variable (C_k) and set it as the parent of both A_i and A_j. We have to estimate the values of C_k for each object via EM algorithm. Also we have to remove A_i and A_j and add C_k to the active set (step 1.2). But, if the chosen pair contains one latent variable and one attribute (A_i and C_j) then we only need to add A_i to the child set of C_j and then update the values of C_j by running EM algorithm. Also we have to remove A_i from the active set. Moreover, if the pair contains two latent variables (C_j and C_k), then we will just ignore it and find the next pair with the highest MI. We do these steps, until there are no attributes left in the active set. At this stage, we shall have our bridge model with k partitions, which would be an NB model. Then, at the second and third steps the partitions and attributes structures have to be learnt, respectively. First, we will run a hill-climbing algorithm only on partition variables to find any potential edge between them. Next, we run another hill-climbing algorithm, but this time for attributes, to find the relationships between them. Note that, for partition variables, which are hidden, we first predict their values according to their posterior probabilities and then try to learn a structure for them.

Our method relaxes the constraint on sharing attributes in LTMs by allowing indirect relations between attributes and partitions. Figure 16 shows a possible outcome of the proposed method. Since, both C_1 and C_2 are dependent on A_3 given A_2, we can say that they share A_3. Meanwhile, since the children of each partition are not allowed to have any intersection, the partitions are guaranteed to be novel. The advantages of our proposed method over LTMs can be summarized as follows:

- Incorporating the cluster variable in the computations, so there would be no redundant attributes for any partition.
- Allowing attributes to have relations with each other.
- Allowing partitions to share attributes indirectly.
- The model is allowed to be disconnected (if we choose to learn a forest structure for both partitions and attributes variables).

We scrutinize the benefit of the third advantage with an example. Back to our previous example (Fig. 12), assume that we have another variable "sprinkler direction" which can have two values "towards wall" and "towards grass." So, we can say that "sprinkler" and "wet grass" variables are dependent on each other conditioned on the direction of the sprinkler. Figure 17 shows such a model.

Fig. 16 A possible structure for our proposed method with two partitions and five attributes. The attributes' structure is a Chow-Liu tree

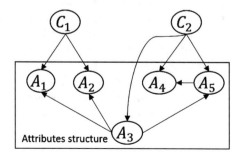

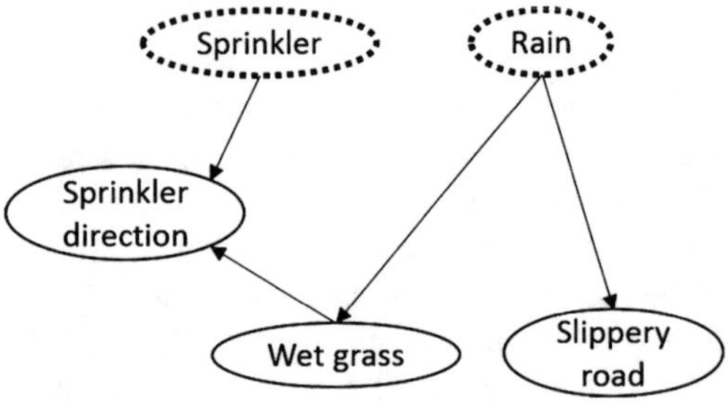

Fig. 17 An example of a
possible model with indirect
dependency between a cluster
variable (Sprinkler) and an
attribute (Wet grass)

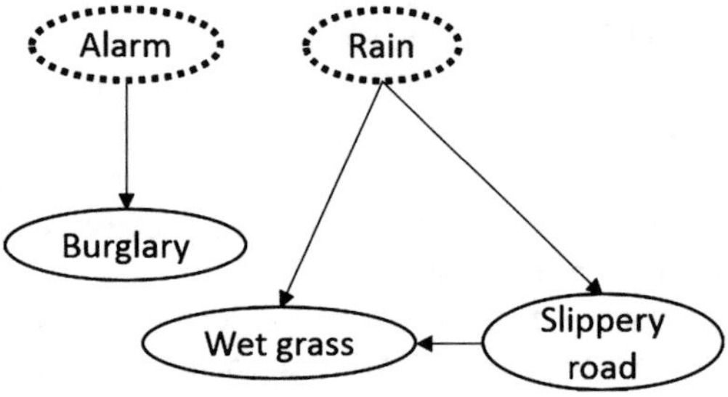

Fig. 18 An example of a
model with completely
independent partitions (Alarm
and Rain)

We can see that "wet grass" will affect both partitions. However, in a LTM model "wet grass" could only affect one of the partitions. Furthermore, the second and fourth advantages will allow the algorithm to cover a wider range of models and be more intuitive. Since the merit of the second advantage is obvious, we only justify the usefulness of the fourth advantage via an example. Consider a problem with three attributes: "slippery road," "wet grass," and "burglary." Maybe the most probable model would be the one in Fig. 18 where two partitions are completely independent of each other (there is no path between them). Again, no LTM algorithm will be able to capture such independencies (alarm and rain), while the proposed method is capable to do so. The reason that LTMs cannot represent such an independent model (Fig. 18) is that, their structure is restricted to a tree, which is always connected. However, one may discuss that removing such a restriction would be easy. Recall that learning LTM structure was possible with two general approaches: search methods and feature selection. Due to their operators (node introduction, node elimination, and neighbor relocation) the former approach will always find a connected model. So, Mourad et al. [13] mentioned latent forest model (LFM) and discussed that one can learn an LTM first and then use some independency test to remove edges between partitions. On the other hand, BI can simply replace its third step (learning Chow-Liu tree) with learning a forest structure. Although these extensions are possible, according to our knowledge, nobody has used them in the LTM literature. In the next section, we are going to show some preliminary results for our second approach and discuss its ability to find meaningful partitions.

5   Preliminary Results

Since a full evaluation of the proposed methods is beyond the goal of this chapter, we only show some preliminary results. First, we will show some results from the first step of our algorithm (constructing the bridge model) to see if it can find meaningful partitions. Next, we will investigate the effect of the second and third steps to find out if these steps improve the structure or not. To test the first step, we used one real dataset and one synthetic dataset.

For our real dataset we choose NBA data, so that we can interpret the meaning of the partitions found according to our basic knowledge of basketball. The data contains the statistics for "all time regular seasons stats."[1] We removed all objects with missing values. It includes 927 players along with 19 attributes. The attributes are related to average performance of players per game. Also, we discretize the attributes into four equal bins. Since we use EM in our approach, each time we run the algorithm we may get slightly different partitions, but some attributes are found to be in the same partition every time. Figure 19 shows those attributes. For example, the attributes of the first partition {OR, DR, TR, BLK} (offensive rebounds, defensive rebounds, total rebounds, and blocks, respectively) are all related to the height of a player which indicates if he is a forward or guard player. Third partition {FG%, 3P%, FT%} (field goal percentage, three point percentage, and free throw percentage, respectively) is obviously related to the shooting accuracy of a player. So we can claim that all introduced partitions are both meaningful and novel (Fig. 20).

Also, in order to achieve a more robust result we have done the experiment merely on the following eight attributes: {3PM, 3PA, 3P%, DR, OR, TR, GP, Min} (three point made, three point attempted, three point percentage, defensive rebounds, offensive rebounds, total rebounds, game played, and minutes played, respectively). In this case we almost found three partitions every time {3PM, 3PA, 3P%}, {DR, OR, TR}, and {GP, Min} which are very intuitive. Note that all three partitions are novel and connote different concepts.

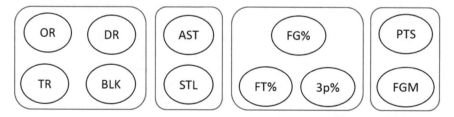

Fig. 19 The attributes which are usually grouped under the same partition for NBA data

---

[1]The data is available on: http://www.stats.nba.com/leaders/alltime/.

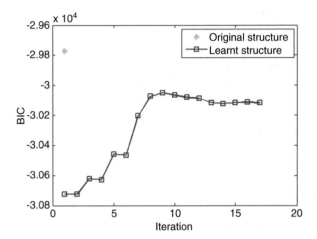

Fig. 20 BIC score of the learnt structures during the second and third steps

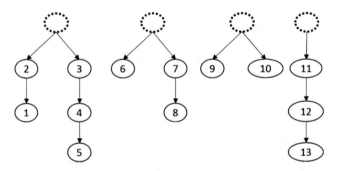

Fig. 21 The original BN for synthetic data with four partitions and 13 attributes

In order to further test the ability of our method, we also used synthetic data. We generated samples from a BN structure with four partitions (Fig. 21). Then we removed the data related to all four partitions and run our algorithm on all 13 observed attributes. The algorithm was able to find five partitions as shown in Fig. 22, which are very close to the original structure. This result is also support the goodness of our method in finding meaningful and novel partitions.

Along with its ability to find meaningful and novel partitions, our algorithm is very time efficient. For NBA data with 19 attributes it only takes 7 min (on average) to find partitions, which is fairly good. We run our algorithm on a computer with Core i7 3.40 GHz Intel CPU.

Now, it is time to see if second and third steps are able to improve the NB model, which is learnt in the first step. So, we have generated 10 random DAGs (each has 12 variables) and we have generated 4000 samples from each one. Next, we hide three variables and run our algorithm on remaining observed variables. In order to measure the effect of the second and third steps, we have compared the

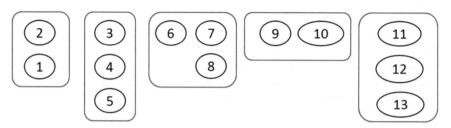

Fig. 22 The partitions found by our second proposed method for the synthetic data

Table 3 Average BIC score in 10 run

Avg initial BIC	Avg final BIC	Avg original BIC
−31137	−30752	−29874

BIC score of NB structure, resulted from the first step, with the BIC score of the final structure. Note that, for the sake of comparison, the BIC score is computed according to the original data (including the values for hidden variables). Table 3 shows the average BIC score for 10 runs. Also, we have measured the original structure's BIC on true data to see how close the resulting structure is to the optimal one. Figure 20 shows a result from one run. As it can be seen, the second and third steps, improve the structure, from the sense of BIC score, over time. Although in some iterations the BIC score may decrease (ninth iteration), but in overall the BIC score will increase in compare to the initial structure and it will get closer to the optimal point, which justifies the usefulness of the second and third steps of our algorithm.

6   Conclusion and Summary

In this chapter, we discussed many model based approaches for both uni- and multi-dimensional clustering and contribute with two new methods for the latter. The aim of model based clustering is to find the best possible model that captures the conditional (in)dependencies exist in the problem and for this purpose, in uni-dimensional clustering, the basic algorithm is BSC and other methods are just a special case of it. However, according to our knowledge, the only graphical model based approach for multi-dimensional clustering is LTMs, which has many limitations due to their assumptions. The main problem of LTMs was their inability to draw an edge between two attributes directly, which will limit their capability of modeling conditional (in)dependencies. Another problem is their inability to model independence partitions and adding attributes to a partition which may be found to be meaningless in the future. Based on these limitations, we proposed two algorithms for multi-dimensional clustering. The first algorithm (Fig. 14) was just a generalization of BSC to multi-dimensional cases and suffers from high time complexity. But, the second method is a combination of MBC and BI (one of the

LTM approaches) methods. We used the idea of grouping edges between variables into three groups from MBCs and used a sibling cluster concept from BI to propose a new method that avoids the mentioned disadvantages of LTMs. Since a comprehensive evaluation and comparison of different methods was beyond the aim of this chapter, we only provide a preliminary evaluation of our second proposed method, which justifies its ability for finding meaningful and novel partitions in an acceptable time.

Acknowledgements This work is funded by a so-called career contract at Linköping University, and by the Swedish Research Council (ref. 2010–4808).

References

1. McLachlan, G., Peel, D.: Finite Mixture Models. Wiley, New York (2004)
2. Peña, J.M., Lozano, J.A., Larrañaga, P.: Learning Bayesian networks for clustering by means of constructive induction. Pattern Recogn. Lett. **20**(11), 1219–1230 (1999)
3. Peña, J.M., Lozano, J.A., Larrañaga, P.: An improved Bayesian structural EM algorithm for learning Bayesian networks for clustering. Pattern Recogn. Lett. **21**(8), 779–786 (2000)
4. Peña, J.M., Lozano, J.A., Larrañaga, P.: Learning recursive Bayesian multinets for data clustering by means of constructive induction. Mach. Learn. **47**(1), 63–89 (2002)
5. Pham, D.T., Ruz, G.A.: Unsupervised training of Bayesian networks for data clustering. Proc. R. Soc. A Math. Phys. Eng. Sci. **465**(2109), 2927–2948 (2009)
6. Van Der Gaag, L.C., De Waal, P.R.: Multi-dimensional Bayesian network classifiers. In: Proceedings of the 3rd European Workshop in Probabilistic Graphical Models, pp. 107–114 (2006)
7. Bielza, C., Li, G., Larrañaga, P.: Multi-dimensional classification with Bayesian networks. Int. J. Approx. Reason. **52**, 705–727 (2011)
8. Sucar, L.E., Bielza, C., Morales, E.F., Hernandez-Leal, P., Zaragoza, J.H., Larrañaga, P.: Multi-label classification with Bayesian network-based chain classifiers. Pattern Recogn. Lett. **41**, 14–22 (2014)
9. Rodríguez, J.D., Lozano, J.A.: Multi-objective learning of multi-dimensional Bayesian classifiers. In: Proceedings of the 8th IEEE International Conference on Hybrid Intelligent Systems, pp. 501–506 (2008)
10. Read, J., Bielza, C., Larrañaga, P.: Multi-dimensional classification with super-classes. IEEE Trans. Knowl. Data Eng. **26**(7), 1720–1733 (2014)
11. Chen, T., Zhang, N.L., Liu, T., Poon, K.M., Wang, Y.: Model-based multidimensional clustering of categorical data. Artif. Intell. **176**(1), 2246–2269 (2012)
12. Liu, T., Zhang, N., Poon, K., Liu, H., Wang, Y.: A novel LTM-based method for multi-partition clustering. In: Proceedings of the 6th European Workshop on Probabilistic Graphical Models, pp. 203–210 (2012)
13. Mourad, R., Sinoquet, C., Zhang, N.L., Liu, T., Leray, P., et al.: A survey on latent tree models and applications. J. Artif. Intell. Res. **47**, 157–203 (2013)
14. Zhang, N.L.: Hierarchical latent class models for cluster analysis. J. Mach. Learn. Res. **5**, 697–723 (2004)
15. Elidan, G., Lotner, N., Friedman, N., Koller, D.: Discovering hidden variables: A structure-based approach. Neural Inf. Process. Syst. **13**, 479–485 (2000)
16. Elidan, G., Friedman, N.: Learning the dimensionality of hidden variables. In: Proceedings of the 17th Conference on Uncertainty in Artificial Intelligence, pp. 144–151 (2001)

17. McLachlan, G., Krishnan, T.: The EM Algorithm and Extensions, vol. 382. Wiley, New York (2007)
18. Friedman, N.: The Bayesian structural EM algorithm. In: Proceedings of the 14th Conference on Uncertainty in Artificial Intelligence, pp. 129–138 (1998)
19. Mossel, E., Roch, S.: Learning nonsingular phylogenies and hidden markov models. In: Proceedings of the 37th Annual ACM Symposium on Theory of Computing, pp. 366–375 (2005)
20. Darwiche, A.: Modeling and Reasoning with Bayesian Networks. Cambridge University Press, Cambridge (2009)
21. Santafé, G., Lozano, J.A., Larrañaga, P.: Bayesian model averaging of naive Bayes for clustering. IEEE Trans. Syst. Man Cybern. B Cybern. 36(5), 1149–1161 (2006)
22. Santafé, G., Lozano, J.A., Larrañaga, P.: Bayesian model averaging of TAN models for clustering. In: 3rd European Workshop on Probabilistic Graphical Models, pp. 271–278 (2006)
23. Neapolitan, R.E.: Learning Bayesian Networks, vol. 38. Prentice Hall, Upper Saddle River (2004)
24. Friedman, N., Geiger, D., Goldszmidt, M.: Bayesian network classifiers. Mach. Learn. 29(2–3), 131–163 (1997)
25. Ramoni, M., Sebastiani, P.: Learning Bayesian networks from incomplete databases. In: Proceedings of the 13th Conference on Uncertainty in Artificial Intelligence, pp. 401–408 (1997)
26. Thiesson, B., Meek, C., Chickering, D.M., Heckerman, D.: Learning mixtures of DAG models. In: Proceedings of the 14th Conference on Uncertainty in Artificial Intelligence, pp. 504–513 (1998)
27. Geiger, D., Heckerman, D.: Knowledge representation and inference in similarity networks and Bayesian multinets. Artif. Intell. 82(1), 45–74 (1996)
28. Galimberti, G., Soffritti, G.: Model-based methods to identify multiple cluster structures in a data set. Comput. Stat. Data Anal. 52(1), 520–536 (2007)
29. Guan, Y., Dy, J.G., Niu, D., Ghahramani, Z.: Variational inference for nonparametric multiple clustering. In: Proceedings of the Workshop on Discovering, Summarizing and Using Multiple Clusterings (2010)
30. Herman, G., Zhang, B., Wang, Y., Ye, G., Chen, F.: Mutual information-based method for selecting informative feature sets. Pattern Recogn. 46(12), 3315–3327 (2013)
31. Zhang, N.L., Kocka, T.: Efficient learning of hierarchical latent class models. In: Proceedings of the 16th IEEE International Conference on Tools with Artificial Intelligence, pp. 585–593 (2004)
32. Poon, L., Zhang, N.L., Chen, T., Wang, Y.: Variable selection in model-based clustering: to do or to facilitate. In: Proceedings of the 27th International Conference on Machine Learning, pp. 887–894 (2010)
33. Poon, L.K., Zhang, N.L., Liu, T., Liu, A.H.: Model-based clustering of high-dimensional data: variable selection versus facet determination. Int. J. Approx. Reason. 54(1), 196–215 (2013)
34. Liu, T.-F., Zhang, N.L., Chen, P., Liu, A.H., Poon, L.K., Wang, Y.: Greedy learning of latent tree models for multidimensional clustering. Mach. Learn. 98, 301–330 (2013)
35. Wang, Y., Zhang, N.L., Chen, T.: Latent tree models and approximate inference in Bayesian networks. J. Artif. Intell. Res., 879–900 (2008)
36. Harmeling, S., Williams, C.K.: Greedy learning of binary latent trees. IEEE Trans. Pattern Anal. Mach. Intell. 33(6), 1087–1097 (2011)
37. Zaragoza, J.C., Sucar, L.E., Morales, E.F.: A two-step method to learn multidimensional Bayesian network classifiers based on mutual information measures. In: Proceedings of Florida Artificial Intelligence Research Society Conference (2011)
38. Zaragoza, J.H., Sucar, L.E., Morales, E.F., Bielza, C., Larrañaga, P.: Bayesian chain classifiers for multidimensional classification. In: Proceedings of the International Joint Conference on Artificial Intelligence, vol. 11, pp. 2192–2197 (2011)

39. Cheng, W., Hüllermeier, E., Dembczynski, K.J.: Bayes optimal multilabel classification via probabilistic classifier chains. In: Proceedings of the 27th International Conference on Machine Learning, pp. 279–286 (2010)
40. Borchani, H., Bielza, C., Martínez-Martín, P., Larrañaga, P.: Predicting EQ-5D from the Parkinson's disease questionnaire PDQ-8 using multi-dimensional Bayesian network classifiers. Biomed. Eng. Appl. Basis Commun. **26**(1), 1450015 (2014)
41. Mihaljevic, B., Bielza, C., Benavides-Piccione, R., DeFelipe, J., Larrañaga, P.: Multi-dimensional classification of GABAergic interneurons with Bayesian network-modeled label uncertainty. Front. Comput. Neurosci. **8**, 150 (2014)
42. Borchani, H., Bielza, C., Toro, C., Larrañaga, P.: Predicting human immunodeficiency virus inhibitors using multi-dimensional Bayesian network classifiers. Artif. Intell. Med. **57**(3), 219–229 (2013)
43. Borchani, H., Bielza, C., Martínez-Martín, P., Larrañaga, P.: Markov blanket-based approach for learning multi-dimensional Bayesian network classifiers: an application to predict the European quality of life-5Dimensions (EQ-5D) from the 39-item Parkinson's disease questionnaire (PDQ-39). J. Biomed. Inform. **45**, 1175–1184 (2012)
44. de Waal, P.R., van der Gaag, L.C.: Inference and learning in multi-dimensional Bayesian network classifiers. In: Proceedings of the 9th European Conference on Symbolic and Quantitative Approaches to Reasoning with Uncertainty, 501–511 (2007)

A Radial Basis Function Neural Network Training Mechanism for Pattern Classification Tasks

Antonios D. Niros and George E. Tsekouras

Abstract This chapter proposes a radial basis function network learning approach for classification problems that combines hierarchical fuzzy clustering and particle swarm optimization (PSO) with discriminant analysis to elaborate on an effective design of radial basis function neural network classifier. To eliminate the redundant information, the training data are pre-processed to create a partition of the feature space into a number of fuzzy subspaces. The center elements of the subspaces are considered as a new data set which is further clustered by means of a weighted clustering scheme. The obtained cluster centers coincide with the centers of the network's basis functions. The method of PSO is used to estimate the neuron connecting weights involved in the learning process. The proposed classifier is applied to three machine learning data sets, and its results are compared to other relative approaches that exist in the literature.

Keywords Radial basis function neural networks • Hierarchical fuzzy clustering • Particle swarm optimization • Discriminant analysis

1 Introduction

Radial basis function (RBF) networks have been used in various cases such as classification [1–4], system identification [5], function approximation [6–8], nonlinear systems modeling [1, 9, 10], and pattern recognition [11–15]. The challenges in designing efficient RBF neural networks refer to the network's parameter estimation procedure. The parameters are the centers and the widths of the basis functions, and the neuron connection weights, as well. An approach able to efficiently train RBF networks is the implementation of fuzzy cluster analysis [6, 9, 16]. Fuzzy clustering attempts to identify the underlying structure of the training data, and then generates a distribution of the RBFs that better describe this structure. This distribution is

A.D. Niros (✉) • G.E. Tsekouras
Department of Cultural Technology and Communication, University of the Aegean, University Hill, 81100 Mytilene, Greece
e-mail: aneiros@aegean.gr

© Springer International Publishing Switzerland 2016
M.E. Celebi, K. Aydin (eds.), *Unsupervised Learning Algorithms*,
DOI 10.1007/978-3-319-24211-8_8

established by determining the appropriate values for the center elements and widths of the basis functions. Finally, to establish the input–output relationships, optimal values for the connection weights can be extracted in terms of a wide range of iterative optimization methods, such as back-propagation [1, 17], gradient descent [1], and particle swarm optimization (PSO) [4, 5, 18–25].

In this chapter we propose a three-staged approach to elaborate on an effective design of RBF neural network for classification problems. The first step uses fuzzy partitions to dismember the feature space into a number of subspaces. Since these subspaces come from direct processing of the available data, they code all the necessary information related to the data distribution in the input space. Therefore, in order to reduce the computational efforts we choose to elaborate them (instead of using directly the training data). To accomplish this task, we assign weights to each fuzzy subspace. The representatives (i.e. center elements) along with the corresponding weights of the aforementioned fuzzy subspaces are treated as a new data set, which is partitioned in terms of a weighted version of the fuzzy c-means. This process yields the network's basis function centers. The network's widths are determined by means of the centers. Finally, the connection weights are provided by applying the PSO.

The material is presented as follows. Section 2 describes the standard topology of a typical RBF neural network. In Sect. 3 we analytically describe the proposed methodology. The simulations experiments take place in Sect. 4. Finally, the chapter concludes in Sect. 5.

2   RBF Neural Network

The basic topology of an RBF network consists in sequence of an input layer, a hidden layer, and a linear processing unit forming the output layer. The standard topology of the RBF network is depicted in Fig. 1.

The hidden layer consists of a number of nodes, each of which corresponds to an RBF. Herein, the RBF will also be referred to as kernel function or simply kernel. The set of input–output data pairs is denoted as

$$Z = \{(x_k, y_k) \in R^p x R^s : 1 \leq k \leq N\} \tag{1}$$

with $x_k = [x_{k1}, x_{k2}, \ldots, x_{kp}]^T$ is the k-th input vector and $y_k \in \{1, 2, \ldots, s\}$ is the class the vector x_k belongs to.

In the case of the Gaussian type of RBFs, the basis functions have the form,

$$h_i(x_k) = \exp\left(-\frac{\|x_k - v_i\|^2}{\sigma_i^2}\right) \tag{2}$$

where v_i $(1 \leq i \leq c)$ are the centers, σ_i $(1 \leq i \leq c)$ the respective widths, and c the number of hidden nodes. The estimated output of the network utilizes the linear regression functional:

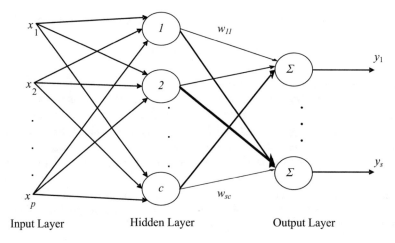

Fig. 1 Standard topology of a multi-input multi-output RBF neural network

$$\tilde{y}_j\left(x_k\right) = \sum_{i=1}^{c} w_{ji} h_i\left(x_k\right) \tag{3}$$

where w_{ji} is the connection weight of the i-th hidden node and j output node ($j = 1, \ldots, s$). The training procedure of RBF networks aims to obtain a set of values for the kernel parameters (v_i, σ_i) and the connecting weights w_{ji}. Usually, the connecting weights are estimated through a regression-based analysis.

3 Particle Swarm Optimization

PSO has been widely used on numerical optimization tasks, in particular on strongly nonlinear problems [5, 19]. The basic design element is the particle, which is a real valued vector $p \in \mathfrak{R}^r$ that represents a possible, yet complete, solution of the problem at hand. The set of particles forms the swarm. In this chapter the swarm size is denoted as M. Every particle p_i ($1 \leq i \leq M$) is assigned a velocity vector $v_i \in \mathfrak{R}^r$. In the t-th iteration, each particle informs others that make up the respective group of informants $Q_i(t)$. As p_i^{best} we symbolize the position of the lowest value of the fitness function obtained so far by the particle p_i. The position associated with the lowest value of fitness function obtained so far by all particles belonging to $Q_i(t)$ is denoted as $p_j^{\text{best}}(t)$. Then, the velocity v_i is calculated as:

$$
\begin{aligned}
v_i\left(t+1\right) = {} & \vartheta v_i(t) + f_1 V\left(0, 1\right) \circ \left(p_i^{\text{best}}(t) - p_i(t)\right) \\
& + f_2 V\left(0, 1\right) \circ \left(p_j^{\text{best}}(t) - p_i(t)\right)
\end{aligned}
\tag{4}
$$

and the position of each particle is updated according to the next learning rule,

$$p_i(t+1) = p_i(t) + v_i(t+1) \tag{5}$$

In the above equations, $\circ$ stands for the point-wise vector multiplication, $V(0,1)$ is a function that returns an r-th dimensional vector whose coordinates are numbers randomly generated by a uniform distribution in $[0, 1]$, f_1 and f_2 are constant positive numbers called the cognitive and social parameters and ϑ is also a positive constant called the inertia factor. Point-wise (or element-wise) vector multiplication is the element-by-element multiplication of two vectors. In accordance with the existing literature, typical values for these parameters are [26]: $\vartheta \in [0, 1]$, while f_1 and f_2 should be around 1.5. In this paper, for all the experiments the value 2 was chosen for the parameters f_1 and f_2. For the inertia factor ϑ, a pseudo-random selection in $(0.5,1)$ took place for all the simulations.

4 RBF Network Training Algorithm

In this section we analytically describe the proposed algorithm, the flow sheet of which is illustrated in Fig. 2.

In view of this figure, three steps can be distinguished. In the first step we define an ordinary fuzzy partition in the input space and pre-process the data in terms of a suitable unsupervised learning process to extract a number of fuzzy subspaces. The second elaborates the resulting fuzzy subspaces by means of a weighted version

Fig. 2 The sequential steps of the proposed algorithm

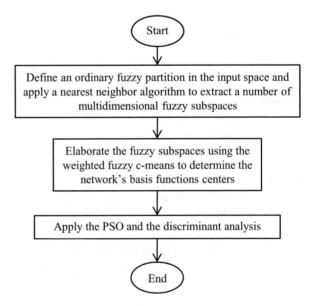

of the fuzzy c-means model. The outcome of this clustering process is the basis function centers. Finally the last step involves the PSO and the discriminant analysis in order to infer the output classes. The detailed presentation of the above steps is provided by the next subsections.

4.1 Extraction of the Multidimensional Fuzzy Subspaces

Let us partition each universe of discourse X_j $(1 \leq j \leq p)$ into q_j symmetric triangular fuzzy sets $A_j^1, A_j^2, \ldots, A_j^{q_j}$ with membership functions of the form

$$A(x) = \begin{cases} 1 - \frac{|x-\alpha|}{\delta\alpha}, & \text{if } x \in [\alpha - \delta\alpha, \alpha + \delta\alpha] \\ 0, & \text{otherwise} \end{cases} \tag{6}$$

where α is the center element and $\delta\alpha$ is the width. Each fuzzy set A_j^l is described as $A_j^l = \left\{ a_j^l, \delta_j^l \right\}$. This procedure will create $L = \prod_{j=1}^p q_j$ multidimensional fuzzy sets in the feature space X. The set V_ℓ $(1 \leq \ell \leq L)$ is composed by p one-dimensional fuzzy sets, $V_\ell = \{A_1^\ell, A_2^\ell, \ldots, A_p^\ell\} = \{a^\ell, \delta a^\ell\}$, with $a^\ell = \left[a_1^\ell, a_2^\ell, \ldots, a_p^\ell \right]$ and $\delta a^\ell = \left[\delta a_1^\ell, \delta a_2^\ell, \ldots, \delta a_p^\ell \right]$. Figure 3 depicts a fuzzy partition in the two-dimensional space.

The matching degree between the training vector x_k and the ℓth multidimensional fuzzy set is [27]:

$$V_\ell(x_k) = 1 - rd^\ell(x_k) \tag{7}$$

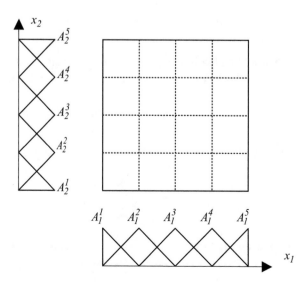

Fig. 3 Fuzzy partition in a two-dimensional input space

where $rd^{\ell}(x_k)$ is the relative Euclidean distance [27]:

$$
rd^{\ell}(x_k) = \begin{cases} \dfrac{\left[\displaystyle\sum_{j=1}^{p}\left(\alpha_j^{\ell}-x_{kj}\right)^2\right]^{1/2}}{\left[\displaystyle\sum_{j=1}^{p}\left(\delta\alpha_j^{\ell}\right)^2\right]^{1/2}} & if \ \left[\displaystyle\sum_{j=1}^{p}\left(\alpha_j^{\ell}-x_{kj}\right)^2\right]^{1/2} \leq \left[\displaystyle\sum_{j=1}^{p}\left(\delta\alpha_j^{\ell}\right)^2\right]^{1/2} \\[6pt] 1 & otherwise \end{cases}
$$

(8)

The application of Eqs. (7) and (8) produces a spherical fuzzy subspace $C^{\ell} = \{a^{\ell}, \rho^{\ell}\}$ centered at $a^{\ell} = [a_1^{\ell}, a_2^{\ell}, \ldots, a_p^{\ell}]$ with radius $\rho^{\ell} = \sqrt{\sum_{j=1}^{p}\left(\delta\alpha_j^{\ell}\right)^2}$. As it was shown in [28], we can select the fuzzy subspaces that better describe the training data set according to the following algorithm:

Algorithm 1

Step 1) Set $k = 1$ and $n = 1$. Using (7)–(8) generate the first fuzzy subspace $C^n = \{a^n, \rho^n\}$.

Step 2) For $k = 2$ to N do

$$rd_{\min}(x_k) = \min_{1 \leq i \leq n}\{rd^i(x_k)\}$$

If $rd_{\min}(x_k) < 1$ then

Assign the x_k to the fuzzy subspace that correspond to the $rd_{\min}(x_k)$.

else

Set $n = n + 1$ and using Eqs. (7)–(8) generate the nth fuzzy subspace $C^n = \{a^n, \rho^n\}$.

endif

endfor

The above algorithm dismembers the input space into n fuzzy subspaces denoted as C^l with $l = 1, 2, \ldots, n$. The fuzzy cardinality of the lth fuzzy subspace is:

$$\aleph\left(C^l\right) = \sum_{k=1}^{N} V_l(x_k) \quad (1 \leq l \leq n) \tag{9}$$

To this end we define the relative fuzzy cardinality of the C^l as:

$$r_l = \frac{\aleph\left(C^l\right)}{\sum_{j=1}^{n}\aleph\left(C^j\right)} \quad (1 \leq l \leq n) \tag{10}$$

Note that the radii ρ^{ℓ} $(1 \leq \ell \leq L)$ were used to extract the corresponding cardinalities. Therefore, they will not be used any more. Instead, each fuzzy subspace will be represented by its center element a_l and the respective relative cardinality r_l. Since $r_l \in [0, 1]$ it can be viewed as the weight of significance of the subspace C_l.

4.2 Estimation of the Network's Basis Function Parameters

The standard methodology to determine the basis functions centers of the RBF network is the utilization of cluster analysis. In this section, we attempt to cluster the fuzzy subspaces resulting from the previous step. In general, the arsenal to cluster fuzzy data consists of a wide range of algorithmic tools [6, 29, 30]. In this section we employ the weighted fuzzy c-means method that was developed in [30, 31]. To implement this clustering scheme, we recall that each fuzzy subspace can be represented by its center element and the weight of significance. Therefore, the pairs $\{(a_l, r_l) : 1 \le l \le n\}$ constitute a new data set, which we intend to cluster by using the aforementioned cluster analysis in order to obtain c fuzzy clusters in the input space with centers $v_1, v_2, \ldots, v_c$. The objective function has the following form [30]:

$$J = \sum_{l=1}^{n} \sum_{i=1}^{c} r_l \, u_{il}^m \, \|a_l - v_i\|^2 \tag{11}$$

where u_{il} is the membership degree and m the fuzziness parameter.

The task is to minimize J under the constraint:

$$\sum_{i=1}^{c} u_{il} = 1 \quad \forall l \tag{12}$$

The membership degrees and cluster centers that solve the above optimization problem are provided by the following equations:

$$u_{il} = \frac{1}{\sum_{j=1}^{c} \left(\dfrac{\|a_l - v_i\|}{\|a_l - v_j\|} \right)^{2/(m-1)}} \tag{13}$$

$$v_i = \frac{\sum_{l=1}^{n} r_l u_{ik}^m a_l}{\sum_{l=1}^{n} r_l u_{il}^m}, \quad (1 \le l \le n) \tag{14}$$

The weighted fuzzy c-means is carried out by iteratively applying the Eqs. (13) and (14) until convergence. The interesting reader in referred to [30, 31] for a more detailed presentation of the clustering process. As mentioned, the results of this clustering approach is a set of cluster centers $\{v_1, v_2, \ldots, v_c\}$, which coincide with the network's RBFs center elements. In order to calculate the kernel width of the i-th RBF, we use the approach developed by Niros and Tsekouras [32]. Specifically, the widths of the network's basis functions are as follows:

$$\sigma_i = \frac{2 * d^i_{max}}{3} \quad (1 \leq i \leq c) \tag{15}$$

The distance function d^i_{max} is:

$$d^i_{max} = \left\{ \max_{1 \leq l \leq n} \{|| x_l - v_i ||^2\} : x_l \in C_i \text{ such that } u_{il} \geq \tau \right\} \tag{16}$$

where C_i is the i-th fuzzy cluster, u_{il} is the membership degree of the l-th training data vector to the i-th final fuzzy cluster, and τ a small positive number such that $\tau \in (0, 1)$.

4.3 Discriminant Analysis and PSO Implementation

Given that the number of classes is denoted as s, the outputs of the network when the input is the vector x_k $(1 \leq k \leq N)$ are denoted as $y_1(x_k), y_2(x_k), \ldots, y_s(x_k)$. By defining the vectors:

$$w_j = \left[w_{j1} \ w_{j2} \ \cdots \ w_{jc} \right]^T \quad (1 \leq j \leq s) \tag{17}$$

and

$$Hx_k = \left[h_1(x_k) \ h_2(x_k) \ \cdots \ h_c(x_k) \right]^T \quad (1 \leq j \leq s)$$

the classifier assigns the vector x_k to the class j_k when the subsequent condition is fulfilled,

$$f_{j_k}(x_k) = \max_{1 \leq \ell \leq s} \{f_l(x_k)\} \tag{18}$$

where f_j is the discriminant function, which is defined in terms of the inner product operation,

$$f_j(x_k) = w_j \cdot Hx_k \tag{19}$$

According to the above analysis, the classifier is viewed as a network that computes s discriminant functions and for the current input selects the class that appears to have the largest value of all the rest of the discriminant functions. Note that the network's

w_{j1}	w_{j2}	...	w_{jc}

Fig. 4 The structure of the particle in the PSO algorithm for the jth class

Table 1 Parameter setting

$\lambda = 0.71$	$f_2 = 2$
$q_j = 21$	Swarm size $M = 20$
$\tau = 0.001$	Number of intervals $= 5$
$f_1 = 2$	$\theta = 0.7$

Table 2 Machine learning data sets used

Data set	Number of classes	Number of inputs	Number of patterns (data)
WDBC	2	30	569
Wine	3	13	178
Pima	2	8	768

output will typically give a positive weight to the RBF neurons that belong to its category, and a negative weight to the others.

The estimation of the connection weights is carried out by the PSO algorithm. Due to the appearance of s outputs, the PSO is implemented separately with respect to each output class. For the jth class the structure of the particle is depicted in Fig. 4.

5 Evaluation Experiments

The experiments are conducted by using a split of the available data set into 60–40 % training and testing subsets, namely 60 % of the whole patterns are selected randomly for training and the remaining patterns are used for testing purposes.

We used a computer with a dual core CPU (i3-3120M) at 2.50 GHz with 4GB Ram Memory and the MATLAB software. Table 1 reports the parameter values that remain constant.

We consider several data sets concerning classification problems taken from the Machine Learning UCI repository provided by the website: http://archive.ics.uci.edu/ml/datasets.html.

Table 2 summarizes the pertinent details of the data set such as the number of features and number of patterns.

Table 3 Classification rates (%) for the WDBC data set

Number of hidden nodes	Training data set	Testing data set
2	76.83 ± 0.01	76.32 ± 0.01
4	86.22 ± 0.01	85.96 ± 0.01
5	89.74 ± 0.29	89.91 ± 0.35
6	88.27 ± 0.01	89.92 ± 0.01

Table 4 Comparison of the average performances for the WDBC data set

Classifier model	Classification rate (%)
Bayes Net [33]	95.81
DigaNN [34]	97.9
FSM [35]	98.3
MLP [36]	85.92
MPANN [37]	98.1
RBF2 [38]	97.13
RVM [39]	97.2
SVM [40]	96.68
Proposed RBF (12 hidden nodes)	97.92 ± 1.11

Fig. 5 Mean values of the fitness function obtained by the PSO for the WDBC data set ($c = 12$)

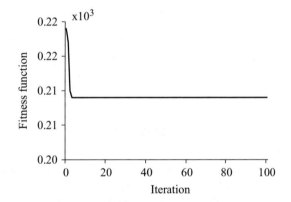

5.1 WDBC Data Set

In this experiment, we are concerned with data of high dimensionality (see Table 2). Initially, we evenly partitioned each input universe of discourse into 21 fuzzy sets of the form (6). By applying Algorithm 1 we obtained $n = 31$ fuzzy subspaces.

Table 3 displays the results when the number of hidden nodes is equal to 2, 4, 5, and 6. Table 4 provides a comparison study where the average classification rates obtained for several models are shown. According to this table, the proposed classifier is quantified as 97.92 reported for 12 hidden nodes.

Finally, Fig. 5 describes the convergence capabilities of the PSO when the number of hidden node is 1, which seems to be quite smooth.

Table 5 Classification rates (%) for the wine data set

Number of hidden nodes	Training data set	Testing data set
2	61.68 ± 0.01	74.65 ± 0.01
4	62.14 ± 0.01	74.83 ± 0.01
6	68.41 ± 2.34	74.04 ± 3.00
9	69.82 ± 1.68	80.04 ± 2.15
12	78.12 ± 2.34	90.12 ± 2.53

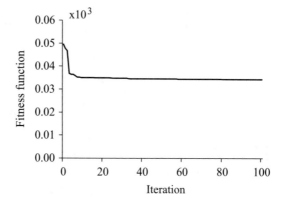

Fig. 6 Mean values of the fitness function J obtained by the PSO for the example 2 ($c = 12$)

5.2 Wine Data Set

Initially, we evenly partitioned each input universe of discourse into 21 fuzzy sets of the form (6). By applying Algorithm 1 we obtained $n = 22$ initial fuzzy subspaces.

Table 5 shows the final results for various number of hidden nodes raining and testing data. Note that the best performance comes in the case of 12 hidden nodes with a 90.12 % classification rate in the testing data case. Figure 6 describes the convergence capabilities of the PSO when the number of hidden node is 12.

5.3 Pima Indians Diabetes Data Set

In this example, each universe of discourse was partitioned into 21 symmetric triangular fuzzy sets. The implementation of Algorithm 1 gave $n = 25$ multidimensional fuzzy subspaces.

Table 6 shows the result when the number of RBF hidden nodes is 4, 6, 8, 10, and 12. The best performance value is 75.97 and it is reported in case of 12 hidden nodes. Table 7 shows the average classification rate obtained for several models and the proposed RBF classifier shows the best performance. Figure 7 describes the convergence capabilities of the PSO when the number of hidden node is 12.

Table 6 Classification rates for the pima data set with 4, 6, 8, 10, and 12 hidden nodes

Number of hidden nodes	Training data set	Testing data set
4	57.57 ± 0.01	53.09 ± 0.01
6	57.82 ± 0.01	55.92 ± 0.01
8	62.08 ± 0.29	59.86 ± 0.35
10	70.15 ± 1.90	65.06 ± 1.13
12	70.98 ± 0.66	68.16 ± 0.58

Table 7 Comparison of the average performance of classifiers for the pima Indians diabetes data set

Classifier model	Classification rate (%)
MLP [41]	73.10
RBF1 [42]	76.48
RBF2 [42]	76.96
RVM [43]	74.83
SVM [40]	74.72
Proposed RBF (14 hidden nodes)	75.97 ± 1.03

Fig. 7 Mean values of the fitness function J obtained by the PSO for the example 3 ($c = 12$)

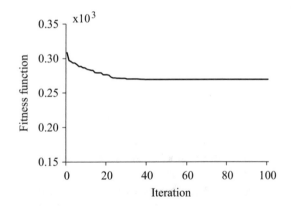

6 Conclusion

In this chapter, we have discussed a new method for classification problems based on hierarchical fuzzy clustering and PSO. The proposed RBF classifier is a trainable mechanism consisting of three steps. The first step includes an ordinary fuzzy partition defined in the input space and a pre-processing unit which elaborates the data in terms of a suitable unsupervised learning process to extract a number of fuzzy subspaces. The second step clusters the fuzzy subspaces resulting from the previous step by means of a weighted version of the fuzzy c-means model. The outcome of this clustering process is the basis function centers. Finally in the last step, the PSO is implemented in order to infer the network's output. The method is applied to three machine learning data sets. As shown in the results of the experimental study, the proposed classifier exhibits a highly reliable performance and can be considered as an effective tool in two-class or multi-class pattern classification.

References

1. Haykin, S.: Neural Networks: A Comprehensive Foundation. Prentice Hall, Upper Saddle River (1999)
2. Schwenker, F., Kestler, H.A., Palm, G.: Three learning phases for radial-basis-function networks. Neural Netw. **14**, 439–458 (2001)
3. Xie, T., Yu, H., Hewlett, J., Rózycki, P., Wilamowski, B.: Fast and efficient second-order method for training radial basis function networks. IEEE Trans. Neural Netw. **23**(4), 609–619 (2012)
4. Chen, Y., Peng, L., Abraham, A.: Hierarchical radial basis function neural networks for classification problems. Advances in Neural Networks. Lecture Notes in Computer Science, vol. 3971, pp. 873–879 (2006)
5. Chen, Y., Yang, B., Zhou, J.: Automatic Design of Hierarchical RBF Networks for System Identification. PRICAI'06, Lecture Notes on Artificial Intelligence, **4099**, 1191–1195 (2006)
6. González, J., Rojas, I., Pomares, H., Ortega, J., Prieto, A.: A new clustering technique for function approximation. IEEE Trans. Neural Netw. **13**(1), 132–142 (2002)
7. Hou, M., Han, X.: Constructive approximation to multivariate function by decay RBF neural network. IEEE Trans. Neural Netw. **21**(9), 1517–1523 (2010)
8. Huan, H.X., Hien, D.T.T., Tue, H.H.: Efficient algorithm for training interpolation RBF networks with equally spaced nodes. IEEE Trans. Neural Netw. **22**(6), 982–988 (2011)
9. Babuska, R., Verbruggen, H.: Neuro-fuzzy methods for nonlinear system identification. Annu. Rev. Control **27**, 73–85 (2004)
10. Behloul, F., Lelieveldt, B.P.F., Boudraa, A., Reiber, J.H.C.: Optimal design of radial basis function neural networks for fuzzy-rule extraction in high dimensional data. Pattern Recognit. **35**, 659–675 (2002)
11. Haddadnia, J.H., Faez, K., Ahmadi, M.: A fuzzy hybrid algorithm for radial basis function neural network application in human face recognition. Pattern Recognit. **36**, 1187–1202 (2003)
12. Wong, Y.W., Seng, K.P., Ang, L.M.: Radial basis function neural network with incremental learning for face recognition. IEEE Trans. Syst. Man. Cybern. B Cybern. **41**(4), 940–949 (2011)
13. Valdez, F., Melin, P., Castillo, O.: Evolutionary method combining particle swarm optimisation and genetic algorithms using fuzzy logic for parameter adaptation and aggregation: the case neural network optimisation for face recognition. Int. J. Artif. Intell. Soft Comput. **2**(1–2), 77–102 (2010)
14. Theodoridis, S., Koutroumbas, K.: Pattern Recognition, 2nd edn. Elsevier, San Diego (2003)
15. Bishop, C.M.: Neural Networks for Pattern Recognition. Clarendon, Oxford (1995)
16. Staiano, A., Tagliaferri, R., Pedrycz, W.: Improving RBF networks performance in regression tasks by means of a supervised fuzzy clustering. Neurocomputing **69**, 1570–1581 (2006)
17. Park, H.S., Pedrycz, W., Oh, S.K.: Granular neural networks and their development through context-based clustering and adjustable dimensionality of receptive fields. IEEE Trans. Neural Netw. **20**(10), 1604–1616 (2009)
18. Fathi, V., Montazer Gh, A.: An improvement in RBF learning algorithm based on PSO for real time applications. Neurocomputing **111**, 169–176 (2013)
19. Carlisle, A., Dozier, G.: Adapting particle swarm optimization to dynamic environments. In: Proceedings of International Conference on Artificial Intelligence, pp. 429–434 (2000)
20. Clerc, M.: Particle Swarm Optimization. ISTE, London (2006)
21. Clerc, M., Kennedy, J.: The particle swarm – explosion, stability, and convergence in a multidimensional complex space. IEEE Trans. Evol. Comput. **6**(1), 58–73 (2002)
22. Oh, S.K., Kima, W.D., Pedrycz, W., Park, B.J.: Polynomial-based radial basis function neural networks (P-RBF NNs) realized with the aid of particle swarm optimization. Fuzzy Sets Syst. **163**, 54–77 (2011) (Elsevier)
23. Tsekouras, G.E., Tsimikas, J.: On training RBF neural networks using input-output fuzzy clustering and particle swarm optimization. Fuzzy Sets Syst. **221**, 65–89 (2013)

24. Qasem, S.N., Shamsuddin, S.M.: Improving performance of radial basis function network with particle swarm optimization. IEEE International Conference on Systems, Man and Cybernetics, pp. 534–540 (2009)
25. Qasem, S.N., Shamsuddin, S.M.: Radial basis function network based on time variant multi-objective particle swarm optimization for medical diseases diagnosis. Appl. Soft Comput. **11**, 1427–1438 (2011)
26. Chen, S., Hong, X., Harris, C.J.: Particle swarm optimization aided orthogonal forward regression for unified data modeling. IEEE Trans. Evol. Comput. **14**(4), 477–499 (2010)
27. Linkens, D.A., Chen, M.Y.: Input selection and partition validation for fuzzy modelling using neural network. Fuzzy Sets Syst. **107**, 299–308 (1999)
28. Tsekouras, G.E., Sarimveis, H., Bafas, G.: A simple algorithm for training fuzzy systems using input–output data. Adv. Eng. Softw. **34**(5), 247–259 (2003)
29. Looney, G.G.: Fuzzy connectivity clustering with radial basis kernel functions. Fuzzy Sets Syst. **160**, 1868–1885 (2009)
30. Tsekouras, G.E.: On the use of the weighted fuzzy c-means in fuzzy modeling. Adv. Eng. Softw. **36**, 287–300 (2005)
31. Tsekouras, G.E., Sarimveis, H., Kavakli, E., Bafas, G.: A hierarchical fuzzy-clustering approach to fuzzy modeling. Fuzzy Sets Syst. **150**(2), 245–266 (2005)
32. Niros, A.D., Tsekouras, G.E.: On training radial basis function neural networks using optimal fuzzy clustering. In: 17th IEEE Mediterranean Conference on Control and Automation, pp. 395–400 (2009)
33. Jensen, F.: An Introduction to Bayesian Networks. UCL Press/Springer, Berlin, Germany (1996)
34. Anagnostopoulos, I., Maglogiannis, I.: Neural network-based diagnostic and prognostic estimations in breast cancer micro scopic instances. Med. Biol. Eng. Comput. **44**(9), 773–784 (2006)
35. Adamczak, R., Duch, W., Jankowski, N.: New developments in the feature space mapping model. In: Third Conference on Neural Networks and Their Applications, Kule, October 1997, pp. 65–70 (1997)
36. Duda, R.O., Hart, P.E.: Pattern Classification and Scene Analysis. Wiley, New York (2002)
37. Abbass, H.A.: An evolutionary artificial neural networks approach for breast cancer diagnosis. Artif. Intell. Med. **25**(3), 265–281 (2002)
38. Young, Z.R.: A novel radial basis function neural network for discriminant analysis. IEEE Trans. Neural Netw. **17**(3), 604–612 (2006)
39. Tipping, M.E.: The relevance vector machine. Adv. Neural Inf. Process. Syst. **12**, 652–658 (2000)
40. Vapnik, V.: The Nature of Statistical Learning Theory. Springer, New York (1995)
41. Aeberhard, S., Coomans, D., de Vel, O.: Comparison of classifiers in high dimensional settings, Department of Computer Science and Department of Mathematics and Statistics, James Cook. University of North Queensland, Technical Report 92-02 (1992)
42. Kotsiantis, S.B., Pintelas, P.E.: Logitboost of simple Bayesian classifier. Informatica **29**, 53–59 (2005)
43. Lim, T., Loh, W., Shih, Y.: A comparison of prediction accuracy, complexity, and training time of thirty three old and new classification algorithms. Mach. Learn. **40**, 203–228 (2000)

A Survey of Constrained Clustering

Derya Dinler and Mustafa Kemal Tural

Abstract Traditional data mining methods for clustering only use unlabeled data objects as input. The aim of such methods is to find a partition of these unlabeled data objects in order to discover the underlying structure of the data. In some cases, there may be some prior knowledge about the data in the form of (a few number of) labels or constraints. Performing traditional clustering methods by ignoring the prior knowledge may result in extracting irrelevant information for the user. Constrained clustering, i.e., clustering with side information or semi-supervised clustering, addresses this problem by incorporating prior knowledge into the clustering process to discover relevant information from the data. In this chapter, a survey of advances in the area of constrained clustering will be presented. Different types of prior knowledge considered in the literature, and clustering approaches that make use of this prior knowledge will be reviewed.

1 Introduction

Data mining, i.e., science of extracting new and useful information from large data sets, has gained a lot of attention recently among scientists and in society as a whole due to the wide availability of huge data sets and the increased power of computers. According to [47], the amount of data produced worldwide during the year 2002 and stored on paper, film, magnetic media, and optical devises was estimated to be between 3,400,000 and 5,600,000 terabytes. These numbers are about twice the size of the data produced during the year 1999 [46]. The amount of data that has exploded year over year introduced new challenges and led to continued innovation in data storage and data mining techniques. A fundamental categorization in data mining techniques is:

1. Supervised learning
2. Unsupervised learning.

D. Dinler (✉) • M.K. Tural
Industrial Engineering Department, Middle East Technical University, 06800 Inonu Bulvari, Ankara, Turkey
e-mail: dinler@metu.edu.tr; tural@metu.edu.tr

© Springer International Publishing Switzerland 2016   207
M.E. Celebi, K. Aydin (eds.), *Unsupervised Learning Algorithms*,
DOI 10.1007/978-3-319-24211-8_9

In the former one, the data is split into training and test data. Observations in the training data are provided with desired output values (labels) which are used to construct a function that predicts the output values of the observations in the test data. Classification and regression are good examples of supervised learning problems. Typically, supervised learning algorithms require a large training data set, which is not always available or may be very costly to acquire. In the latter one, however, there is no learning from cases (no training data), but instead one tries to find intrinsic "natural" structures in unlabeled data. Clustering is one of the most famous unsupervised learning problems. As unsupervised learning methods are completely unguided, the structure extracted from the data may not always be relevant to the analyst or user.

In some cases, the data analyst may have a priori (domain) knowledge about the underlying structure of the data. With completely unlabeled data or a few number of labeled data, the supervised learning techniques may not be suitable. One way of learning from such data would be to completely ignore the prior knowledge and apply unsupervised learning methods. This may, however, result in extracting irrelevant structure from the data. Another way of learning from such data is by means of semi-supervised learning which incorporates the prior knowledge to the learning process to improve the quality of the result. Note that this type of learning is at the intersection of (completely) supervised and unsupervised learning.

In the last two decades, semi-supervised clustering which is also known as constrained clustering or clustering with side information has attracted several researchers as it has been observed that even with a small amount of prior knowledge, the clustering performance can be improved and running time of the process can be decreased significantly [62]. A survey of advances in the area of constrained clustering will be presented in this chapter. The survey will cover different types of prior knowledge considered in the literature, and clustering approaches that make use of this knowledge. In Sect. 2, we introduce unsupervised clustering and review some of the widely used unsupervised clustering algorithms which form the basis of several constrained clustering algorithms. In Sect. 3, we provide a review of the constrained clustering literature. Finally, we conclude the chapter with Sect. 4.

2   Unsupervised Clustering

Clustering can be defined as partitioning of unlabeled observations into groups (clusters) so that the observations in a group will be similar to each other with respect to some similarity criteria (i.e., high intra-cluster similarity) and different from the observations in other groups (i.e., high inter-cluster dissimilarity). It aims to find meaningful or useful clusters which are defined according to the goals of the data analyst. In some clustering problems, the number of clusters to be formed, k, is given as a parameter. In some others, however, k may be unknown. Clustering has been employed in many disciplines such as statistics, biology, marketing, and medicine. It has mainly been used for three purposes:

- to discover the underlying structure of the data,
- to find natural classification of data objects by identifying similarities among them, and
- to organize and summarize the data [38].

In some cases, clustering can be used as a preprocessing tool for other tasks such as regression, principal component analysis, and association analysis. See, for instance, [23] in which the authors used clustering techniques to reduce complexity in solving automated planning problems.

Typically, clustering problems are considered as optimization problems with various objectives and solved by exact algorithms, approximation methods, or heuristics [3]. Clustering methods can be broadly divided into two categories: hierarchical and partitional methods. Hierarchical clustering algorithms build nested clusters by merging (agglomerative methods) or splitting (divisive methods) clusters successively. The hierarchy of clusters is represented with a tree known as dendrogram. Partitional clustering algorithms construct a one-level clustering of the observations without a nested structure. It should be noted that in some clustering algorithms all of the data is studied at once whereas in others data objects are taken into consideration one by one. Algorithms falling in the former category are known as the batch algorithms, whereas algorithms in the latter category are known as incremental algorithms. For online applications where the input arrives piece by piece, incremental algorithms are better suited.

Several different criteria have been used in cluster analysis. Among them, the minimum sum-of-squares is the most popular one.

2.1 Minimum Sum-of-Squares Clustering

Consider a data set consisting of n data objects in the d-dimensional Euclidean space, $X = \{x_i\}$, $i = 1, \ldots, n$, $x_i \in R^d$. The Minimum Sum-of-Squares Clustering (MSSC) problem is the problem of partitioning the objects into a set of a predefined number, say k, of clusters, $C = \{c_j\}$, $j = 1, \ldots, k$, so as to minimize the sum of squared Euclidean distances between the data objects and the centers of the clusters (representatives) they belong to. The problem can be formulated as follows:

$$\text{minimize} \sum_{j=1}^{k} \sum_{x_i \in c_j} \left\| x_i - \mu_j \right\|^2, \tag{1}$$

where μ_j represents the center of the jth cluster and $\{c_1, \ldots, c_k\}$ a partition of X.

Note that the center of a cluster may not coincide with any of the data objects belonging to that cluster. In the MSSC problem, a hard assignment of the data objects to clusters is made, i.e., every object is assigned to exactly one cluster. On the other hand, a soft assignment would assign for each object and each cluster a probability that the object belongs to the cluster.

The MSSC problem is known to be NP-hard in general [50] and therefore heuristic algorithms have been widely employed. K-means algorithm and its variants are among the most commonly used heuristics proposed for the MSSC problem.

2.1.1 K-Means Algorithm

K-means algorithm proposed by Macqueen in 1967 is the most famous partitional (batch) clustering algorithm [48]. Because of its easy to implement nature, simplicity, efficiency, and empirical success, its framework has also been commonly used in algorithms developed for constrained clustering problems. The steps of the K-means algorithm are based on two simple observations. First, if the centers of the clusters are known, then each data object is assigned to the cluster whose center is closest to the object. Second, if all the data objects belonging to a cluster are known, then the center of the cluster is computed by averaging all data objects in the cluster.

The basic steps of the K-means algorithm are given below.

Algorithm 1: K-means

1: Initialization: Start with initial cluster centers (seed), $\mu_j, j = 1, \ldots, k$.
   repeat
2:   Assignment: Assign each data object x_i to the closest cluster j^*, and let $x_i \in c_{j^*}$, where
     $j^* = \text{argmin}_j \|x_i - \mu_j\|$.
3:   Update: Update the cluster centers by averaging the data objects assigned to them, i.e.,
     $\mu_j = \frac{\sum_{x_i \in c_j} x_i}{|c_j|}$.
   until Convergence is achieved.
4: **return** Partition $\{c_1, \ldots, c_k\}$ of X.

The algorithm finds a local minimum and the convergence can be slow. It may return some empty clusters. The final partition obtained with the K-means algorithm is highly depended on the initialization step. To overcome this problem, the K-means algorithm may be initialized with multiple different sets of initial centers and the solution with the minimum objective function value may be returned to the user. One way to initialize the K-means algorithm is by randomly choosing the initial centers. Several initialization heuristics have also been proposed with the aim of starting with a "good" set of initial centers, see, for example, [14, 15]. The runtime of a single iteration of the K-means algorithm is equal to $O(nkd)$.

Due to the known issues with the standard K-means algorithm, several variants of the algorithm have been proposed over the years [35, 36, 44].

2.2 *Agglomerative Hierarchical Clustering*

Traditional agglomerative hierarchical clustering algorithms begin with each data object in a separate cluster and in a progressive manner merge two clusters that are the closest to reduce the number of clusters by one until all the data objects are in a single cluster.

The steps of a basic agglomerative hierarchical clustering algorithm are as follows:

Algorithm 2: Agglomerative hierarchical clustering

1: Let $c_i = \{x_i\}$, for every $i = 1, 2, \ldots, n$ and let $C = \{c_1, c_2, \ldots, c_n\}$
    for $k = n$ to 1 **do**
2:     $Dendrogram(k) = C$
     Let $(u, v) = \text{argmin}_{i \neq j: \ c_i, c_j \in C} \ \rho\left(c_i, c_j\right)$
     $c_u = c_u \bigcup c_v$ and $C = C \setminus c_v$.
    end for

Here $\rho\left(c_i, c_j\right)$ represents the distance between the clusters c_i and c_j and (u, v) the arguments of the closest two clusters. Distance between two clusters can be measured in several different ways. In single linkage clustering, the distance between two clusters is measured by the minimum distance between the data objects in the two clusters, i.e.,

$$\rho\left(c_i, c_j\right) = \min_{x \in c_i, \ y \in c_j} d(x, y), \tag{2}$$

where $d(x, y)$ represents the distance between two data objects x and y.

In complete linkage clustering, distance between clusters is measured by the two most distant objects rather than the closest ones. Average linkage clustering uses, on the other hand, the average distance between the objects in the two clusters. There are several other ways to measure distances between groups, see, for instance, [66].

The result of a hierarchical clustering algorithm is usually depicted as a tree known as dendrogram. The root node of the dendrogram represents all the data objects as a single cluster and the leaf nodes represent each object as a single cluster. The height of a node in a dendrogram is proportional to the value of the clustering criterion at this node. By cutting the dendrogram with a horizontal line at the appropriate level, any desired number of clusters can be obtained and the height of the line represents the value of the clustering criterion. For more on hierarchical clustering algorithms and complexity issues, we refer the reader to [27, 49, 53].

2.3 *COBWEB*

COBWEB is an incremental hierarchical clustering algorithm that was proposed by Fisher in 1987 [26]. It is used to obtain a hierarchy (classification tree) of data objects with nominal features.

In this algorithm, clusters are represented by d probability distributions (one distribution for each dimension or feature) instead of cluster centers. When a data object is available to cluster, algorithm performs one of the four operators:

1. Insert operator: insert the new object to an existing cluster (node),
2. Create operator: create a new cluster which will include the new object,
3. Merge operator: merge two of the existing children (clusters) of a node and place the new object in the resulting hierarchy, and
4. Split operator: split the objects of a cluster and place the new object in the resulting hierarchy.

While deciding which one of these operators will be performed, the algorithm uses the concept of category utility proposed by Gluck and Corter in 1985 [31] which takes both intra-cluster similarity and inter-cluster dissimilarity into account. Category utility of a hierarchical partition is defined as follows:

$$CU(c_1, \ldots, c_k) = \frac{1}{k} \left(\left(\sum_{j=1}^{k} \frac{|c_j|}{n} \sum_{d} \sum_{t \in V(d)} P(D_d = t|c_j)^2 \right) - \sum_{d} \sum_{t \in V(d)} P(D_d = t)^2 \right),$$

(3)

where $\{c_1, c_2, \ldots, c_n\}$ are the nodes in the hierarchy, D_d is one of the dimensions (features), and $V(d)$ is the set of discrete values that feature d can take. $P(D_d = t|c_j)$ is the probability that the dth feature of a data object in cluster c_j is equal to t and $P(D_d = t)$ is the probability that the dth feature of a data object is equal to t.

3   Constrained Clustering

In some cases, there may be available some prior information about the underlying cluster structure of the data in the form of constraints or a few labeled data [34]. Such prior knowledge generally arises from expert opinion, user feedback, or the needs of the problem owner [9]. Incorporating such background knowledge into the clustering process and hence allowing the user to guide the process in a manner toward a "better" partitioning of the data is known as constrained clustering and has been the subject of extensive research recently.

Constrained clustering has been used in several domains and applications. In gene clustering based on gene expression data obtained with DNA microarrays, databases of co-occurrence data have been used to generate constraints forcing that certain genes must be in the same cluster (must-link constraints) [25, 56, 64].

In some agricultural areas, each farmer cultivates a large number of small and dispersed land parcels. This has several disadvantages. A solution to this problem would be land consolidation which refers to the process that the farmers surrender their dispersed parcels in order to receive a more continuous equivalent land area. Land consolidation is clearly a clustering problem with several constraints. Firstly, neighboring parcels should be assigned to a farmer. Secondly, the total land area of a farmer should not change much after consolidation. Thirdly, the quality of soil of each farmer's land should also not change by too much. There are other constraints about the geometry of the land a farmer receives, such as it should not be a continuous long but narrow land [11].

In text clustering, the goal is to automatically categorize a large number of text documents into smaller and manageable groups (clusters) based on their content. The user may specify that some documents should be clustered into the same group as they have similar contents or have the same authors (must-link constraints) and/or some documents should be separated from each other due to the differences in their subjects (cannot-link constraints) [37].

For more applications of constrained clustering, we refer the reader to the survey written by Davidson and Basu [18].

Methods proposed for constrained clustering can be divided into three categories; namely search based (also known as constraint based), distance based (also known as similarity based), hybrid (also known as search and distance based) methods [71]. In search based methods, clustering algorithms are modified to incorporate the prior knowledge into the clustering task. In other words, the solution space to be searched is adjusted according to the constraints. Common techniques in search based methods are modifying the objective function by adding penalty terms for unsatisfied constraints, enforcing constraints to be satisfied and using prior knowledge to initialize clusters.

In distance based methods, an existing clustering method is generally used but the distance measure of the method is modified in accordance with the prior knowledge. The distance measure is adjusted in such a way that data objects that should be placed in the same cluster will be closer to each other while data objects that should be placed in different clusters will be farther away from each other. Hybrid methods integrate search based and distance based methods. They benefit from the advantages of both and generally perform better than the individual methods.

In constrained clustering, the prior knowledge can appear in the form of labeled data or constraints. In the former one, labels of some of the data objects are known, but the amount of available information (number of known labels) may not be sufficient to perform classification. In such a case one can perform clustering instead of classification incorporating the known labels in some way to the clustering procedure. Instead of having labeled data, there may be some constraints on the data objects or clusters. For example, there may be constraints on pairs of data objects in the form of pairwise must-link and cannot-link relations. This type of knowledge is generally more practical. Getting the true labels of the data objects may require too much effort or may be costly, while whether pairs of data objects belong to the same cluster or different clusters can be easily specified by an expert.

Constraints can be considered as hard or soft. If they are considered as hard constraints, they should be satisfied in the final partition of the data objects. A problem with using such constraints is that there may not be any partition at all satisfying all the constraints. In the soft constraint case, however, constraint violations are allowed with associated violation costs.

Constraints encountered in clustering problems can also be categorized as instance-level and cluster-level constraints. Instance-level constraints are pairwise must-link and cannot-link constraints on some pairs of data objects. A must-link constraint enforces that two objects must be placed in the same cluster. On the other hand, a cannot-link constraint enforces that two objects cannot be placed in the same cluster. Must-link constraints form an equivalence relation. The resulting equivalence classes are sometimes called as chunklets [58]. Each chunklet consists of data objects that are known to be in the same cluster. Hence the data can initially be partitioned into chunklets in the presence of must-link constraints.

It should be noted that prior information in the form of pairwise constraints is weaker than the prior information in the form of labeled data [40]. While labeled data can easily be transformed into pairwise constraints, the labels of data objects cannot be inferred from pairwise constraints.

In addition to instance-level constraints, there can be several other constraints in a constrained clustering problem. Balancing constraints would force the sizes of the clusters to be comparable. More generally, given data objects with associated positive weights, one may restrict the total weight of the data objects in each cluster (this is known as the capacitated clustering problem). In particular, if all the weights are one, then the size of each cluster is restricted. There could be constraints that put lower or upper bounds on the radii of the clusters. Given an initial clustering of the data, one may want to obtain a clustering that is "different" from the initial one. After this point, we will call all such constraints, i.e., all the constraints different from instance-level constraints, as cluster-level constraints.

3.1 Constrained Clustering with Labeled Data

3.1.1 Search Based Methods

Constrained clustering with limited number of labeled data can be considered as a multi-objective optimization problem with objectives of maximizing intra-cluster similarity, maximizing inter-cluster dissimilarity and minimizing cluster impurity which is a measure of the consistency between the partition and the prior knowledge (labels).

Basu et al. [4] develop two variants of the K-means algorithm for the clustering problem with labeled data; namely, Seeded-K-means and Constrained-K-means. In the Seeded-K-means algorithm, labeled data is used only in the initialization step of the K-means algorithm. The authors assume that at least one labeled data is available for each of the k clusters. By partitioning the labeled data, k groups are obtained and initial cluster centers are computed by taking the mean of only the labeled data objects in each cluster. Then, assignment and update steps are repeated until convergence. In the Seeded-K-means algorithm, labels are considered as soft constraints. This is because as the labels are not taken into account after the initialization, they are subject to change in later steps. In the Constrained-K-means algorithm, labeled data are considered both in the initialization step and in the assignment steps of the K-means algorithm. The initialization step of the Constrained-K-means is the same as that of the Seeded-K-means. In the Constrained-K-means, only unlabeled data objects are reassigned during the assignment step, but the clusters of labeled data objects are never changed after the initialization. So labels are considered as hard constraints here. When the initial labels are noise free, i.e., there are not inconsistencies in prior knowledge, Constrained-K-means can be preferred. If there is noise in initial labels, then Seeded-K-means may be helpful in reducing the effect of noise by allowing the clustering algorithm to change initial labels. Experimental studies show that both of the proposed algorithms perform better than the K-means algorithm. Also, the algorithms outperform COP-K-means proposed in [63] (see Sect. 3.2.1 for COP-K-means). In his dissertation, Basu provides more discussion on both of the algorithms. For more details, see [3].

For the same problem, Demiriz et al. [22] proposed an objective function that is a linear combination of cluster dispersion (intra-cluster similarity and/or inter-cluster dissimilarity) and cluster impurity. As a cluster impurity measure, the authors used the Gini Index (see, for instance, [13]). As a dispersion measure, the authors considered two alternatives:

1. mean square error: this is the objective function used in the MSSC problem divided by the number of data objects,
2. Davies-Bouldin index [21]: this is an index of dispersion taking both intra-cluster similarity and inter-cluster dissimilarity into account.

As the resulting objective functions are highly nonlinear with many local optima, the authors propose a genetic algorithm (GA), a nature-inspired metaheuristic, for the problem. In their algorithm, authors use operators (crossover, mutation, etc.) from GAlib which is a general purpose GA library instead of defining new GA operators. Besides operators used, another main aspect in GAs is the representation of a solution. In this study, authors used a $k \times d$ matrix in which each row contains the values of the d coordinates of the corresponding cluster center. Since this representation does not depend on n, the proposed GA is scalable to large data sets. Based on the experimental studies, it is shown that constrained clustering performs better than unsupervised clustering. When dispersion measures are investigated separately, using Davies-Bouldin index results in better performance than mean square error since it finds more compact and well-separated clusters.

3.1.2 Distance Based Methods

Constrained clustering problem with labeled data can also be handled in two stages instead of considering it as a multi-objective optimization problem. In the first stage, cluster impurity is considered. In this stage, data objects are transformed into a new space using the prior knowledge on hand. This transformation is done in such a way that objects having the same label will be closer to each other and objects with different labels will be farther away from each other in the new space. This first stage is called as distance metric learning. After the metric is learnt, cluster dispersion is considered in the second stage where traditional clustering algorithms are generally employed.

In [70], authors propose a parametric distance learning method for the clustering problem with labeled data. Data objects are transformed into a new Euclidean space in which the Euclidean distance between two data objects shows the dissimilarity between them. This transformation is performed in such a way that the square of the Euclidean distance between data objects provided with labels in the new space will be equal to the dissimilarity of them in the original space which is calculated as

$$\delta_{ij} = \begin{cases} 1 - e^{-\frac{\|x_i - x_j\|^2}{\beta}} & \text{if labels of } x_i \text{ and } x_j \text{ are the same ,} \\ 1 + e^{-\frac{\|x_i - x_j\|^2}{\beta}} & \text{otherwise,} \end{cases} \tag{4}$$

where $\beta > 0$ is a width parameter and δ_{ij} is a measure of dissimilarity between x_i and x_j in the original space. Such a transformation can be achieved by multidimensional scaling method. As this method may be intractable for large data sets, authors find a regression mapping from the original space to the new space using the prior knowledge (labeled data) to be able to map the unlabeled data objects into the new space. In the new space, the authors then used the K-means algorithm for clustering. Experimental studies show that the method they propose outperforms the traditional K-means algorithm.

3.2 Constrained Clustering with Instance-Level Constraints

3.2.1 Search Based Methods

Clustering with constraints can be considered as a multi-objective optimization problem as in the case with clustering in the presence of labeled data. In addition to cluster dispersion, a measure of constraint violation is used as another objective instead of using cluster impurity. If constraints are seen as hard constraints, then the constraint violation is set to zero.

Wagstaff and Cardie [62] study the clustering problem with must-link and cannot-link constraints. They consider the constraints as hard constraints. To enforce the constraints to be satisfied, authors propose a modified version of COBWEB, called as COP-COBWEB. The main difference of their algorithm from COBWEB is that COP-COBWEB is a partitional clustering algorithm that would return a single level of the dendrogram in COBWEB (level after the root node) rather than a hierarchy in the absence of constraints.

The steps of COP-COBWEB can be described as follows. When a new data object, x_i, arrives, first a must-link check is done. If there is a data object, x_j, that is in one of the existing clusters and that has to be in the same cluster with x_i, then x_i

is placed into the cluster where x_j belongs to and the category utility of the resulting partition is recorded (call it as CU_{ML}). Then, the split operator is applied to the cluster that x_i now belongs to and let CU_{S1} be the resulting category utility. If CU_{ML} is greater than CU_{S1}, then do not split and go to the next data object and otherwise do the splitting and go to the next data object.

If x_i is not inserted into a cluster in the must-link check step, then x_i is inserted in every possible cluster after a cannot-link check. The best two resulting clusters (in terms of the category utility) are recorded (call the resulting category utilities as CU_{I1} and CU_{I2}, respectively). These two clusters are merged if possible after a cannot-link check and the category utility of the resulting partition is recorded (let it be CU_M). Also with the create operator a new cluster is formed containing only x_i and the category utility of the resulting partition is recorded (say CU_C). Furthermore try splitting the best partition obtained with the insert operator that had the category utility of CU_{I1} and record the category utility of the resulting partition (say CU_{S2}). Choose the partition having the largest category utility out of CU_{I1}, CU_M, CU_C, and CU_{S2} and go to the next data object.

Authors experimentally show that including even a small number of constraints into the clustering process increases the clustering accuracy and decreases the running time since it reduces the solution space to be searched. Also, authors discover which constraint type is more beneficial for which type of clustering problem. For example, they observe experimentally that cannot-link constraints result in better accuracy for the data sets for which unsupervised clustering leads in less number of clusters than the true number of clusters. On the contrary, for the data sets for which unsupervised clustering leads in too many or true number of clusters, including must-link constraints is better than including cannot-link constraints.

In [63], authors consider the same problem considered in [62]. They modify the K-means algorithm. Authors take the transitive closure over constraints and used the whole derived set of constraints. For example, if x_i must-link to x_j which cannot-link to x_k, then we know that x_i cannot-link to x_k and this is added to the constraint set. Also if x_i must-link to x_j which must-link to x_k, then we know that x_i must-link to x_k as well and again this is added to the constraint set.

The algorithm proposed in this study, called as COP-K-means, uses the same initialization and update steps of the K-means algorithm but differs from it in the assignment step. In the assignment step, the data objects are considered one by one. For each data object x_i, the closest appropriate cluster is found and the object is placed in that cluster. A cluster is appropriate for a data object x_i, if

1. there exists a data object in the cluster that must-link to x_i, and
2. there does not exist any data object in the cluster that cannot-link to x_i.

Note that the assignment step of the algorithm is order-dependent. If the algorithm cannot find an appropriate cluster for any x_i, then the algorithm stops with no partition to return. The running time of a single iteration of the algorithm is $O(ndk + n^2)$ where n is the number of data objects, d is the number of features (dimension) each data object has, and k is the number of clusters. Authors compare performance of COP-K-means with COP-COBWEB developed in [62]. For data sets for which k is known, both of the algorithms result in similar amount of improvement in clustering accuracy over the original K-means. For data set for which k is unknown, COP-K-means is quite better than the original K-means in terms of clustering accuracy. Also, COP-K-means is better than the original K-means in determining the correct value of the number of clusters which is found by solving both algorithms for many times with different random initializations for different k values.

In her dissertation, Wagstaff provides more discussions on COP-COBWEB and COP-K-means [61]. In addition to these two algorithms which incorporate pairwise must-link and cannot-link constraints as hard constraints into the clustering process, Wagstaff proposes another algorithm which handles pairwise constraints as soft constraints. This algorithm is a modified version of the K-means and called as SCOP-K-means which incorporates violation costs into the objective function to be minimized in the assignment step. Initialization and update steps of the algorithm are the same as the K-means. In the assignment step, each data object is assigned to the cluster which contributes to the objective function the least. Experimental studies show that SCOP-K-means outperforms K-means in terms of accuracy.

Basu et al. in [6] propose PC-K-means, a modified version of the K-means, for clustering with pairwise must-link and cannot-link constraints. Similar to [61], authors consider the constraints as soft constraints. In the algorithm, prior knowledge is used in the initialization step and in the assignment step through a modified objective function. The objective function of the PC-K-means includes constraint violation terms in addition to the objective function of the MSSC problem. To initialize cluster centers, chunklets of size ≥ 2 are used. If the number of such chunklets is equal to k, the centers of them are used as initial centers. If the number of such chunklets is greater than k, then the largest k of them (in terms of size) are used in the initialization. If the number of such chunklets is less than k, then using all the chunklets less than k centers are initialized. The remaining centers are initialized randomly (with an exception). In the assignment step, each data object is assigned to the cluster minimizing the contribution to the objective function. This step is highly order-dependent. Authors considered random ordering of the objects in the assignment step in their computational experiments. Update step of the algorithm is the same as that of the original K-means. Authors prove that PC-K-means converges to a local optimum.

The authors also propose a two-phased method for actively selecting informative constraints. Their method uses farthest-first traversal scheme which aims to select a number of data objects that are far from each other. The active learning method proposed assumes that we can ask a given number of queries, say Q, to a noiseless oracle where the input is a pair of data objects and in return the oracle states whether there is a must-link or cannot-link constraint between the objects. In the first phase, called as the explore phase, the algorithm starts with selecting a data object at random and placing it in the first neighborhood. At the beginning of each explore step, we have a certain number of non-empty neighborhoods. At this point, the algorithm finds the data object that is farthest from the data objects so far traversed. Then by querying, the new data object is placed in an existing neighborhood if at any time a must-link constraint is returned by the oracle. Otherwise, i.e., the new object is cannot-linked to all existing neighborhoods, it is placed in a new neighborhood. The explore phase continues until k neighborhoods are formed or all Q queries are used up. In the latter case the number of neighborhoods formed may be less than k. In the former case, however, if there are still left some unused queries, the algorithm moves to the second phase, called as consolidate phase.

At the beginning of the consolidate phase we have k neighborhoods and certain number of queries to be used. The consolidate phase starts with estimating the centers of the neighborhoods. Then a data object that is not in any of the neighborhoods is selected at random and the distance between this new object and the centers of the neighborhoods are computed. Starting with the closest center, queries will be formed by taking a point from the neighborhood. In at most $(k-1)$ queries, the new object is placed in one of the neighborhoods.

Experimental studies show that the learning curve for the clustering process is significantly steeper with the proposed active learning method. In other words, number of constraints to be included in the clustering process to improve accuracy is much smaller with the proposed method in comparison with random selection of the constraints. It is also experimentally shown that the consolidate phase makes the clustering accuracy even better when compared with the active learning method that includes only the explore phase.

Basu [3] uses the concept of hidden Markov Random Field (HMRF) to solve the same problem considered in [6, 61]. An HMRF is a Markov Random Field in which some of the random variables are hidden (unobservable). A Markov Random Field consists of an undirected graph whose nodes represent random variables and a set of potential functions taking maximal cliques on the graph as input and returning non-negative real numbers. Edges between two random variables represent the dependencies between them. Generally, an HMRF model consists of an observable set of random variables, an unobservable set of random variables, an unobservable set of model parameters, and an observable set of constraints. For the clustering problem with constraints, these components are data objects to be clustered, labels of the data objects, cluster centers and pairwise constraints, respectively. Taking advantage of the resemblance between HMRF models and constrained clustering, the objective function used in this study is defined in the framework of HMRF. To solve the problem the author proposes an iterative K-means like algorithm, called as HMRF-K-means. Initialization step of the algorithm is the same with that of the PC-K-means [6]. After the initialization, the assignment step takes place. In the original K-means, assigning each data point to nearest cluster center minimizes the objective function. But in the framework of HMRF, cluster centers are dependent to each other and assigning data points to clusters is computationally intractable. In the assignment step, the author uses a greedy iterated conditional modes (ICM) approach proposed by Zhang et al. in 2001 [69]. Data points are assigned to clusters that minimize the contribution to the objective function in random order. After all objects are reassigned, the process of random reassigning is repeated until no change occurs in assignments for the two successive iterations. In addition to the ICM method, the author tries two global methods for the assignment step; namely, belief propagation and linear programming relaxation. These methods outperform greedy method when the amount of prior knowledge is small but as this amount increases global methods become computationally expensive and quality of greedy and global methods become comparable. So, greedy method is a wise choice for the assignment step in such cases. Based on the data points assigned to clusters, cluster means are updated. It is proved that the algorithm converges to a local optimum. Experimental studies show that the HMRF-K-means outperforms the original K-means.

Yu and Shi [68] consider the inclusion of must-link constraints in image segmentation and propose a graph-theoretic solution method. In the method, each data object (pixel) is considered as a node in a graph. The weight associated with each edge represents the similarity between the nodes connected by the edge. The authors then formulate the image segmentation problem with must-link constraints as a node partitioning problem in which edges within a partition (cluster) should have a high total weight (high similarity) and edges across partitions should have a low total weight (low similarity). Normalized cuts criterion proposed by Shi and Malik in 2000 [59] considers both of these objectives. By using this criterion the authors model the problem as a constrained optimization problem. Relaxing the discrete assignment constraints, authors showed that the problem resulted in a constrained eigenvalue problem. Authors then find the optimal solution of the constrained eigenvalue problem by eigenvalue decomposition from which a near global optimum solution is obtained for the discrete problem. The algorithm proposed works reasonably well with enough number of must-link constraints. On the other hand, when there are only a few such constraints, the final partition may not be as it is desired. In this case, considering the constraints as soft, the authors modify their algorithm that produces a better partitioning of the data.

In [20], Davidson and Ravi consider the clustering problem in the presence of must-link and cannot-link constraints. They propose a modified version of the K-means algorithm by adding constraint violation terms to the objective function to penalize violated constraints. Their algorithm called as the constrained vector quantization error (CVQE) algorithm therefore considers the instance-level constraints as soft constraints. Initialization step of the CVQE algorithm is the same as that of the K-means. In the assignment step, each pair of data objects that form a constraint are assigned to the clusters in such a way that the objective function is minimized. This takes $O(k^2)$ time per constraint. The remaining data objects are assigned as in the K-means. In the update step, the update rule is computed by taking the first derivative of the objective function and setting it to zero. Computational studies show that inclusion of constraints results in better clustering accuracy and faster convergence.

The advantage of algorithms considering constraints as soft over algorithms considering constraints as hard is that the former usually better handles noisy constraints. Algorithms in which the constraints are taken as hard may not find a feasible partition if there is noise with the constraints. While algorithms in [62, 63] may end up with no partition at all, the algorithms in [3, 6, 20, 61, 68] find a partition in every case with some amount of constraint violations.

The authors in [51] consider the use of instance-level constraints in an agglomerative hierarchical clustering algorithm. The authors first form the chunklets by using the must-link constraints. This initial partition of the data is used in the initial step of their algorithm. Rather than starting with clusters consisting of individual data objects, the authors start with the chunklets and hence making sure that the must-link constraints are all satisfied. As the initial clusters are never split, this makes sure that in the final partition, all of the must-link constraints will be satisfied (so they are considered as hard constraints). An important aspect of agglomerative hierarchical clustering algorithms is the way the distance is measured between the clusters. The distance function used by the authors consists of two terms: a cluster dispersion measure and a constraint violation term for the cannot-link constraints. They use two different dispersion measures; namely the centroid method and the Ward method. The centroid method measures the distance between two clusters as the distance between their centers. Given a cluster c_i and its center μ_i, let $E(c_i) = \sum_{x \in c_i} \|x - \mu_i\|^2$. In the Ward method, the distance between two clusters c_i and c_j is computed as $E(c_i \bigcup c_j) - E(c_i) - E(c_j)$. Starting with the chunklets, the algorithm proposed combines two closest clusters until the desired number of clusters is obtained. The authors also compare the results from the centroid method and the Ward method with and without the Gaussian kernel. They show computationally that their methods outperform the COP-K-means algorithm.

3.2.2 Distance Based Methods

Similar to the clustering problem with labeled data, the clustering problem with instance-level constraints can be handled in two stages. In the first stage, a new distance metric which brings must linked data objects closer and pushes cannot linked data objects apart is defined. A measure of dispersion as a function of this newly defined distance metric is then used in the second stage for clustering.

Klein et al. [40] state that just using the provided instance-level constraints in learning a distance metric that will help to reveal the desired partition of the data may result in missing some spatial information. If there is must-link between x_i and x_j then the data objects at the neighborhoods of x_i and x_j should also be put into the same cluster intuitively. The authors propose a distance measure reflecting both spatial information that instance-level constraints imply and the provided instance-level information. With this distance measure, it is aimed that must-linked objects will be close to each other while cannot-linked objects will be far from each other. Also, with the new distance measure, if x_i and x_j are close to each other, a data object close to x_i will be enforced to be close to x_j, and if x_i and x_j are far apart, a data object close to x_i will be imposed to be far from x_j. The authors use a complete-link hierarchical agglomerative clustering which takes a proximity matrix showing the pairwise proximities of data objects and merges two clusters based on their proximities. To incorporate must-link constraints and the spatial information implied by them into the proximity matrix, the authors run all-pairs-shortest-paths algorithm on the original proximity matrix. By this way, a new proximity matrix that is faithful to the original proximity matrix is obtained. Then, to include the cannot-link constraints, the distance between cannot-linked objects is set to the maximum distance between two data objects in the problem plus 1. The algorithm called as constrained complete-link, CCL, starts with all data objects in different clusters and merges clusters in order of proximities until one cluster including all data objects is obtained. Experimental studies show that CCL outperforms the COP-K-means [63] which does not consider spatial information implied by pairwise constraints.

To obtain a relevant partition of the data objects, distance metric used in the clustering process should reflect the relationships between data objects very well. If two data objects are stated as similar (i.e., there is a must-link constraint between them), the distance between these objects should be small while if they are stated as dissimilar (i.e., there is a cannot-link constraint between them), the distance between them should be high. To find a distance metric satisfying these relationships, Xing et al. [65] propose the following convex optimization model.

$$\text{minimize}_A \sum_{(x_i,x_j)\in ML} (x_i - x_j)^T A(x_i - x_j)$$

$$\text{subject to} \qquad\qquad\qquad\qquad\qquad\qquad\qquad\qquad (5)$$

$$\sum_{(x_i,x_j)\in CL} \sqrt{(x_i - x_j)^T A(x_i - x_j)} \geq 1,$$

where A is a positive semi-definite matrix, ML is the set of must-link constraints, and CL is the set of cannot-link constraints. Finding a positive-semi-definite matrix A by solving the optimization problem above is equivalent to mapping data objects $x_i, i = 1, 2, \ldots, n$ to $y_i = A^{1/2} x_i, i = 1, 2, \ldots, n$ and using the Euclidean distance between y_i's. If A is restricted to be the identity matrix, then the final distance metric is the standard Euclidean norm. If it is restricted to be diagonal, then this means that different weights are given to different dimensions (features). In this case, the above model can be solved by the Newton–Raphson method. If A is a full matrix, authors propose an efficient method using gradient ascent and iterative projections. With the transformed objects in the new space, authors use the COP-K-means [63]. Experimental results show that using such a mapping in the COP-K-means outperforms both traditional K-means and the COP-K-means. Also, as it is expected, increasing the amount of prior knowledge leads to a better distance metric learning and so better clustering accuracy is achieved.

In [39], Kamvar et al. study the clustering problem involving labeled data and/or instance-level constraints. Their problem can be considered as a clustering problem with just instance-level constraints as they transform the labeled data (if there exist any in the problem) into the instance-level constraints. They propose a spectral clustering method for the problem. In spectral clustering methods, the input is an $n \times n$ similarity matrix, where n is the number of data objects to be clustered. Each entity of this matrix is a value between 0 and 1 showing the amount of pairwise similarity between corresponding data objects. By using eigenvalues and eigenvectors of the normalized similarity matrix, blocks corresponding to the clusters are found in the similarity matrix. In this study, the authors incorporate instance-level constraints into the similarity matrix by replacing similarity entity corresponding to the two must-linked data objects with 1 and two cannot-linked data objects with 0. The authors then find the k largest eigenvectors of the normalized similarity matrix (say e_j, $j = 1, 2, \ldots, k$). After the normalization of the rows of the $n \times k$ matrix $Y = [e_1 \ e_2 \ \ldots \ e_k]$, each row of Y is considered as a data object in R^k. To cluster these transformed data objects, the authors use the traditional K-means algorithm. The empirical results show that the proposed spectral clustering method performs better than the traditional K-means algorithm. It should be noted that the main purpose of the paper is classification. After the clusters are found by the proposed spectral learning method, each data object attains a label based on the prior information. Then, class labels are determined according to the labels of majority in the classes.

Bar-Hillel et al. [2] propose a distance learning method for the clustering problem with must-link constraints. The authors use relative component analysis (RCA) proposed by Shental et al. in 2002 [57]. RCA tries to find reasons of the unwanted variability present in the data and decrease that variability by transforming the data objects into a new space in which the desired structure in the data is more apparent. This transformation is done in such a way that high weights are given to the relevant dimensions (features) while low weights are assigned to the irrelevant dimensions. The relevant dimensions are determined by using chunklets. Given the set of must-link constraints (say ML), RCA starts with calculating the center of each chunklet of size ≥ 2 and computing the covariance matrix C of all centered data objects in these chunklets. After the transformation of the data objects into a new space by whitening transformation, $C^{-1/2}$, COP-K-means [63] is performed. Authors compare their method with the traditional K-means, COP-K-means without any distance learning, COP-K-means with distance learning proposed in [65], expectation-maximization (EM) and constrained EM. The proposed algorithm significantly outperforms the K-means. Also, it performs similar to or better than COP-K-means with distance learning method proposed in [65].

Similar to [2], Chang and Yeung [16] address the constrained clustering problem with only must-link constraints. They propose a distance metric learning method transforming the data objects into a different space. For each pair of data objects, (x_r, x_s), in the set ML which is the set of all pairwise must-link constraints, the vector $(x_s - x_r)$ is linearly transformed to $A_r(x_s - x_r) + c_r$ where A_r is a $d \times d$ matrix and c_r is a d dimensional vector (d is the number of features). Then, every data object, x_i, in the neighborhood of x_r is transformed by the following formula:

$$y_i = x_i + (A_r - I)x_i + b_r, \tag{6}$$

where $b_r = (I - A_r)x_r + c_r$. The authors assume that a data object may belong to more than one neighborhood. So, the transformation formula becomes

$$y_i = x_i + \sum_{(x_r,x_s)\in ML} I_{ri}((A_r - I)x_i + b_r), \tag{7}$$

where I_{ri} is an indicator function showing whether x_i belongs to the neighborhood of x_r or not. The problem here is to estimate $|ML|$ different A_r and c_r. In other words, $O(|ML|d^2)$ transformation parameters are required. To decrease this amount to $|ML|d$, authors use Gaussian neighborhood functions. They try to minimize the sum of squared Euclidean distances between must-linked data objects in the new space and penalty for the degree of transformation. To solve this nonlinear objective function of the distance learning process, authors use an iterative procedure. In the initialization step of the procedure, y_i's are set to x_i's. In each iteration, transformation parameters are optimized for given x_i's and y_i's are then updated by using newly found transformation parameters. Procedure continues until convergence is achieved. To solve the optimization problem in each iteration, authors use two methods: gradient method and iterative majorization method. After transforming data objects into the new space, authors apply K-means and compare their method with traditional K-means, COP-K-means [63] and K-means with distance learning method proposed in [2]. In the experimental studies, the effectiveness of the method is verified as it outperforms the other methods for most of the data sets tried.

3.2.3   Search and Distance Based Methods

Search and distance based methods are hybrid methods. They benefit from the advantages of both search based and distance based methods and generally perform better than the individual methods. Lately, noticing this potential, some authors proposed such hybrid methods.

Basu et al. [5] propose an algorithm called MPC-K-means which is the unification of PC-K-means [6] and a distance learning method. There are two components of the objective function of the PC-K-means; namely, constraint violation terms and the objective function of the MSSC problem. The constraint violation terms are directly proportional with the "seriousness" of the violation. For example, putting must-linked data objects close to each other into different clusters is more serious than putting must-linked data objects that are relatively far from each other into different clusters. The distance calculation in the objective function is parameterized with a positive-definite matrix A which corresponds to feature weighting as follows:

$$\sum_{j=1}^{k}\sum_{x_i\in c_j}\left\|x_i - \mu_j\right\|_A^2 = \sum_{j=1}^{k}\sum_{x_i\in c_j}(x_i - \mu_j)^T A(x_i - \mu_j), \tag{8}$$

where μ_j represents the center of the jth cluster. By this way, the distance metric learning is incorporated into the clustering process. In most of the studies using distance metric learning, distance metric is determined before the clustering algorithm by using only the prior knowledge. In this study, data for which there is no prior information also affects the metric learning process. To initialize cluster centers, the method used in [6] is followed. In the assignment step of the algorithm, data objects are assigned to the clusters that result in least contribution to the objective function. Then, as in K-means, cluster centers are updated by taking average of the data objects assigned to them. Moreover, positive-definite matrix used in distance calculation is updated by taking partial derivatives of the objective function and setting them to zero. The assignment and update steps alternate until convergence. In the experimental studies, authors ablate MPC-K-means as M-K-means that only uses metric learning and does not consider violation terms in the objective function and PC-K-means that only considers pairwise constraint violations in the objective function and does not learn a distance metric. They compare performances of MPC-K-means, PC-K-means, M-K-means, and traditional K-means. Results show that supporting search based method with distance metric learning results in the best clustering accuracy. In other words, MPC-K-means outperforms PC-K-means and M-K-means. But, since distance learning task requires considerable amount of prior knowledge, search based methods (like PC-K-means) are better suited when there is a small amount of prior knowledge. However, if there is enough prior knowledge to learn a distance metric, unified approaches become favorable. By ablating MPC-K-means, authors had a chance to compare search based and distance based methods for constrained clustering in the presence of instance-level constraints to discover their relative strengths and weaknesses.

In [10], Bilenko et al. make minor modifications on the MPC-K-means [5]. While penalty terms included in the objective function for constraint violations are the weights associated with the constraints in [5], the penalty terms are defined as functions in this study. Besides that, in experimental studies, authors compare their algorithm with Supervised-K-means in which the assignment step is supervised in the light of the available constraints in addition to comparisons made in [5]. Results show that the updated MPC-K-means outperforms all of the methods including Supervised-K-means.

Basu et al. propose another hybrid method [7]. Different than their previous work which is developed for just Euclidean distance and presented in [5], the method in this study can be used for any Bregman divergence. Also, it can be used for directional similarity measures. As in [3], authors define the objective function of the problem in the framework of HMRF. The potential function used in the objective function includes penalty terms for must-link and cannot-link constraint violations. These penalty terms reflect the aim of learning the underlying distance measure. For example, if two cannot-linked data objects are near to each other with respect to the distance measure in use and are put into the same cluster, the distance measure should be modified to put these objects far from each other. After the modelling, to solve the problem author proposes an iterative K-means like algorithm. The initialization step of the algorithm is the same with that of HMRF-K-means [3]. In the assignment step, each data object is assigned to the cluster minimizing the contribution to the objective function of the algorithm. In the update step, in addition to the update of the cluster centers based on the data points assigned to them, the distance measure is modified to reduce the objective function. It is proved that the algorithm converges to a local optimum. Authors explain the details of the steps of their algorithm for two distance measures; namely cosine similarity and Kullback–Leibler divergence. Moreover, they give details of the steps of their algorithm for the squared Euclidean distance in [8]. Experimental studies show that clustering with constraints results in better clustering accuracy and the proposed algorithm outperforms the original K-means algorithm.

The methods incorporating distance learning generally learn the distance in the light of prior knowledge before the clustering process. The distance metric learnt is then used without any modification during clustering. In [5, 7, 8, 10], authors incorporate distance metric learning into the clustering process that modifies the distance metric in each iteration.

Law et al. [42, 43] also address the problem of clustering with instance-level constraints. Most of the methods mentioned so far consider the constraints as correct and consistent. In this study, authors consider the constraints as random variables between 0 and 1 which reflects the certainty of the constraints. If there is a must-link constraint between data objects x_i and x_j, the constraint that z_i and z_j (which are the cluster labels that x_i and x_j belongs to, respectively) are equal to l, $l \in \{1, 2, \ldots, k\}$ is given a probability. Having values between 0 and 1 for constraints actually means the use of soft constraints. If the probability that z_i and z_j are equal to l is set to 1, the must-link between x_i and x_j is considered as hard. Authors build a graphical model with these random variables. Since there are hidden variables (labels of data objects) in their model, they consider a missing data problem and propose an EM algorithm. In each maximization step, posterior probabilities are re-estimated. That re-estimation corresponds to the distance metric learning in the algorithm. Experimental studies show that the proposed method is beneficial even with small number of constraints. The method outperforms the hard constraint case and is more robust to noisy data as it allows constraint violations. Moreover, the proposed method performs better than the method with no constraints.

3.3 Constrained Clustering with Cluster-Level Constraints

Constrained clustering with cluster-level constraints did not receive too much attention as clustering with instance-level constraints in data mining community. Constraints like forcing the sizes of the groups (clusters) to be comparable and putting lower or upper bounds on the radii or diameter of the groups are generally used in facility location problems. A facility location problem is the problem of finding locations of a pre-determined number of facilities to serve the customers so as to optimize a certain objective like a function of distances between facilities and customers assigned to them. If we consider the facility locations as cluster centers, customers as data objects and the objective to be optimized in the facility location problem as a cluster distortion measure, facility location problem with such constraints and the clustering with cluster-level constraints are very alike. Interested readers are referred to [24] in which Drezner proposes two heuristics and an optimal algorithm for the p-center problem encountered in the facility location literature which can be considered as clustering with upper bounds on the radii of the clusters.

The methods proposed for clustering with cluster-level constraints are search based methods. A distance metric learning in such problems is not applicable as there is no aim of bringing some data objects closer to each other while pulling some others apart. In the paper by Davidson and Ravi [19], instance-level constraints and two types of cluster-level constraints are used in a search based agglomerative hierarchical clustering algorithm. The cluster-level constraints considered are

- δ-constraint that enforces a distance of at least δ between any two data objects that are not in the same cluster.
- ϵ-constraint that requires that for any data object if there is any other object in the same cluster, then there should be at least one object which is at most ϵ away from it.

The proposed algorithm starts with the chunklets formed using the must-link constraints as the initial partition of the data. For every pair of mergeable clusters (taking the constraints into account), distance is computed. Two closest clusters are then merged. Clusters are iteratively merged unless every pair of clusters is non-mergeable due to the constraints. At this point, the algorithm stops. The authors investigate the benefits of using constraints empirically for agglomerative hierarchical clustering. To improve the efficiency of agglomerative clustering, they also introduce a new constraint called as the γ-constraint which enforces that two clusters whose centers are more than γ away cannot be merged.

In addition, the authors explore the feasibility of agglomerative hierarchical clustering under combinations of the instance-level and δ, ϵ constraints. See Sect. 3.4 for the details.

In the constrained clustering literature, there is a body of research focusing on partitioning the data into clusters of approximately equal size. This has applications, for example, in direct marketing campaigns where one may want to partition customers into segments of roughly equal size so that same number of sales representatives can be allocated to each segment [67].

An approach for obtaining balanced clusters would be using an agglomerative hierarchical clustering algorithm and removing those clusters having a certain size from consideration during the remaining merge steps [1]. The authors in [1] state that this approach may significantly reduce the quality of the partition and does not scale well.

In [30], the authors consider clustering of sensor nodes in a distributed sensor network where each cluster has exactly one master node. To be able to evenly distribute the load on all the master nodes, they consider the problem of finding balanced clusters of sensor nodes while minimizing the total distance between the sensor nodes and master nodes. The problem is formulated as a minimum cost flow (MCF) problem and solved to optimality. In some sensor network applications, sensors in a cluster talk to each other. In such applications, the distance between all pairs of sensor nodes in a cluster should be less than a given threshold. The authors also considered this problem, where the maximum diameter of the clusters is minimized and proposed an algorithm for its solution. Here the diameter of a cluster is defined as the maximum distance between pairs of objects in the cluster.

The problem of finding balanced clusters is also considered in [1]. They propose an algorithm that is based on sampling of the data objects. The steps of their algorithm are

- sampling of a small representative subset of the data objects,
- clustering of the sampled data objects,
- populating and refining the clusters while keeping them balanced.

The authors prove the complexity of their framework and show its efficacy on several data sets including high-dimensional ones.

Balancing constraints can be seen as a special case of size constraints which put constraints on the size of each cluster (or on the total weight of the data objects in each cluster in the case with weights on the data objects). This is known as the capacitated clustering problem. Zhu et al. [73] consider the problem of clustering with size constraints where number of data objects in each cluster is known a priori. They propose a heuristic algorithm to transform the problem into a 0–1 integer linear programming problem.

The clustering problem with size constraints is also considered in [28]. Here the authors propose a modified K-means algorithm (initialization and the assignment steps are modified) that takes the size constraint of each cluster into account.

Bradley et al. [12] draw attention to a drawback of the K-means which may end up with poor local optima containing empty clusters or clusters with very few data points especially when it is used with high-dimensional data ($d \geq 10$) and high number of clusters ($k \geq 20$). To overcome this drawback, the K-means is generally initialized with different starting points and the best partition among different initializations is reported to the user. In this study, to prevent empty clusters or clusters with a few data object to appear, authors add a size constraint for each cluster. Their constraints enforce that c_j should have at least τ_j data objects in the final partition. Authors propose an iterative K-means like algorithm to solve the problem. The initialization and update steps of the algorithm are the same with those of the K-means. In the assignment step, algorithm solves the following optimization problem.

$$\text{minimize } J(C) = \sum_{j=1}^{k} \sum_{x_i \in c_j} T_{ij} \left\| x_i - \mu_j \right\|^2$$

subject to

$$\sum_{i=1}^{n} T_{ij} \geq \tau_j \quad \forall j \tag{9}$$

$$\sum_{j=1}^{k} T_{ij} = 1 \quad \forall i$$

$$T_{ij} \in \{0, 1\},$$

where T_{ij} takes value 1 if x_i is assigned to c_j or 0 otherwise and μ_j represents the center of c_j. This model can be considered as a MCF problem in which the data objects are supply nodes, cluster centers are demand nodes and distances between data objects and cluster centers are costs associated with arcs. Thanks to the unimodularity of the problem, relaxing the integrality constraints, leads in optimal solution in which $T_{ij} \in \{0, 1\}$. In the assignment step of the algorithm, above LP is solved by assuming that μ_j's are fixed. Experimental studies show that the constrained version of the K-means converges to better local optima than the original K-means for the same starting points. Also, it better summarizes the data. Other solution approaches and applications of the capacitated clustering problem can be found in [17, 29, 52].

In [60], Tung et al. consider a cluster-level constraint called as the existential constraint which can be thought of as a generalization of the constraints considered in [12]. Let X' be a subset of X, the set of all data objects. The elements of X' are called as pivot elements. An existential constraint enforces that each cluster c_j must include at least τ_j pivot elements. When $X' = X$ the existential constraint is the same as the size constraint considered in [12]. The algorithm proposed in the paper is a graph based algorithm having two phases: pivot moving and deadlock resolution. Authors prove that finding an optimal solution for both of the phases are NP-hard. Let's consider a graph whose nodes are all possible k-partitions of X, nodes satisfying existential constraints are called as valid nodes while others are called as invalid nodes. There is an edge between two nodes if and only if these two solutions differ in the assignment of just one pivot element. Algorithm starts with randomly selecting a valid node (feasible solution) and moves to a neighboring solution which has a refined partition. When there is no such neighbor, algorithm performs deadlock resolution phase to escape the local solution. As it is expected, when n is too high, the number of possible k-partitions of X will be too high. This means that the graph to be searched in the algorithm is of very large scale and efficient search may not be possible. Authors also propose a scaling method using micro-clustering methodology to overcome such situations. Micro-clustering reduces the size of the data by pre-clustering some of them into micro-clusters in which the maximum radius is constrained. The proposed algorithm for constrained clustering and the scaling method are both evaluated with a computational study.

Gonzales [32] considers the clustering problem that minimizes the maximum inter-cluster distance. He proposes an approximation algorithm which has $O(kn)$ time complexity where k is the number of clusters and n is the number of data objects. The algorithm consists of an initialization and $(k - 1)$ expanding steps. In the initialization step, all data objects are assigned to the first cluster, c_1, and one of them is selected randomly as head (center) of the cluster. In the hth expanding step, cluster c_{h+1} is constructed. The data object that has a maximal distance to its cluster center becomes the center of c_{h+1}. Then, every data object in previous clusters $\{c_1, \ldots, c_h\}$ for which distance to the cluster center they belong to is more than the distance to the center of c_{h+1} is moved from its cluster to c_{h+1}. Author proves that the algorithm guarantees to find a solution that is at most 2 times worse than the optimal solution. Moreover, it is proven that this upper bound 2 is the best possible approximation bound if $P \neq NP$.

The clustering algorithms discussed so far partition data objects by taking intra-cluster similarities and inter-cluster dissimilarities. Similarities and dissimilarities are based on features only. Contiguity constrained clustering takes the spatial information of the data objects into account in addition to feature information [45]. Contiguity constraints are usually handled in three ways:

- ignoring the contiguity information completely during the clustering and assessing the final partition by investigation,
- embedding the contiguity information into the similarity/dissimilarity matrix (or distances),
- utilizing a contiguity matrix, which must be consulted before merging the clusters.

Murtgah [54] provides a review of contiguity constrained clustering algorithms. He states that contiguity information is generally provided in two ways. In the first way, a binary $n \times n$ matrix where n is the number of data objects is used to describe contiguity. Here the entry of the matrix in the intersection of the ith row and jth column takes the value 1 if and only if x_i and x_j are contiguous. In the second way, a dissimilarity matrix involving continuous contiguity values are used instead of binary values. The main difference between the algorithms for clustering problems with contiguity information is in their clustering criteria. Author discusses algorithms under categorization based on their application areas. See [54] for more on the contiguity constrained clustering.

3.4 Feasibility Issues

In this section, we review the complexity of the clustering feasibility problem under constraints that is the problem of finding a feasible partition of the data satisfying all of the given constraints. Four types of constraints have been considered in the literature [19, 20, 40]; namely must-link, cannot-link, δ, and ϵ constraints. Two types of feasibility problems are considered:

- Given a value of k, does there exist a feasible partition of the data into k clusters satisfying all the given constraints?
- Given the constraints, does there exist a feasible partition of the data (into any number of clusters) satisfying all the constraints?

The complexity results are due to [19, 20, 40]. Table 1 summarizes the complexity of the clustering feasibility problem under different combinations of the constraints and is taken from [19].

In general, the feasibility problem with only must-link constraints can be solved in polynomial time whether k is given or not. The feasibility problem with only cannot-link constraints is polynomial time solvable when k is not given, but is NP-complete in general when k (≥ 3) is specified. The feasibility problem with only δ constraint or only ϵ constraint can be solved in polynomial time in both cases.

Table 1 Complexity of the clustering feasibility problem

Constraint combination	k given	k not given
Cannot-link	NP-complete [20, 40]	P [19]
Must-link and δ	P [20]	P [19]
Must-link and ϵ	NP-complete [20]	P [19]
δ and ϵ	P [20]	P [19]
Must-link, cannot-link, δ and ϵ	NP-complete [20]	NP-complete [19]

Note that, in general, when a combination of ϵ constraint and must-link constraints are given, the feasibility problem becomes NP-complete for a fixed value of k even though the problem is easy if only one of the constraint types is given.

3.5 Related Studies

Using constraints in classification problems as well as clustering problems has gained a lot of interest recently. In some classification problems, data labeling may be expensive although data collection is easy [41]. As in remote sensing, collecting images of Earth surface with high resolution is easy but stating whether there is water or snow in the images may not be so easy. But an expert can readily state that some of the regions are very alike and they should be labeled with the same label or some of the regions should be labeled with different labels since they are too dissimilar. In other words, having must-link or cannot-link constraints is usually easier than getting the labels. In such situations, semi-supervised classification, i.e., constrained classification can be performed.

As mentioned in Sect. 3.2.2, Kamvar et al. [39] use semi-supervised clustering for semi-supervised classification. Like [39], Lange et al. also use semi-supervised clustering for semi-supervised classification in [41].

Interested readers are referred to [33, 55, 72] which are good surveys on semi-supervised classification.

4 Conclusion

The need for reasonable grouping of data objects naturally appears in many domains like social sciences, computer science, business and marketing, and medicine. Therefore, clustering is very popular among scientist and analysts. However, traditional clustering methods may extract irrelevant information from data since they only use unlabeled data objects as input. This natural difficulty of clustering can be handled by incorporating some amount of prior knowledge into the process. In many domains, such a prior knowledge is available in the form of labeled data objects and/or constraints. In this chapter, we have addressed the clustering in the light of prior knowledge, reviewed some of the developments in the area of constrained clustering, and discussed the selected solution approaches from the literature. The studies in this area show that the inclusion of even a small amount of prior knowledge results in improvements of clustering performance and decrease in running time of the algorithms.

References

1. Banerjee, A., Ghosh, J.: Scalable clustering algorithms with balancing constraints. Data Min. Knowl. Disc. **13**(3), 365–395 (2006). doi:10.1007/s10618-006-0040-z
2. Bar-Hillel, A., Hertz, T., Shental, N., Weinshall, D.: Learning distance functions using equivalence relations. In: Proceedings of 20th International Conference on Machine Learning, vol. 3, pp. 11–18 (2003)
3. Basu, S.: Semi-supervised clustering: probabilistic models, algorithms and experiments. Ph.D. thesis, Austin, TX (2005). AAI3187658
4. Basu, S., Banerjee, A., Mooney, R.: Semi-supervised clustering by seeding. In: Proceedings of 19th International Conference on Machine Learning, pp. 19–26. Citeseer (2002)
5. Basu, S., Bilenko, M., Mooney, R.J.: Comparing and unifying search-based and similarity-based approaches to semi-supervised clustering. In: Proceedings of the ICML-2003 Workshop on the Continuum from Labeled to Unlabeled Data in Machine Learning and Data Mining Systems, pp. 42–49. Citeseer (2003)
6. Basu, S., Banerjee, A., Mooney, R.J.: Active semi-supervision for pairwise constrained clustering. In: Proceedings of the SIAM International Conference on Data Mining, vol. 4, pp. 333–344. SIAM (2004)
7. Basu, S., Bilenko, M., Mooney, R.J.: A probabilistic framework for semi-supervised clustering. In: Proceedings of 10th ACM SIGKDD International Conference on Knowledge Discovery and Data Mining, pp. 59–68. ACM (2004)
8. Basu, S., Bilenko, M., Banerjee, A., Mooney, R.J.: Probabilistic semi-supervised clustering with constraints. In: Chapelle, O., Schölkopf, B., Zien, A. (eds.) Semi-Supervised Learning, pp. 73–102. The MIT Press, Cambridge (2006)
9. Basu, S., Davidson, I., Wagstaff, K. (eds.): Constrained Clustering: Advances in Algorithms, Theory, and Applications. CRC Press, Boca Raton (2009)
10. Bilenko, M., Basu, S., Mooney, R.J.: Integrating constraints and metric learning in semi-supervised clustering. In: Proceedings of 21st International Conference on Machine Learning, pp. 11–18. ACM (2004)
11. Borgwardt, S., Brieden, A., Gritzmann, P.: Geometric clustering for the consolidation of farmland and woodland. Math. Intell. **36**(2), 37–44 (2014). doi:10.1007/s00283-014-9448-2. http://www.dx.doi.org/10.1007/s00283-014-9448-2
12. Bradley, P.S., Bennett, K.P., Demiriz, A.: Constrained k-means clustering. Tech. Rep., Microsoft Corporation (2000). http://www.machinelearning102.pbworks.com/f/Constrained KMeanstr-2000-65.pdf
13. Breiman, L., Friedman, J., Stone, C.J., Olshen, R.A.: Classification and Regression Trees. Chapman and Hall, New York (1984)
14. Celebi, M.E., Kingravi, H.A.: Deterministic initialization of the k-means algorithm using hierarchical clustering. Int. J. Pattern Recogn. Artif. Intell. **26**(07), 1250,018 (2012)
15. Celebi, M.E., Kingravi, H.A., Vela, P.A.: A comparative study of efficient initialization methods for the k-means clustering algorithm. Expert Syst. Appl. **40**(1), 200–210 (2013)
16. Chang, H., Yeung, D.Y.: Locally linear metric adaptation for semi-supervised clustering. In: Proceedings of 21st International Conference on Machine Learning, pp. 153–160. ACM (2004)

17. Chou, C.A., Chaovalitwongse, W.A., Berger-Wolf, T.Y., DasGupta, B., Ashley, M.V.: Capacitated clustering problem in computational biology. Comput. Oper. Res. **39**(3), 609–619 (2012)
18. Davidson, I., Basu, S.: A survey of clustering with instance level constraints. ACM Trans. Knowl. Discov. Data **1**, 1–41 (2007)
19. Davidson, I., Ravi, S.: Agglomerative hierarchical clustering with constraints: theoretical and empirical results. In: Jorge, A.M., Torgo, L., Brazdil, P., Camacho, R., Gama, J. (eds.) Knowledge Discovery in Databases: PKDD 2005, pp. 59–70. Springer, Berlin/Heidelberg (2005)
20. Davidson, I., Ravi, S.: Clustering with constraints: feasibility issues and the k-means algorithm. In: Proceedings of 2005 SIAM International Conference on Data Mining, pp. 138–149. SIAM (2005)
21. Davies, D.L., Bouldin, D.W.: A cluster separation measure. IEEE Trans. Pattern Anal. Mach. Intell. **1**(2), 224–227 (1979)
22. Demiriz, A., Bennett, K.P., Embrechts, M.J.: Semi-supervised clustering using genetic algorithms. In: Proceedings of Artificial Neural Networks in Engineering - ANNIE'99, pp. 809–814 (1999)
23. Dicken, L., Levine, J.: Applying clustering techniques to reduce complexity in automated planning domains. In: Proceedings of 11th International Conference on Intelligent Data Engineering and Automated Learning – IDEAL 2010, vol. 6283, pp. 186–193 (2010)
24. Drezner, Z.: The p-centre problem-heuristic and optimal algorithms. J. Oper. Res. Soc. **35**, 741–748 (1984)
25. Eisen, M.B., Spellman, P.T., Brown, P.O., Botstein, D.: Cluster analysis and display of genome-wide expression patterns. In: Proceedings of the National Academy of Sciences of the United States of America, vol. 95, pp. 14,863–14,868. National Academy Sciences (1998)
26. Fisher, D.H.: Knowledge acquisition via incremental conceptual clustering. Mach. Learn. **2**(2), 139–172 (1987)
27. Gan, G., Ma, C., Wu, J.: Data Clustering: Theory, Algorithms, and Applications (ASA-SIAM Series on Statistics and Applied Probability), SIAM, Philadelphia, ASA, Alexandria, VA (2007)
28. Ganganath, N., Cheng, C.T., Chi, K.T.: Data clustering with cluster size constraints using a modified k-means algorithm. In: Proceedings of 2014 International Conference on Cyber-Enabled Distributed Computing and Knowledge Discovery, pp. 158–161. IEEE (2014)
29. Geetha, S., Poonthalir, G., Vanathi, P.: Improved k-means algorithm for capacitated clustering problem. INFOCOMP J. Comput. Sci. **8**(4), 52–59 (2009)
30. Ghiasi, S., Srivastava, A., Yang, X., Sarrafzadeh, M.: Optimal energy aware clustering in sensor networks. Sensors **2**(7), 258–269 (2002)
31. Gluck, M.A., Corter, J.E.: Information uncertainty and the utility of categories. In: Proceedings of 7th Annual Conference of Cognitive Science Society, pp. 283–287. Lawrence Erlbaum (1985)
32. Gonzalez, T.F.: Clustering to minimize the maximum intercluster distance. Theor. Comput. Sci. **38**, 293–306 (1985)
33. Gordon, A.D.: A survey of constrained classification. Comput. Stat. Data Anal. **21**(1), 17–29 (1996)
34. Grira, N., Crucianu, M., Boujemaa, N.: Unsupervised and semi-supervised clustering: a brief survey. In: A Review of Machine Learning Techniques for Processing Multimedia Content, Report of the MUSCLE European Network of Excellence (2004)
35. Hansen, P., Mladenovic, N.: J-means: a new local search heuristic for minimum sum of squares clustering. Pattern Recogn. **34**(2), 405–413 (2001)
36. Howard, R.: Classifying a population into homogeneous groups. In: Lawrence, J.R. (ed.) Operational Research in the Social Sciences. Tavistock Publication, London (1966)
37. Huang, Y., Mitchell, T.M.: Text clustering with extended user feedback. In: Proceedings of 29th Annual International ACM SIGIR Conference on Research and Development in Information Retrieval, pp. 413–420. ACM (2006)
38. Jain, A.K.: Data clustering: 50 years beyond k-means. Pattern Recogn. Lett. **31**(8), 651–666 (2010)

39. Kamvar, S.D., Klein, D., Manning, C.D.: Spectral learning. In: Proceedings of 18th International Joint Conference of Artificial Intelligence, pp. 561–566. Stanford InfoLab (2003)
40. Klein, D., Kamvar, S.D., Manning, C.D.: From instance-level constraints to space-level constraints: making the most of prior knowledge in data clustering. In: Proceedings of 19th International Conference on Machine Learning, pp. 307–314. Stanford (2002)
41. Lange, T., Law, M.H.C., Jain, A.K., Buhmann, J.M.: Learning with constrained and unlabeled data. In: Proceedings of IEEE Computer Society Conference on Computer Vision and Pattern Recognition, vol. 1, pp. 731–738. IEEE (2005)
42. Law, M.H.C., Topchy, A., Jain, A.K.: Clustering with soft and group constraints. In: Fred, A., Caeli, T.M., Duin, R.P.W., Campilho, A.C., Ridder, D. (eds.) Structural, Syntactic, and Statistical Pattern Recognition, pp. 662–670. Springer, Berlin/Heidelberg (2004)
43. Law, M.H.C., Topchy, A.P., Jain, A.K.: Model based clustering with probabilistic constraints. In: Proceedings of 2005 SIAM International Conference on Data Mining, pp. 641–645. Citeseer (2005)
44. Likas, A., Vlassis, N., Verbeek, J.J.: The global k-means clustering algorithm. Pattern Recogn. **36**(2), 451–461 (2003)
45. Luo, Z.: Clustering under spatial contiguity constraint: a penalized k-means method. Tech. Rep., Department of Statistics, Penn State University (2001)
46. Lyman, P., Varian, H.R.: How much information 2000? (2000). http://www.groups.ischool. berkeley.edu/archive/how-much-info
47. Lyman, P., Varian, H.R.: How much information 2003? (2003). http://www2.sims.berkeley. edu/research/projects/how-much-info-2003
48. MacQueen, J.: Some methods for classification and analysis of multivariate observations. In: Proceedings of 5th Berkeley Symposium on Mathematical Statistics and Probability, vol. 1, pp. 281–297, Oakland, CA (1967)
49. Manning, C.D., Raghavan, P., Schütze, H.: Introduction to Information Retrieval. Cambridge University Press, Cambridge, England (2008)
50. Megiddo, N., Supowit, K.J.: On the complexity of some common geometric location problems. SIAM J. Comput. **13**(1), 182–196 (1984)
51. Miyamoto, S., Terami, A.: Constrained agglomerative hierarchical clustering algorithms with penalties. In: Proceedings of 2011 IEEE International Conference on Fuzzy Systems, pp. 422–427. IEEE (2011)
52. Mulvey, J.M., Beck, M.P.: Solving capacitated clustering problems. Eur. J. Oper. Res. **18**(3), 339–348 (1984)
53. Murtagh, F.: A survey of recent advances in hierarchical clustering algorithms. Comput. J. **26**(4), 354–359 (1983)
54. Murtagh, F.: A survey of algorithms for contiguity-constrained clustering and related problems. Comput. J. **28**(1), 82–88 (1985)
55. Pise, N.N., Kulkarni, P.: A survey of semi-supervised learning methods. In: Proceedings of 2008 International Conference on Computational Intelligence and Security, vol. 2, pp. 30–34. IEEE (2008)
56. Segal, E., Shapira, M., Regev, A., Pe'er, D., Botstein, D., Koller, D., Friedman, N.: Module networks: identifying regulatory modules and their condition-specific regulators from gene expression data. Nat. Genet. **34**(2), 166–176 (2003)
57. Shental, N., Hertz, T., Weinshall, D., Pavel, M.: Adjustment learning and relevant component analysis. In: Heyden, A., Sparr, G., Nielsen, M., Johansen, P. (eds.) Computer Vision—ECCV 2002, pp. 776–790. Springer, Berlin/Heidelberg (2002)
58. Shental, N., Bar-Hillel, A., Hertz, T., Weinshall, D.: Computing gaussian mixture models with em using equivalence constraints. In: Proceedings of the Advances in Neural Information Processing Systems 16, vol. 16, pp. 465–472. MIT Press (2004)
59. Shi, J., Malik, J.: Normalized cuts and image segmentation. IEEE Trans. Pattern Anal. Mach. Intell. **22**(8), 888–905 (2000)

60. Tung, A.K.H., Han, J., Lakshmanan, L.V.S., Ng, R.T.: Constraint-based clustering in large databases. In: Bussche, J.V., Vianu, V. (eds.) Database Theory, pp. 405–419. Springer, Berlin/Heidelberg (2001)
61. Wagstaff, K.L.: Intelligent clustering with instance-level constraints. Ph.D. thesis (2002)
62. Wagstaff, K., Cardie, C.: Clustering with instance-level constraints. In: Proceedings of 17th International Conference on Machine Learning, pp. 1103–1110. Standford (2000)
63. Wagstaff, K., Cardie, C., Rogers, S., Schroedl, S.: Constrained k-means clustering with background knowledge. In: Proceedings of 18th International Conference on Machine Learning, vol. 1, pp. 577–584. Williams College (2001)
64. Xenarios, I., Fernandez, E., Salwinski, L., Duan, X.J., Thompson, M.J., Marcotte, E.M., Eisenberg, D.: Dip: the database of interacting proteins: 2001 update. Nucleic Acids Res. 29(1), 239–241 (2001)
65. Xing, E.P., Jordan, M.I., Russell, S., Ng, A.Y.: Distance metric learning with application to clustering with side-information. In: Proceedings of the Advances in Neural Information Processing Systems 15, pp. 505–512. MIT Press (2002)
66. Xu, R., Wunsch, D.: Clustering, vol. 10. Wiley, Hoboken, New Jersey (2008)
67. Yang, Y., Padmanabhan, B.: Segmenting customer transactions using a pattern-based clustering approach. In: Proceedings of 3rd IEEE International Conference on Data Mining, pp. 411–418. IEEE (2003)
68. Yu, S.X., Shi, J.: Segmentation given partial grouping constraints. IEEE Trans. Pattern Anal. Mach. Intell. 26(2), 173–183 (2004)
69. Zhang, Y., Brady, M., Smith, S.: Segmentation of brain mr images through a hidden Markov random field model and the expectation-maximization algorithm. IEEE Trans. Med. Imaging 20(1), 45–57 (2001)
70. Zhang, Z., Kwok, J.T., Yeung, D.Y.: Parametric distance metric learning with label information. In: Proceedings of the International Joint Conference on Artificial Intelligence, pp. 1450–1452 (2003)
71. Zhigang, C., Xuan, L., Fan, Y.: Constrained k-means with external information. In: Proceedings of 8th International Conference on Computer Science & Education, pp. 490–493. IEEE (2013)
72. Zhu, X.: Semi-supervised learning literature survey (2008). URL http://www.pages.cs.wisc.edu/~jerryzhu/research/ssl/semireview.html
73. Zhu, S., Wang, D., Li, T.: Data clustering with size constraints. Knowl.-Based Syst. 23(8), 883–889 (2010)

An Overview of the Use of Clustering for Data Privacy

Vicenç Torra, Guillermo Navarro-Arribas, and Klara Stokes

Abstract In this chapter we review some of our results related to the use of clustering in the area of data privacy. The paper gives a brief overview of data privacy and, more specifically, on data driven methods for data privacy and discusses where clustering can be applied in this setting. We discuss the role of clustering in the definition of masking methods, and on the calculation of information loss and data utility.

Keywords Data privacy • Clustering • Fuzzy clustering • Information loss • Microaggregation

1 Introduction

Data privacy has emerged as an important area of research in the last years due to the increasing amount of information available that contains sensitive data from people and companies. Privacy preserving data mining (PPDM) and statistical disclosure control (SDC) are the two areas that study methods and tools to ensure that disclosure does not take place.

Methods for data privacy can be classified into different categories, and there exist different approaches for this classification. One of them is according to the information on the type of calculation that the receptor of the data (a third party) will apply to the data. Under this categorization we can distinguish between computation-driven, data-driven and result-driven approaches.

Computation-driven methods are defined taking into account which is the analysis to be applied to the data. Data-driven methods are defined when the detailed analysis is unknown. Result-driven focuses on the sensitivity of the outcomes of the analysis. In this paper we focus on data-driven methods.

V. Torra (✉) • K. Stokes
University of Skövde, Skövde, Sweden
e-mail: vtorra@his.se; klara.stokes@his.se

G. Navarro-Arribas
Universitat Autònoma de Barcelona, Barcelona, Spain
e-mail: guillermo.navarro@uab.cat

© Springer International Publishing Switzerland 2016
M.E. Celebi, K. Aydin (eds.), *Unsupervised Learning Algorithms*,
DOI 10.1007/978-3-319-24211-8_10

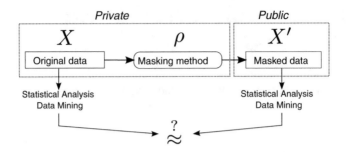

Fig. 1 Common scenario for data-driven protection methods

Data-driven methods for databases typically consist of modifying a database reducing its quality so that sensitive information is not disclosed. The modification should be in a way that the analyses on the modified data are similar to the analysis on the original data. Formally, if X is the original information, we have a method ρ such that when applied to X leads to a file X' that is quite similar to X but with less disclosure risk. Methods ρ of this characteristics are known as masking methods. Figure 1 shows the typical scenario of data-driven protection methods.

Three main topics of research are of interest for data-driven methods. They are (1) masking methods (this is to answer which are the effective methods for data protection), (2) disclosure risk measures (how we evaluate that the modified database X' is appropriate to ensure confidentiality), (3) information loss measures (how we evaluate that the perturbation is not too high to make analysis useless).

Masking methods ρ can be classified into three main categories. The first one corresponds to methods that modify the original data introducing some kind of error. That is, $X' = X + \epsilon$. In these methods, records in X' will contain some incorrect information. For example, salaries of individuals are lower or larger than the real ones. This category corresponds to perturbative methods and includes noise addition, microaggregation, and rank swapping. The second class which corresponds to non-perturbative methods is defined by methods that do not produce erroneous data but change the level of detail. For example, salaries can be replaced by intervals, and cities by counties or regions. No correct value is replaced by an incorrect one. Generalization and suppression are the typical examples of non-perturbative methods. The third category corresponds to synthetic data generators. That is, the original data is replaced by synthetic data which follow a certain model that approximates the original data.

In this chapter we focus on data-driven approaches and we review masking methods based on clustering and information loss measures based on clustering. Clustering has an important role in both the definition of masking methods and the measure of information loss. More specifically, microaggregation is a well-known perturbative masking method based on clustering that we will discuss in Sect. 2. In addition, clustering has been used extensively to evaluate the quality of protected data. We will discuss clustering to measure information loss in Sect. 3.

In addition to the topics explained in this chapter, clustering has also been studied in computation-driven approaches. That is, when we know that the third party will cluster the data set. In this framework, the typical scenario is that a few data owners (e.g., companies) want to apply a clustering algorithm to their data but without sharing their records. To do so, a cryptographic protocol is established so that the resulting set of clusters are computed without revealing the original records to the other data owners. Different algorithms exist. Some algorithms presume that the data is vertically partitioned and others that it is horizontally partitioned. That is, data owners have information on different variables from the same people or the same variables from different people. See, e.g., [60] for details.

2 Clustering to Define Masking Methods

As explained in the previous section, masking methods are functions that introduce some distortion to the data in order to protect sensitive information.

2.1 Clustering in Microaggregation

Microaggregation [10, 13, 19] is one of the methods for data protection. It has been proven [14–16] to be effective for data protection as it permits us to obtain a good trade-off between information loss and disclosure risk.

Given a data set, microaggregation consists of building small clusters and then replacing the data by the cluster representatives. Privacy is achieved because we require that each cluster contains at least k records where k is a parameter of the method. The larger the k, the more privacy we have. Nevertheless, a large k also implies a large information loss. Because of that, microaggregation algorithms try to find a good trade-off between privacy and information loss by means of an appropriate value for k. Formally, this method is defined by the following optimization problem. In this definition we have that $x \in X$ are the records, p_i is the centroid of the ith cluster, and $\chi_i(x) = 1$ represents that record x is assigned to the ith cluster. The application of the algorithms requires that we have a distance function between records and cluster centers, and the value k which is the minimum number of records in a cluster. We denote as $d(x_j, p_i)$ the distance between record x_j and centroid p_i. Equation (1) shows the formalization microaggregation as an optimization problem, minimizing the distance between records of a given cluster with their centroid, subject to the constraint imposed by the k parameter regarding the size of the clusters.

Table 1 Example of microaggregation

				(b) Masked microdata with microaggregation for $k = 3$	

(a) Original microdata					
Id	Age	Income	Id	Age	Income
885	24	21,000.00	–	25.00	20,166.67
795	31	19,500.00	–	25.00	20,166.67
295	32	22,000.00	–	38.00	31,595.00
058	57	43,480.00	–	52.33	41,916.67
732	49	39,220.00	–	52.33	41,916.67
925	43	32,285.00	–	38.00	31,595.00
465	39	40,500.00	–	38.00	31,595.00
321	20	20,000.00	–	25.00	20,166.67
223	51	43,050.00	–	52.33	41,916.67

$$\text{Minimize} \quad \sum_{i=1}^{c} \sum_{j=1}^{n} \chi_i(x_j)(d(x_j, p_i))^2 \tag{1}$$

$$\text{Subject to} \quad \sum_{i=1}^{c} \chi_i(x_j) = 1 \text{ for all } j = 1, \ldots, n$$

$$2k \geq \sum_{j=1}^{n} \chi_i(x_j) \geq k \text{ for all } i = 1, \ldots, c$$

$$\chi_i(x_j) \in \{0, 1\}$$

Table 1 shows a simple example of microaggregation applied to numerical continuous attributes. The resulting masked table, composed of 3 clusters, is 3-anonymous. As it is a common practice in data privacy, identifiers are removed.

Microaggregation algorithms have been proven to be NP-hard problems [43] except for the case of a single variable (univariate microaggregation). A polynomial algorithm exists for this problem [24] and for some variants (e.g., univariate microaggregation with data suppression [29]).

Microaggregation was originally defined for data represented as records on a set of numerical variables [10], and later extended to categorical variables [55]. Currently, there are extensions and variations for other types of structures as search and access logs, time series, documents, and graphs.

All heuristic algorithms for microaggregation follow the same pattern. First, data is clustered so that records are assigned to clusters and each cluster has at least k records. This is the clustering step and for this purpose a distance is needed in the space of the original data. Then, a cluster representative is selected from the cluster. For this purpose aggregation operators [58] are typically used. When data is numerical it is usual to use the mean while other operators as the median and the mode are used for non-numerical data. Finally, the original data is replaced by the cluster representative.

Documents, or, in general, categorical information that can be interpreted semantically permits us to consider semantic versions of microaggregation. Note that as clustering algorithms are based on distances, and that it is usual to consider different types of distances. When data is categorical, we can use syntactic distances but also distances based on the semantics between terms. Semantic distances based on Wordnet [21] and on the Open Directory Project [11] have been considered in microaggregation. This is discussed in more detail below.

Microaggregation is related to k-anonymity [45, 54] as the application of microaggregation to a data set considering all the variables at the same time with a certain given k will satisfy k-anonymity.

2.2   Clustering for Graphs: Microaggregation and k-Anonymity

The underlying structure of a social network is a graph, where nodes represent the individuals in the network and the edges their connections. In addition to the connectivity, both the nodes and the edges can contain additional information about the individuals and their relationships.

Masking methods for data protection for graphs can be classified using the same classes that exist for data files. There are perturbative methods that, e.g., modify the graph adding and removing edges and vertices. In addition, non-perturbative methods reduce the graph into a kind of supernodes which in some sense generalize the connections between the original nodes.

The similarities and differences between k-anonymity for graphs and k-anonymity for standard files are discussed in [53]. This discussion follows the arguments in [62] to state the difficulty of working with graphs. However, in general, as [53] points out, every type of data has its own peculiarities.

Different masking methods for graphs consider different types of attacks. There are methods [34] that presume that the information available to an intruder is the number of connections of a node (e.g., the number of friends in a social network). This corresponds to the degree of the nodes. Others assume that the intruder knows a subgraph (some relationships between nodes) or, in general, a certain type of query on the graph [25]. See also, [22, 62] for other types of definitions. Stokes and Torra [53] reviews reidentification and k-anonymity definitions for graphs.

Given a graph $G = (V, E)$ where V are the nodes and E the edges of a graph ($E \subseteq V \times V$), [53] defines k-anonymity for graphs in terms of the neighbors of a node. The set of neighbors of a node $v \in V$ is defined as

$$N(v) := \{u \in V : (u, v) \in E\}.$$

Then, k-anonymity for graphs is defined as follows, see [53].

Definition 1. Let $G = (V, E)$ be a graph; then, we say that G is k-anonymous if for any vertex v_1 in V, there are at least k distinct vertices $\{v_i\}_{i=1}^{k}$ in V, such that $N(v_i) = N(v_1)$ for all $i \in \{1, \ldots, k\}$.

This definition can be applied when the intruder knows (some of) the neighbors of a node.

Clustering algorithms have been used in masking methods for graphs. Standard clustering algorithms for numerical data can be used for ensuring k-degree anonymity (i.e., that a given degree sequence is k-anonymous). In contrast, specific algorithms for graphs have been used to cluster the nodes of a graph to build k-anonymous graphs. In this case the goal is to build a graph that is k-anonymous in the sense of Definition 1. Figure 2 gives a small example of a 3-anonymous graph whose adjacency matrix is given in Table 2. It can be seen that for each node (each row) there are other 2 which have the same neighbors. Most algorithms for the clustering of social networks are centralized. That is, it is assumed that they are applied by the data owner who has all the data. Stokes [49] presents a distributed approach of message passing type.

Stokes and Torra [52] discusses the difference between two different approaches for graph partitioning: direct and indirect partitioning.

- **Direct partitioning.** This consists of partitioning the original matrix that represents the graph. This implies that clusters are built gathering together sets of nodes that have a good connectivity among themselves. This approach does not permit us to distinguish between the different roles of the vertices in well-connected regions.

Fig. 2 Example of a simple 3-anonymous graph

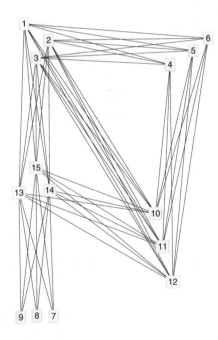

Table 2 Adjacency table of a simple 3-anonymous graph

0	0	0	1	1	1	0	0	0	1	1	1	1	1	1
0	0	0	1	1	1	0	0	0	1	1	1	1	1	1
0	0	0	1	1	1	0	0	0	1	1	1	1	1	1
1	1	1	0	0	0	0	0	0	1	1	1	0	0	0
1	1	1	0	0	0	0	0	0	1	1	1	0	0	0
1	1	1	0	0	0	0	0	0	1	1	1	0	0	0
0	0	0	0	0	0	0	0	0	0	0	0	1	1	1
0	0	0	0	0	0	0	0	0	0	0	0	1	1	1
0	0	0	0	0	0	0	0	0	0	0	0	1	1	1
1	1	1	1	1	1	0	0	0	0	0	0	1	1	1
1	1	1	1	1	1	0	0	0	0	0	0	1	1	1
1	1	1	1	1	1	0	0	0	0	0	0	1	1	1
1	1	1	0	0	0	1	1	1	1	1	1	0	0	0
1	1	1	0	0	0	1	1	1	1	1	1	0	0	0
1	1	1	0	0	0	1	1	1	1	1	1	0	0	0

- Indirect partitioning. In this case, the partitioning algorithm is not applied to the graph (the original matrix) but to a similarity matrix computed from the graph. That is, given a similarity function S we build a matrix $M_S : V \times V \rightarrow [0, 1]$ defining $M_s(V_1, V_2) = S(V_1, V_2)$. Then, we partition M_S.

It is clear that while the direct partitioning gathers in the same class connected nodes, the indirect partitioning gathers in the same class nodes that are similar.

2.3 Attacks on Microaggregation

When a data set is protected using microaggregation and taking all the variables at the same time in the clustering, the microaggregated file satisfies k-anonymity. In this case, attacks for k-anonymity are of relevance. They are [9, 32] the homogeneity and the background attacks.

Nevertheless, microaggregation is also applied to subsets of variables. This is used to decrease information loss at the cost of some disclosure risk. This implies that k-anonymity is not guaranteed. Table 3 illustrates the application of optimal microaggregation to each variable independently.

In this case, intruders can use the values of the masked file as well as their own information to attack the data set. In this example we can see that there are two unique records in the masked data set. The real k-anonymity of the protected file is one. Here, we use real k-anonymity as first defined in [41]. So, effective re-identification attacks can be done to this file. In general, as microaggregation modifies the original data, re-identification does not need to be straightforward, and we may have some records for which the nearest masked one is the correct link but others for which is not true. Nevertheless, intersection attacks are possible combining the information the intruder has for each of the variables.

Table 3 Example of microaggregation

(a) Original microdata

(b) Masked microdata with microaggregation for $k = 3$ applying optimal microaggregation to each variable

Id	Age	Income	Id	Age	Income
885	24	21,000.00	–	25.00	20,166.67
795	31	19,500.00	–	25.00	20,166.67
295	32	22,000.00	–	38.00	31,168.33
058	57	43,480.00	–	52.33	42,343.33
732	49	39,220.00	–	52.33	31,168.33
925	43	32,285.00	–	38.00	31,168.33
465	39	40,500.00	–	38.00	42,343.33
321	20	20,000.00	–	25.00	20,166.67
223	51	43,050.00	–	52.33	42,343.33

This type of intersection attack was first considered in [57] and later in [40, 42]. These latter works show empirically that some microaggregation methods fail to protect the data file.

This type of attacks are related to the idea of transparency in data privacy. We have transparency when a release of a file goes with all the information on how the data is produced. This includes information on the masking method applied as well as its parameters.

2.4 Fuzzy Clustering for Microaggregation

Most methods for microaggregation are based on crisp clustering methods. In order to avoid some of the attacks mentioned in the previous section fuzzy clustering [5, 37, 38] was introduced in [17] (see also [18, 56, 57]) as the clustering algorithm for microaggregation. Recall that the idea behind fuzzy clustering is that records can belong to more than one cluster.

In this approach, the assignment of records to clusters is not deterministic. Instead, it is done probabilistically according to a probability distribution. This probability can be proportional to the membership degrees of records to clusters or just uniformly distributed for clusters with membership above a certain threshold.

The goal of using fuzzy clustering and the random selection is to avoid intersection attacks when the different variables are considered. In addition, an intruder cannot be sure that the nearest masked record will be the one that correctly matches to the one in its own database.

2.5   Clustering for Masking Data Streams

The application of microaggregation directly to mask data streams is usually not recommended. If microaggregation is applied using a sliding (or tumbler) window, thus applying microaggregation by parts of the stream, the result might be very poor in terms of information loss. Moreover, if not done carefully, e.g. by allowing re-computation of centroids after publication, it can be vulnerable to inference attacks through intersection [6, 51].

Stream clustering methods for data privacy differ from common stream clustering techniques in several points. Most notably, the main objective of the masking method is to produce the masked output, not the partition or structure of the clustering. This makes methods based on corsets, or in general techniques that require adjusting the clusters parameters as the stream is processed, not suitable for masking. Several techniques have been developed with this constraints in mind [6, 8, 44, 59]. These are streaming clustering methods that can implement k-anonymity in data streams, while avoiding disclosure from intersection of clusters.

Some ideas behind stream masking based on clustering have been extended to support fully dynamic data. Allowing the deletion of already masked data imposes an important threat. For instance if a cluster is left with less than k elements, these elements need to be protected. This protection is difficult since it has to prevent inferences. There are some works providing dynamic clustering and microaggregation as a masking method [39, 61], which still present some important drawbacks, for example, regarding information loss. The protection of dynamic data for publication is still an open research field.

2.6   Masking Very Large Data Sets

Although stream masking methods discussed in the previous section can be used to mask high volumes of data, there are specific approaches to deal with this problem without the restrictions imposed by streaming data. These proposals improve generic microaggregation algorithms which need to access the whole data set during the masking process repetitively.

Some efficiency improvements can be achieved by projecting the data into one dimension and performing an optimal microaggregation [23], or by using specific data structures [26, 31]. Other approaches define an initial partition of the data in order to apply microaggregation in each part separately [46, 47].

Very efficient microaggregation can also be achieved by defining the clustering using k-nearest neighbors searches [48]. Laszlo and Mukherjee [30] is another recent approach based on local search.

Note that common microaggregation algorithms such as MDAV [13] present a complexity of $O(n^2)$ (where n is the number of records), which is unaffordable for very large data sets. The previously cited works reduce this complexity at least for some specific cases.

2.7 Masking Through Semantic Clustering

As previously mentioned microaggregation can be defined for categorical data exploiting their semantics. This is very convenient for data privacy since precisely the semantics of the data is the important part to be preserved when masking data. These methods usually achieve a better compromise between privacy and information loss than syntactic approaches.

Microaggregation can be defined in terms of a semantic distance and a semantic aggregation operator to compute the clusters representatives. An example is to use an ontology such as Wordnet to define the distance and to aggregate words or synsets by means of generalization [1, 33, 36]. Note that here generalization is from the point of view of semantics (e.g., dog and cat are generalized into pets) and the dictionary can be used for this purpose.

This approach can be extended to deal with document vectors (algebraic representation of documents widely used in information retrieval and text mining) providing anonymous document vector spaces using microaggregation by clustering the vectors [39]. Although it is not a semantic microaggregation strictly speaking, spherical microaggregation has also been introduced to deal with document vectors [2].

Semantic microaggregation has also been applied to the anonymization of query logs from a search engine [4, 20]. In this case the Open Directory Project is used to semantically categorize queries (based on their actual results), and semantic distances are computed over those categories for clustering user queries. The semantic anonymization of set valued data has also been treated in [3].

2.8 Clustering in Other Masking Methods

Clustering has been also used to define other masking methods. It is worth to mention its application to build data models that are accurate on subdomains. For example in [50] data is clustered in a first step and then masked data is generated within the clusters. In this way, properties of the data at the cluster level can be preserved. For example, as microaggregation preserves mean, means will be preserved in the clusters if microaggregation is used for masking data in the second step. Similarly, if we use rank swapping in the second step, as rank swapping preserves frequencies, frequencies will be preserved in the clusters. Domingo-Ferrer and González-Nicolás [12] follows a different approach, it uses microaggregation in the first step, and then a synthetic data generator for the second step. In this way, the first step ensures a certain privacy level through the selection of the value of k. When $k = 1$ the original data is retrieved. So, there is no information loss and the risk is maximal. When $k = |X|$, protection is maximal and X is replaced by data according to the synthetic data generator for the full data set X. Note that this is different to what we obtain with microaggregation directly applied to the file. In such case, for $k = |X|$ we have that all records are replaced by the mean of the whole file X (i.e., all masked records are equal to the mean of X).

3   Clustering to Measure Information Loss

Information loss depends on the data use. That is, on the analysis or function that the user intends to apply to the data. Naturally, the results of an analysis are different when data sets are different. Therefore, the analysis on the protected data set and on the original data set are different. The more perturbation the masking method applies to the data, the larger the difference between the original and the protected data set, and the larger the information loss.

Information loss measures can be formalized as follows. If f is the analysis to be applied to the data, X the original data set and X' the protected one obtained as the application of a masking method ρ to X (i.e., $X' = \rho(X)$), the information loss for a particular use f is the measure

$$IL(X, X') = \text{divergence}(f(X), f(X'))$$

where divergence is a function that quantifies the difference between $f(X)$ and $f(X')$. Naturally, divergence$(Y, Y)=0$.

Different types of analysis f have been considered in the literature. Clustering is one of them. Both crisp and fuzzy clustering have been considered, and the corresponding information loss has been measured. For example, information loss for k-means and fuzzy c-means have been measured for a few masking methods as, e.g., microaggregation.

In the case of crisp clustering, divergence is a function that needs to consider the results of the clustering algorithm on the original file (i.e., $f(X)$) and on the protected file (i.e., $f(X')$). In this case, these results $f(X)$ and $f(X')$ are two partitions of the elements in X. Therefore any distance or similarity measure on pairs of partitions can be used to define the divergence. For example, we can use the Rand or the Jaccard index for this purpose. When $f(X)$ is a partition $\Pi = \{\pi_1, \ldots, \pi_n\}$ and $f(X')$ is the partition $\Pi' = \{\pi'_1, \ldots, \pi'_n\}$, the Rand and Jaccard indices are defined by:

Rand index:

$$RI(\Pi, \Pi') = (r + u)/(r + s + t + u)$$

Jaccard Index:

$$JI(\Pi, \Pi') = r/(r + s + t)$$

Adjusted Rand Index: This is a correction of the Rand index so that the expectation of the index for partitions with equal number of objects is 0. This adjustment was done assuming generalized hypergeometric distribution as the model of randomness. That is,

$$ARI(\Pi, \Pi') = \frac{r - \exp}{\max - \exp}$$

where $\exp = (np(\Pi)np(\Pi'))/(n(n-1)/2)$ and where $\max = 0.5(np(\Pi) + np(\Pi'))$.

In these indices, r, s, t, u, and $np(\Pi)$ are defined as follows:

- r is the number of pairs (a, b) where a and b are in the same cluster in Π and in Π';
- s is the number of pairs where a and b are in the same cluster in Π but not in Π';
- t is the number of pairs where a and b are in the same cluster in Π' but not in Π;
- u is the number of pairs where a and b are in different clusters in both partitions.
- $np(\Pi)$ is the number of pairs within clusters in the partition π.

The Rand index is 1 when the two partitions are equal and 0 when the difference is maximal. Therefore, we can define

$$IL_{RI}(X, X') = \text{divergence}_{RI}(f(X), f(X')) = 1 - RI(f(X), f(X')).$$

The Adjusted Rand Index has the same behavior as the Rand Index, but has an expected value of zero and can take negative values. Therefore, we can also use in this case:

$$IL_{ARI}(X, X') = \text{divergence}_{ARI}(f(X), f(X')) = 1 - ARI(f(X), f(X')).$$

The Jaccard index can be used in the same way.

In the case of fuzzy clustering, $f(X)$ and $f(X')$ will be fuzzy partitions. Therefore, we can use here distances and generalization of these indices for fuzzy partitions. See [7, 27] for details.

These indices have been used, e.g., in [28] to compare the results of masking methods with respect to the use of clustering.

Examples of the use of clustering as an information loss measure in the particular case of semantic-based masking methods can be found in [4, 35].

4   Conclusion

In this chapter we have discussed the application of clustering in data privacy. We have seen that clustering is applied in the definition of masking methods and also at the time of computing information loss.

In the context of data privacy protection and evaluation, clustering has an important role. Moreover, it still presents some open research problems and there is room for improvement in existing approaches both in protection and evaluation methods.

Acknowledgements Partial support by the Spanish MEC (projects TIN2011-27076-C03-03 and TIN2014-55243-P) is acknowledged.

References

1. Abril, D., Navarro-Arribas, G., Torra, V.: Towards semantic microaggregation of categorical data for confidential documents. Modeling Decisions for Artificial Intelligence. Lecture Notes in Computer Science, vol. 6408, pp. 266–276. Springer, Heidelberg (2010)
2. Abril, D., Navarro-Arribas, G., Torra, V.: Spherical microaggregation: Anonymizing sparse vector spaces. Comput. Secur. **49**, 28–44 (2015)
3. Batet, M., Erola, A., Sánchez, D., Castellà-Roca, J.: Semantic anonymisation of set-valued data. In: Proceedings of the 6th International Conference on Agents and Artificial Intelligence (ICAART) vol. 1, pp. 102–112 (2014)
4. Batet, M., Erola, A., Sánchez D., Castellà-Roca, J.: Utility preserving query log anonymization via semantic microaggregation. Inf. Sci. **242**, 49–63 (2013)
5. Bezdek, J.C.: Pattern Recognition with Fuzzy Objective Function Algorithms. Plenum, New York (1981)
6. Byun, J.-W., Sohn, Y., Bertino, E., Li, N.: Secure anonymization for incremental datasets. In: Secure Data Management. Lecture Notes in Computer Science, pp. 48–63. Springer, Heidelberg (2006)
7. Campello, R.J.G.B.: A fuzzy extension of the Rand index and other related indexes for clustering and classification assessment. Pattern Recogn. Lett. **28**(7), 833–841 (2007)
8. Cao, J., Carminati, B., Ferrari, E., Tan, K.-L.: CASTLE: continuously anonymizing data streams. IEEE Trans. Dependable Secure Comput. **8**, 337–352 (2011)
9. De Capitani di Vimercati, S., Foresti, S., Livraga, G., Samarati, P.: Data privacy: definitions and techniques. Int. J. Uncertainty Fuzziness Knowledge Based Syst. **20**(6), 793–817 (2012)
10. Defays, D., Nanopoulos, P.: Panels of enterprises and confidentiality: the small aggregates method. In: Proceeding of the 1992 Symposium on Design and Analysis of Longitudinal Surveys, pp. 195–204. Statistics Canada (1993)
11. DMOZ: The Open Directory Project. www.dmoz.org (2015)
12. Domingo-Ferrer, J., González-Nicolás, U.: Hybrid microdata using microaggregation. Inf. Sci. **180**, 2834–2844 (2010)
13. Domingo-Ferrer, J., Mateo-Sanz, J.M.: Practical data-oriented microaggregation for statistical disclosure control. IEEE Trans. Knowl. Data Eng. **14**(1), 189–201 (2002)
14. Domingo-Ferrer, J., Mateo-Sanz, J.M., Torra, V.: Comparing SDC methods for microdata on the basis of information loss and disclosure risk. In: Pre-proceedings of ETK-NTTS'2001 (Eurostat, ISBN 92-894-1176-5), vol. 2, pp. 807–826. Creta, Greece (2001)
15. Domingo-Ferrer, J., Torra, V.: Disclosure control methods and information loss for microdata. In: Doyle, P., Lane, J.I., Theeuwes, J.J.M., Zayatz, L. (eds.) Confidentiality, Disclosure, and Data Access: Theory and Practical Applications for Statistical Agencies, pp. 91–110. Elsevier (2001)
16. Domingo-Ferrer, J., Torra, V.: A quantitative comparison of disclosure control methods for microdata. In: Doyle, P., Lane, J.I., Theeuwes, J.J.M., Zayatz, L. (eds.) Confidentiality, Disclosure and Data Access: Theory and Practical Applications for Statistical Agencies, pp. 111–134. North-Holland, Amsterdam, The Netherlands (2001)

17. Domingo-Ferrer, J., Torra, V.: Towards fuzzy c-means based microaggregation. In: Grzegorzewski, P., Hryniewicz, O., Gil, M.A. (eds.) Soft Methods in Probability and Statistics, pp. 289–294. Physica, Heidelberg (2002)
18. Domingo-Ferrer, J., Torra, V.: Fuzzy microaggregation for microdata protection. J. Adv. Comput. Intell. Intell. Inform. 7(2), 153–159 (2003)
19. Domingo-Ferrer, J., Torra, V.: Ordinal, continuous and heterogeneous k-anonymity through microaggregation. Data Min. Knowl. Disc. 11(2), 195–212 (2005)
20. Erola, A., Castellà-Roca, J., Navarro-Arribas, G., Torra, V.: Semantic microaggregation for the anonymization of query logs using the open directory project. SORT Stat. Oper. Res. 35, Trans. 41–58 (2011)
21. Fellbaum, C. (ed.): WordNet: An Electronic Lexical Database. MIT, Cambridge 1998
22. Feder, T., Nabar, S.U., Terzi, E.: Anonymizing graphs. CoRR abs/0810.5578 (2008)
23. Ghinita, G., Karras, P., Kalnis, P., Mamoulis, N.: Fast data anonymization with low information loss. In: Proceedings of the 33rd International Conference Very Large Data Bases, pp. 758–769 (2007)
24. Hansen, S.L., Mukherjee, S.: A polynomial algorithm for optimal univariate microaggregation. IEEE Trans. Knowl. Data Eng. 15(4), 1043–1044 (2003)
25. Hay, M., Miklau, G., Jensen, D.: Anonymizing social networks. In: Proceedings of the VLDB Endowment (2008)
26. Hore, B., Jammalamadaka, R.C., Mehrotra, S.: Flexible anonymization for privacy preserving data publishing: a systematic search based approach. In: Proceedings of the 7th SIAM International Conference on Data Mining (2007)
27. Hüllermeier, E., Rifqi, M.: A fuzzy variant of the rand index for comparing clustering structures. In: Proceedings of IFSA-EUSFLAT (2009)
28. Ladra, S., Torra, V.: On the comparison of generic information loss measures and cluster-specific ones. Int. J. Uncertainty Fuzziness Knowledge Based Syst. 16(1) 107–120 (2008)
29. Laszlo, M., Mukherjee, S.: Optimal univariate microaggregation with data suppression. J. Syst. Softw. 86, 677–682 (2013)
30. Laszlo, M., Mukherjee, S.: Iterated local search for microaggregation. J. Syst. Softw. 100, 15–26 (2015)
31. LeFevre, K., DeWitt, D.J., Ramakrishnan, R.: Mondrian multidimensional k-anonymity. In: Proceedings of International Conference on Data Engineering (2006)
32. Li, N., Li, T., Venkatasubramanian, S.: T-closeness: privacy beyond k-anonymity and l-diversity. In: Proceedings of the IEEE ICDE (2007)
33. Liu, J., Wang, K.: Anonymizing bag-valued sparse data by semantic similarity-based clustering. Knowl. Inf. Syst. 35, 435–461 (2013)
34. Liu, K., Terzi, E.: Towards identity anonymization on graphs. In: Proceeding of the SIGMOD (2008)
35. Martínez, S., Sánchez, D., Valls, A., Batet, M.: Privacy protection of textual attributes through a semantic-based masking method. Inf. Fusion 13(4), 304–314 (2012)
36. Martínez, S., Sánchez, D., Valls, A.: Semantic adaptive microaggregation of categorical microdata. Comput. Secur. 31(5), 653–672 (2012)
37. Miyamoto, S.: Introduction to Fuzzy Clustering (in Japanese). Morikita, Tokyo (1999)
38. Miyamoto, S., Ichihashi, H., Honda, K.: Algorithms for Fuzzy Clustering. Springer, Berlin (2008)
39. Navarro-Arribas, G., Abril, D., Torra, V.: Dynamic anonymous index for confidential data. Data Privacy Management and Autonomous Spontaneous Security. Lecture Notes in Computer Science, vol. 8247, pp. 362–368. Springer Berlin Heidelberg, Germany (2014)
40. Nin, J., Herranz, J., Torra, V.: On the disclosure risk of multivariate microaggregation. Data Knowl. Eng. 67, 399–412 (2008)
41. Nin, J., Herranz, J., Torra, V.: How to Group Attributes in Multivariate Microaggregation. Int. J. Uncertainty Fuzziness Knowledge Based Syst. 16(1), 121–138 (2008)
42. Nin, J., Torra, V.: Analysis of the univariate microaggregation disclosure risk. N. Gener. Comput. 27, 177–194 (2009)

43. Oganian, A., Domingo-Ferrer, J.: On the complexity of optimal microaggregation for statistical disclosure control. Stat. J. U. N. Econ. Comm. Eur. **18**(4), 345–353 (2001)
44. Pei, J., Xu, J., Wang, Z., Wang, W., Wang, K.: Maintaining K-anonymity against incremental updates. In: Proceedings of the 19th International Conference on Scientific and Statistical Database Management, 2007 (SSBDM, 2007), pp. 5–5 (2007)
45. Samarati, P.: Protecting respondents identities in microdata release. IEEE Trans. Knowl. Data Eng. **13**, 1010–1027 (2001)
46. Solanas, A., Martínez-Balleste, A., Domingo-Ferrer, J., Mateo-Sanz, J.M.: A 2d-tree-based blocking method for microaggregating very large data sets. In: The First International Conference on Availability, Reliability and Security (ARES) (2006)
47. Solanas, A., Pietro, R.D.: A linear-time multivariate micro-aggregation for privacy protection in uniform very large data sets. Modeling Decisions for Artificial Intelligence. Lecture Notes in Computer Science, pp. 203–214. Springer, Heidelberg (2008)
48. Solé, M., Muntés-Mulero, V., Nin, J.: Efficient microaggregation techniques for large numerical data volumes. Int. J. Inf. Secur. **11**, 253–267 (2012)
49. Stokes, K.: Graph k-anonymity through k-means and as modular decomposition. In: Proceedings of the NordSec 2013. Lecture Notes in Computer Science, vol. 8208, pp. 263–278. (2013)
50. Stokes, K., Torra, V.: n-Confusion: a generalization of k-anonymity. In: Proceedings of the 5th International Workshop on Privacy and Anonymity in the Information Society (PAIS). Berlin, Germany (2012)
51. Stokes,K., Torra, V.: Multiple releases of k-anonymous data sets and k-anonymous relational databases. Int. J. Uncertainty Fuzziness Knowledge Based Syst. **20**(06), 839–853 (2012)
52. Stokes, K., Torra, V.: On some clustering approaches for graphs. In: Proceeding of the IEEE International Conference on Fuzzy Systems (FUZZ-IEEE 2011) (ISBN 978-1-4244-7315-1), pp. 409–415. Taipei, Taiwan (2011)
53. Stokes, K., Torra, V.: Reidentification and k-anonymity: a model for disclosure risk in graphs. Soft. Comput. **16**(10), 1657–1670 (2012)
54. Sweeney, L.: k-anonymity: a model for protecting privacy. Int. J. Uncertainty Fuzziness Knowledge Based Syst. **10**, 557–570 (2002)
55. Torra, V.: Microaggregation for categorical variables: a median based approach. In: Proceeding of the Privacy in Statistical Databases (PSD 2004). Lecture Notes in Computer Science, vol. 3050, pp. 162–174 (2004)
56. Torra, V. (2015) A fuzzy microaggregation algorithm using fuzzy c-means, Proc. CCIA 2015, Volume 277: Artificial Intelligence Research and Development, IOS Press, 214–223 DOI: 10.3233/978-1-61499-578-4-214
57. Torra, V., Miyamoto, S.: Evaluating fuzzy clustering algorithms for microdata protection. Privacy in Statistical Databases. Lecture Notes in Computer Science, vol. 3050, pp. 175–186 (2004)
58. Torra, V., Narukawa, Y.: Modeling Decisions: Information Fusion and Aggregation Operators. Springer, Heidelberg (2007)
59. Truta, T.M., Campan, A.: K-anonymization incremental maintenance and optimization techniques. In: Proceeding of the 2007 ACM Symposium on Applied Computing, pp. 380–387 (2007)
60. Vaidya, J., Clifton, C., Zhu, M.: Privacy Preserving Data Mining. Springer, New York (2006)
61. Xiao, X., Tao, Y.: M-invariance: towards privacy preserving re-publication of dynamic datasets. In: Proceedings of the 2007 ACM SIGMOD International Conference on Management of Data, SIGMOD 2007, pp. 689–700. ACM (2007)
62. Zhou, B., Pei. J.: Preserving privacy in social networks against neighborhood attacks. In: Proceeding of the ICDE 2008 (2008)

Nonlinear Clustering: Methods and Applications

Chang-Dong Wang and Jian-Huang Lai

Abstract As a fundamental classification method for pattern recognition, data clustering plays an important role in various fields such as computer science, medical science, social science, and economics. According to the data distribution of clusters, data clustering problem can be categorized into linearly separable clustering and nonlinearly separable clustering. Due to the complex manifold of the real-world data, nonlinearly separable clustering is one of most popular and widely studied clustering problems. This chapter reviews nonlinear clustering algorithms from four viewpoints, namely kernel-based clustering, multi-exemplar model, graph-based method, and support vector clustering (SVC). Accordingly, this chapter reviews four nonlinear clustering methods, namely conscience on-line learning (COLL) for kernel-based clustering, multi-exemplar affinity propagation (MEAP), graph-based multi-prototype competitive learning (GMPCL), and position regularized support vector clustering (PSVC), and demonstrates their applications in computer vision such as digital image clustering, video segmentation, and color image segmentation.

1 Introduction

Data clustering plays an indispensable role in various fields such as computer science, medical science, social science, and economics [47]. According to the data distribution of clusters, data clustering problem can be categorized into linearly separable clustering and nonlinearly separable clustering. Nonlinearly separable clustering means that the data set to be partitioned contains at least one cluster with concave boundaries or even of arbitrary shapes. Figure 1 illustrates two typical

C.-D. Wang
School of Mobile Information Engineering, Sun Yat-sen University, Zhuhai, People's Republic of China
e-mail: changdongwang@hotmail.com

J.-H. Lai (✉)
School of Information Science and Technology, Sun Yat-sen University, Guangzhou, People's Republic of China
e-mail: stsljh@mail.sysu.edu.cn

© Springer International Publishing Switzerland 2016
M.E. Celebi, K. Aydin (eds.), *Unsupervised Learning Algorithms*,
DOI 10.1007/978-3-319-24211-8_11

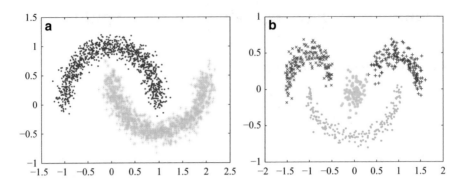

Fig. 1 Two typical data sets consisting of nonlinear separable clusters. (**a**) Two moons data set;
(**b**) smile face data set

data sets consisting of nonlinear separable clusters. Due to the complex manifold
of the real-world data, it is important to identify the nonlinear clusters. However,
classical linear separable clustering methods such as k-means cannot meet the need
of identifying nonlinear clusters. In this chapter, we will review several nonlinear
clustering works from four viewpoints, namely kernel-based clustering, multi-
exemplar model, graph-based method and support vector clustering (SVC), and
analyze their applications in computer vision such as digital image clustering, video
segmentation, and color image segmentation. They are summarized as follows.

1. To address the ill-initialization problem in kernel-based clustering, Sect. 2 first
   of all reviews a conscience on-line learning (COLL) for kernel-based clustering [40, 44]. Kernel-based clustering is one of the most widely used nonlinear
   clustering methods. However, classical kernel-based clustering methods, such
   as kernel k-means [31], suffer from one shortcoming that their clustering
   results seriously degenerate due to ill-initialization [6, 7]. The advantage of
   the conscience mechanism is that, by decreasing the winning frequency of the
   frequent winners, all the prototypes would converge to the optimal solution
   quickly, such that the degeneration problem due to ill-initialization can be
   successfully avoided. By conducting experiments in video segmentation, we
   validate the effectiveness of the COLL algorithm in the applications of computer
   vision.
2. From the viewpoint of multi-exemplar model, Sect. 3 reviews a novel multi-
   exemplar clustering method, termed multi-exemplar affinity propagation
   (MEAP) [43]. In the multi-exemplar model, each cluster is represented by an
   automatically determined number of exemplars and a super-exemplar. Each data
   point is assigned to the most suitable exemplar and each exemplar is assigned to
   the most suitable super-exemplar. The model aims to maximize the sum of all
   similarities between data points and the corresponding exemplars *plus* the sum
   of all linkages between exemplars and the corresponding super-exemplars.
   To solve this NP-hard problem, we use the max-sum belief propagation,

producing clusters insensitive to initialization and converged to the neighborhood maximum. Experimental results in natural image categorization validate the effectiveness of MEAP in computer vision application.

3. Based on the advantages of graph method, multi-prototype representation and competitive learning, Sect. 4 reviews a novel nonlinear clustering method, namely graph-based multi-prototype competitive learning (GMPCL) [45]. Based on the graph method, the GMPCL method firstly generates an initial coarse clustering. That is, we first construct a graph with vertex energy vector from the data set; a core-point set is computed according to the vertex energy, from which we can obtain an initial clustering by constructing connected components via neighbor connectivity. Then, a multi-prototype competitive learning is designed to refine this initial clustering so as to generate accurate clustering results. Experimental results in automatic color image segmentation validate the effectiveness of GMPCL and show that GMPCL provides a useful tool for computer vision.

4. To eliminate the influence of the trade-off parameter C to SVC [3], Sect. 5 reviews a position regularized support vector clustering (PSVC) [39]. SVC is a nonlinear clustering method developed from support vector domain description (SVDD). Compared with other clustering methods, the advantages of SVC include the abilities to discover clusters of arbitrary shapes and to deal with outliers. However, the value of the trade-off parameter directly determines the size of sphere and therefore influences the data distribution of sphere surface, i.e., the clustering results of SVC. By assigning each data point with a position based weighting, rather than using the same trade-off parameter for all data points, the PSVC algorithm can adaptively generate a suitable sphere, leading to accurate clustering results.

In the discussion and conclusion section, we will discuss the issues associated with these methods, as well as the relation between these methods and other nonlinear clustering methods.

2   COLL for Kernel-Based Clustering

In this section, we review a COLL for kernel-based clustering [40, 44], which is designed to address the ill-initialization problem in kernel-based clustering. The advantage of the conscience mechanism is that, by decreasing the winning frequency of the frequent winners, all the prototypes would converge to the optimal solution quickly, such that the degeneration problem due to ill-initialization can be successfully avoided. Experimental results in video segmentation have validated the effectiveness of the COLL algorithm in the applications of computer vision.

2.1   Problem Formulation

Given an unlabelled data set $\mathscr{X} = \{\mathbf{x}_1, \ldots, \mathbf{x}_n\}$ of n data points in $\mathbb{R}^d$ which is projected into a kernel space $\mathscr{Y}$ by a mapping ϕ, and the number of clusters c, we wish to find an assignment of each data point to one of c clusters, such that in the kernel space $\mathscr{Y}$, the within-cluster similarity is high and the between-cluster one is low. That is, we seek a map

$$v : \mathscr{X} \to \{1, \ldots, c\} \tag{1}$$

to optimize [32]

$$v = \arg\min_{v} \left\{ \sum_{i,j: v_i = v_j} \| \phi(\mathbf{x}_i) - \phi(\mathbf{x}_j) \|^2 - \lambda \sum_{i,j: v_i \neq v_j} \| \phi(\mathbf{x}_i) - \phi(\mathbf{x}_j) \|^2 \right\}, \tag{2}$$

where $\lambda > 0$ is some parameter and we use the short notation $v_i = v(\mathbf{x}_i)$.

Theorem 1. *The optimization criterion (2) is equivalent to the criterion*

$$v = \arg\min_{v} \sum_{i=1}^{n} \| \phi(\mathbf{x}_i) - \boldsymbol{\mu}_{v_i} \|^2, \tag{3}$$

where $\boldsymbol{\mu}_k$ is the mean of data points assigned to cluster k

$$\boldsymbol{\mu}_k = \frac{1}{|v^{-1}(k)|} \sum_{i \in v^{-1}(k)} \phi(\mathbf{x}_i), \ \forall k = 1, \ldots, c \tag{4}$$

and v_i satisfies

$$v_i = \arg\min_{k=1,\ldots,c} \| \phi(\mathbf{x}_i) - \boldsymbol{\mu}_k \|^2, \ \forall i = 1, \ldots, n. \tag{5}$$

Proof. See [32].   □

Thus the goal of kernel clustering is to solve the optimization problem in (3). The objective term

$$\sum_{i=1}^{n} \| \phi(\mathbf{x}_i) - \boldsymbol{\mu}_{v_i} \|^2 \tag{6}$$

is known as the *distortion error* [4]. Ideally, all possible assignments of the data into clusters should be tested and the best one with smallest *distortion error* selected. This procedure is unfortunately computationally infeasible in even a very small data set, since the number of all possible partitions of a data set grows exponentially with the number of data points. Hence, efficient algorithms are required.

In practice, the mapping function ϕ is often not known or hard to obtain and the dimensionality of $\mathcal{Y}$ is quite high. The feature space $\mathcal{Y}$ is characterized by the kernel function κ and corresponding kernel matrix K [32].

Definition 1. A kernel is a function κ, such that $\kappa(\mathbf{x}, \mathbf{z}) = \langle \phi(\mathbf{x}), \phi(\mathbf{z}) \rangle$ for all $\mathbf{x}, \mathbf{z} \in \mathcal{X}$, where ϕ is a mapping from $\mathcal{X}$ to an (inner product) feature space $\mathcal{Y}$. A kernel matrix is a square matrix $K \in \mathbb{R}^{n \times n}$ such that $K_{i,j} = \kappa(\mathbf{x}_i, \mathbf{x}_j)$ for some $\mathbf{x}_1, \ldots, \mathbf{x}_n \in \mathcal{X}$ and some kernel function κ.

Thus for an efficient approach, the computation procedure using only the kernel matrix is also required.

2.2 Batch Kernel k-Means and Issues

The k-means [26] algorithm is one of most popular iterative methods for solving the optimization problem (3). It begins by initializing a random assignment v and seeks to minimize the *distortion error* by iteratively updating the assignment v

$$v_i \leftarrow \arg\min_{k=1,\ldots,c} \| \phi(\mathbf{x}_i) - \boldsymbol{\mu}_k \|^2, \ \forall i = 1, \ldots, n \tag{7}$$

and the prototypes $\boldsymbol{\mu}$

$$\boldsymbol{\mu}_k \leftarrow \frac{1}{|v^{-1}(k)|} \sum_{i \in v^{-1}(k)} \phi(\mathbf{x}_i), \ \forall k = 1, \ldots, c, \tag{8}$$

until all prototypes converge or the number of iterations reaches a prespecified value $t_{\max}$. Let $\hat{\boldsymbol{\mu}}$ denote the old prototypes before the t-th iteration, the convergence of all prototypes is characterized by the *convergence criterion*

$$\mathfrak{e}^{\phi} = \sum_{k=1}^{c} \| \boldsymbol{\mu}_k - \hat{\boldsymbol{\mu}}_k \|^2 \le \epsilon, \tag{9}$$

where ϵ is some very small positive value, e.g., 10^{-4}.

Since in practice, only the kernel matrix K is available, the updating of assignment (7) is computed based on the kernel trick [30]

$$v_i \leftarrow \arg\min_{k=1,\ldots,c} \| \phi(\mathbf{x}_i) - \boldsymbol{\mu}_k \|^2$$

$$= \arg\min_{k=1,\ldots,c} \left\{ K_{i,i} + \frac{\sum_{h \in v^{-1}(k)} \sum_{l \in v^{-1}(k)} K_{h,l}}{|v^{-1}(k)|^2} - \frac{2 \sum_{j \in v^{-1}(k)} K_{i,j}}{|v^{-1}(k)|} \right\}. \tag{10}$$

Then the updated prototypes $\boldsymbol{\mu}$ are implicitly expressed by the assignment v, which is further used in the next iteration. Let $\hat{v}$ denote the old assignment before the t-th iteration, the *convergence criterion* (9) is computed as

$$e^{\phi} = \sum_{k=1}^{c} \| \boldsymbol{\mu}_k - \hat{\boldsymbol{\mu}}_k \|^2$$

$$= \sum_{k=1}^{c} \left\{ \frac{\sum_{h \in v^{-1}(k)} \sum_{l \in v^{-1}(k)} K_{h,l}}{|v^{-1}(k)|^2} \right.$$

$$\left. + \frac{\sum_{h \in \hat{v}^{-1}(k)} \sum_{l \in \hat{v}^{-1}(k)} K_{h,l}}{|\hat{v}^{-1}(k)|^2} - \frac{2 \sum_{h \in v^{-1}(k)} \sum_{l \in \hat{v}^{-1}(k)} K_{h,l}}{|v^{-1}(k)||\hat{v}^{-1}(k)|} \right\}. \quad (11)$$

The above procedures lead to the well-known kernel k-means [31]. Given the appropriate initialization, it can find the sub-optimal solution (local minima).

Despite great success in practice, it is quite sensitive to the initial positions of cluster prototypes, leading to degenerate local minima.

Example 1. In Fig. 2b, $\boldsymbol{\mu}_2$ is ill-initialized relatively far away from any points such that it will get no chance to be assigned with points and updated in the iterations of k-means. As a result, the clustering result by k-means is trapped in degenerate local minima as shown in Fig. 2c, where $\boldsymbol{\mu}_2$ is assigned with no point.

To overcome this problem, current k-means algorithm and its variants usually run many times with different initial prototypes and select the result with the smallest *distortion error* (6) [36], which are, however, computationally too expensive. The COLL method is an efficient and effective approach to the optimization problem (3), which can output smaller *distortion error* that is insensitive to the initialization.

2.3   Conscience On-Line Learning

In this section, we will introduce COLL to overcome the above issue.

2.3.1   The COLL Model

Let n_k denote the cumulative winning number of the k-th prototype, and $f_k = n_k / \sum_{l=1}^{c} n_l$ the corresponding winning frequency. In the beginning, they are initialized as

$$n_k = |v^{-1}(k)|, f_k = n_k/n, \forall k = 1, \ldots, c. \quad (12)$$

The *COLL* is performed as follows: initialize the same random assignment v as k-means, and iteratively update the assignment v and the prototypes $\boldsymbol{\mu}$ based on the *frequency sensitive (conscience) on-line learning* rule. That is, in the t-th iteration, for one randomly taken data point $\phi(\mathbf{x}_i)$, *select* the winning prototype $\boldsymbol{\mu}_{v_i}$ guided by the winning frequency f_k, i.e.,

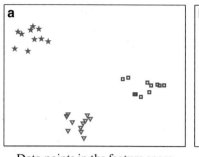

Data points in the feature space

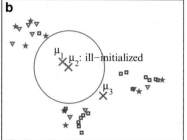

A random but ill initialization

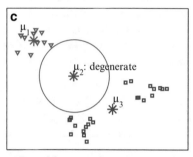

Kernel k-means degenerate result

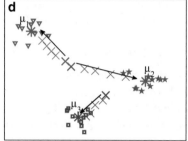

COLL procedure and result

Fig. 2 Illustration of ill-initialization and comparison of kernel k-means and COLL. Different clusters are plotted in *different colors* and *marks*, and the initial prototypes are plotted in "×" while the final in "*". (**a**) Data points in the feature space. (**b**) A random but ill initialization. Each prototype is computed as the mean of points randomly assigned to the corresponding cluster. μ_2 is ill-initialized, i.e., it is relatively far away from any points. (**c**) The degenerate result by kernel k-means. The ill-initialization makes the final μ_2 assigned with no point. (**d**) The procedure and result of COLL. The updating procedure of prototypes is plotted in thinner "×". The *conscience* mechanism successfully makes μ_2 win in the iterations, leading to a satisfying result

conscience based winner selection rule:

$$v_i = \arg \min_{k=1,\dots,c} \{f_k \parallel \phi(\mathbf{x}_i) - \boldsymbol{\mu}_k \parallel^2\}, \tag{13}$$

and *update* the winner $\boldsymbol{\mu}_{v_i}$ with learning rate η_t, i.e.,
on-line winner updating rule:

$$\boldsymbol{\mu}_{v_i} \leftarrow \boldsymbol{\mu}_{v_i} + \eta_t \left(\phi(\mathbf{x}_i) - \boldsymbol{\mu}_{v_i} \right), \tag{14}$$

as well as update the **winning frequency**

$$n_{v_i} \leftarrow n_{v_i} + 1, f_k = n_k \Big/ \sum_{l=1}^{c} n_l, \forall k = 1, \dots, c. \tag{15}$$

The iteration procedure continues until all prototypes converge or the number of iterations reaches a prespecified value t_{max}. The *convergence criterion* identical to that of k-means (9) is used. The learning rates $\{\eta_t\}$ satisfy conditions [4]:

$$\lim_{t \to \infty} \eta_t = 0, \qquad \sum_{t=1}^{\infty} \eta_t = \infty, \qquad \sum_{t=1}^{\infty} \eta_t^2 < \infty. \tag{16}$$

In practice, $\eta_t = const/t$, where *const* is some small constant, e.g., 1.

By reducing the winning rate of the frequent winners according to (13), the goal of the *conscience* mechanism is to bring all the prototypes available into the solution quickly and to bias the competitive process so that each prototype can win the competition with almost the same probability. In this way, it is insensitive to the ill-initialization, and thus prevents the result from being trapped into degenerate local minima; meanwhile converges much faster than other iterative methods [8].

Example 2. The same ill-initialization as kernel k-means in Example 1 is used by COLL. However, since the winning frequency of the ill-initialized μ_2 will become smaller in the later iterations, it gets the chance to win according to (13). Finally, an almost perfect clustering with small distortion error is obtained as shown in Fig. 2d.

2.3.2 The Computation of COLL

As discussed before, any effective algorithm in the kernel space must compute with only the kernel matrix K. To this end, we devise an efficient framework for computation of COLL based on a novel representation of prototypes termed *prototype descriptor*.

Let $\bar{\mathbb{R}}^{c \times (n+1)}$ denote the $c \times (n+1)$ matrix space satisfying $\forall A \in \bar{\mathbb{R}}^{c \times (n+1)}$, $A_{:,n+1} \geq 0$, i.e., the last column of matrix A is nonnegative. We define the *prototype descriptor* based on *kernel trick* as follows.

Definition 2 (Prototype Descriptor). A prototype descriptor is a matrix $W^\phi \in \bar{\mathbb{R}}^{c \times (n+1)}$, such that the k-th row represents prototype μ_k by

$$W^\phi_{k,i} = \langle \mu_k, \phi(\mathbf{x}_i) \rangle, \forall i = 1, \ldots, n, \ W^\phi_{k,n+1} = \langle \mu_k, \mu_k \rangle, \tag{17}$$

i.e.,

$$W^\phi = \begin{pmatrix} \langle \mu_1, \phi(\mathbf{x}_1) \rangle & \cdots & \langle \mu_1, \phi(\mathbf{x}_n) \rangle & \langle \mu_1, \mu_1 \rangle \\ \langle \mu_2, \phi(\mathbf{x}_1) \rangle & \cdots & \langle \mu_2, \phi(\mathbf{x}_n) \rangle & \langle \mu_2, \mu_2 \rangle \\ \vdots & \vdots & \vdots & \vdots \\ \langle \mu_c, \phi(\mathbf{x}_1) \rangle & \cdots & \langle \mu_c, \phi(\mathbf{x}_n) \rangle & \langle \mu_c, \mu_c \rangle \end{pmatrix}. \tag{18}$$

With this definition, the computation of the *distortion error* (6) now becomes:

$$\sum_{i=1}^{n} \| \phi(\mathbf{x}_i) - \boldsymbol{\mu}_{v_i} \|^2 = \sum_{i=1}^{n} \left(K_{i,i} + W^{\phi}_{v_i,n+1} - 2W^{\phi}_{v_i,i} \right). \tag{19}$$

Let's consider the computation of four ingredients of the COLL model.

Theorem 2 (Initialization). *The random initialization can be realized in the way of*

$$W^{\phi}_{:,1:n} = AK, \qquad W^{\phi}_{:,n+1} = diag(AKA^{\top}) \tag{20}$$

where $\mathrm{diag}(M)$ *denotes the main diagonal of a matrix* M *and the positive matrix* $A = [A_{k,i}]^{c\times n} \in R^{c\times n}_{+}$ *has the form*

$$A_{k,i} = \begin{cases} \frac{1}{|v^{-1}(k)|} & \text{if } i \in v^{-1}(k) \\ 0 & \text{otherwise.} \end{cases} \tag{21}$$

That is, the matrix A *reflects the initial assignment* v.

Proof. Assume the assignment is randomly initialized as v. Substitute the computation of the prototypes (4) to the definition of $W^{\phi}_{k,:}$ in (17), we get

$$W^{\phi}_{k,i} = \langle \boldsymbol{\mu}_k, \phi(\mathbf{x}_i) \rangle, \forall i = 1, \dots, n$$

$$= \left\langle \frac{\sum_{j \in v^{-1}(k)} \phi(\mathbf{x}_j)}{|v^{-1}(k)|}, \phi(\mathbf{x}_i) \right\rangle$$

$$= \sum_{j \in v^{-1}(k)} \frac{1}{|v^{-1}(k)|} K_{j,i}$$

$$= \sum_{j=1}^{n} A_{k,j} K_{j,i}$$

$$= A_{k,:} K_{:,i} \tag{22}$$

$$W^{\phi}_{k,n+1} = \langle \boldsymbol{\mu}_k, \boldsymbol{\mu}_k \rangle$$

$$= \left\langle \frac{\sum_{h \in v^{-1}(k)} \phi(\mathbf{x}_h)}{|v^{-1}(k)|}, \frac{\sum_{l \in v^{-1}(k)} \phi(\mathbf{x}_l)}{|v^{-1}(k)|} \right\rangle$$

$$= \sum_{h \in v^{-1}(k)} \sum_{l \in v^{-1}(k)} \frac{1}{|v^{-1}(k)|^2} K_{h,l}$$

$$= \sum_{h=1}^{n} \sum_{l=1}^{n} A_{k,h} A_{k,l} K_{h,l}$$

$$= A_{k,:} K A_{k,:}^{\top}. \tag{23}$$

Thus we obtain the initialization of W^{ϕ} as (20). The proof is finished. □

Theorem 3 (Conscience Based Winner Selection Rule). *The conscience based winner selection rule (13) can be realized in the way of*

$$v_i = \arg \min_{k=1,\ldots,c} \{f_k \cdot (K_{i,i} + W^{\phi}_{k,n+1} - 2W^{\phi}_{k,i})\}. \tag{24}$$

Proof. Consider the winner *selection* rule, i.e., (13), one can get

$$v_i = \arg \min_{k=1,\ldots,c} \{f_k \parallel \phi(\mathbf{x}_i) - \boldsymbol{\mu}_k \parallel^2\}$$

$$= \arg \min_{k=1,\ldots,c} \{f_k \cdot (K_{i,i} + W^{\phi}_{k,n+1} - 2W^{\phi}_{k,i})\}. \tag{25}$$

Thus we get the formula required. □

Theorem 4 (On-Line Winner Updating Rule). *The on-line winner updating rule (14) can be realized in the way of*

$$W^{\phi}_{v_i,j} \leftarrow \begin{cases} (1 - \eta_t) W^{\phi}_{v_i,j} + \eta_t K_{i,j} & j = 1, \ldots, n, \\[2mm] (1 - \eta_t)^2 W^{\phi}_{v_i,j} + \eta_t^2 K_{i,i} + 2(1 - \eta_t) \eta_t W^{\phi}_{v_i,i} & j = n+1. \end{cases} \tag{26}$$

Proof. Although we do not know exactly the expression of $\boldsymbol{\mu}_{v_i}$, however we can simply take $\boldsymbol{\mu}_{v_i}$ as a symbol of this prototype and denote its updated one as $\hat{\boldsymbol{\mu}}_{v_i}$. Substitute the on-line winner *updating* rule (14) to the winning prototype $W^{\phi}_{v_i,:}$, we have

$$W^{\phi}_{v_i,j} \leftarrow \langle \hat{\boldsymbol{\mu}}_{v_i}, \phi(\mathbf{x}_j) \rangle \quad \forall j = 1, \ldots, n$$

$$= \langle \boldsymbol{\mu}_{v_i} + \eta_t(\phi(\mathbf{x}_i) - \boldsymbol{\mu}_{v_i}), \phi(\mathbf{x}_j) \rangle$$

$$= (1 - \eta_t) \langle \boldsymbol{\mu}_{v_i}, \phi(\mathbf{x}_j) \rangle + \eta_t \langle \phi(\mathbf{x}_i), \phi(\mathbf{x}_j) \rangle, \tag{27}$$

$$W^{\phi}_{v_i,j} \leftarrow \langle \hat{\boldsymbol{\mu}}_{v_i}, \hat{\boldsymbol{\mu}}_{v_i} \rangle \quad j = n+1$$

$$. \quad = \langle \boldsymbol{\mu}_{v_i} + \eta_t(\phi(\mathbf{x}_i) - \boldsymbol{\mu}_{v_i}), \boldsymbol{\mu}_{v_i} + \eta_t(\phi(\mathbf{x}_i) - \boldsymbol{\mu}_{v_i}) \rangle$$

$$= (1 - \eta_t)^2 \langle \boldsymbol{\mu}_{v_i}, \boldsymbol{\mu}_{v_i} \rangle + \eta_t^2 \langle \phi(\mathbf{x}_i), \phi(\mathbf{x}_i) \rangle + 2(1 - \eta_t) \eta_t \langle \boldsymbol{\mu}_{v_i}, \phi(\mathbf{x}_i) \rangle. \tag{28}$$

Then we get the on-line winner *updating* rule as (26). □

It is a bit complicated to compute the *convergence criterion* without explicit expression of $\{\boldsymbol{\mu}_1, \ldots, \boldsymbol{\mu}_c\}$. Notice that, in **one** iteration, each point $\phi(\mathbf{x}_i)$ is assigned to **one and only one** winning prototype. Let array $\pi^k = [\pi_1^k, \pi_2^k, \ldots, \pi_{m_k}^k]$ store the indices of m_k *ordered* points assigned to the k-th prototype in **one** iteration. For instance, if $\phi(\mathbf{x}_1), \phi(\mathbf{x}_{32}), \phi(\mathbf{x}_8), \phi(\mathbf{x}_{20}), \phi(\mathbf{x}_{15})$ are 5 *ordered* points assigned to the 2-nd prototype in the t-th iteration, then the index array of the 2-nd prototype is $\pi^2 = [\pi_1^2, \pi_2^2, \ldots, \pi_{m_2}^2] = [1, 32, 8, 20, 15]$ with $\pi_1^2 = 1, \pi_2^2 = 32, \pi_3^2 = 8, \pi_4^2 = 20, \pi_5^2 = 15$ and $m_2 = 5$. The following lemma formulates the cumulative update of the k-th prototype based on the array $\pi^k = [\pi_1^k, \pi_2^k, \ldots, \pi_{m_k}^k]$.

Lemma 1. *In the t-th iteration, the relationship between the updated prototype $\boldsymbol{\mu}_k$ and the old $\hat{\boldsymbol{\mu}}_k$ is:*

$$\boldsymbol{\mu}_k = (1 - \eta_t)^{m_k} \hat{\boldsymbol{\mu}}_k + \sum_{l=1}^{m_k} (1 - \eta_t)^{m_k - l} \eta_t \phi(\mathbf{x}_{\pi_l^k}), \tag{29}$$

where array $\pi^k = [\pi_1^k, \pi_2^k, \ldots, \pi_{m_k}^k]$ stores the indices of m_k ordered points assigned to the k-th prototype in this iteration.

Proof. To prove this relationship, we use the Principle of Mathematical Induction. One can easily verify that (29) is true for $m_k = 1$ directly from (14),

$$\boldsymbol{\mu}_k = \hat{\boldsymbol{\mu}}_k + \eta_t \left(\phi(\mathbf{x}_{\pi_1^k}) - \hat{\boldsymbol{\mu}}_k \right)$$

$$= (1 - \eta_t)^1 \hat{\boldsymbol{\mu}}_k + \sum_{l=1}^{1} (1 - \eta_t)^0 \eta_t \phi(\mathbf{x}_{\pi_l^k}). \tag{30}$$

Assume that it is true for $m_k = m$, that is, for the first m *ordered* points,

$$\boldsymbol{\mu}_k = (1 - \eta_t)^m \hat{\boldsymbol{\mu}}_k + \sum_{l=1}^{m} (1 - \eta_t)^{m-l} \eta_t \phi(\mathbf{x}_{\pi_l^k}). \tag{31}$$

Then for $m_k = m + 1$, i.e., the $(m + 1)$-th point, from (14) we have

$$\boldsymbol{\mu}_k = \boldsymbol{\mu}_k + \eta_t \left(\phi(\mathbf{x}_{\pi_{m_k}^k}) - \boldsymbol{\mu}_k \right)$$
$$= (1 - \eta_t)\boldsymbol{\mu}_k + \eta_t \phi(\mathbf{x}_{\pi_{m_k}^k})$$
$$= (1 - \eta_t) \left((1 - \eta_t)^m \hat{\boldsymbol{\mu}}_k + \sum_{l=1}^{m} (1 - \eta_t)^{m-l} \eta_t \phi(\mathbf{x}_{\pi_l^k}) \right) + \eta_t \phi(\mathbf{x}_{\pi_{m_k}^k})$$
$$= (1 - \eta_t)^{m+1} \hat{\boldsymbol{\mu}}_k + \sum_{l=1}^{m+1} (1 - \eta_t)^{m+1-l} \eta_t \phi(\mathbf{x}_{\pi_l^k}). \tag{32}$$

This expression shows that (29) is true for $m_k = m + 1$. Therefore, by mathematical induction, it is true for all positive integers m_k. $\qquad\square$

Theorem 5 (Convergence Criterion). *The convergence criterion can be computed by*

$$\mathfrak{e}^{\phi} = \sum_{k=1}^{c} \left(\left(1 - \frac{1}{(1-\eta_t)^{m_k}} \right)^2 W_{k,n+1}^{\phi} \right) + \eta_t^2 \sum_{k=1}^{c} \sum_{h=1}^{m_k} \sum_{l=1}^{m_k} \frac{K_{\pi_h^k, \pi_l^k}}{(1-\eta_t)^{h+l}}$$

$$+ 2\eta_t \sum_{k=1}^{c} \left(\left(1 - \frac{1}{(1-\eta_t)^{m_k}} \right) \sum_{l=1}^{m_k} \frac{W_{k,\pi_l^k}^{\phi}}{(1-\eta_t)^l} \right). \tag{33}$$

Proof. According to Lemma 1, the old $\hat{\boldsymbol{\mu}}_k$ can be retained from the updated $\boldsymbol{\mu}_k$ as

$$\hat{\boldsymbol{\mu}}_k = \frac{\boldsymbol{\mu}_k}{(1-\eta_t)^{m_k}} - \eta_t \sum_{l=1}^{m_k} \frac{\phi(\mathbf{x}_{\pi_l^k})}{(1-\eta_t)^l}. \tag{34}$$

Substitute it to $\mathfrak{e}^{\phi} = \sum_{k=1}^{c} \| \boldsymbol{\mu}_k - \hat{\boldsymbol{\mu}}_k \|^2$, we have

$$\mathfrak{e}^{\phi} = \sum_{k=1}^{c} \| \boldsymbol{\mu}_k - \left(\frac{\boldsymbol{\mu}_k}{(1-\eta_t)^{m_k}} - \eta_t \sum_{l=1}^{m_k} \frac{\phi(\mathbf{x}_{\pi_l^k})}{(1-\eta_t)^l} \right) \|^2$$

$$= \sum_{k=1}^{c} \left(\left(1 - \frac{1}{(1-\eta_t)^{m_k}} \right)^2 \langle \boldsymbol{\mu}_k, \boldsymbol{\mu}_k \rangle \right) + \eta_t^2 \sum_{k=1}^{c} \sum_{h=1}^{m_k} \sum_{l=1}^{m_k} \frac{\langle \phi(\mathbf{x}_{\pi_h^k}), \phi(\mathbf{x}_{\pi_l^k}) \rangle}{(1-\eta_t)^{h+l}}$$

$$+ 2\eta_t \sum_{k=1}^{c} \left(\left(1 - \frac{1}{(1-\eta_t)^{m_k}} \right) \sum_{l=1}^{m_k} \frac{\langle \boldsymbol{\mu}_k, \phi(\mathbf{x}_{\pi_l^k}) \rangle}{(1-\eta_t)^l} \right). \tag{35}$$

Thus $\mathfrak{e}^{\phi}$ can be computed by (33). This ends the proof. $\qquad\square$

For clarification, Algorithm 1 summaries the *COLL* method. Figure 3 compares the clustering results generated by kernel k-means and COLL on one nonlinearly separable data set.

2.3.3   Computational Complexity

The computation of the COLL method consists of two parts: initialization of W^{ϕ} and iterations to update W^{ϕ}. From (20), the initialization of W^{ϕ} takes $O(cn^2)$ operations. For each iteration, the computational complexity is $O(n(c + (n+1)) + (c + n^2 + n))$. Since $O(n(c + (n+1)))$ operations are needed to perform the iteration (for each point, $O(c)$ to select a winner and $O(n+1)$ to update the winner, there being n points) and $O(c + n^2 + n)$ operations are needed to compute the *convergence criterion* $\mathfrak{e}^{\phi}$ (the first term of (33) taking $O(c)$ operations, the second term at most $O(n^2)$ operations, and the third term $O(n)$ operations). Assume the iteration number is $t_{\max}$, since

---

Algorithm 1: Conscience on-line learning (COLL)

1: **Input:** kernel matrix $K \in \mathbb{R}^{n \times n}$, c, $\{\eta_t\}$, ϵ, t_{max}.

2: **Output:** cluster assignment v subject to (3).

3: Randomly initialize assignment v and set $t = 0$;
   Initialize the winning frequency $\{f_k\}$ by (12);
   Initialize *prototype descriptor* $W^{\phi} \in \mathbb{R}^{c \times (n+1)}$ by (20).
   repeat

4:    Get c empty index arrays $\{\pi^k = \emptyset : k = 1, \ldots, c\}$ and a random permutation
      $\{I_1, \ldots, I_n : I_i \in \{1, \ldots, n\}$, s.t. $I_i \neq I_j, \forall i \neq j\}$, set $t = t + 1$.
      for $l = 1, \ldots, n$ **do**

5:       **Select** the winning prototype $W^{\phi}_{v_i,:}$ of the i-th point ($i = I_l$) by (24), and append i to the
         v_i-th index array π^{v_i}.

6:       **Update** the winning prototype $W^{\phi}_{v_i,:}$ with learning rate η_t by (26) and the winning
         frequency by (15).
      end for

7:    Compute $\mathbf{e}^{\phi}$ via (33).
   until $\mathbf{e}^{\phi} \leq \epsilon$ or $t \geq t_{max}$

8: Obtain the cluster assignment v_i by (24), $\forall i = 1, \ldots, n$.

---

in general $1 < c < n$, the computational complexity for iteration procedure is $O\left(t_{\max}\left(n(c + (n+1)) + (c + n^2 + n)\right)\right) = O\left(t_{\max}n^2\right)$. Consequently the total computational complexity of the COLL method is $O\left(cn^2 + t_{\max}n^2\right) = O\left(\max(c, t_{\max})n^2\right)$, which is the same as that of kernel k-means if the same number of iterations is obtained. However, due to the *conscience* mechanism [8] and on-line learning rule [4], the COLL achieves faster convergence rate than its counterpart. Thus fewer iterations are needed for COLL to converge. This is especially beneficial in large-scale data clustering.

2.4 Experiments and Applications

In this section, we report the experimental results in the application of video clustering (automatic segmentation). Video clustering plays an important role in automatic video summarization/abstraction as a preprocessing step [35]. Consider a video sequence in which the camera is fading/switching/cutting among a number of scenes, the goal of automatic video clustering is to cluster the video frames according to the different scenes. The gray-scale values of the raw pixels were used as the feature vector for each frame. For one video sequence, the frames $\{\mathbf{f}_i \in \mathbb{R}^d\}_{i=1}^n$ are taken as the data set, where $d = $ width $\times$ height and n is the length of the video sequence. We selected the 11 video sequences from the open-video website [16], which are 11 segments of the whole "NASA 25th Anniversary Show" with $d = 320 \times 240 = 76800$ and n (i.e., the duration of the sequence) varying from one sequence to another.

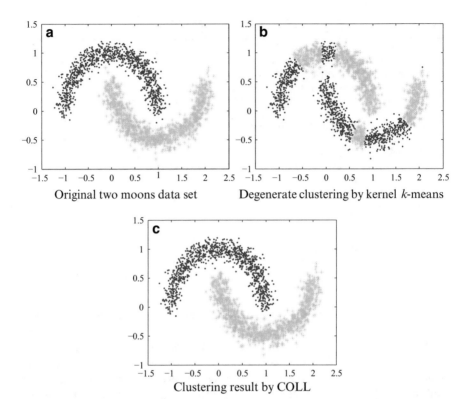

Original two moons data set

Degenerate clustering by kernel *k*-means

Clustering result by COLL

Fig. 3 Original two moons data set and clustering results by kernel *k*-means and the COLL method. (**a**) Original two moons data set. (**b**) Kernel *k*-means result. It is obvious that degenerate local optimum is obtained due to the sensitivity to ill-initialization. (**c**) Clustering result by COLL with the same initialization. An almost perfect clustering is obtained with only a few points being erroneously assigned

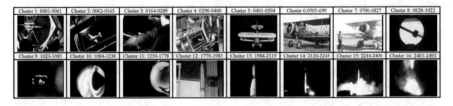

Fig. 4 Clustering frames of the video sequence ANNI002 into 16 scenes using COLL. A satisfying segmentation has been achieved

Figure 4 illustrates the clustering result of one video sequence "NASA 25th Anniversary Show, Segment 2" (ANNI002) by COLL. 2492 frames have been clustered into 16 scenes. Except for the frames from 400 to 405 and 694 to 701 as well as the last two clusters, where the separation boundaries are not so clear, satisfactory segmentation has been obtained. For comparison, "ground truth"

Table 1 The means (running 10 times) of NMI and computational time in seconds on the 11 video sequences, The bold values denote the best results

Video sequence (#frames)	Kernel k-means [31]		Global kernel k-means [36]		COLL	
	NMI	Time	NMI	Time	NMI	Time
ANNI001 (914)	0.781	72.2	0.801	94.0	**0.851**	**70.4**
ANNI002 (2492)	0.705	94.7	0.721	126.4	**0.741**	**89.0**
ANNI003 (4265)	0.712	102.2	0.739	139.2	**0.762**	**99.5**
ANNI004 (3897)	0.731	98.3	0.750	121.6	**0.759**	**93.6**
ANNI005 (11361)	0.645	152.2	0.656	173.3	**0.680**	**141.2**
ANNI006 (16588)	0.622	193.0	0.638	255.5	**0.642**	**182.3**
ANNI007 (1588)	0.727	81.1	0.740	136.7	**0.770**	**79.1**
ANNI008 (2773)	0.749	95.9	0.771	119.0	**0.794**	**81.5**
ANNI009 (12304)	0.727	167.0	0.763	184.4	**0.781**	**160.4**
ANNI010 (30363)	0.661	257.2	0.709	426.4	**0.734**	**249.0**
ANNI011 (1987)	0.738	85.4	0.749	142.7	**0.785**	**83.7**

#frames denote the number of frames of each video sequence

segmentation of each video sequence has been manually obtained, through which NMI values are computed to compare COLL with kernel k-means and global kernel k-means. Table 1 lists the average values (running 10 times) of NMI and computational time in seconds on the 11 video sequences. Additionally, the length of each video sequence (i.e., number of the frames) is also listed. The results in terms of average values of NMI and computational time reveal that COLL generates the best segmentation among the compared methods with the least computational time, which is a significant improvement.

3   Multi-exemplar Affinity Propagation

Multi-exemplar is another method for solving nonlinear clustering problem. From the viewpoint of multi-exemplar model, this section reviews a novel multi-exemplar clustering method, termed MEAP [43]. In the multi-exemplar model, each cluster is represented by an automatically determined number of exemplars and a super-exemplar. Each data point is assigned to the most suitable exemplar and each exemplar is assigned to the most suitable super-exemplar. The model aims to maximize the sum of all similarities between data points and the corresponding exemplars *plus* the sum of all linkages between exemplars and the corresponding super-exemplars. To solve this NP-hard problem, we use the max-sum belief propagation, producing clusters insensitive to initialization and converged to the neighborhood maximum. Experimental results in natural image categorization validate the effectiveness of MEAP in computer vision application.

3.1 Affinity Propagation

Affinity propagation is a single-exemplar clustering algorithm using max-sum (max-product) belief propagation to obtain good exemplars. Given a user-defined similarity matrix $[s_{ij}]_{N \times N}$ of N points, it aims at searching for a valid configuration of labels $\mathbf{c} = [c_1, \ldots, c_N]$ to maximize the following objective function [10],

$$\mathscr{S}(\mathbf{c}) = \sum_{i=1}^{N} s_{ic_i} + \sum_{k=1}^{N} \delta_k(\mathbf{c}), \tag{36}$$

where $\delta_k(\mathbf{c})$ is an *exemplar-consistency* constraint such that if some data point i has selected k as its exemplar, i.e., $c_i = k$, then data point k must select itself as an exemplar, i.e., $c_k = k$,

$$\delta_k(\mathbf{c}) = \begin{cases} -\infty & \text{if } c_k \neq k \text{ but } \exists i : c_i = k \\ 0 & \text{otherwise.} \end{cases} \tag{37}$$

AP is an optimized max-sum belief propagation algorithm over the factor graph in Fig. 5. It begins by simultaneously considering all data points as potential exemplars, and recursively transmits real-valued messages between data points, until high-quality exemplars emerge. There are two kinds of messages, which are $r(i, k)$, sent from point i to the candidate exemplar k, reflecting the accumulated evidence for how well-suited point k is to serve as the exemplar for point i, and $a(i, k)$, sent from the candidate exemplar k to point i, reflecting the accumulated evidence for how appropriate it would be for point i to choose point k as its exemplar (Fig. 5). They are initialized as zero, and updated, respectively, as follows:

$$r(i, k) \leftarrow s_{ik} - \max_{j \neq k}[s_{ij} + a(i, j)] \tag{38}$$

$$a(i, k) \leftarrow \min\left[0, r(k, k) + \sum_{i' \notin \{i, k\}} \max[0, r(i', k)]\right], k \neq i \tag{39}$$

$$a(k, k) \leftarrow \sum_{i' \neq k} \max[0, r(i', k)]. \tag{40}$$

Fig. 5 Factor graph of the AP method and its messages

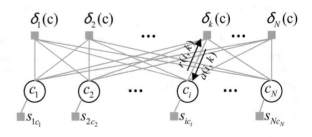

The assignment vector $\mathbf{c} = [c_1, \ldots, c_N]$ is computed as $c_i = \arg\max_j[a(i,j)+r(i,j)]$ after convergence.

3.2 Multi-exemplar Affinity Propagation

Let $[s_{ij}]_{N\times N}$ be a user-defined similarity matrix with s_{ij} measuring the similarity between point i and the potential exemplar j, and $[l_{ij}]_{N\times N}$ a linkage matrix with l_{ij} measuring the linkage between exemplar i and its potential super-exemplar j. We develop a multi-exemplar model that seeks two mappings, $\psi_1 : \{1, \ldots, N\} \rightarrow \{1, \ldots, N\}$ assigning data point i to exemplar $\psi_1(i)$, and $\psi_2 : \{\psi_1(1), \ldots, \psi_1(N)\} \rightarrow \{\psi_1(1), \ldots, \psi_1(N)\}$ assigning exemplar $\psi_1(i)$ to super-exemplar $\psi_2(\psi_1(i)) = (\psi_2 \circ \psi_1)(i)$, where $\circ$ denotes the function composition. The goal is to maximize the sum $\mathscr{S}_1$ of all similarities between data points and the corresponding exemplars *plus* the sum $\mathscr{S}_2$ of all linkages between exemplars and the corresponding super-exemplars.

3.2.1 The Model

Let $C = [c_{ij}]_{N\times N}$ be an assignment matrix, where the non-diagonal elements $c_{ij} \in \{0, 1\}(j \neq i)$ denote that point j is the exemplar of point i if $c_{ij} = 1$, i.e. $\psi_1(i) = j$, and the diagonal elements $c_{ii} \in \{0, \ldots, N\}$ denote that exemplar c_{ii} is the super-exemplar of exemplar i if $c_{ii} \in \{1, \ldots, N\}$, i.e. $\psi_2(i) = c_{ii}$,

$$c_{ij} = \begin{cases} 1 & \text{if } j \text{ is an exemplar of } i \\ 0 & \text{otherwise} \end{cases} \quad \forall i \neq j, \tag{41}$$

$$c_{ii} = \begin{cases} k \in \{1, \ldots, N\} & \text{if } k \text{ is a super} - \text{exemplar of } i \\ 0 & \text{if } i \text{ is not an exemplar.} \end{cases} \tag{42}$$

The sum of all similarities between data points and the corresponding exemplars, i.e., $\mathscr{S}_1$, and the sum of all linkages between exemplars and the corresponding super-exemplars, i.e., $\mathscr{S}_2$, can be respectively expressed as

$$\mathscr{S}_1 = \sum_{i=1}^{N}\sum_{j=1}^{N} s_{ij} \cdot [c_{ij} \neq 0], \quad \mathscr{S}_2 = \sum_{i=1}^{N} l_{ic_{ii}} \cdot [c_{ii} \neq 0], \tag{43}$$

where $[\cdot]$ is the Iverson notation with [true] $= 1$ and [false] $= 0$. We define a function matrix $[S_{ij}(c_{ij})]_{N\times N}$ with non-diagonal elements incorporating the similarities s_{ij} between data point i and the potential exemplar j, and diagonal elements incorporating the exemplar preference s_{ii} *plus* the linkage $l_{ic_{ii}}$ between exemplar i and its super-exemplar c_{ii}. That is,

$$S_{ij}(c_{ij}) = \begin{cases} s_{ij} & \text{if } i \neq j \& c_{ij} \neq 0 \\ s_{ii} + l_{ic_{ii}} & \text{if } i = j \& c_{ii} \neq 0 \\ 0 & \text{otherwise.} \end{cases} \tag{44}$$

We have $\mathscr{S}_1 + \mathscr{S}_2 = \sum_{i=1}^{N} \sum_{j=1}^{N} S_{ij}(c_{ij})$. The valid assignment matrix C must satisfy the following three constraints:

1. *Exemplar's "1-of-N" constraint* [15]: Each data point i must be assigned to *exactly one* exemplar,

$$I_i(c_{i1}, \dots, c_{iN}) = \begin{cases} -\infty & \text{if } \sum_{j=1}^{N} [c_{ij} \neq 0] \neq 1 \\ 0 & \text{otherwise.} \end{cases} \tag{45}$$

2. *Exemplar consistency* constraint [15]: If there exists a data point i selecting data point j as its exemplar, then data point j must be an exemplar itself,

$$E_j(c_{1j}, \dots, c_{Nj}) = \begin{cases} -\infty & \text{if } c_{jj} = 0 \text{ but } \exists i : c_{ij} = 1 \\ 0 & \text{otherwise.} \end{cases} \tag{46}$$

3. *Super-exemplar consistency* constraint: If some exemplar i has chosen exemplar k as its super-exemplar, i.e., $c_{ii} = k$, then k must be a super-exemplar itself,

$$F_k(c_{11}, \dots, c_{NN}) = \begin{cases} -\infty & \text{if } c_{kk} \neq k \text{ but } \exists i : c_{ii} = k \\ 0 & \text{otherwise.} \end{cases} \tag{47}$$

The goal of the multi-exemplar model is to maximize the following objective function

$$\mathscr{S}(C) = \mathscr{S}_1 + \mathscr{S}_2 + \text{three constraints}$$

$$= \sum_{i=1}^{N} \sum_{j=1}^{N} S_{ij}(c_{ij}) + \sum_{i=1}^{N} I_i(c_{i1}, \dots, c_{iN})$$

$$+ \sum_{j=1}^{N} E_j(c_{1j}, \dots, c_{Nj}) + \sum_{k=1}^{N} F_k(c_{11}, \dots, c_{NN}). \tag{48}$$

Figure 6 illustrates the multi-exemplar model. The data points, exemplars and super-exemplars form a two-layer structure by the two mappings ψ_1 and ψ_2. The lower-layer is modeled by the mapping ψ_1. The sum of all similarities between data points and the corresponding exemplars, i.e. $\mathscr{S}_1$, is used to measure the *within-subcluster* compactness. The higher-layer is modeled by the mapping ψ_2. The sum of all linkages between exemplars and the corresponding super-exemplars, i.e. $\mathscr{S}_2$,

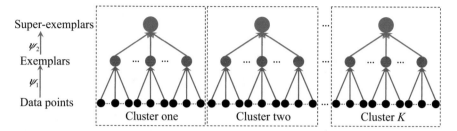

Fig. 6 Illustration of the multi-exemplar model. The mapping ψ_1 assigns each data point to the most appropriate exemplar and the mapping ψ_2 assigns each exemplar to the most appropriate super-exemplar. This model can effectively characterize clusters consisting of multiple subclusters

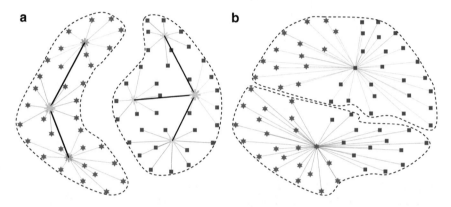

Fig. 7 MEAP vs. AP. The two ground-truth categories are plotted by *red hexagrams* and *blue squares*, respectively. The decision boundaries are plotted by *dash curves*. Each point is assigned to the most appropriate exemplar by *thin lines*. In MEAP, each exemplar (small "*") is assigned to its most appropriate super-exemplar (large "*") by *thick lines*. (**a**) MEAP; (**b**) AP

is used to measure the *within-cluster* compactness. From the single-exemplar theory, maximizing the within-cluster similarity automatically maximizes the between-cluster separation [32]. Therefore, the appropriate multi-exemplar model should be that both the *within-subcluster* compactness and the *within-cluster* compactness are maximized. Maximizing $\mathscr{S}_1 + \mathscr{S}_2$ under the constraints of producing valid clusters (i.e., I, E, F) makes the model effectively characterize clusters consisting of multiple subclusters. Figure 7 compares MEAP with AP in one synthetic data set. From the viewpoint of the maximum margin clustering [49], MEAP finds better decision boundaries than AP, i.e., larger margins are obtained.

3.2.2 Optimization

Exactly searching for an optimal assignment matrix that maximizes the objective function (48) is NP-hard, since the multi-exemplar model is a generalization of

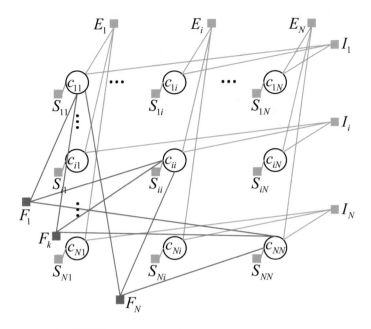

Fig. 8 Factor graph of MEAP

the single-exemplar one, the optimization of which is proved to be NP-hard [14]. To this end, the max-sum belief propagation is utilized, which is a local-message-passing algorithm guaranteed to converge to the neighborhood maximum [46]. The factor graph is shown in Fig. 8. There are seven types of messages passing between variable nodes and function nodes as shown in Fig. 9. In the max-sum algorithm, the message updating involves either a message from a variable to each adjacent function or that from a function to each adjacent variable. The message from a variable to a function sums together the messages from all adjacent functions except the one receiving the message [19],

$$\mu_{x \to f}(x) \leftarrow \sum_{h \in \text{ne}(x) \setminus \{f\}} \mu_{h \to x}(x), \tag{49}$$

where $\text{ne}(x)$ denotes the set of adjacent functions of variable x. The message from a function to a variable involves a maximization over all arguments of the function except the variable receiving the message [19],

$$\mu_{f \to x}(x) \leftarrow \max_{X \setminus \{x\}} \left[f(X) + \sum_{y \in X \setminus \{x\}} \mu_{y \to f}(y) \right], \tag{50}$$

Fig. 9 Messages of MEAP

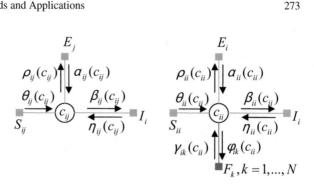

where $X = \text{ne}(f)$ is the set of arguments of function f. Since the messages associated with the non-diagonal variables (i.e., $c_{ij}, i \neq j$) and the diagonal variables (i.e., c_{ii}) are quite different, we will discuss them separately.

1. **Messages of Non-diagonal Elements**

    As shown on the left of Fig. 9, there are five types of messages associated with $c_{ij}, i \neq j$ as follows ($m = 0, 1$):

$$\theta_{ij}(m) = \mu_{S_{ij} \to c_{ij}}(m) = S_{ij}(m) \tag{51}$$

$$\rho_{ij}(m) = \mu_{c_{ij} \to E_j}(m) = \theta_{ij}(m) + \eta_{ij}(m) \tag{52}$$

$$\alpha_{ij}(m) = \mu_{E_j \to c_{ij}}(m)$$

$$= \max_{c_{i'j}: i' \neq i} \left[E_j(c_{1j}, \ldots, c_{Nj}) + \sum_{i'' \neq i} \rho_{i''j}(c_{i''j}) \right] \tag{53}$$

$$\beta_{ij}(m) = \mu_{c_{ij} \to I_i}(m) = \theta_{ij}(m) + \alpha_{ij}(m) \tag{54}$$

$$\eta_{ij}(m) = \mu_{I_i \to c_{ij}}(m)$$

$$= \max_{c_{ij'}: j' \neq j} \left[I_i(c_{i1}, \ldots, c_{iN}) + \sum_{j'' \neq j} \beta_{ij''}(c_{ij''}) \right]. \tag{55}$$

According to (44) and (51), these messages take as input the similarity $s_{ij}, i \neq j$. Consequently, these messages reflect the accumulated evidence for deciding the partial mapping ψ_1 from data point i to the potential exemplar j ($j \neq i$). That is, $m = c_{ij} = 1$ implies that $\psi_1(i) = j$, and $m = c_{ij} = 0$ implies that $\psi_1(i) \neq j$.

2. **Messages of Diagonal Elements**

There are seven types of messages associated with $c_{ij}, i = j$ (the right of Fig. 9) as follows ($m = 0, \ldots, N$):

$$\theta_{ii}(m) = \mu_{S_{ii} \to c_{ii}}(m) = S_{ii}(m) \tag{56}$$

$$\rho_{ii}(m) = \mu_{c_{ii} \to E_i}(m)$$

$$= \theta_{ii}(m) + \eta_{ii}(m) + \sum_{k=1}^{N} \gamma_{ik}(m) \tag{57}$$

$$\alpha_{ii}(m) = \mu_{E_i \to c_{ii}}(m)$$

$$= \max_{c_{i'i}:i' \neq i} \left[E_i(c_{1i}, \ldots, c_{Ni}) + \sum_{i'' \neq i} \rho_{i''i}(c_{i''i}) \right] \tag{58}$$

$$\beta_{ii}(m) = \mu_{c_{ii} \to I_i}(m)$$

$$= \theta_{ii}(m) + \alpha_{ii}(m) + \sum_{k=1}^{N} \gamma_{ik}(m) \tag{59}$$

$$\eta_{ii}(m) = \mu_{I_i \to c_{ii}}(m)$$

$$= \max_{c_{ii'}:i' \neq i} \left[I_i(c_{i1}, \ldots, c_{iN}) + \sum_{i'' \neq i} \beta_{ii''}(c_{ii''}) \right] \tag{60}$$

$$\phi_{ik}(m) = \mu_{c_{ii} \to F_k}(m)$$

$$= \theta_{ii}(m) + \alpha_{ii}(m) + \eta_{ii}(m) + \sum_{k' \neq k} \gamma_{ik'}(m) \tag{61}$$

$$\gamma_{ik}(m) = \mu_{F_k \to c_{ii}}(m)$$

$$= \max_{c_{i'i'}:i' \neq i} \left[F_k(c_{11}, \ldots, c_{NN}) + \sum_{j \neq i} \phi_{jk}(c_{jj}) \right]. \tag{62}$$

According to (44) and (56), these messages take as input both of the exemplar preference s_{ii} and the linkage $l_{ic_{ii}}$ between exemplars and super-exemplars. Consequently, these messages reflect the accumulated evidence for simultaneously deciding the partial mapping ψ_1 from point i to itself and the mapping ψ_2 from exemplar i to the potential super-exemplar k (k can be either $k = i$ or $k \neq i$). That is, $m = c_{ii} \in \{1, \ldots, N\}$ implies that $\psi_1(i) = i, \psi_2(i) = c_{ii}$, and $m = c_{ii} = 0$ implies that $\psi_1(i) \neq i$ and i is not in the domain of ψ_2.

3. **Simplified Messages**

By applying some mathematical tricks used in [14, 15], we can obtain the simplified messages as follows:

$$i \neq j$$

$$\tilde{\rho}_{ij} \leftarrow s_{ij} - \max \left[\max_{j' \notin \{j,i\}} [s_{ij'} + \tilde{\alpha}_{ij'}], \max_{m \in \{1, \ldots, N\}} [l_{im} + \tilde{\gamma}_{im}] + s_{ii} + \tilde{\alpha}_{ii} \right] \tag{63}$$

$$\tilde{\alpha}_{ij} \leftarrow \min\left[0, \max_{m\in\{1,\ldots,N\}} \tilde{\rho}_j^m + \sum_{i'\notin\{i,j\}} \max[0, \tilde{\rho}_{i'j}]\right] \tag{64}$$

$$\forall i = 1,\ldots,N, \ k = 1,\ldots,N$$

$$\tilde{\rho}_i^k \leftarrow s_{ii} + l_{ik} - \max_{i'\neq i}[s_{ii'} + \tilde{\alpha}_{ii'}] + \tilde{\gamma}_{ik} \tag{65}$$

$$\tilde{\alpha}_{ii} \leftarrow \sum_{i'\neq i} \max\left[0, \tilde{\rho}_{i'i}\right] \tag{66}$$

$$\tilde{\phi}_{ik} \leftarrow \min\left[l_{ik} - \max_{m\neq k}[l_{im} + \tilde{\gamma}_{im}], \tilde{\alpha}_{ii} + \tilde{\rho}_i^k - \tilde{\gamma}_{ik}\right] \tag{67}$$

$$\tilde{\gamma}_{kk} \leftarrow \sum_{i'\neq i} \max\left[0, \tilde{\phi}_{i'k}\right] \tag{68}$$

$$\tilde{\gamma}_{ik} \leftarrow \min\left[0, \tilde{\phi}_{kk} + \sum_{i'\notin\{i,k\}} \max[0, \tilde{\phi}_{i'k}]\right], k \neq i, \tag{69}$$

where $\tilde{\rho}_{ij} = \rho_{ij}(1) - \rho_{ij}(0)$, $\tilde{\alpha}_{ij} = \alpha_{ij}(1) - \alpha_{ij}(0)$, $\tilde{\rho}_i^k = \rho_{ii}(k) - \rho_{ii}(0)$, $\tilde{\alpha}_{ii} = \alpha_{ii}(k) - \alpha_{ii}(0)$, $\tilde{\phi}_{ik} = \phi_{ik}(k) - \max_{m\neq k}\phi_{ik}(m)$, $\tilde{\gamma}_{ik} = \gamma_{ik}(k) - \gamma_{ik}(m : m \neq k)$, and β, η are eliminated. The values of all messages $\tilde{\rho}, \tilde{\alpha}, \tilde{\phi}$, and $\tilde{\gamma}$ are initialized as zero and updated via Eqs. (63)–(69). The message-updating procedure may be terminated after the local decisions stay constant for some number of iterations t_{conv}, or after a fixed number of iterations t_{max}.

4. Computing Assignment Matrix

To estimate the value of an element c_{ij}, we sum together all incoming messages to c_{ij} and take the value $\hat{c}_{ij}$ that maximizes the sum. That is,

$$\hat{c}_{ij} = \arg\max_{c_{ij}} \left[\theta_{ij}(c_{ij}) + \alpha_{ij}(c_{ij}) + \eta_{ij}(c_{ij})\right]$$

$$= \begin{cases} 1 & \text{if } \tilde{\alpha}_{ij} + \tilde{\rho}_{ij} \geq 0 \\ 0 & \text{otherwise} \end{cases} \forall i \neq j, \tag{70}$$

$$\hat{c}_{ii} = \arg\max_{c_{ii}} \left[\theta_{ii}(c_{ii}) + \alpha_{ii}(c_{ii}) + \eta_{ii}(c_{ii}) + \sum_{k=1}^{N} \gamma_{ik}(c_{ii})\right]$$

$$= \begin{cases} \arg\max_k \tilde{\rho}_i^k & \text{if } \tilde{\alpha}_{ii} + \max_k \tilde{\rho}_i^k \geq 0 \\ 0 & \text{otherwise.} \end{cases} \tag{71}$$

The derivation can be found in the supplementary material. From the assignment matrix, we obtain the two mappings ψ_1 and ψ_2, and thus generate the clustering labels $\{(\psi_2 \circ \psi_1)(1), \ldots, (\psi_2 \circ \psi_1)(N)\}$ of N data points.

3.3   Experiments and Applications

In this section, we report the applications in image categorization on the JAFFE data set.

The JAFFE database [25] contains 213 images of 7 facial expressions posed by 10 Japanese females. The 7 facial expressions include 6 basic facial expressions, i.e., happiness, sadness, surprise, anger, disgust, fear, *plus* 1 neutral expression. There are 3 or 4 examples for each expression per person. Images of the same person in different facial expressions should be taken as in distinct subclasses [52, 53]. The goal is to cluster the 213 images into 10 groups according to identity.

We plot in Fig. 10 the complete results on the JAFFE data set by MEAP when setting the preferences at the median of the similarities. The exemplars and super-exemplars are bounded by thin and thick red lines, respectively. Each exemplar is

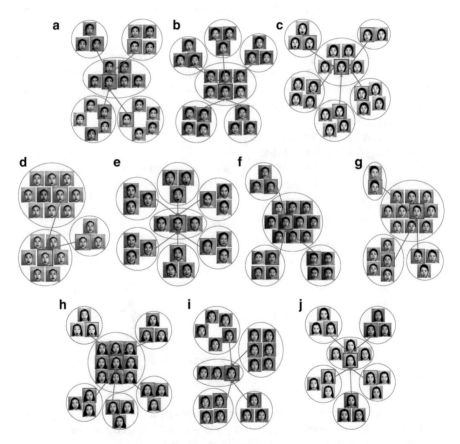

Fig. 10 Clusters learned by MEAP on the JAFFE data set. (**a**) KA cluster; (**b**) KL cluster; (**c**) KM cluster; (**d**) KR cluster; (**e**) MK cluster; (**f**) NA cluster; (**g**) NM cluster; (**h**) TM cluster; (**i**) UY cluster; (**j**) YM cluster

connected to the corresponding super-exemplar by the blue arrow. In this setting, although the AP method results in a moderate number of clusters [14], yet it over-segments the images of ten subjects into more clusters than the actual, with each cluster corresponding to a facial expression. When setting the preferences at the minimum of similarities, leading to a small number of clusters in AP, some images are misclassified by AP due to the presence of facial expression. However, MEAP can overcome the problem and almost perfectly group images of the same subject into one cluster by assigning exemplars of the same subject to the corresponding super-exemplar. Among 213 images, only 3 images belonging to the YM category are misclassified to the TM category, leading to a 98.6 % classification rate. The reason is that the exemplar of these 3 images is more similar to the super-exemplar of the TM category than YM. On the other hand, the AP algorithm misclassifies 36 images, resulting in a 83.1 % classification rate.

4   Graph-Based Multi-prototype Competitive Learning

Based on the advantages of graph method, multi-prototype representation, and competitive learning, this section reviews a novel nonlinear clustering method, namely GMPCL [45]. Based on the graph method, the GMPCL method firstly generates an initial coarse clustering. That is, we first construct a graph with vertex energy vector from the data set; a core-point set is computed according to the vertex energy, from which we can obtain an initial clustering by constructing connected components via neighbor connectivity. Then, a multi-prototype competitive learning is designed to refine this initial clustering so as to generate accurate clustering results. Experimental results in automatic color image segmentation validate the effectiveness of GMPCL and show that GMPCL provides a useful tool for computer vision.

4.1   Graph-Based Initial Clustering

Given a data set $\mathscr{D} = \{\mathbf{x}_1, \ldots, \mathbf{x}_n\}$ of n points in $\mathbb{R}^d$, the first step of the graph-based algorithm is to construct a graph $G^{\mathbf{e}} = (\mathscr{V}, A, \mathbf{e})$. The vertex set $\mathscr{V}$ contains one node for each sample in $\mathscr{D}$. The affinity matrix $A = [A_{ij}]_{n \times n}$ is defined as

$$A_{ij} \triangleq \begin{cases} \exp(-\| \mathbf{x}_i - \mathbf{x}_j \|^2) & \text{if } \mathbf{x}_i \in \mathscr{N}_k(\mathbf{x}_j) \wedge \mathbf{x}_j \in \mathscr{N}_k(\mathbf{x}_i) \\ 0 & \text{otherwise,} \end{cases} \tag{72}$$

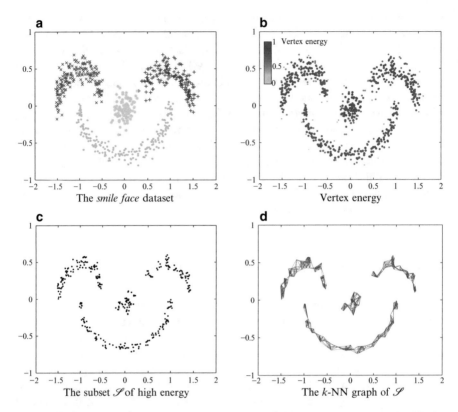

Fig. 11 *Smile face* example. (**a**) The *smile face* data set, different classes are plotted in *different markers*. (**b**) The vertex energy of *smile face*, the color and size of each point is marked according to its vertex energy. (**c**) The subset $\mathscr{S}$ comprising vertices of high energy (*core points*). (**d**) The k-NN graph of $\mathscr{S}$ with $k = 21$

where $\mathscr{N}_k(\mathbf{x}_i)$ denotes the set consisting of k nearest neighbors of $\mathbf{x}_i$. The vertex energy vector $\mathbf{e} = [e_1, \ldots, e_n]^\top$ is defined as

$$e_i \triangleq \log_2\left(1 + \frac{\sum_j A_{ij}}{\max_{l=1,\ldots,n} \sum_j A_{lj}}\right), i = 1, \ldots, n. \tag{73}$$

The component $e_i \in [0, 1]$ is the vertex energy of $\mathbf{x}_i$, which measures how "important" $\mathbf{x}_i$ is.

Figure 11a shows a *smile face* data set, and Fig. 11b plots its vertex energy. In [12], the density is measured simply by the number of points within the ϵ-neighborhood of one point. However, the vertex energy defined in (73) takes into account the correlations between all data points, which results in a global estimate of the vertex energy. Although both of them can be used to discover arbitrarily shaped clusters, the vertex energy is more suitable for datasets containing clusters of differing densities.

A subset $\mathscr{S}$ consisting of the vertices of higher energy is obtained (e.g., Fig. 11c), which is termed *core-point* set.

Definition 3. Given a graph $G^{\mathrm{e}} = (\mathscr{V}, A, \mathbf{e})$ and a percentage ρ, the core-point set $\mathscr{S}$ is defined as $\mathscr{S} \triangleq \{\mathbf{x}_i | e_i \geq \zeta\}$ with $\zeta \in [0, 1]$ such that $|\mathscr{S}|/|\mathscr{V}| = \rho$.

The core-point-connectivity of any two core points $\mathbf{p}$ and $\mathbf{q}$ in $\mathscr{S}$ is defined as follows.

Definition 4 (Core-Point-Connectivity). Two core points $\mathbf{p}$ and $\mathbf{q}$ in $\mathscr{S}$ are core-point-connected w.r.t. k (denoted as $\mathbf{p} \bowtie_k^{\mathscr{S}} \mathbf{q}$) if there exists a chain of core points $\mathbf{p}_1, \ldots, \mathbf{p}_m$, $\mathbf{p}_1 = \mathbf{p}$, $\mathbf{p}_m = \mathbf{q}$ such that $\mathbf{p}_{i+1} \in \mathscr{N}_k(\mathbf{p}_i) \bigcap \mathscr{S}$ and $\mathbf{p}_i \in \mathscr{N}_k(\mathbf{p}_{i+1}) \bigcap \mathscr{S}$.

From the viewpoint of density-based clustering [11, 12], the core-point-connectivity separates $\mathscr{S}$ into some natural subgroups, as shown in Fig. 11d, which are defined as connected components as follows.

Definition 5. A set of c connected components $\{\mathscr{I}_1, \ldots, \mathscr{I}_c\}$ is obtained by separating the core-point set $\mathscr{S}$ w.r.t. k, such that $\forall i \neq j, \mathscr{I}_i \bigcap \mathscr{I}_j = \emptyset$, $\mathscr{S} = \bigcup_{i=1}^c \mathscr{I}_i$, and $\forall \mathbf{p}, \mathbf{q} \in \mathscr{I}_i$, $\mathbf{p} \bowtie_k^{\mathscr{S}} \mathbf{q}$, while $\forall \mathbf{p} \in \mathscr{I}_i, \forall \mathbf{q} \in \mathscr{I}_j, i \neq j, \mathbf{p} \bowtie_k^{\mathscr{S}} \mathbf{q}$ does not hold.

The connected components $\{\mathscr{I}_1, \ldots, \mathscr{I}_c\}$ are taken as initial clusters which will be further refined via *multi-prototype* competitive learning.

4.2 Multi-prototype Competitive Learning

The initial clusters $\{\mathscr{I}_1, \ldots, \mathscr{I}_c\}$ obtained in the first phase take into account only data points of higher energy, and the remaining data points are not assigned with cluster labels. Therefore, the output of the first phase is only a coarse clustering that requires further refinement. Rather than directly assigning the unlabeled data points to the *core points* as in [11], this section employs classical competitive learning to refine the initial clustering and assign cluster labels to all data points. Experimental results show that the approach can obtain at least 9.8 % improvement over the direct assignment.

Since the data set is nonlinearly separable, a nonlinear cluster with concave boundaries would always exist, which cannot be characterized by a *single* prototype that produces *convex* boundaries [24]. However, *multiple* prototypes produce subregions of the Voronoi diagram which can approximately characterize one cluster of an arbitrary shape. Therefore, we represent each cluster by *multiple* prototypes.

Every point in $\mathscr{I}_j$ can be taken as one of the initial prototypes representing the j-th cluster $\mathscr{C}_j$. But there is no need of using so many prototypes to represent one cluster and some of them are more appropriate and more effective than others. These points should be as few as possible to lower the computational complexity of *multi-prototype* competitive learning, meanwhile be scattered in the whole space of the initial cluster in order to suitably characterize the corresponding cluster. Affinity propagation [14] can generate suitable prototypes to represent an input data set without preselecting the number of prototypes. In experiments, the representative points are obtained by applying affinity propagation to each $\mathscr{I}_j$. The similarity $s(\mathbf{x}_i, \mathbf{x}_{i'})$ between $\mathbf{x}_i, \mathbf{x}_{i'} \in \mathscr{I}_j$ is set to $- \parallel \mathbf{x}_i - \mathbf{x}_{i'} \parallel^2$ and the preferences are set to the median of the similarities, which outputs p_j suitable *multi-prototypes* $\mathbf{w}_j^1, \ldots, \mathbf{w}_j^{p_j}$. In this way, we obtain an initial *multi-prototype* set

$$\mathscr{W} = \{\underbrace{\mathbf{w}_1^1, \ldots, \mathbf{w}_1^{p_1}}_{\text{represent } \mathscr{C}_1}, \underbrace{\mathbf{w}_2^1, \ldots, \mathbf{w}_2^{p_2}}_{\text{represent } \mathscr{C}_2}, \ldots, \underbrace{\mathbf{w}_c^1, \ldots, \mathbf{w}_c^{p_c}}_{\text{represent } \mathscr{C}_c}\}. \tag{74}$$

We use the index notation ω_j^q to denote the *multi-prototype* $\mathbf{w}_j^q$. That is, referring to the ω_j^q-th *multi-prototype* is equivalent to mentioning $\mathbf{w}_j^q$, and $\omega = \{\omega_1^1, \ldots, \omega_1^{p_1}, \omega_2^1, \ldots, \omega_2^{p_2}, \ldots, \omega_c^1, \ldots, \omega_c^{p_c}\}$.

After the initial *multi-prototype* set $\mathscr{W}$ is obtained, classical competitive learning is performed to iteratively update the *multi-prototypes* such that the *multi-prototype objective* function is minimized:

$$J(\mathscr{W}) = \sum_{i=1}^n \parallel \mathbf{x}_i - \mathbf{w}_{v_i}^{v_i} \parallel^2, \tag{75}$$

where $\mathbf{w}_{v_i}^{v_i}$ satisfies $\omega_{v_i}^{v_i} = \arg\min_{\omega_j^q \in \omega} \parallel \mathbf{x}_i - \mathbf{w}_j^q \parallel^2$, i.e., the winning *multi-prototype* of $\mathbf{x}_i$ that is nearest to $\mathbf{x}_i$. For each randomly taken $\mathbf{x}_i$, the winning *multi-prototype* $\omega_{v_i}^{v_i}$ is selected via the **winner selection rule**:

$$\omega_{v_i}^{v_i} = \arg\min_{\omega_j^q \in \omega} \parallel \mathbf{x}_i - \mathbf{w}_j^q \parallel^2, \tag{76}$$

and is updated by the **winner update rule**:

$$\mathbf{w}_{v_i}^{v_i} \leftarrow \mathbf{w}_{v_i}^{v_i} + \eta_t(\mathbf{x}_i - \mathbf{w}_{v_i}^{v_i}) \tag{77}$$

with learning rates $\{\eta_t\}$ satisfying [4]: $\lim_{t \to \infty} \eta_t = 0$, $\sum_{t=1}^{\infty} \eta_t = \infty$, $\sum_{t=1}^{\infty} \eta_t^2 < \infty$. In practice, $\eta_t = const/t$, where *const* is a small constant, e.g., 0.5.

Figure 12a illustrates the procedure of updating the winning *multi-prototype*. The converged *multi-prototype* set $\mathscr{W}$ and the corresponding Voronoi diagram are shown in Fig. 12b. The *multi-prototypes* representing different clusters are plotted in different markers. The *piecewise linear separator* consists of hyperplanes shared by two subregions which are induced by the *multi-prototypes* representing different clusters. This *piecewise linear separator* is used to identify nonlinearly separable clusters, as shown in Fig. 12c.

Algorithm 2 summarizes the *GMPCL* method.

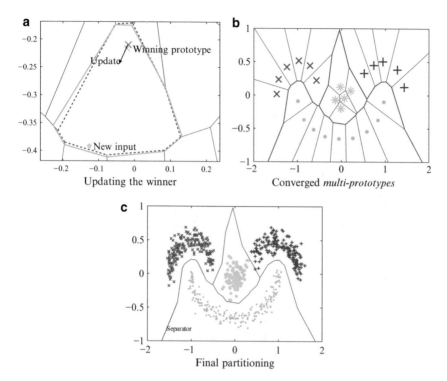

Fig. 12 The winner update and the clustering result of *smile face*. (**a**) The procedure of updating the winning *multi-prototype*, during which both the winning *multi-prototype* and the corresponding lines in Voronoi diagram move slightly: *red* "×" marker and *dashed lines* before update, while *green* "+" marker and *lines* after update. (**b**) The converged *multi-prototypes* and the corresponding Voronoi diagram, the *multi-prototypes* representing different clusters are plotted in *different markers*, and the *piecewise linear separator* is plotted in *red*. (**c**) The final clusters separated by the *piecewise linear separator*

4.3 Fast GMPCL

High-dimensional clustering applications, such as video clustering, are character-ized by a high computational load, which is mainly due to the *redundant* calculation of the distances between high-dimensional points in the update procedure of competitive learning. To overcome this problem, an approach similar to the kernel trick [31] is considered. First, an *inner product* matrix $M = [M_{i,j}]_{n \times n}$ of the data set $\mathscr{D}$ is computed such that $M_{i,j} = \langle \mathbf{x}_i, \mathbf{x}_j \rangle$. Then the computation of $\| \mathbf{x}_i - \mathbf{x}_j \|^2$ is efficiently accomplished by $\| \mathbf{x}_i - \mathbf{x}_j \|^2 = M_{i,i} + M_{j,j} - 2M_{i,j}$. Thus, the *redundant* high-dimensional computation is avoided. Unfortunately, it cannot be directly applied in competitive learning due to the incremental update rule. Since the winning *multi-prototype* $\mathbf{w}_{v_i}^{v_i}$ is updated by $\mathbf{w}_{v_i}^{v_i} \leftarrow \mathbf{w}_{v_i}^{v_i} + \eta_t(\mathbf{x}_i - \mathbf{w}_{v_i}^{v_i})$, it is unlikely

---

Algorithm 2: GMPCL

1: **Input:** $\mathcal{D} = \{x_1, \ldots, x_n\}$, p, t_{max}, $\{\eta_t\}$, ϵ.
2: Construct a graph $G^e = (\mathcal{V}, A, e)$ and initialize a clustering $\{\mathcal{I}_1, \ldots, \mathcal{I}_c\}$.
3: Initialize a *multi-prototype* set $\mathcal{W}$, set $t = 0$.
    repeat
4:    Randomly permute $\{x_1, \ldots, x_n\}$, $\hat{\mathcal{W}} = \mathcal{W}$, $t \leftarrow t + 1$.
    for $i = 1, \ldots, n$ **do**
5:      Select a winning *multi-prototype* $\omega_{v_i}^{v_i}$ by (76).
6:      Update $\mathbf{w}_{v_i}^{v_i}$ with learning rate η_t by (77).
    end for
7:    Compute $L = \| \mathcal{W} - \hat{\mathcal{W}} \|^2 = \sum_{\omega_j^q \in \omega} \| \mathbf{w}_j^q - \hat{\mathbf{w}}_j^q \|^2$.
    until $L \leq \epsilon$ or $t \geq t_{max}$
8: **Output:** clusters
    $\{\mathcal{C}_1, \ldots, \mathcal{C}_c : x_i \in \mathcal{C}_{v_i}, \text{ s.t. } \omega_{v_i}^{v_i} = \arg\min_{\omega_j^q \in \omega} \| x_i - \mathbf{w}_j^q \|^2, \forall i = 1, \ldots, n\}$.

---

that the updated $\mathbf{w}_{v_i}^{v_i}$ satisfies $\mathbf{w}_{v_i}^{v_i} \in \mathcal{D}$. No pre-computed distance $\| x_i - \mathbf{w}_{v_i}^{v_i} \|^2$ is available for calculating (76).

In [40], a *prototype descriptor* W^ψ is designed to represent c prototypes $\{\mu_1, \ldots, \mu_c\}$ in the kernel space induced by a mapping ψ. The *prototype descriptor* W^ψ is a $c \times (n + 1)$ matrix, whose rows represent prototypes as the inner products between a prototype and the data points, as well as the squared length of the prototype. That is,

$$W^\psi = \begin{pmatrix} \langle \mu_1, \psi(x_1) \rangle & \cdots & \langle \mu_1, \psi(x_n) \rangle & \langle \mu_1, \mu_1 \rangle \\ \langle \mu_2, \psi(x_1) \rangle & \cdots & \langle \mu_2, \psi(x_n) \rangle & \langle \mu_2, \mu_2 \rangle \\ \vdots & \vdots & \vdots & \vdots \\ \langle \mu_c, \psi(x_1) \rangle & \cdots & \langle \mu_c, \psi(x_n) \rangle & \langle \mu_c, \mu_c \rangle \end{pmatrix}. \tag{78}$$

Competitive learning in the kernel space becomes a process of updating W^ψ. *Multi-prototype descriptor* is developed, which is a row-block matrix independent of the dimensionality, and extend GMPCL to deal with high-dimensional clustering.

4.3.1 Inner Product Based Computation

According to the initialization of *multi-prototypes*, the initial $\mathcal{W}$ satisfies $\mathcal{W} \subset \mathcal{D}$. The *multi-prototype descriptor* is defined as follows.

Definition 6 (Multi-prototype Descriptor). A multi-prototype descriptor is a row-block matrix W of size $|\mathcal{W}| \times (n + 1)$,

$$W = \begin{pmatrix} W_1 \\ \vdots \\ W_c \end{pmatrix} \qquad (79)$$

such that the j-th block W_j represents $\mathscr{C}_j$, and the q-th row of W_j, i.e. $W_{j,:}^q$, represents $\mathbf{w}_j^q$ by

$$W_{j,i}^q = \langle \mathbf{w}_j^q, \mathbf{x}_i \rangle, i = 1, \ldots, n, \ W_{j,n+1}^q = \langle \mathbf{w}_j^q, \mathbf{w}_j^q \rangle, \qquad (80)$$

where $W_{j,i}^q$ denotes the i-th column of $W_{j,:}^q$. That is,

$$\begin{pmatrix}
\langle \mathbf{w}_1^1, \mathbf{x}_1 \rangle & \langle \mathbf{w}_1^1, \mathbf{x}_2 \rangle & \cdots & \langle \mathbf{w}_1^1, \mathbf{x}_n \rangle & \langle \mathbf{w}_1^1, \mathbf{w}_1^1 \rangle \\
\langle \mathbf{w}_1^2, \mathbf{x}_1 \rangle & \langle \mathbf{w}_1^2, \mathbf{x}_2 \rangle & \cdots & \langle \mathbf{w}_1^2, \mathbf{x}_n \rangle & \langle \mathbf{w}_1^2, \mathbf{w}_1^2 \rangle \\
\vdots & \vdots & \vdots & \vdots & \vdots \\
\langle \mathbf{w}_1^{p_1}, \mathbf{x}_1 \rangle & \langle \mathbf{w}_1^{p_1}, \mathbf{x}_2 \rangle & \cdots & \langle \mathbf{w}_1^{p_1}, \mathbf{x}_n \rangle & \langle \mathbf{w}_1^{p_1}, \mathbf{w}_1^{p_1} \rangle \\
\vdots & \vdots & \vdots & \vdots & \vdots \\
\langle \mathbf{w}_c^1, \mathbf{x}_1 \rangle & \langle \mathbf{w}_c^1, \mathbf{x}_2 \rangle & \cdots & \langle \mathbf{w}_c^1, \mathbf{x}_n \rangle & \langle \mathbf{w}_c^1, \mathbf{w}_c^1 \rangle \\
\langle \mathbf{w}_c^2, \mathbf{x}_1 \rangle & \langle \mathbf{w}_c^2, \mathbf{x}_2 \rangle & \cdots & \langle \mathbf{w}_c^2, \mathbf{x}_n \rangle & \langle \mathbf{w}_c^2, \mathbf{w}_c^2 \rangle \\
\vdots & \vdots & \vdots & \vdots & \vdots \\
\langle \mathbf{w}_c^{p_c}, \mathbf{x}_1 \rangle & \langle \mathbf{w}_c^{p_c}, \mathbf{x}_2 \rangle & \cdots & \langle \mathbf{w}_c^{p_c}, \mathbf{x}_n \rangle & \langle \mathbf{w}_c^{p_c}, \mathbf{w}_c^{p_c} \rangle
\end{pmatrix}.$$

Using the ω-notation, the ω_j^q-th row, i.e. $W_{j,:}^q$, represents the ω_j^q-th multi-prototype, i.e. $\mathbf{w}_j^q$.

The initial *multi-prototype descriptor* W is obtained as a sub-matrix of M. In Algorithm 2, three key procedures of *multi-prototype* competitive learning are involved with the *redundant* computation of distances, which are the winning *multi-prototype* selection, the winner update, and the computation of the sum of prototype update, i.e. L. Based on the *multi-prototype descriptor*, we implement these procedures whose computational complexity is independent of the dimensionality. For detailed proofs of these theorems and lemmas, readers are encouraged to refer to the [40].

Theorem 6 (Winner Selection Rule). *The selection of the winning multi-prototype $\omega_{v_i}^{v_i}$ of $\mathbf{x}_i$ can be realized by*

$$\omega_{v_i}^{v_i} = \arg\min_{\omega_j^q \in \omega} \left(W_{j,n+1}^q - 2 W_{j,i}^q \right). \qquad (81)$$

Theorem 7 (Winner Update Rule). *The update of the winning multi-prototype $\omega_{v_i}^{v_i}$ of $\mathbf{x}_i$ can be realized by*

$$
W_{v_i,j}^{v_i} \leftarrow
\begin{cases}
(1 - \eta_t) W_{v_i,j}^{v_i} + \eta_t M_{i,j} & \text{if } j = 1, \ldots, n, \\[2ex]
(1 - \eta_t)^2 W_{v_i,j}^{v_i} + \eta_t^2 M_{i,i} + 2(1 - \eta_t)\eta_t W_{v_i,i}^{v_i} & \text{if } j = n+1.
\end{cases}
\tag{82}
$$

Similar to [40], in **one** iteration of competitive learning, each data point $\mathbf{x}_i$ is assigned to **exactly one** *multi-prototype*. Let the index array $\pi_j^q = [\pi_j^q(1), \pi_j^q(2), \ldots, \pi_j^q(m_j^q)]$ store the indices of m_j^q *ordered* data points assigned to the ω_j^q-th *multi-prototype* in **one** iteration. For instance, if $\mathbf{x}_1, \mathbf{x}_{32}, \mathbf{x}_8, \mathbf{x}_{20}, \mathbf{x}_{15}$ are 5 *ordered* data points assigned to the ω_3^2-th *multi-prototype* in the t-th iteration, then the index array $\pi_3^2 = [\pi_3^2(1), \pi_3^2(2), \ldots, \pi_3^2(m_3^2)] = [1, 32, 8, 20, 15]$ with $\pi_3^2(1) = 1, \pi_3^2(2) = 32, \pi_3^2(3) = 8, \pi_3^2(4) = 20, \pi_3^2(5) = 15$ and $m_3^2 = 5$. The following lemma formulates the cumulative update of the ω_j^q-th *multi-prototype* based on the index array π_j^q.

Lemma 2. *In the t-th iteration, the relation between the updated multi-prototype $\mathbf{w}_j^q$ and the old $\hat{\mathbf{w}}_j^q$ is:*

$$
\mathbf{w}_j^q = (1 - \eta_t)^{m_j^q} \hat{\mathbf{w}}_j^q + \eta_t \sum_{l=1}^{m_j^q} (1 - \eta_t)^{m_j^q - l} \mathbf{x}_{\pi_j^q(l)}.
\tag{83}
$$

Theorem 8 (Iteration Stopping Criteria). *The iteration stopping criteria of multi-prototype competitive learning can be realized by $L \leq \epsilon$ or $t \geq t_{max}$, where L is computed as*

$$
L = \sum_{\omega_j^q \in \omega} \left(\left(1 - \frac{1}{(1 - \eta_t)^{m_j^q}}\right)^2 W_{j,n+1}^q \right) + \eta_t^2 \sum_{\omega_j^q \in \omega} \sum_{h=1}^{m_j^q} \sum_{l=1}^{m_j^q} \frac{M_{\pi_j^q(h), \pi_j^q(l)}}{(1 - \eta_t)^{h+l}}
$$

$$
+ 2\eta_t \sum_{\omega_j^q \in \omega} \left(\left(1 - \frac{1}{(1 - \eta_t)^{m_j^q}}\right) \sum_{l=1}^{m_j^q} \frac{W_{j,\pi_j^q(l)}^q}{(1 - \eta_t)^l} \right).
\tag{84}
$$

4.3.2 FGMPCL in High Dimension

Based on the *multi-prototype descriptor W* and the above theorems, a *fast* GMPCL (FGMPCL) is proposed for high-dimensional clustering, which is summarized in Algorithm 3. According to the three theorems, it is easy to prove that FGMPCL generates the same clustering as GMPCL. The analysis of the asymptotic computational complexity reveals that, in high-dimensional clustering, i.e. when $d \gg n$, FGMPCL can save $O\left(t_{max} n(d|\mathcal{W}| - n)\right)$ computations.

Algorithm 3: FGMPCL

1: **Input:** $\mathscr{D} = \{\mathbf{x}_1, \ldots, \mathbf{x}_n\}$, ρ, t_{max}, $\{\eta_t\}$, ϵ.
2: Compute an *inner product* matrix $M = [\langle \mathbf{x}_i, \mathbf{x}_j \rangle]_{n \times n}$.
3: The same graph-based initial clustering step as GMPCL is performed except directly taking values from M.
4: Initialize a *multi-prototype descriptor* W, set $t = 0$.
   repeat
5:     Randomly permute $\{\mathbf{x}_1, \ldots, \mathbf{x}_n\}$, initialize $|\mathscr{W}|$ empty index arrays $\{\pi_j^q = \emptyset : \omega_j^q \in \omega\}$, $t \leftarrow t + 1$.
       for $i = 1, \ldots, n$ **do**
6:         Select a winning *multi-prototype* $\omega_{v_i}^{v_i}$ by (81), and append i to the index array $\pi_{v_i}^{v_i}$.
7:         Update $W_{v_i,:}^{v_i}$ (i.e., the $\omega_{v_i}^{v_i}$-th *multi-prototype*) with learning rate η_t by (82).
       end for
8:     Compute L by (84).
   until $L \leq \epsilon$ or $t \geq t_{max}$
9: **Output:** clusters
   $\{\mathscr{C}_1, \ldots, \mathscr{C}_c : \mathbf{x}_i \in \mathscr{C}_{v_i}, \text{ s.t. } \omega_{v_i}^{v_i} = \arg\min_{\omega_j^q \in \omega} \left(W_{j,n+1}^q - 2W_{j,i}^q \right), \forall i = 1, \ldots, n\}.$

First, we should notice that one computation of the distances or the inner products between all data points is unavoidable, which takes $O(n^2 d)$ operations. Our goal here is to eliminate the *redundant* computation occurring in the procedure of the *multi-prototype* update.

The initialization of the *multi-prototype descriptor* W only takes $O(|\mathscr{W}|(n + 1))$ operations (taking $|\mathscr{W}|(n + 1)$ entries from the matrix M). For the t-th iteration, the initialization takes $O(|\mathscr{W}| + 1 + n)$ operations ($O(|\mathscr{W}|)$ for initializing $|\mathscr{W}|$ empty index arrays, $O(1)$ for increasing $t = t + 1$ and $O(n)$ for randomly permuting the data set). There are n data points, and each takes $O(|\mathscr{W}|)$ operations to select a winner by (81) and $O(n + 1)$ to update the winner by (82). Thus $O(n(n + 1 + |\mathscr{W}|))$ operations are needed for the update procedure in one iteration. The computation of (84) takes $O(|\mathscr{W}| + n^2 + n)$ operations (the first term is $O(|\mathscr{W}|)$, the second term is $O(n^2)$ and the third term is $O(n)$). Therefore, in one iteration, the total computational complexity is $O(|\mathscr{W}| + 1 + n) + O(n(n + 1 + |\mathscr{W}|)) + O(|\mathscr{W}| + n^2 + n) = O(n^2)$, since $|\mathscr{W}| < n$. Assume that the iteration number reaches the maximum iteration number t_{max}, the computational complexity of iteration procedure is $O(t_{max} n^2)$. Therefore, in FGMPCL, the computational complexity of multi-prototype competitive learning is $O(|\mathscr{W}|(n + 1)) + O(t_{max} n^2) = O(t_{max} n^2)$.

However, in GMPCL, the computational complexity of multi-prototype competitive learning is $O(t_{max} nd|\mathscr{W}|)$. In high-dimensional clustering where $d \gg n$, it is easy to see that $O(t_{max} nd|\mathscr{W}|) \gg O(t_{max} n^2 |\mathscr{W}|) > O(t_{max} n^2)$. The fast version can save $O(t_{max} nd|\mathscr{W}|) - O(t_{max} n^2) = O(t_{max} n(d|\mathscr{W}| - n))$ computations when $d \gg n$.

In practice, we suggest making a choice between GMPCL and FGMPCL before performing clustering based on the relation between n and d. If $d \gg n$, FGMPCL is preferable; otherwise, use GMPCL to perform clustering.

4.4 Experiments and Applications

In this section, we applied GMPCL to automatic color image segmentation. Our intention here is to demonstrate that GMPCL has the capability of finding visually appealing structures in real color images. The experiment was performed on images from the Berkeley Segmentation data set (BSDS)[1] [28]. BSDS contains 300 images of a wide variety of natural scenes, as well as "ground truth" segmentations produced by humans [13], aiming at providing an empirical basis for research on image segmentation and boundary detection. The size of each image is either 480×320 or 320×480, which is too large for directly computing 153,600 pixel feature vectors. So we resized the images by 0.4 into either 192×129 or 129×192. We used the 3-D vectors of color features for each pixel as the feature vectors to segment a color image. Since the L*a*b color is designed to approximate human vision and suitable for interpreting the real world [17], the coordinates in the L*a*b* color space were used as the features. Thus for each image we obtained a data set $\mathscr{D} = \{\mathbf{x}_i \in \mathbb{R}^3 : i = 1, \ldots, 24768\}$. Before applying clustering, smoothing was performed using the averaging filter of size 3×3 to avoid the over-segmentation caused by local color variants.

Figure 13 displays some of the segmentation results obtained by GMPCL, without any further postprocessing. Table 2 lists the means and variances of PRI, NPRI and NMI, and the average computational time in seconds on 300 images from BSDS. The compared methods include Rival penalized competitive learning (RPCL) [48], kernel k-means (k$k$means) [31], Ncut [33], Graclus [9], and gPb-owt-ucm [1]. The results show that although GMPCL is not the best of all compared methods, it is ranked in the second place and is comparable to the best algorithm. Please note that, gPb-owt-ucm is coupled to a high-performance contour detector proposed in [27], meanwhile the other algorithms only rely on the color information to perform image segmentation.

5 Position Regularized Support Vector Clustering

SVC is a nonlinear clustering method developed from SVDD. Compared with other clustering methods, the advantages of SVC include the abilities to discover clusters of arbitrary shapes and to deal with outliers. However, the value of the trade-off parameter directly determines the size of sphere and therefore influences the data distribution of sphere surface, i.e., the clustering results of SVC. To eliminate the influence of the trade-off parameter C to SVC [3], this section reviews a PSVC [39].

---

[1] www.eecs.berkeley.edu/Research/Projects/CS/vision/grouping/segbench/.

Fig. 13 Some of the image segmentation results by GMPCL. The first column displays the original images and the second column displays the segmentation results. Each segment (cluster) is painted with its mean color

Table 2 The means and variances of PRI, NPRI and NMI, and the average computational time in seconds on images from BSDS, The bold values denote the best results

Methods	PRI	NPRI	NMI	Time
RPCL [48]	0.739 ± 0.020	0.701 ± 0.022	0.607 ± 0.025	89.17
kkmeans [31]	0.743 ± 0.014	0.723 ± 0.016	0.626 ± 0.020	74.05
Ncut [33]	0.741 ± 0.015	0.722 ± 0.017	0.623 ± 0.022	28.32
Graclus [9]	0.751 ± 0.011	0.742 ± 0.014	0.625 ± 0.015	19.24
GMPCL	$\mathbf{0.758 \pm 0.009}$	$\mathbf{0.753 \pm 0.010}$	$\mathbf{0.642 \pm 0.012}$	**18.53**
gPb-owt-ucm [1]	0.760 ± 0.007	0.756 ± 0.007	0.651 ± 0.008	21.09

5.1 Background

This section briefly reviews SVC [3] as background.

5.1.1 Support Vector Domain Description

In domain description, the task is to give a description of a set of objects, which should cover the positive objects and reject the negative ones in the object space [34]. SVDD is a sphere shaped data description. By using a nonlinear transformation to map the data from the input data space into a high-dimensional kernel space, SVDD can obtain a very flexible and accurate data description relying on only a small number of SVs.

Given a data set $\mathscr{X} = \{\mathbf{x}_i \in \mathbb{R}^d | i = 1, \ldots, N\}$ consisting of N data points and a nonlinear transformation ϕ from the input space to a Gaussian kernel feature space,[2] we look for the smallest sphere that encloses most of the mapped data points in the feature space. Using the center $\boldsymbol{\mu}$ and radius R to describe the sphere, we minimize

$$F(R, \boldsymbol{\mu}, \xi_i) = R^2 + C \sum_{i=1}^{N} \xi_i, \qquad (85)$$

under the constraints

$$\|\phi(\mathbf{x}_i) - \boldsymbol{\mu}\|^2 \leq R^2 + \xi_i, \forall i = 1, \ldots, N, \qquad (86)$$

where the trade-off parameter C gives the trade-off between the volume of the sphere and the accuracy of data description, and $\xi_i \geq 0$ are some slack variables allowing

---

[2]Although any kernel space works here, as discussed in [3, 34], Gaussian kernels provide more tight contour representations. Therefore, Gaussian kernels with appropriate kernel parameters are used in most SVDD-based approaches, e.g., [3, 5, 21, 22, 34, 41, 42, 50].

for soft boundaries. By introducing the Lagrangian, we have

$$\mathbb{L}(R, \boldsymbol{\mu}, \xi_i, \beta_i, \alpha_i) = R^2 + C \sum_{i=1}^{N} \xi_i - \sum_{i=1}^{N} \beta_i (R^2 + \xi_i - \|\phi(\mathbf{x}_i) - \boldsymbol{\mu}\|^2) - \sum_{i=1}^{N} \alpha_i \xi_i,$$
(87)

where $\beta_i \geq 0$, $\alpha_i \geq 0$ are Lagrange multipliers. By eliminating the variables R, $\boldsymbol{\mu}$, ξ_i and α_i, the Lagrangian can be turned into the Wolfe dual form

$$\max_{\beta_i} W = \sum_{i=1}^{N} \beta_i K(\mathbf{x}_i, \mathbf{x}_i) - \sum_{i,j=1}^{N} \beta_i \beta_j K(\mathbf{x}_i, \mathbf{x}_j)$$

$$\text{subject to } \sum_{i=1}^{N} \beta_i = 1, \ 0 \leq \beta_i \leq C, \forall i = 1, \dots, N,$$
(88)

where the dot products $\phi(\mathbf{x}_i) \cdot \phi(\mathbf{x}_j)$ are replaced by the corresponding Gaussian kernel $K(\mathbf{x}_i, \mathbf{x}_j) = \exp(-q\|\mathbf{x}_i - \mathbf{x}_j\|^2)$ with the width parameter q.

According to the values of Lagrange multipliers $\beta_i, i = 1, \dots, N$, the data points are classified into three types:

- Inner point (IP): $\beta_i = 0$, which lies *inside* the sphere surface.
- Support vector (SV): $0 < \beta_i < C$, which lies *on* the sphere surface.
- Bounded support vector (BSV): $\beta_i = C$, which lies *outside* the sphere surface.

It is obvious that setting $C \geq 1$ will result in no BSV.

The kernel radius function, defined by the Euclidian distance of $\phi(\mathbf{x})$ from $\boldsymbol{\mu}$, is given by

$$R(\mathbf{x}) = \|\phi(\mathbf{x}) - \boldsymbol{\mu}\|$$

$$= \sqrt{1 - 2\sum_{i=1}^{N} \beta_i K(\mathbf{x}_i, \mathbf{x}) + \sum_{i,j=1}^{N} \beta_i \beta_j K(\mathbf{x}_i, \mathbf{x}_j)}.$$
(89)

The radius of the sphere is defined as

$$R = \max\{R(\mathbf{x}_i)|\mathbf{x}_i \text{ is a SV, i.e.} 0 < \beta_i < C\}.$$
(90)

Ideally, all the SVs should have the same $R(\mathbf{x}_i)$. However, due to the numerical problem, they may be slightly different. A practical strategy is to use their maximum value as the radius. The contours enclosing most of the data points in the data space are defined by the set $\{\mathbf{x}|R(\mathbf{x}) = R\}$. In the original SVDD algorithm [34], data points lying outside the sphere surface, namely BSVs, are taken as outliers, from which outlier detection can be derived.

5.1.2 Support Vector Clustering

The SVC algorithm is directly based on the previously described SVDD by adding
a cluster labeling step after obtaining the domain description of the data set [3].
By mapping the data back to the input data space, the sphere surface corresponds
to contours in the data space, which separate the data into several components,
each enclosing a cluster of data points. Constructing connected components is
accomplished by computing an adjacency matrix of the data set according to the
sphere.

To this end, an adjacency matrix $A = [A_{ij}]^{N \times N}$ is computed as

$$A_{ij} = \begin{cases} 1 & \text{if } \forall \mathbf{y} \text{ on the line segment connecting } \mathbf{x}_i \text{ and } \mathbf{x}_j, R(\mathbf{y}) \leq R, \\ 0 & \text{otherwise.} \end{cases} \tag{91}$$

Checking the line segment is implemented by sampling a number of points (usually
ten points are used) [3]. Clusters are defined as the connected components of the
graph induced by the adjacency matrix A. BSVs are unclassified by this procedure
since their feature space images always lie outside the sphere. Each of them is
directly assigned to the closest cluster [3].

5.2 Position Regularized Support Vector Clustering

One inherent drawback associated with the SVC algorithm is that the constructed
sphere is very sensitive to the selection of the trade-off parameter C. Especially, by
directly controlling the penalty term $C \sum_{i=1}^{N} \xi_i$ in the objective function (85), the
volume of the sphere and the accuracy of data description strongly rely on the value
of C.

Figure 14 illustrates the sensitivity of SVC to the value of C on a data set
consisting of 200 data points with the best kernel parameter q. The results show that
when different values of C are used, different spheres (including SVs, BSVs, and
contours) are generated but none of them can accurately discover the four classes of
this data distribution.

1. When using a small C, too many BSVs are generated. Some non-outliers are
   erroneously taken as BSVs, which divide the class containing these data points
   into at least two clusters, as shown in Fig. 14b.
2. When using a large C, too few BSVs are generated. Some outliers are incorrectly
   taken as IPs, which connect two classes (that should be separated by these
   outliers) into one cluster, as shown in Fig. 14e, f.
3. Even using some "median" C, say, $C = 0.2$ (Fig. 14c) and $C = 0.3$ (Fig. 14d),
   which produces an appropriate number of SVs and BSVs, there are still some
   erroneously detected SVs and BSVs, leading to the incorrect identification of
   classes.

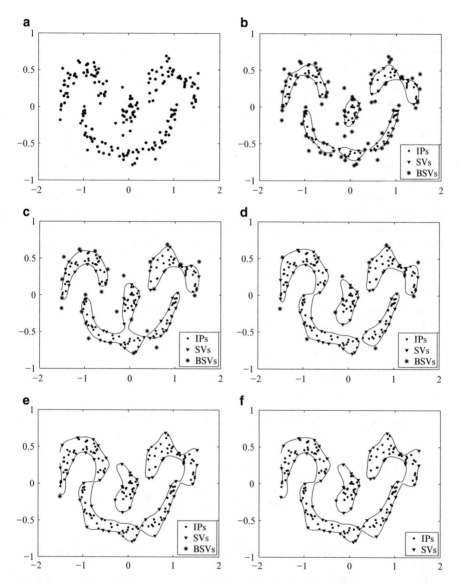

Fig. 14 The spheres generated by SVC on a synthetic data set with the best kernel parameter $q = 13$ using different C. (**a**) The data set of 200 data points. (**b**) The sphere by SVC with $C = 0.1$. (**c**) The sphere by SVC with $C = 0.2$. (**d**) The sphere by SVC with $C = 0.3$. (**e**) The sphere by SVC with $C = 0.4$. (**f**) The sphere by SVC with $C \geq 0.5$

The main reason is that, by using the same global trade-off parameter C for all data points, the likelihood of each data point to be an outlier is taken the same, which is actually not true due to the change of the density distribution or the relation between data points.

According to the above analysis, a distinct trade-off parameter should be used for each data point, reflecting the possibility of each data point to be outside the sphere surface. To address the above issue, one can assign a weighting to each data point so as to represent the confidence of the corresponding data point to be an outlier. In the PSVC algorithm, the weighting W_i is defined to be inversely proportional to the distance between the feature space image of $\mathbf{x}_i$ (i.e., $\phi(\mathbf{x}_i)$) and the mean of feature space images (i.e. $\frac{1}{N}\sum_{j=1}^{N}\phi(\mathbf{x}_j)$).

The underlying rationale is that, in kernel space, if the feature space image of a data point is too distant from the mean of feature space images, which implies that the corresponding data point is relatively isolated from the rest data points in the input data space, this data point should be more likely considered as an outlier. Additionally, since the larger the slack variable ξ_i, the more likely the data point corresponds to an outlier, the corresponding slack variables ξ_i of the isolated data points should be larger than those of the remaining data points. Consequently, if we replace C with the corresponding W_i in the objective function (85), the relatively isolated data points should have smaller weightings W_i. This weighting is called position-based weighting and the corresponding SVC with such weighting is called PSVC.

To compute the position-based weighting, we first compute a kernel distance vector $D^\phi = [D_l^\phi | l = 1, \dots, N]$ from the kernel matrix K,

$$
\begin{aligned}
D_l^\phi &= \left\| \phi(\mathbf{x}_l) - \frac{1}{N}\sum_{j=1}^{N}\phi(\mathbf{x}_j) \right\|^2 \\
&= K(\mathbf{x}_l, \mathbf{x}_l) + \frac{1}{N^2}\sum_{i,j=1}^{N}K(\mathbf{x}_i, \mathbf{x}_j) - \frac{2}{N}\sum_{j=1}^{N}K(\mathbf{x}_l, \mathbf{x}_j), \quad \forall l = 1, \dots, N.
\end{aligned} \tag{92}
$$

The weighting $W_i, \forall i = 1, \dots, N$ is computed as

$$
W_i = \max_{l=1,\dots,N}\{D_l^\phi\} - D_i^\phi, \quad W_i = \frac{W_i}{\max_{l=1,\dots,N} W_l}, \tag{93}
$$

which is inversely proportional to the distance D_i^ϕ and normalized to be no larger than 1.

Figure 15 illustrates the concept of weighting. The farther away one feature space image $\phi(\mathbf{x}_i)$ from the mean $\frac{1}{N}\sum_{j=1}^{N}\phi(\mathbf{x}_j)$ is, the smaller the corresponding weighting W_i is. The weightings $W_i, i = 1, \dots, N$ are used to regularize the sphere volume in place of the trade-off parameter C shared by all data points. That is, we minimize the radius

$$
F(R, \boldsymbol{\mu}, \xi_i) = R^2 + \sum_{i=1}^{N} W_i \xi_i, \tag{94}
$$

Fig. 15 Illustration of the position-based weighting in kernel space. The mean vector of the kernel space images of data points is shown as a bold start. The size of each data point is marked according to its weighting

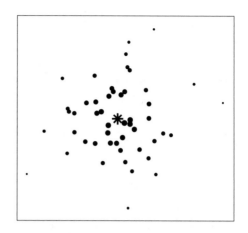

under the constraints

$$\|\phi(\mathbf{x}_i) - \boldsymbol{\mu}\|^2 \le R^2 + \xi_i, \forall i = 1, \ldots, N. \tag{95}$$

Unlike in the original SVC, in the objective function (94), each weighting W_i, respectively, regularizes the likelihood for the corresponding data point $\mathbf{x}_i$ to be an outlier. The smaller W_i is, the larger ξ_i is. The slack variable ξ_i is used to generate soft boundaries and BSV. We will show later that when $\xi_i > 0$, $\mathbf{x}_i$ is taken as a BSV. Therefore, the weighting mechanism adaptively regularizes each data point to be an outlier or not.

The main difference between the weighting mechanism and the one proposed in [51] is that the weighting of [51] is computed according to the Mahalanobis distance of the corresponding factor analyzer; meanwhile the weighting is computed according to the position of the feature space image in the kernel space. The weighting of [51] strongly relies on the results of the mixture of factor analyzers (MFA). Different number of analyzers leads to different weightings, and estimating the number of analyzers is itself a difficult task. In addition, the PSVC algorithm does not require performing MFA to learn the weighting and is therefore more efficient than its counterpart.

By introducing the Lagrangian for (94), we have

$$\mathbb{L}(R, \boldsymbol{\mu}, \xi_i, \beta_i, \alpha_i) = R^2 + \sum_{i=1}^{N} W_i \xi_i - \sum_{i=1}^{N} \beta_i (R^2 + \xi_i - \|\phi(\mathbf{x}_i) - \boldsymbol{\mu}\|^2) - \sum_{i=1}^{N} \alpha_i \xi_i, \tag{96}$$

where $\beta_i \ge 0$, $\alpha_i \ge 0$ are Lagrange multipliers. Setting to 0 the derivative of $\mathbb{L}$ w.r.t. R, $\boldsymbol{\mu}$ and ξ_i, respectively, leads to

$$\sum_{i=1}^{N} \beta_i = 1, \quad \boldsymbol{\mu} = \sum_{i=1}^{N} \beta_i \phi(\mathbf{x}_i), \quad \beta_i = W_i - \alpha_i. \tag{97}$$

The KKT complementarity conditions result in

$$\alpha_i \xi_i = 0, \quad \beta_i(R^2 + \xi_i - \|\phi(\mathbf{x}_i) - \boldsymbol{\mu}\|^2) = 0. \tag{98}$$

By eliminating the variables R, $\boldsymbol{\mu}$, ξ_i, and α_i, the Lagrangian (96) can be turned into the Wolfe dual form

$$\max_{\beta_i} \mathbb{W} = \sum_{i=1}^{N} \beta_i K(\mathbf{x}_i, \mathbf{x}_i) - \sum_{i,j=1}^{N} \beta_i \beta_j K(\mathbf{x}_i, \mathbf{x}_j)$$

$$\text{subject to } \sum_{i=1}^{N} \beta_i = 1, \ 0 \le \beta_i \le W_i, \forall i = 1, \ldots, N. \tag{99}$$

Please notice that, the upper bounds for Lagrange multipliers $\beta_i, i = 1, \ldots, N$ are no longer the same. Instead, each of them is, respectively, controlled by the corresponding weighting. According to (93) and (99), one side effect of the position-based weighting is the increase in the influence of the kernel parameter. This is because not only the construction of kernel space but also the computation of weighting, as well as the upper bounds for Lagrange multipliers, rely on the kernel parameter. However, as we will show later, the most suitable kernel parameter can be estimated based on stability arguments.

According to the values of Lagrange multipliers β_i and the corresponding weightings $W_i, i = 1, \ldots, N$, the data points are classified into three types:

- Inner point (IP): $\beta_i = 0$, which lies *inside* the sphere surface. That is, $\beta_i = 0$ implies that $\xi_i = 0$ according to $\beta_i = W_i - \alpha_i$ and $\alpha_i \xi_i = 0$. Therefore, $\|\phi(\mathbf{x}_i) - \boldsymbol{\mu}\|^2 \le R^2 + \xi_i = R^2$.
- Support vector (SV): $0 < \beta_i < W_i$, which lies *on* the sphere surface. That is, $\beta_i < W_i$ implies that $\xi_i = 0$ according to $\beta_i = W_i - \alpha_i$ and $\alpha_i \xi_i = 0$; and $\beta_i > 0$ implies that $\|\phi(\mathbf{x}_i) - \boldsymbol{\mu}\|^2 = R^2 + \xi_i$ according to $\beta_i(R^2 + \xi_i - \|\phi(\mathbf{x}_i) - \boldsymbol{\mu}\|^2) = 0$. Therefore, $\|\phi(\mathbf{x}_i) - \boldsymbol{\mu}\|^2 = R^2 + \xi_i = R^2$.
- BSV: $\beta_i = W_i$, which lies *outside* the sphere surface. That is, $\beta_i = W_i$ implies that $\xi_i \ge 0$ according to $\beta_i = W_i - \alpha_i$ and $\alpha_i \xi_i = 0$, and that $\|\phi(\mathbf{x}_i) - \boldsymbol{\mu}\|^2 = R^2 + \xi_i$ according to $\beta_i(R^2 + \xi_i - \|\phi(\mathbf{x}_i) - \boldsymbol{\mu}\|^2) = 0$. Therefore, $\|\phi(\mathbf{x}_i) - \boldsymbol{\mu}\|^2 = R^2 + \xi_i \ge R^2$.

The kernel radius function $R(\mathbf{x}_i)$, the radius of the sphere and the contours enclosing most of the data points in the data space are defined the same as the original SVDD. This sphere shaped domain description is termed PSVDD.

After obtaining PSVDD, we follow the same post-processing steps of SVC, i.e., constructing the adjacency matrix A via (91) and defining clusters as the connected components of the graph induced by the adjacency matrix A, leading to the new PSVC algorithm.

Figure 16 illustrates some key concepts of PSVC. By comparing the weightings W_i (Fig. 16a) and the resulting multipliers $\beta_i, i = 1, \ldots, N$ (Fig. 16b), the sphere structure is obtained, which accurately discovers the clusters in the data distribution (Fig. 16c) and therefore generates a correct clustering (Fig. 16d).

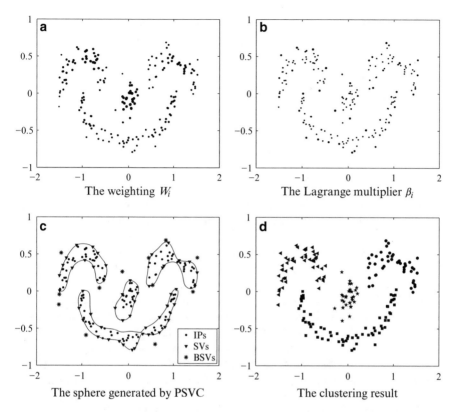

Fig. 16 PSVC demonstration. The same kernel space as Fig. 14 is used, i.e., the same data set with the same kernel parameter q. In (**a**) and (**b**), the size of each data point is marked according to its weighting and Lagrange multiplier, respectively. In (**d**), different clusters are plotted in *different markers*

5.3 Experiments and Applications

In this section, we report the performance improvement achieved by PSVC over its original counterpart SVC.

Seven data sets from the UCI repository [2] are used in comparative experiments, which are summarized in Table 3. In sccts (the synthetic control chart time series data set), four statistical features of each series are used as the feature vector, i.e., the mean, the standard deviation, the skewness, and the kurtosis. The original Wisconsin's breast cancer database consists of 699 samples and after removing 16 samples with missing values, the database consists of 683 samples belonging to 2 classes, benign and malignant tumors. In pendigits (the pen-based recognition of handwritten digit data set), we use a subset consisting of 3000 samples belonging to 3 classes, digit 1, digit 2, and digit 3. The mfeat (the multiple features data set)

Table 3 Summary of the data sets used in comparative experiments

Data set	Number of samples	Dimension	Number of classes
Wine	178	13	3
Glass	214	9	6
Sccts	600	4	6
Wisconsin	683	10	2
Pendigits	3000	16	3
Mfeat	2000	649	10
Isolet	900	617	3

consists of 2000 samples belonging to 10 digit classes. In isolet (the isolated letter speech recognition data set), a subset consisting of 900 samples belonging to classes A, B and C, is used.

In this comparison, the best kernel parameters q for SVC and PSVC, respectively are used. Figure 17 compares the performances of SVC and PSVC when the best kernel parameters q (for SVC and PSVC, respectively) are used on each data set. The clustering results obtained by SVC are plotted as a function of the trade-off parameter C, meanwhile the clustering result obtained by PSVC is plotted as a baseline independent of C since there is no such trade-off parameter in PSVC. The comparative results reveal that, in most cases, no matter how the trade-off parameter is adjusted, SVC is not comparable with PSVC. The exception is that, on the wine data set, when q is set to 0.2 (the best for SVC), SVC with C set to 0.4 can generate higher NMI than PSVC. On data sets such as wine, mfeat, and isolet, when too small values of C are used, e.g., $C = 0.1$ and $C = 0.2$, SVC will generate 0 NMI value, which indicates that the entire data set is partitioned into only one cluster according to the definition of NMI. Additionally, on all data sets, the fact that SVC generates the dramatically changing NMI with different C also reveals that SVC is quite sensitive to the selection of the trade-off parameter.

We also compare the performances of PSVC with three recently improved variants of SVC, including LSVC [51], VM-SVC [38], and DDM-SVC [20]. All the three variants are proposed for improving the clustering performance of SVC. For each algorithm, the best clustering result (i.e., the highest NMI) is reported on each data set by adjusting parameters under the constraint of generating the actual number of clusters.

The comparative results are listed in Table 4, which show that the PSVC algorithm significantly outperforms the compared methods in terms of clustering accuracy (i.e., NMI) and computational time. Especially, on the isolet data set, PSVC has obtained an NMI higher than 0.9, achieving a significantly 37% improvement over its counterparts. On the Wisconsin data set, although all the compared algorithms have got NMI values as high as 0.94, the PSVC algorithm is still better than the other three algorithms by achieving an NMI value higher than 0.96. These comparative results demonstrate the effectiveness of the PSVC algorithm.

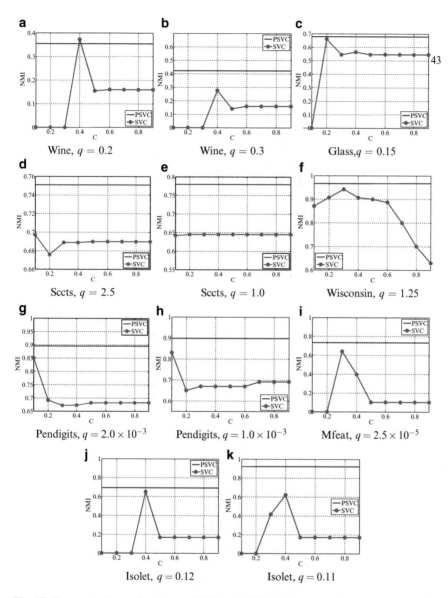

Fig. 17 Comparing the performances of SVC and PSVC using the best kernel parameter q for SVC and PSVC, respectively. For each data set, the clustering results generated by SVC are plotted as a function of C. (**a**) Wine, $q = 0.2$ best for SVC; (**b**) Wine, $q = 0.3$ best for PSVC; (**c**) Glass, $q = 0.15$ best for both; (**d**) Sccts, $q = 2.5$ best for SVC; (**e**) Sccts, $q = 1.0$ best for PSVC; (**f**) Wisconsin, $q = 1.25$ for both; (**g**) Pendigits, $q = 2.0 \times 10^{-3}$ best for SVC; (**h**) Pendigits, $q = 1.0 \times 10^{-3}$ best for PSVC; (**i**) Mfeat, $q = 2.5 \times 10^{-5}$ best for both; (**j**) Isolet, $q = 0.12$ best for SVC; (**k**) Isolet, $q = 0.11$ best for PSVC

Table 4 Comparative results

Data sets	SVC [3]		LSVC [51]		VM-SVC [38]		DDM-SVC [20]		PSVC	
	NMI	Time	NMI	Time	NMI	Time	NMI	Time	NMI	Time
Wine	0.373	9.74×10^0	0.418	8.31×10^0	0.413	4.27×10^0	0.410	3.43×10^0	**0.423**	$\mathbf{0.99 \times 10^0}$
Glass	0.661	3.31×10^1	0.665	2.79×10^1	0.666	1.33×10^1	0.668	1.23×10^1	**0.679**	$\mathbf{3.10 \times 10^0}$
Sccts	0.697	8.11×10^2	0.738	6.43×10^2	0.751	3.34×10^2	0.745	3.23×10^2	**0.780**	$\mathbf{8.65 \times 10^1}$
Wisconsin	0.942	9.07×10^2	0.957	6.65×10^2	0.953	3.92×10^2	0.949	3.74×10^2	**0.967**	$\mathbf{9.18 \times 10^1}$
Pendigits	0.850	7.59×10^4	0.863	4.87×10^4	0.864	4.24×10^4	0.869	4.12×10^4	**0.898**	$\mathbf{3.37 \times 10^4}$
Mfeat	0.641	3.01×10^4	0.671	1.94×10^4	0.672	1.88×10^4	0.667	1.73×10^4	**0.734**	$\mathbf{1.44 \times 10^4}$
Isolet	0.650	2.12×10^3	0.671	1.77×10^3	0.670	1.61×10^3	0.670	1.54×10^3	**0.921**	$\mathbf{1.49 \times 10^3}$

On each data set, we report the highest NMI value by each algorithm, and the computational time in seconds used to achieve such good results

6 Conclusion and Discussion

Nonlinear clustering is one of the most important research problems in data clustering. In this chapter, we have reviewed several nonlinear clustering algorithms from the viewpoints of kernel-based clustering, multi-exemplar model, graph-based method, and SVC. In particular, we have reviewed four nonlinear clustering methods, namely COLL for kernel-based clustering, MEAP, GMPCL, and PSVC. Experimental results in computer visions have shown the effectiveness of these methods.

Despite the great attraction of these methods, there still exist some issues needed to be further discussed and addressed. For instance, for the kernel based methods, namely COLL and PSVC, one challenging issue is to select a suitable kernel function to map the data set from the original data space into a high-dimensional kernel space. Obviously, different kernel transformation would lead to different clustering results no matter what clustering mechanism is employed. Usually but most naively, this issue is addressed by trial and error, which is, however, not applicable due to the lack of prior information. Another strategy in the unsupervised scenario is to select the suitable kernel function by using stability arguments—one chooses the kernel parameter such that the resulting domain descriptions are "most stable" [37, 39].

For the multi-exemplar/multi-prototype methods, they have the same observation, i.e. applicable to only a special case of nonlinearly separable clusters with multiple sub-clusters. This assumption implies that they only focus on a very small portion of a nonlinear problem, where the data points are relatively homogenous. In many applications, such as image categorization, face categorization, multi-font optical character recognition, and handwritten digit classification, each cluster may contain several sub-clusters. For instance, in the natural scene categorization experiments, a scene category often contains multiple "themes," e.g., the street scene may contain themes like "road," "car," "pedestrian," "building," etc. Similarly, in the face categorization experiments, images of the same person in different facial expressions should be taken as in distinct sub-clusters. In the applications of optical character recognition and handwritten digit classification, the cluster representing a letter or a digit could be composed of several sub-clusters, each corresponding to a different style or font. The data containing multiple sub-clusters obviously cannot be represented by a single exemplar/prototype, which however is suitable for the multi-exemplar/multi-prototype model. Therefore, apart from the above reviewed multi-exemplar/multi-prototype methods, some other strategies have been developed for sub-cluster cases. For instance, in [23], Lin et al. developed a cohesion-based self-merging clustering method which combines partitional and hierarchical clustering methods by first partitioning the input data into several small sub-clusters and then merging the sub-clusters based on cohesion in a hierarchical way.

Apart from the above nonlinear clustering methods, there are also some other widely used approaches such as density based clustering (e.g., the well-known

DBSCAN [12]), ensemble clustering (e.g., Link-based ensemble clustering [18]), and clustering by fast search and find of density peaks [29]. The basic idea of density based clustering methods, say, DBSCAN, is to group together data points that are close to each other with high density, and mark as outliers data points that lie alone in low-density regions. The ensemble clustering aggregates several input data clusterings to generate a single output clustering, which can be taken as a space transformation from nonlinear space to linear space. The clustering by fast search and find of density peaks works by using the assumption that cluster centers are characterized by a higher density than their neighbors and by a relatively large distance from points with higher densities, which can be viewed as a combination of both density and distance. From the above discussion, it is clear that nonlinear clustering has been and will continue to be a hot research topic.

References

1. Arbeláez, P., Maire, M., Fowlkes, C., Malik, J.: From contours to regions: an empirical evaluation. In: Proceedings of the 2009 IEEE Computer Society Conference on Computer Vision and Pattern Recognition, pp. 2294–2301 (2009)
2. Asuncion, A., Newman, D.: UCI machine learning repository. http://www.ics.uci.edu/~mlearn/MLRepository.html (2007)
3. Ben-Hur, A., Horn, D., Siegelmann, H.T., Vapnik, V.: Support vector clustering. J. Mach. Learn. Res. **2**, 125–137 (2001)
4. Bishop, C.M.: Pattern Recognition and Machine Learning. Springer, Berlin (2006)
5. Camastra, F., Verri, A.: A novel kernel method for clustering. IEEE Trans. Pattern Anal. Mach. Intell. **27**(5), 801–805 (2005)
6. Celebi, M.E., Kingravi, H.A.: Deterministic initialization of the k-means algorithm using hierarchical clustering. Int. J. Pattern Recogn. Artif. Intell. **26**(7) 1–23 (2012)
7. Celebi, M.E., Kingravi, H.A., Vela, P.A.: A comparative study of efficient initialization methods for the k-means clustering algorithm. Expert Syst. Appl. **40**, 200–210 (2013)
8. DeSieno, D.: Adding a conscience to competitive learning. In: IEEE International Conference on Neural Networks (1988)
9. Dhillon, I.S., Guan, Y., Kulis, B.: Weighted graph cuts without eigenvectors: a multilevel approach. IEEE Trans. Pattern Anal. Mach. Intell. **29**(11), 1944–1957 (2007). http://www.cs.utexas.edu/users/dml/Software/graclus.html
10. Dueck, D.: Affinity propagation: clustering data by passing messages. Ph.D. thesis, University of Toronto (2009)
11. Ertöz, L., Steinbach, M., Kumar, V.: Finding clusters of different sizes, shapes, and densities in noisy, high dimensional data. In: Proceedings of the 3rd SIAM International Conference on Data Mining, pp. 47–58 (2003)
12. Ester, M., Kriegel, H.P., Sander, J., Xu, X.: A density-based algorithm for discovering clusters in large spatial databases with noise. In: Proceedings of the 2nd International Conference on Knowledge Discovery and Data Mining, pp. 226–231 (1996)
13. Fowlkes, C., Martin, D., Malik, J.: Local figure-ground cues are valid for natural images. J. Vis. **7**(8), 2, 1–9 (2007)
14. Frey, B.J., Dueck, D.: Clustering by passing messages between data points. Science **315**, 972–976 (2007)
15. Givoni, I.E., Frey, B.J.: A binary variable model for affinity propagation. Neural Comput. **21**(6), 1589–1600 (2009)

16. http://www.open-video.org. The Open Video Project is managed at the Interaction Design Laboratory, at the School of Information and Library Science, University of North Carolina at Chapel Hill (2015)
17. Hunter, R.S.: Photoelectric color difference meter. J. Opt. Soc. Am. **48**(12), 985–993 (1958)
18. Iam-On, N., Boongoen, T., Garrett, S., Price, C.: A link-based approach to the cluster ensemble problem. IEEE Trans. Pattern Anal. Mach. Intell. **33**(12), 2396–2409 (2011)
19. Kschischang, F.R., Frey, B.J., Loeliger, H.A.: Factor graphs and the sum-product algorithm. IEEE Trans. Inf. Theory **47**(2), 498–519 (2001)
20. Lee, D., Lee, J.: Dynamic dissimilarity measure for support-based clustering. IEEE Trans. Knowl. Data Eng. **22**(6), 900–905 (2010)
21. Lee, J., Lee, D.: An improved cluster labeling method for support vector clustering. IEEE Trans. Pattern Anal. Mach. Intell. **27**(3), 461–464 (2005)
22. Lee, J., Lee, D.: Dynamic characterization of cluster structures for robust and inductive support vector clustering. IEEE Trans. Pattern Anal. Mach. Intell. **28**(11), 1869–1874 (2006)
23. Lin, C.R., Chen, M.S.: Combining partitional and hierarchical algorithms for robust and efficient data clustering with cohesion self-merging. IEEE Trans. Knowl. Data Eng. **17**(2), 145–159 (2005)
24. Liu, M., Jiang, X., Kot, A.C.: A multi-prototype clustering algorithm. Pattern Recogn. **42**, 689–698 (2009)
25. Lyons, M.J., Akamatsu, S., Kamachi, M., Gyoba, J.: Coding facial expressions with gabor wavelets. In: Proceedings of the 3rd IEEE International Conference on Automatic Face and Gesture Recognition, pp. 200–205 (1998)
26. MacQueen, J.: Some methods for classification and analysis of multivariate observations. In: Proceedings of the 15th Berkeley Symposium on Mathematical Statistics and Probability, vol. 1, pp. 281–297. University of California Press, Berkeley (1967)
27. Maire, M., Arbeláez, P., Fowlkes, C., Malik, J.: Using contours to detect and localize junctions in natural images. In: Proceedings of the 2008 IEEE Computer Society Conference on Computer Vision and Pattern Recognition, pp. 1–8 (2008)
28. Martin, D., Fowlkes, C., Tal, D., Malik, J.: A database of human segmented natural images and its application to evaluating segmentation algorithms and measuring ecological statistics. In: Proceedings of the 8th International Conference on Computer Vision, vol. 2, pp. 416–423 (2001). doi:10.1109/ICCV.2001.937655
29. Rodriguez, A., Laio, A.: Clustering by fast search and find of density peaks. Science **344**(6191), 1492–1496 (2014)
30. Schölkopf, B.: The kernel trick for distances. In: Advances in Neural Information Processing Systems 301–307 (2000)
31. Schölkopf, B., Smola, A., Müller, K.R.: Nonlinear component analysis as a kernel eigenvalue problem. Neural Comput. **10**, 1299–1319 (1998)
32. Shawe-Taylor, J., Cristianini, N.: Kernel Methods for Pattern Analysis. Cambridge University Press, Cambridge (2004)
33. Shi, J., Malik, J.: Normalized cuts and image segmentation. IEEE Trans. Pattern Anal. Mach. Intell. **22**(8), 888–905 (2000). http://www.cis.upenn.edu/~jshi/software/
34. Tax, D.M., Duin, R.P.: Support vector domain description. Pattern Recogn. Lett. **20**, 1191–1199 (1999)
35. Truong, B.T., Venkatesh, S.: Video abstraction: a systematic review and classification. ACM Trans. Multimed. Comput. Commun. Appl. **3**(1), 1–37 (2007)
36. Tzortzis, G.F., Likas, A.C.: The global kernel k-means algorithms for clustering in feature space. IEEE Trans. Neural Netw. **20**(7), 1181–1194 (2009)
37. von Luxburg, U.: Clustering stability: an overview. Found. Trends Mach. Learn. **2**, 235–274 (2009)
38. Wang, J.S., Chiang, J.C.: A cluster validity measure with outlier detection for support vector clustering. IEEE Trans. Syst. Man Cybern. Part B Cybern. **38**(1), 78–89 (2008)
39. Wang, C.D., Lai, J.H.: Position regularized support vector domain description. Pattern Recogn. **46**, 875–884 (2013)

40. Wang, C.D., Lai, J.H., Zhu, J.Y.: A conscience on-line learning approach for kernel-based clustering. In: Proceedings of the 10th International Conference on Data Mining, pp. 531–540 (2010)
41. Wang, C.D., Lai, J.H., Huang, D.: Incremental support vector clustering. In: Proceedings of the ICDM 2011 Workshop on Large Scale Visual Analytics, pp. 839–846 (2011)
42. Wang, C.D., Lai, J.H., Huang, D., Zheng, W.S.: SVStream: a support vector based algorithm for clustering data streams. IEEE Trans. Knowl. Data Eng. $25(6)$, 1410–1424 (2013)
43. Wang, C.D., Lai, J.H., Suen, C.Y., Zhu, J.Y.: Multi-exemplar affinity propagation. IEEE Trans. Pattern Anal. Mach. Intell. $35(9)$, 2223–2237 (2013)
44. Wang, C.D., Lai, J.H., Zhu, J.Y.: Conscience online learning: an efficient approach for robust kernel-based clustering. Knowl. Inf. Syst. $31(1)$, 79–104 (2012)
45. Wang, C.D., Lai, J.H., Zhu, J.Y.: Graph-based multiprototype competitive learning and its applications. IEEE Trans. Syst. Man Cybern. C Appl. Rev. $42(6)$, 934–946 (2012)
46. Weiss, Y., Freeman, W.T.: On the optimality of solutions of the max-product belief-propagation algorithm in arbitrary graphs. IEEE Trans. Inf. Theory $47(2)$, 736–744 (2001)
47. Xu, R., Wunsch II, D.: Survey of clustering algorithms. IEEE Trans. Neural Netw. $16(3)$, 645–678 (2005). doi:10.1109/TNN.2005.845141
48. Xu, L., Krzyżak, A., Oja, E.: Rival penalized competitive learning for clustering analysis, RBF net, and curve detection. IEEE Trans. Neural Netw. $4(4)$, 636–649 (1993)
49. Xu, L., Neufeld, J., Larson, B., Schuurmans, D.: Maximum margin clustering. In: NIPS 17 (2004)
50. Yang, J., Estivill-Castro, V., Chalup, S.K.: Support vector clustering through proximity graph modelling. In: Proceedings of the 9th International Conference on Neural Information Processing (2002)
51. Yankov, D., Keogh, E., Kan, K.F.: Locally constrained support vector clustering. In: Proceedings of the 7th International Conference on Data Mining, pp. 715–720 (2007)
52. Zhou, S.K., Chellappa, R.: Multiple-exemplar discriminant analysis for face recognition. In: ICPR, pp. 191–194 (2004)
53. Zhu, M., Martinez, A.M.: Subclass discriminant analysis. IEEE Trans. Pattern Anal. Mach. Intell. $28(8)$, 1274–1286 (2006)

Swarm Intelligence-Based Clustering Algorithms: A Survey

Tülin İnkaya, Sinan Kayalıgil, and Nur Evin Özdemirel

Abstract Swarm intelligence (SI) is an artificial intelligence technique that depends on the collective properties emerging from multi-agents in a swarm. In this work, the SI-based algorithms for hard (crisp) clustering are reviewed. They are studied in five groups: particle swarm optimization, ant colony optimization, ant-based sorting, hybrid algorithms, and other SI-based algorithms. Agents are the key elements of the SI-based algorithms, as they determine how the solutions are generated and directly affect the exploration and exploitation capabilities of the search procedure. Hence, a new classification scheme is proposed for the SI-based clustering algorithms according to the agent representation. We elaborate on which representation schemes are used in different algorithm categories. We also examine how the SI-based algorithms, together with the representation schemes, address the challenging characteristics of the clustering problem such as multiple objectives, unknown number of clusters, arbitrary-shaped clusters, data types, constraints, and scalability. The pros and cons of each representation scheme are discussed. Finally, future research directions are suggested.

1 Introduction

Clustering aims at discovering the natural groups in a set of points. It forms groups of points based on a similarity measure so that the similarities between the points inside the same group are high, and the similarities between the points in different groups are low [34]. It can be used as an exploratory tool to discover the hidden and potentially useful groups in a data set. It also helps simplification and

T. İnkaya (✉)
Industrial Engineering Department, Uludağ University, Görükle, 16059, Bursa, Turkey
e-mail: tinkaya@uludag.edu.tr

S. Kayalıgil • N.E. Özdemirel
Industrial Engineering Department, Middle East Technical University,
Çankaya, 06800, Ankara, Turkey
e-mail: skayali@metu.edu.tr; nurevin@metu.edu.tr

© Springer International Publishing Switzerland 2016
M.E. Celebi, K. Aydin (eds.), *Unsupervised Learning Algorithms*,
DOI 10.1007/978-3-319-24211-8_12

303

data compression through generalizations. It has applications in many real-world problems such as image segmentation and clustering [8, 53], gene expression [30], geographic information systems [3], market segmentation [13], and so on.

Clustering is an unsupervised learning method, in which natural groupings are explored without any guidance and use of external information. For instance, the number of clusters in a data set is in general unknown a priori. The attributes may include various types of data, such as numerical, ordinal, and categorical. Hence, the similarity/dissimilarity function should be selected according to the type(s) of the attributes. Other complications with the clustering problem may be arbitrary-shaped clusters and density variations. The existence of such patterns complicates the evaluation of clustering quality. In addition to these, domain specific constraints and handling large data sets are some other challenging issues in the clustering problem.

Clustering problem has been studied in various domains including statistics, pattern recognition, management information systems, optimization, and artificial intelligence. Hence, a vast amount of literature has been accumulated. Basically, clustering approaches can be classified as partitional, hierarchical, density-based, graph-based, probabilistic, metaheuristic, and fuzzy approaches. Partitional approaches divide the data set into a number of groups according to an objective function [10]. Hierarchical approaches find the nested clusters recursively, and build a dendogram. A hierarchical algorithm works either in agglomerative (bottom-up) or divisive (top-down) mode. Density-based approaches form a cluster from a group of points in a dense region surrounded by less dense regions. Graph-based approaches are built upon the connectivity idea in graphs, and a cluster corresponds to a set of connected components. Probabilistic approaches assume that clusters originate from certain probability distributions and try to estimate the associative parameters. Metaheuristic approaches are stochastic search methods that consider clustering as an optimization problem. Typically, they are used to obtain the optimal clustering based on a criterion. Fuzzy clustering approaches determine the membership values when the cluster boundaries are not clear cut.

The flexibility of metaheuristic approaches makes them a promising tool for clustering. In particular, Swarm Intelligence (SI) has become an emerging solution approach for clustering. SI is an artificial intelligence technique that depends on the collective property emerging from multi-agents in a swarm. Agents in the swarm can collectively discover the attractive regions, i.e. core regions of clusters, through sharing density and connectivity information. Moreover, SI is a robust, scalable, and easily distributed tool. Motivated by these, SI has attracted attention of the clustering community, and it has become a widely used solution method. In recent years, several articles have been presented in the intersection of clustering and SI-based approaches. Some of them are ant colony optimization (ACO), ant-based sorting (ABS), particle swarm optimization (PSO), artificial bee colony (ABC), and so on.

There are several review articles about SI-based approaches for clustering [1, 2, 28, 50, 57]. Handl and Meyer [28] focus on the ant-based clustering algorithms, and study them in two categories: (1) methods that directly mimic the real ant colonies such as patch sorting and annular sorting and (2) ant-based optimization methods including ACO and its derivatives. They also discuss other types of swarm-based clustering methods such as PSO and information flocking. Abraham et al. [1] review the clustering algorithms based on ACO and PSO. Rana et al. [57] and Alam et al. [2] provide reviews of PSO-based clustering. Alam et al. [2] classify the PSO clustering algorithms into two categories: (1) algorithms in which PSO is directly used for clustering and (2) algorithms in which PSO is hybridized with another clustering method. Also, [50] study nature-inspired algorithms including evolutionary algorithms, simulated annealing (SA), genetic algorithms (GA), PSO, and ACO. However, their scope is limited to partitional clustering.

The studies in the intersection of data mining and SI are reviewed by Grosan et al. [24] and Martens et al. [48]. Grosan et al. [24] consider the early studies of PSO and ACO in data mining. Martens et al. [48] classify the SI-based data mining approaches into two categories: effective search and data organizing approaches. In effective search approaches, the agents move in the solution space and construct the solution. ACO and PSO are studied in this group. In data organizing approaches, the agents move the data points located in a low-dimensional space to accomplish the clustering task or to find a mapping solution. ABS is considered in this group.

In this paper, we study SI-based approaches for hard (crisp) clustering in five categories: ACO, ABS, PSO, hybrid approaches, and other SI-based approaches. ACO, ABS, and PSO are the well-known and widely used SI approaches that use the collective behavior of ants and birds. In the last decade, new SI-based approaches have emerged such as ABC, wasp swarm optimization (WSO), firefly algorithm (FA), and so on. We examine these algorithms as other SI-based algorithms. Moreover, SI-based approaches are combined with other metaheuristics and AI methods. Even two SI-based approaches can work together. These are considered as hybrid approaches.

Different from the previous review studies, we propose a new classification scheme for the SI-based clustering approaches according to agent representation. Agent representation is a key element, as it determines how solutions are generated. It also directly affects the exploration and exploitation capabilities of the search procedure. Taking into account this key role, we discuss the use of each agent representation scheme. The advantages and disadvantages of representation schemes are elaborated. Finally, we examine which challenging characteristics of the clustering problem are addressed by each representation scheme.

The rest of the chapter is organized as follows. Section 2 provides a definition of the clustering problem including its challenging issues. Section 3 briefly explains each SI-based approach. The new classification scheme based on agent representation is introduced in Sect. 4. The main properties of each representation scheme are explained. In addition, the studies using each representation scheme are discussed. Section 5 includes discussions about the representation schemes, SI-based algorithms, and challenging issues in the clustering problem. Finally, we conclude and give future research directions in Sect. 6.

2 The Clustering Problem

Given a data set $X = \{x_1, .., x_i, .., x_n\}$, which includes n data points each defined with d attributes, clustering forms k groups such that data points in the same cluster are similar to each other, and data points in different clusters are dissimilar.

The aim is to obtain compact, connected, and well-separated clusters. Compactness is related to intra-cluster similarity. It implies that the points in the same cluster should be similar/close to each other. Compactness measures are classified into two categories, namely representative point-based and edge-based measures. Representative point-based measures minimize the total dissimilarity/distance between each cluster member and a point that represents the cluster (centroid, medoid, and so on). When the number of clusters is given a priori, these measures usually lead to spherical clusters, and elongated or spiral clusters can hardly be obtained [19, 25]. The edge-based measures use the pairwise distances of the data points in the same cluster, for example, the sum of pairwise distances within each cluster, or the maximum edge length in the connected graph of a cluster. Edge-based approaches are more powerful than the other methods in handling arbitrary shapes.

Connectivity is the linkage between the data points in close proximity and having similarities. Data points in a certain vicinity of a given data point form the neighbors of that point, and a point and its neighbors should be assigned to the same cluster. For example, [29] calculate the connectivity objective for k-nearest neighbors as the degree of the neighboring data points placed in the same cluster.

Separation implies inter-cluster dissimilarity, that is, data points from different clusters should be dissimilar. Total inter-cluster distances between cluster representatives, single-link, average-link, and complete-link are some commonly used separation measures in the literature. For instance, single-link calculates the dissimilarity between a pair of clusters as the distance between the closest (most similar) points in the two clusters, whereas complete-link takes into account the most distant (most dissimilar) points. In average-link, the average distance between all point pairs in two clusters is used.

In this chapter, we consider the following challenging characteristics of the clustering problem.

(a) *Multiple objectives*: It is possible to conceptualize the aim of clustering, however it is difficult to combine and optimize the three objectives, namely compactness, connectivity, and separation, simultaneously. To the best of our knowledge, there is not a generally accepted clustering objective or measure that fits all data sets. The data set characteristics (types of attributes, shapes, and densities of clusters) and the application field determine the choice of the objective functions to be used in clustering.

(b) *Unknown number of clusters*: In real-world applications such as those in image segmentation, bioinformatics (gene clustering), and geographic information systems, the number of clusters is typically unknown, and there are no given class labels. Instead, one should extract it from the data set as a major piece of knowledge. Some approaches use a validity index to find the number of clusters

[5, 6]. After the execution of the clustering algorithm for different numbers of clusters, the one with the best validity index is selected. However, there is no validity index that works well for all data sets. There are also studies that assume the number of clusters is given. For instance, in a facility location problem, the number of facilities that will serve clusters of customers is typically pre-determined. Then, the task is to determine the facility locations and to assign customers to these facilities.

(c) *Arbitrary-shaped clusters*: A data set may include clusters of any size and shape such as ellipsoids, elongated structures, and concentric shapes. Density variations within and between clusters make the clustering task even more difficult. Arbitrary-shaped clusters are in particular seen in geographical infor-mation systems, image segmentation, geology, and earth sciences [19, 35]. The success in extraction of arbitrary-shaped clusters with density variations depends on the clustering objective function and the similarity/distance function used. In addition to the compactness and separation objectives, the proximity and connectivity relations are also crucial in discovering these clusters.

(d) *Data type*: A data set may include two types of attributes: (1) qualitative or cat-egorical (binary, nominal, or ordinal), (2) quantitative or numerical (continuous, discrete, or interval). The selection of the similarity/distance measure depends on the attribute type. When all the attributes are quantitative (e.g., height, weight), Manhattan, Euclidean, and Minkowski distances can be used [27]. In order to calculate the dissimilarities in qualitative data (e.g., color, rank), some well-known similarity measures are simple matching coefficient, Jaccard and Hamming distances [27]. If there exists a mixture of quantitative and qualitative attributes in a data set, the general similarity coefficient by Gower [23] and the generalized Minkowski distance by Ichino and Yaguchi [32] can be used. Note that, throughout the text, the terms similarity/dissimilarity measure and distance are used interchangeably.

(e) *Constraints*: Constrained clustering is a semi-supervised learning technique, in which additional conditions should be satisfied while grouping similar points into clusters. Constraints can be interpreted as a priori domain knowledge, and they improve the clustering accuracy and efficiency. However, additional mechanisms are needed in a clustering algorithm to handle such constraints. Constraints can be classified into three categories: cluster-level, attribute-level, and instance-level [7]. Cluster-level constraints include capacity, minimum clus-ter separation, maximum/minimum cluster diameter, and ϵ-distance neighbor within cluster. Attribute-level constraints are order preferences and constraints on the attribute values. Instance-level (relational) constraints are partial labels and pairwise relationships such as must-link (ML) and cannot-link (CL) constraints.

(f) *Scalability*: Improvements in the data storage and computing enable working with large amounts of data sets. However, as the number of data points and the number of attributes increase, data analysis requires more memory and computing time. For this reason, clustering algorithms that are capable of handling large data sets within a reasonable amount of time are needed.

3   Overview of the Swarm Intelligence-Based Approaches

A swarm is composed of simple agents. Even though there is a decentralized control mechanism in the swarm, it can perform complex tasks (such as finding food sources and protection from enemies) through sharing of information and experience among its members and interacting with its environment. Working principles of a swarm are dictated by the behavior of autonomous agents, the communication methods among the agents and with the environment, and moves of the agents. In the SI-based algorithms, these concepts correspond to agent representation, neighborhood definition, decision rule, exploration and exploitation mechanisms.

The starting point in the design of an SI-based algorithm is the agent representation; other characteristics of SI are built upon this ground. An agent may represent a set of variables in the solution, or it may work as a means to construct the solution. Hence, agent representation is directly related with the solutions obtained by the algorithm and the search procedure. The representation must allow all feasible solutions to be reachable during the search. Another important design element is the neighborhood definition. It is related with the locality of the agents, and defines their next possible moves during the solution construction. Next, a solution component is selected using a decision rule, which is often a stochastic choice mechanism. Together with the neighborhood properties, this rule determines the exploration and exploitation properties of the algorithm. Exploration ensures that the search space is examined efficiently, and the algorithm does not prematurely converge to a local optimal solution. This is usually provided by the randomness in the search process. Exploitation, however, is based on the experience of the swarm. It helps investigate the promising regions of the search space thoroughly. There should be a balance between exploration and exploitation for the success of the algorithm.

We study the SI-based algorithms in five categories: PSO, ACO, ABS, hybrid approaches, and other SI-based approaches. In the hybrid approaches, the aim is to strengthen an SI-based algorithm with the exploration and exploitation properties of other search methods such as other metaheuristics, AI approaches, or another SI-based algorithm. There are also studies that combine SI-based algorithms with local search algorithms such as k-means, k-harmonic mean, and so on. These local search algorithms are deterministic search procedures, and improve the exploitation property of the SI-based algorithm.

In this section, we briefly explain the basic characteristics of each SI-based algorithm. The hybrid approaches are derivatives of SI-based algorithms, therefore they are discussed in Sect. 4.

3.1   Particle Swarm Optimization

PSO was introduced by Kennedy and Eberhart [42]. It is inspired from the social behavior of a bird flock. Each particle represents a solution, and particles in the swarm fly in the search space with velocities updated according to the experiences they gained throughout the search. The aim is to discover better places in the search space. Let $x_i(t)$ and $v_i(t)$ denote the position and velocity of particle i in iteration t, respectively. The new velocity and new position of the particle are calculated according to the following equations:

$$v_i(t+1) = w \times v_i(t) + c1 \times rand1 \times (pbest_i(t) - x_i(t)) + c2 \tag{1}$$
$$\times rand\,2 \times (gbest(t) - x_i(t))$$

$$x_i(t+1) = x_i(t) + v_i(t+1) \tag{2}$$

where w denotes the inertia weight, $c1$ and $c2$ are the acceleration coefficients (learning factors), $rand1$ and $rand2$ are random numbers between 0 and 1, $pbest_i(t)$ and $gbest(t)$ denote the personal best solution of particle i and the global best solution for the swarm, respectively. Random functions in the velocity update ensure exploration, whereas the use of personal and global best solutions directs the particles to the promising regions of the search space.

The outline of the PSO algorithm is presented in Table 1.

3.2 Ant Colony Optimization

ACO was introduced by Dorigo et al. [16] and Dorigo et al. [17]. It is inspired by the foraging behavior of ants. Ants search for food on the ground, and during their search, they deposit a substance called pheromone. This smelling substance is used by the other ants to direct the search. Promising paths have higher pheromone concentration, and more ants are directed towards these paths. In this manner, ants can find the food sources by indirect communication between them.

Each ant constructs a solution at each iteration. In each step of construction, the ant selects a solution component from its neighborhood according to a decision rule.

Table 1 The outline of the PSO algorithm

Set parameters and initialize a population of particles with random positions and velocities in the search space.

While (termination conditions are not met)

    For each particle i do

        Update the velocity of particle i according to Eq. (1).

        Update the position of particle i according to Eq. (2).

        Calculate the fitness value of particle i.

        Update $pbest_i$ and $gbest$.

    End for

End while

The decision rule depends on the pheromone concentration on the solution components as well as some problem specific heuristic information. For instance, the ant chooses solution component i according to the following probability.

$$p_i = \frac{(\tau_i)^\alpha (\eta_i)^\beta}{\sum\limits_{k \in N(i)} (\tau_k)^\alpha (\eta_k)^\beta} \tag{3}$$

where τ_i denotes the pheromone concentration of solution component i, η_i is the heuristic information indicating the quality of solution component i, α and β determine the relative effects of pheromone and quality, and $N(i)$ shows the neighborhood of component i. As seen in Eq. (4) below, all components used in a good solution receive the same pheromone deposit. However, not every component in a good solution has an equally high quality, hence η_i is used to reflect the quality differences among the solution components.

After ants construct their solutions, the pheromone concentration of each solution component is updated according to Eqs. (4) and (5).

$$\tau_i = (1 - \rho)\tau_i + \rho \Delta \tau_i \tag{4}$$

$$\Delta \tau_i = \sum_{k \in S | i \in k} w_k f_k \tag{5}$$

where $0 < \rho < 1$ is the rate of pheromone evaporation, S is the set of selected solutions such that they contain solution component i, f_k defines the quality of solution constructed by ant k, and w_k is the weight of solution constructed by ant k. There are different rules for selecting solutions to be used for the update, ranging from all solutions generated in the current iteration to the best solution found in the history of the search.

The pheromone deposit quantified in Eq. (5) helps the ant colony intensify the search towards the good solution components. On the other hand, the evaporation of pheromone in the first term of Eq. (4) and probabilistic selection of the solution components according to Eq. (3) increase the exploration capabilities of ACO.

The outline of the ACO algorithm is presented in Table 2.

Table 2 The outline of the ACO algorithm

Set parameters, and initialize pheromone trails.
While (termination conditions are not met)
For each ant k do
Construct a solution by selecting solution components using Eq. (4).
Calculate the fitness value.
Perform local search.
End for
Update the pheromone concentrations according to Eqs. (5) and (6).
End while

3.3 Ant-Based Sorting

ABS was introduced by Deneubourg et al. [15]. It is inspired from the gathering and sorting behaviors of ants such as corpse clustering, brood sorting, and nest building. Ants work as if they were forming a topographic map. An ant picks up a point in the search space and drops it off near the points similar to it. The pick-up and drop-off operations are performed based on the similarity within the surrounding neighborhood.

Lumer and Faieta [47] generalized this method for clustering. In their model, data points are scattered in a search space, which is split into grids. Ants walk through the grids and move the data points. The probabilities that an ant picks up and drops off point i are denoted as $p_{pick}(i)$ and $p_{drop}(i)$, respectively, and they are calculated as follows.

$$p_{pick}(i) = \left(\frac{k_1}{k_1 + f(i)} \right)^2 \tag{6}$$

$$p_{drop}(i) = \begin{cases} 2f(i), & \text{if } f(i) < k_2 \\ 1, & \text{if } f(i) \geq k_2 \end{cases} \tag{7}$$

where k_1 and k_2 are constants, and $f(i)$ is the density dependent function of point i defined as:

$$f(i) = \begin{cases} \dfrac{1}{s^2} \sum_{j \in N(i)} (1 - d_{ij}/\alpha), & \text{if } f(i) > 0 \\ 0, & \text{otherwise.} \end{cases} \tag{8}$$

Here s^2 is the size of the local neighborhood $N(i)$ (vision field), d_{ij} is the distance between points i and j, and α is the dissimilarity scaling parameter.

An ant has a higher pick-up (drop-off) probability for a dissimilar (similar) point compared to a similar (dissimilar) point. Hence, these probabilistic pick-up and drop-off operations support the exploration and exploitation properties of ABS.

The ABS algorithm is outlined in Table 3.

Table 3 The outline of the ABS algorithm

Randomly scatter the data points and ants on the toroidal grid.
While (termination conditions are not met)
For each ant k do
Calculate the similarity between each point and its neighbors.
Pick-up or drop-off a point according to Eqs. (7) and (8).
Move the ant to a randomly selected empty neighboring grid.
End for
End while

3.4 Other Swarm Intelligence-Based Metaheuristics

In recent years, many new nature-inspired algorithms have been proposed based on the collective behavior of bees, wasps, fish, termites, and so on. For the clustering problem, we examine the ABC, WSO, firefly (FA), and artificial fish (AF) algorithms as the other SI-based (OSIB) algorithms.

ABC was introduced by Karaboga [40], and it is based on the foraging behavior of the honeybees. In ABC, there are three groups of bees including employed bees, onlookers, and scouts. Each employed bee visits a potential food source. Onlooker bees choose promising food sources among them. When an employed bee abandons a food source, it becomes a scout, and starts to search for a new food source randomly. In WSO, wasps share limited resources according to their importance to the entire swarm and social status. In FA, fireflies have a luminescence quality, namely luciferin. They emit light proportional to this value, and each firefly is attracted by the brighter glow of other neighboring fireflies. AF is based on the shoal memberships. The fish shoals are open groups. That is, when shoal encounters are observed, individuals can choose neighboring fish with similar phenotypic properties such as color, size, and species.

Note that there are other population-based methods inspired by nature. However, they do not include the collective behavior of the swarm. For instance, genetic algorithms are based on evolving generations through crossover and mutation. These methods are out of our scope, and they are not included in this review.

4 Classification of the Swarm Intelligence-Based Algorithms for Clustering

Taking into account the importance of the agent representation, we classify the SI-based clustering algorithms into four groups: (1) data point-to-cluster assignment, (2) cluster representatives, (3) direct point-agent matching, and (4) search agent. In the data point-to-cluster assignment and cluster representatives schemes, the SI agent keeps information about the clusters. On the other hand, in the direct point-agent matching and search agent schemes, the agents are concerned about data points.

In this section, we provide the main properties of these representation schemes.

4.1 Data Point-to-Cluster Assignment

In this representation, an agent keeps track of the cluster assignments of the points in the data set. The number of clusters should be provided a priori. The size of the solution vector is defined by the number of points in the data set, and the number

Data set ➔ $X = \{1, 2, 3, 4, 5, 6\}$

<u>**Clustering solution**</u>

Cluster 1 ➔ $C_1 = \{1, 2, 3\}$

Cluster 2 ➔ $C_2 = \{4, 5, 6\}$

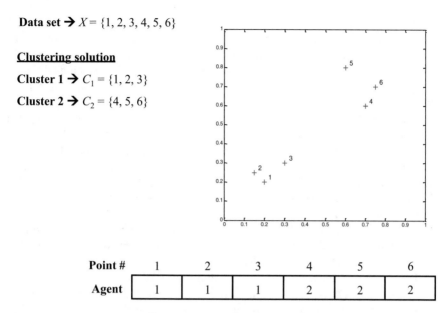

Point #	1	2	3	4	5	6
Agent	1	1	1	2	2	2

Fig. 1 An example for the data point-to-cluster assignment representation

of attributes in the data set has no impact on the representation. For example, in a two-dimensional data set with six points, an agent is a six-dimensional vector. Each dimension shows the clustering assignment of the respective point. An example is provided in Fig. 1.

Jarboui et al. [36] adopt the data point-to-cluster assignment to PSO. PSO has originally been proposed for search in continuous space, so combinatorial PSO is used to switch between the combinatorial and continuous states. A dummy vector takes values from the set $\{-1, 1, 0\}$ according to the solution status, namely the local best cluster, the global best cluster, or none of these, respectively. After the update in each iteration, the new position of the particle is discretized using the dummy vector and a threshold value. The performance of the proposed approach is tested using two objective functions: minimization of within cluster variation and maximization of variance ratio criterion.

In ACO, there are several studies using this representation scheme [11, 58, 61]. The common point of these studies is that they minimize the within cluster variation/distance. Also, each point-to-cluster assignment is associated with a pheromone concentration. However, the pheromone update mechanisms in these algorithms are different. For instance, there are local and global pheromone update mechanisms in [11]. In the local update, each ant simultaneously updates the pheromone concentration on the particular point-to-cluster assignment. In the global update, after all ants build their solutions, the pheromone concentration is updated only for the global best assignments. Runkler [58] updates the pheromone concentration according to the difference between the fitness values of the incumbent solution and the current solution. Shelokar et al. [61] perform the pheromone update using best L solutions out of R solutions. An extended version of this representation proposed by Jiang et al. [38] can handle attribute selection and clustering tasks together. That is, an ant represents not only the cluster assignments, but also the selected attributes. They use two pheromone matrices, one for the attributes selected and another one for point-to-cluster assignments.

In the OSIB category, [59] uses data point-to-cluster assignment scheme together with WSO. The objective is to minimize within cluster distances. Data point-to-cluster assignments are performed using the stochastic tournament selection. Assignment probability is determined according to the distance between the cluster center and the point. An interesting point in this study is that there is no information exchange among the wasps. Hence, in contrast with the SI principles, collective behavior of the swarm is not observed directly.

ABS forms a topographic map, so the number of clusters is not known a priori. Hence, data point-to-cluster assignment scheme is not applied in ABS. This representation is not used in hybrid approaches either.

The main disadvantage of this representation is that the number of clusters should be given a priori. In addition, the studies that use this representation minimize the within cluster distance, and they typically generate spherical clusters.

4.2 Cluster Representatives

In this scheme, an agent shows the representatives of the clusters such as center, median, medoid, and so on. Typically, the number of clusters (representatives) is given, and the size of the solution vector depends on the number of attributes and the number of clusters. This representation scheme has an analogy with the partitional clustering approaches.

A stream of studies adopts this representation to determine the number of clusters. In one of the revised representations, the maximum number of clusters is used. An additional vector shows whether the corresponding cluster representative is selected or not. The length of this additional vector is equal to the maximum number of clusters. In order to determine the number of clusters, these studies use clustering validity indices such as Dunn's index and Davies Bouldin index for fitness calculation. In another revised representation, an additional variable is used to define the number of clusters for each particle.

In Fig. 2, an example agent is given for a data set with six points and two attributes. If the number of clusters is given as two, the agent corresponds to a four-dimensional vector.

In PSO, for the given number of clusters, there are several studies that use cluster representatives scheme [39, 45, 52, 53, 55, 63, 65, 73]. Van der Merwe et al. [65] propose two PSO algorithms for clustering. These are the standard *gbest* PSO algorithm and a combination of PSO and k-means. Both algorithms minimize the average quantization error, i.e. the distance between each point and its cluster

Data set → $X = \{1, 2, 3, 4, 5, 6\}$

Clustering solution

Cluster 1 → $C_1 = \{1, 2, 3\}$

Cluster 2 → $C_2 = \{4, 5, 6\}$

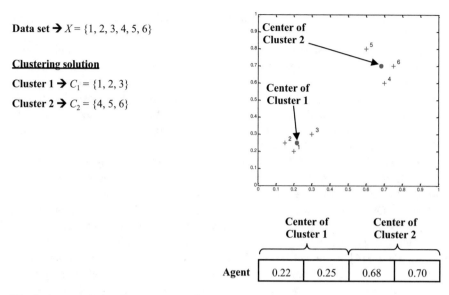

Fig. 2 An example for cluster representatives

representative. In the combined algorithm, k-means seeds the initial swarm. The velocity of a particle is updated according to the particle's previous velocity, its local best, and the swarm's global best. As the particles move, the coordinates of the representative points also change. Omran et al. [52] attempt to consider the multi-objective nature of the clustering algorithm, and use the weighted sum of the maximum average intra-cluster distance and the minimum inter-cluster distance. They test the proposed approach using MRI and satellite data sets. Omran et al. [53] revise the work in [52] by inserting the quantization error term to the objective function. They also use a different version of PSO in which the global best particle is updated by resetting the particle's position to the global best position and using the previous velocity and a random component for search direction. Paterlini and Krink [55] compare the performances of k-means, GA, differential evolution (DE), and PSO. The algorithms give similar results for simple data sets, however population-based heuristics are better for more complex data sets. The performance of DE is superior and more robust compared to the other algorithms, but it has poor performance for noisy data.

A group of studies use PSO together with the well-known local search algorithms to increase the exploitation capability. For example, [39] propose a hybrid algorithm combining Nelder-Mead simplex algorithm and PSO. Nelder-Mead algorithm is used for local search, whereas PSO is used as an exploration tool. The proposed algorithm converges to the global optimal solution with a relatively smaller error rate. Also, it provides an improvement in the number of function evaluations. Yang et al. [73] combine PSO with k-harmonic means (KHM). KHM is used as a local search method, which is applied to the particles in the swarm every eight generations. In a similar manner, [45] apply PSO together with k-means.

Different from the studies discussed so far, [63] consider document clustering. In order to increase the search capability of the algorithm, they introduce an environmental factor for each centroid, which is the mean of the data points belonging to the corresponding centroid. In the first stage of the proposed approach, instead of global best, the environmental factor is used to update the particle's velocity. In the second stage, the particle's velocity is updated according to not only the environmental factor, but also the global best.

A research stream in PSO assumes that the number of clusters is unknown [12, 14, 54]. They start with a set of potential cluster representatives for the maximum number of clusters. For instance, [54] use binary PSO in which a binary particle shows whether the cluster representative is used in the solution or not. The coordinates of the potential representatives are optimized using the k-means algorithm. Clustering solutions are evaluated using Turi's validity index. Instead of binary PSO, [14] use an activation function value for each potential cluster center. That is, a cluster center becomes active if its activation function value is larger than a given threshold. They also adopt a multi-elitist PSO for the clustering problem. In the proposed approach, the global best is replaced with the local best having the highest growth rate. In this way, early convergence is prevented. Moreover, a kernel-induced similarity measure is used instead of the conventional sum-of-squared distances. They use kernelized CS measure as the fitness function. This helps the algorithm extract complex and linearly non-separable clusters. Cura [12] uses an additional variable to define the number of clusters (K_p) for particle p. In each iteration, the particle first moves in dimension K_p. Next, the existing cluster centers of the particle are split or deleted according to the new K_p. Finally, the particle updates the new cluster centers. The proposed approach aims to obtain compact and well-separated clusters by minimizing the difference between total intra-cluster distances and minimum inter-cluster distance.

Lu et al. [46] extend the cluster representatives scheme for variable (attribute) weighting problem. In the proposed approach, there are three swarms: (1) position swarm of variable weights, (2) velocity swarm of variable weights, and (3) swarm of cluster centroids. Position and velocity particles denote the weight set for each cluster. Also, they use comprehensive learning mechanism by applying a crossover between the best position of the weight matrix and one of the other best personal positions in the swarm. Although the proposed approach runs slowly, its performance is superior to those of the other algorithms.

To the best of our knowledge, cluster representatives scheme is not applied in ACO and ABS. However, it is widely used in hybrid approaches [21, 31, 37, 51, 69, 70, 72]. Jiang and Wang [37] combine cooperative co-evolution (CC) framework with bare-bone PSO (BPSO). Different from classical PSO, BPSO updates the position of the particle according to the Gaussian distribution. CC decomposes the clustering problem into subproblems, and BPSO is applied to find the best solution for each subproblem, which represents a single cluster centroid. Each particle in the subproblem is evaluated using the best particles from the other subproblems.

Huang et al. [31] hybridize the continuous version of ACO (ACOR) and PSO. Both ACOR and PSO search for cluster centers. Pheromone-particle table is used to share the new particle solutions generated with PSO and ACOR. Four types of models are introduced: sequential, parallel, sequential with enlarged pheromone-particle table, and global best exchange. In the sequential model, ACOR and PSO share the same set of solutions using pheromone-particle table. For ACOR, pheromone-particle table shows the pheromone values. For PSO, it shows the current solution. In the parallel approach, new ants are generated directly from the pheromone-particle table, without considering the new particles. The sequential update model with the enlarged pheromone-particle table combines the new particles with the pheromone-particle table, and new ants are generated according to the enlarged pheromone-particle table. The global best exchange approach does not have a pheromone-particle table, and the two algorithms share the best solution only. Among the four models, the sequential update with the enlarged pheromone-particle table gives the best average performance.

Firouzi et al. [21] combine PSO, simulated annealing (SA), and k-means. In each iteration, SA searches around the global best, and strengthens the exploration property of PSO. At the end of PSO, k-means is applied. Niknam and Amiri [51] address the lack of search capability for the global best particle, and propose a combination of fuzzy adaptive PSO, ACO, and k-means. The inertia weight and learning factors of PSO are dynamically adjusted using fuzzy if-then rules. Also, for each particle, a different global best position is assigned according to the ACO best path selection method. Yan et al. [72] propose a hybrid version of ABC and GA. Arithmetic crossover is applied to the randomly selected cluster centers (food sources). Xiao et al. [69] combine self-organizing map (SOM) and PSO. SOM is used for clustering, whereas PSO optimizes the weights of SOM. Note that the weights of SOM correspond to the cluster representatives.

Similar to [14, 70] assume that the number of clusters is unknown, and they use an activation function value for each potential cluster center. They combine PSO and DE, and propose the DEPSO algorithm. DEPSO integrates the DE operator into the PSO procedure in order to diversify the PSO. Also, they compare the performances of various clustering validity indices including the silhouette index, Calinski-Harabasz index, CS index, and Davies-Bouldin index. They conclude that the silhouette index gives superior results in most of the data sets for DEPSO, PSO, and DE-based clustering algorithms.

In the OSIB algorithms, an agent usually denotes the cluster representatives. Fathian et al. [20] apply honeybee mating optimization to the clustering problem. New broods are created by exchanging the drone's genes with the queen's (crossover). Worker bees improve the broods by performing local search (royal jelly). Different from the other SI-based algorithms, the bees do not move in the search space; instead, they exchange the gene information. Hence, this algorithm fits more the principles of evolutionary algorithms than those of SI. Karaboga and Ozturk [41] and Zhang et al. [75] apply ABC algorithm to the clustering problem. Zhang et al. [75] study the unsupervised clustering problem, whereas [41] consider the supervised clustering problem, where each data set is partitioned into two subsets for training and testing. In both algorithms, employed and onlooker bees exploit the search space within the given neighborhood. Employed bees turn into scout bees so that they explore different regions of the search space. Zou et al. [78] propose a modified version of ABC, namely the cooperative ABC algorithm. There is a super best solution vector that keeps the best component in each dimension among all populations. The employed and onlooker bees perform their search using the super best solution.

Senthilnath et al. [60] propose a firefly algorithm for supervised clustering. The main idea behind the firefly algorithm is that particles move towards the attractive particles having higher light intensity. Wan et al. [68] apply chaotic ant swarm to clustering. The swarm performs two phases: chaotic phase and organization phase. An organization variable is introduced to achieve self-organization from the chaotic state. This controls the exploration and exploitation properties.

Unless the data set is projected to a transformed feature space, cluster representatives scheme is limited to generating spherical and ellipsoidal clusters. The revised version of this representation can determine the number of clusters, however most of the studies in the literature assume that the number of clusters is given. Moreover, the revised version should couple with an objective function that can optimize the number of clusters.

4.3   Direct Point-Agent Matching

There is a one-to-one correspondence between the agents and the points in this representation. That is, each agent is loaded with a data point, and the number of agents is equal to the number of data points. Agents are scattered in a two-dimensional search space. They move and carry with them the data points such that similar ones are positioned close to each other whereas dissimilar ones are far apart. Eventually, the ants' movements in the search space form a topographic map.

Direct point-agent matching is applied by Picarougne et al. [56] and Veenhuis and Köppen [66] in PSO. Both algorithms have two phases. In the first phase, similar points are gathered together whereas dissimilar ones are set apart using collective behavior of the swarm. In the second phase, clusters are retrieved using a hierarchical algorithm. The number of clusters is assumed to be unknown in both algorithms. The difference between the two algorithms is in the velocity update mechanism. Picarougne et al. [56] use the concept of flock of agents, whereas [66] is an application of PSO. Picarougne et al. [56] update the agent direction using the sum of all the influences in its neighborhood, and the agent speed using the minimum speed and the agent's isolation degree. Hence, agents in a group travel slowly compared to the agents traveling alone. In [66], a particle's new velocity is calculated using its previous velocity, influence factor for the points in its neighborhood, velocity matching factor to the similar points, and avoidance factor to the dissimilar ones.

Direct point-agent matching is widely used in ABS. Xu et al. [71] incorporate ML and CL constraints into the clustering problem. Points with ML constraints are added to the neighborhood with high weights, whereas points with CL constraints are excluded from the neighborhood by defining high negative weights. An ant sleeps or awakens according to a randomly generated number. It performs two types of heuristic walk, namely max-number direction moving and adaptive direction walking. In the max-number direction moving, the ant moves in the direction where most of its neighbors are located. In adaptive direction walking, direction selection is based on the weighted similarity values. The proposed approach gives better results compared to the previous constrained clustering methods.

Zhang et al. [77] introduce a novel ant movement model based on Renyi entropy. In this approach, the local similarity of each point is measured by entropy rather than distance. Also, kernel entropy component analysis (KECA) is used to create rough clusters. Ghosh et al. [22] incorporate aggregation pheromone to ABS in order to gather similar points into closer proximity. The amount of movement for an ant towards a point depends on the intensity of aggregation pheromone deposited by all other ants at that point. They model the pheromone intensity emitted by an ant using Gaussian distribution. Zhang and Cao [76] integrate ABS with kernel principle component analysis (KPCA). KPCA creates a projection for the rough clusters in order to find the non-spherical clusters. Then, the ABS algorithm is applied.

In the OSIB category, [4] introduce a hierarchical ant-based algorithm, which is inspired from the self-assembly behavior of real ants. Ants build a decision tree such that each ant moves in the tree according to the similarity. At the end of the algorithm, each sub-tree, which is directly connected to the support node, forms a cluster. Hence, a cluster retrieval operation is not needed. They apply the proposed approach to several real-world problems such as human skin analysis, web usage mining, and portal site data. Khereddine and Gzara [43] propose a hierarchical divisive clustering algorithm inspired from artificial fish shoals. The proposed algorithm forms a binary tree of clusters. Each time, the cluster with the maximum diameter is selected, and a bi-partitioning algorithm is applied. The direction of the fish is determined according to the similarity of the data point and its neighbors.

The direct agent-point matching is not used in ACO and hybrid approaches.

The common property of the algorithms in this category is that they do not use an explicit objective function or a fitness value. They implicitly maximize the total within-cluster similarity. The main advantage of this representation is that it visualizes the clusters with a topographic map or a decision tree. Moreover, most of the algorithms in this category determine the number of clusters. However, a separate cluster retrieval operation is needed in most of the studies. Another drawback of this representation is that, as the number of points increases, the number of agents increases leading to scalability problems.

4.4 Search Agent

In this representation, agent is a means to carry the similar points to the same neighborhood and the dissimilar points away from each other. In contrast to direct point-agent matching, agents do not have one-to-one correspondence with the data points, and there are fewer agents than points. Agents move around the search space picking up and dropping off the data points. The number of ants in this representation does not depend on the clustering problem parameters such as the number of points, the number of attributes, and the number of clusters.

İnkaya et al. [33] use the search agent representation in ACO. Ants choose the data points they will visit according to the pheromone concentration on the edges connecting pairs of points. A sequence of visits forms a subtour, and each subtour of an ant forms a cluster. The pheromone concentrations on the visited edges are updated in every iteration. In order to handle arbitrary-shaped clusters and unknown number of clusters, they propose two objective functions based on adjusted compactness and relative separation. The algorithm outputs a set of non-dominated solutions for these two criteria.

Lumer and Faieta [47] are the first authors that use search agent representation in ABS. Later, several variations of ABS are introduced. Martin et al. [49] is inspired from formation of ant cemetery activities, and revise the model in [15] with simpler assumptions about ants' behavior. For instance, when an unloaded ant has a body in one of the eight neighboring cells, it is loaded with probability 1.0. If there are several bodies, it chooses one of them at random. After the ant moves at least one step away, it drops its corpse if there are one or more corpses in its eight neighboring cells. The ant drops the corpse in a randomly chosen empty neighboring cell. The authors point out that the simple ants can do the corpse clustering as well. The only difference is that they are slower compared to the intelligent ants.

Vizine et al. [67] develop the Adaptive Ant Clustering Algorithm (A^2CA), which is an extension of the work by Lumer and Faieta [47]. There are two main modifications. Firstly, the vision field is changed adaptively according to a density dependent function. This is called progressive vision. Secondly, a pheromone function is used for each grid. Positions with high pheromone values are attractive for the ants to drop off the points. In contrast, the pick-up probability is inversely proportional to the amount of pheromone at that position. Yang and Kamel [74] propose an aggregated clustering approach. There are several parallel and independent ant colonies, and a queen ant agent. Firstly, clusters are formed by each colony. Secondly, the clustering results are combined by the queen ant using a hypergraph model. Finally, the new similarity matrix calculated by the queen ant is shared with the ant colonies.

Handl et al. [29] also modify the work in [47] with three features: (1) a short-term memory with look ahead capability, that is the next position of the ant is the grid cell with the best match (highest neighborhood function value), (2) increasing the size of the local neighborhood during the algorithm, and (3) a modified neighborhood function and threshold functions. An agglomerative clustering algorithm is used to retrieve the final clusters, and the size of the local neighborhood in the last phase of the ABS becomes the stopping condition for the cluster retrieval process. Hence, the proposed approach finds the number of clusters automatically.

Boryczka [9] propose two revised versions of the algorithm in [47], namely ACA and ACAM. The neighborhood function of ACA depends on the size of the local neighborhood and a dissimilarity scaling parameter (α). Parameter α determines the percentage of points on the grid that are classified as similar, and it is adaptively changed according to the rate of failure for drop-off. In ACAM a new neighborhood scaling parameter is defined using the relationship between the initial and the current size of the local neighborhood for an ant. Also, a cooling procedure is applied for parameter α, where it changes according to the average behavior of the density dependent function over the iterations. Gunes and Uyar [26] propose the parallelized version of ABS. Although clustering quality does not improve, the proposed approach works faster than the traditional ABS.

A recent study by Elkamel et al. [18] introduces a novel ABS algorithm, namely Communicating Ants for Clustering with Backtracking (CACB). CACB is a hierarchical clustering algorithm. It applies backtracking strategy so that the clustering errors in the previous iterations are corrected. Moreover, the use of dynamic aggregation threshold and dynamic data structure with doubly linked list improves the execution time of the algorithm.

In hybrid approaches, [62, 64] use the search agent representation. Tsai et al. [64] hybridize ACO with SA and tournament selection from GA. Ants generate subtours by inserting edges between data points. Pheromone concentration is updated for each edge connecting two points. The closer the distance between the two points is, the higher the pheromone deposit is. An ant selects the next point to visit using the tournament selection. SA ensures the reduction of number of points visited by each ant over the iterations. In the first phase of the algorithm, the edges between similar points become denser in terms of pheromone concentration. In the second phase, edges having higher pheromone density are set as dense, and they are merged using a hierarchical clustering algorithm to form the final clusters. The proposed approach in [62] is similar to the one in [64] except the use of tabu search. During the solution construction, an ant cannot visit the same point repeatedly. This restriction is implicitly imposed in [64] as well.

In the OSIB category, [44] introduce a clustering algorithm based on the API metaheuristic, which is inspired from the prey model of Pachycondyla apicalis ants. Ants generate random points around the nest, i.e. their initial hunting sites. Each ant visits its hunting site, and performs a local search. This is repeated until the ant does not find a prey. Then, the ant starts with another random point close to the nest.

In most of the algorithms, search agent representation helps find the number of clusters. Also, arbitrary-shaped clusters can be extracted. However, some of the algorithms require a cluster retrieval process that runs in isolation from the SI-based algorithm.

5 Discussion

The PSO, ACO, ABS, hybrid, and other SI-based algorithms reviewed in Sect. 4 are summarized in Appendix Tables 6, 7, 8, 9, and 10, respectively. The tables include the features of each study concerning the challenging issues in the clustering problem as discussed in Sect. 2, clustering objectives, agent representation, and application area if any.

In this section we provide a brief discussion of the review results from two perspectives: the agent representation and the challenging issues in clustering.

5.1 *Agent Representation Versus SI-Based Clustering Algorithms*

We summarize the frequency of studies discussed so far in Table 4 according to the agent representation and the type of the SI algorithm. Cluster representatives scheme is the most widely used agent representation in PSO. As PSO has originally been proposed for continuous optimization problems, it becomes a perfect fit for searching the cluster representatives.

An interesting observation is that, there is a single study that uses the data point-to-cluster assignment representation scheme. As this scheme works in the discrete domain, combinatorial PSO is adopted for the clustering problem. The PSO studies discussed so far do not use the search agent representation.

Data point-to-cluster assignment and search agent representations are applied in ACO. ACO is a construction metaheuristic where ants construct a solution from scratch at every iteration. However, cluster representatives and direct point-agent matching representation schemes tend to facilitate solution improvement rather than solution construction. For this reason, these representations have not been used in ACO so far.

In ABS, direct point-agent matching and search agent are the most popular representations. Clusters in ABS are formed using the collective behavior of ants,

Table 4 The frequency of articles by agent representation and SI algorithm

	Agent representation				
	Data point-to-cluster assignment	Cluster representatives	Direct point-agent matching	Search agent	Total
PSO	1	12	2	0	15
ACO	4	0	0	1	5
ABS	0	0	4	8	12
Hybrid	0	7	0	2	9
OSIB	1	6	2	1	10

each carrying or visiting individual data points. These two representation schemes are also based on the individual characteristics of the agents. Hence, they offer a good fit for the gathering and sorting activities of ABS. On the other hand, data point-to-cluster assignment and cluster representatives carry the information about the clusters rather than the points. Hence, they are not as suitable for the working principles of ABS.

Cluster representatives and search agent are the two representation schemes used in hybrid approaches. The cluster representatives scheme takes a partitional approach to the clustering problem, and it is often used together with minimization of the within cluster variance/distance or optimization of a validity index. In order to avoid premature convergence, the strong features of the other metaheuristics and AI techniques are combined with SI. In a similar manner, studies using search agent benefit from the exploration and exploitation capabilities of the other metaheuristics.

In the OSIB algorithms, all four types of representations are used. The most popular one is the cluster representatives scheme. In particular, cluster representatives are used in ABC and its variants, and the firefly algorithm.

5.2 Agent Representation Versus Challenging Issues in Clustering

In this section, we discuss which challenging characteristics of the clustering problem are addressed by each representation scheme. Use of the agent representation schemes in handling the challenging issues in clustering is overviewed in Table 5. The following observations can be made according to Table 5.

(a) *Multiple objectives*

- Most of the literature focuses on a single objective. In particular, the compactness objective, which minimizes the within cluster variance/distance, is widely used. Unless a transformation is applied to the data set, these

Table 5 Use of agent representations in handling the challenging issues in clustering

Clustering characteristic	Agent Representation			
	Data point-to-cluster assignment	Cluster representatives	Direct point-agent matching	Search agent
Multiple objectives		✓		✓
Unknown number of clusters		✓	✓	✓
Arbitrary-shaped clusters		✓	✓	✓
Data type		✓	✓	
Constraints			✓	
Scalability		✓	✓	✓

approaches generate spherical shapes regardless of the representation scheme.

- An early study considers the weighted sum of the two conflicting objectives, namely compactness and separation [53]. Cluster representatives are optimized using PSO. Although this study addresses multiple objectives, a multi-objective optimization method is not used.
- A recent study considers two objectives, namely adjusted compactness and relative separation, separately [33]. In ACO ants are search agents, and they insert edges between pairs of similar points. The proposed approach outputs a set of non-dominated solutions. However, it does not guarantee to generate the entire Pareto-efficient frontier.

(b) *Unknown number of clusters*

- The number of clusters can be extracted in the studies that use the direct point-agent matching and search agent representations. Some of the algorithms in this category form a tree or a map that illustrates the relative position of each data point. An additional cluster retrieval algorithm might be needed to determine the number of clusters. Hierarchical algorithms with thresholds of minimum similarity or maximum dissimilarity are often used for this purpose [29]. This limits the role of SI in finding the final clusters.
- Only a few studies using cluster representatives can extract the number of clusters. In these studies, the maximum number of clusters should be given, and an additional solution vector or variable is defined to control the active cluster representatives [12, 14, 54].

(c) *Arbitrary-shaped clusters*

- Direct point-agent matching and search agent representation schemes do not require any a priori information about the number of clusters or shapes of the clusters. They are flexible tools for extracting the relations among data points. Hence, they can find the arbitrary-shaped clusters, if they are coupled with proper objective(s) and a cluster retrieval process.
- When the agents denote the cluster representatives, and the objective is to minimize the total distance between cluster representatives and data points, the algorithm generates spherical clusters. This deficiency is resolved by the use of kernel-based similarity functions [14].
- In direct point-agent matching, kernel-based similarity measures are also used to extract the arbitrary-shaped clusters [76, 77]. In these algorithms, a cluster retrieval process is executed after the SI algorithm in order to determine the clusters.

(d) *Data type*

- Only a limited number of studies focus on the data sets with non-numerical attributes. For the binary and categorical attributes, Hamming distance is used together with direct agent-point matching [4].

- A couple of studies consider text and document clustering. They use cluster representatives, and apply extended Jaccard coefficient [46] and cosine distance functions [63] to measure the dissimilarities.
- SI-based algorithms are applied to the image and MRI data sets [53, 54]. However, these studies also use Euclidean distance, as the attributes are numerical.

(e) *Constraints*

- The constraint handling mechanisms in SI-based algorithms are limited. To the best of our knowledge, there is only one study that incorporates ML and CL constraints into ABS [71]. They use direct point-agent matching.

(f) *Scalability*

- The most significant work in this category is the parallelization of the ABS algorithm [26].
- Kernel-based approaches produce the rough clusters, so they reduce the runtime of the algorithm [14, 76, 77]. This approach is used in PSO and ABS so far.
- Another approach to handle the scalability is the execution of a preprocessing step for the subclusters [33]. As the subclusters include connected points, only the points on the boundaries of the subclusters are considered for merging and outlier detection operations. A boundary formation algorithm is used to exclude the interior points in the subclusters. The proposed clustering algorithm is executed for the boundary points of the subclusters only.

6 Conclusion

In this study we provide a systematic review for the SI-based hard (crisp) clustering algorithms. Our review includes PSO, ACO, ABS, hybrid, and other SI-based algorithms. Agent representation is the key element for the design of an SI-based algorithm. Motivated by this, we classify the SI-based clustering algorithms into four categories according to agent representation. We examine the capabilities of each representation scheme in handling the challenging characteristics of the clustering problem including multiple objectives, unknown number of clusters, arbitrary-shaped clusters, data types, constraints, and scalability.

As agent representation, search agent and direct point-agent matching schemes can be used when there is no a priori information about the clustering problem. When the data set includes spherical clusters and the number of clusters is known a priori, data point-to-cluster assignment and cluster representatives schemes are candidates for agent representation.

Not all representation schemes are applied to all categories of SI-based algorithms. For instance, ACO works in discrete search space, whereas direct point-agent matching works in continuous space. ABS is based on the moves of ants in the search space, so representation schemes that carry cluster information are not used in ABS. The search agent constructs a solution, whereas PSO is an improvement metaheuristic. Still, the use of different representation schemes in different types of SI-based algorithms is a promising future study. Such adoptions may lead to more effective and efficient search mechanisms for the SI-based algorithms.

Furthermore, SI researchers could pay more attention to the challenging characteristics of the clustering problem. Future research directions can be as follows:

- Current SI-based algorithms use different similarity measures for mixed data types. The development of SI-based mechanisms to handle various data types can be studied further.
- SI-based algorithms for semi-supervised clustering can be a future research direction.
- The development of multi-objective SI-based algorithms for the clustering problem can be a future research direction. In particular, the extraction of Pareto-efficient frontier needs to be explored.
- Another important issue with the real-life clustering problems is the scalability. Hence, scalability should be taken into account during the SI-algorithm design.

Appendix

See Tables 6, 7, 8, 9, and 10.

Table 6 Characteristics of the PSO algorithms for clustering

Article	Problem characteristics						Clustering objective	Particle representation	Application area
	Multi-objective	# of clusters	Arbitrary shapes	Data type	Constraints				
[36]	No	Given	No	Numerical	No		– Minimize within cluster variation (sum of squared Euclidean distances from cluster centers) – Minimize variance ratio	Data point-to-cluster assignment: A particle shows cluster assignments of all points	–
[12]	Yes	Not given	Yes	Numerical	No		Minimize the difference between total intra-cluster distances and minimum inter-cluster distance	Cluster representatives: A particle represents the number of clusters and cluster centers	–
[14]	No	Not given	Yes	Numerical	No		Maximize kernelized CS measure	Cluster representatives: A particle represents the activation function values and cluster centers	–

(continued)

Table 6 (continued)

Article	Problem characteristics						Clustering objective	Particle representation	Application area
	Multi-objective	# of clusters	Arbitrary shapes	Data type	Constraints				
[39]	No	Given	No	Numerical	No		Minimize within cluster distances (sum of Euclidean distances from cluster centers)	Cluster representatives: Cluster centroids are represented by a particle	–
[45]	No	Given	No	Numerical	No		Minimize within cluster variation (sum of squared Euclidean distances from cluster centers)	Cluster representatives: Cluster centroids are represented by a particle	Clustering production orders of a company
[46]	No	Given	No	Numerical and text	No		Minimize within cluster distance	Cluster representatives: Position and velocity particles denote cluster weight sets. Third particle represents cluster centroids	Text clustering
[52]	Yes	Given	Yes	Numerical	No		– Minimize maximum average within cluster distance – Maximize minimum inter-cluster distance	Cluster representatives: Cluster centroids are represented by a particle	MRI and satellite

[53]	Yes	Given	Yes	Numerical	No	– Minimize maximum average within cluster distance – Maximize minimum inter-cluster distance – Minimize quantization error	Cluster representatives: Cluster centroids are represented by a particle	MRI and satellite
[54]	No	Not given	No	Numerical	No	Improve validity index	Cluster representatives: Binary particle denotes whether a cluster centroid is used or not	MRI and satellite
[55]	No	Given	No	Numerical	No	– Minimize trace within criterion – Minimize variance ratio – Minimize Marriott's criterion	Cluster representatives: Cluster centroids are represented by a particle	–
[63]	No	Given	No	Document	No	Minimize average distance between the documents and their closest cluster centroids	Cluster representatives: Cluster centroids are represented by a particle	Document

(continued)

Table 6 (continued)

| Article | Problem characteristics | | | | | Clustering objective | Particle representation | Application area |
	Multi-objective	# of clusters	Arbitrary shapes	Data type	Constraints			
[65]	No	Given	No	Numerical	No	Minimize quantization error	Cluster representatives: Cluster centroids are represented by a particle	–
[73]	No	Given	No	Numerical	No	Minimize harmonic mean of distances from data points to all centers	Cluster representatives: Cluster centroids are represented by a particle	–
[56]	No	Not given	Yes	Numerical	No	Maximize within cluster similarity	Direct point-agent matching: Each point is represented by a particle. Position of particle in two-dimensional space denotes position of point	Facial skin database for women
[66]	No	Not given	Yes	Numerical	No	Maximize within cluster similarity	Direct point-agent matching: Each point is represented by a particle (datoid). Position of particle in two-dimensional space denotes position of datoid	–

Table 7 Characteristics of the ACO algorithms for clustering

Article	Problem characteristics					Clustering objective	Ant representation	Application area
	Multi-objective	# of clusters	Arbitrary shapes	Data type	Constraints			
[11]	No	Given	No	Numerical	No	Minimize within cluster variation (sum of squared Euclidean distances from cluster centers)	Data point-to-cluster assignment: An ant shows cluster assignments of all points	Electronic library
[38]	No	Given	No	Numerical	No	Minimize average quantization error	Data point-to-cluster assignment: An ant shows cluster assignments and selected attributes	–
[58]	No	Given	No	Numerical	No	Minimize within cluster variation (sum of squared Euclidean distances from cluster centers)	Data point-to-cluster assignment: An ant represents cluster assignment of a point	–
[61]	No	Given	No	Numerical	No	Minimize the sum of squared Euclidean distances between each object and the center	Data point-to-cluster assignment: An ant represents cluster assignments of points	–
[33]	Yes	Not given	Yes	Numerical	No	– Adjusted compactness – Relative separation	Search agent: An ant connects pairs of points as it moves	–

Table 8 Characteristics of the ABS algorithms for clustering

Article	Problem characteristics					Clustering objective	Particle representation	Application area
	Multi-objective	# of clusters	Arbitrary shapes	Data type	Constraints			
[22]	No	Given	Yes	Numerical	No	Maximize within cluster similarity and average linkage (implicit)	Direct point-agent matching: Each point is represented by an ant	–
[71]	No	Given	Yes	Numerical	ML and CL	Maximize within cluster similarity (implicit)	Direct point-agent matching: Each point is represented by an ant	–
[76]	No	Not given	Yes	Numerical	No	Minimize within cluster distances (implicit)	Direct point-agent matching: Each point is represented by an ant	–
[77]	No	Not given	Yes	Numerical	No	Maximize within cluster similarity	Direct point-agent matching: Each point is represented by an ant	–
[9]	No	Not given	Yes	Numerical	No	Maximize within cluster similarity (implicit)	Search agent: An ant carries data points	–
[18]	No	Not given	Yes	Numerical	No	Maximize within cluster similarity (implicit)	Search agent: An ant carries data points	Image indexing
[26]	No	Not given	Yes	Numerical	No	Maximize within cluster similarity (Algorithm is adopted from Handl et al. 2006)	Search agent: An ant searches through data points	–

[29]	No	Not given	Yes	Mixed	No	Maximize within cluster similarity (implicit)	Search agent: An ant searches through solution components (points)	–
[47]	No	Not given	Yes	Numerical	No	Minimize within cluster distances (implicit)	Search agent: An ant carries data points	–
[49]	No	Not given	Yes	Numerical	No	Cemetery formation (corpse clustering)	Search agent: An ant carries corpses (points)	–
[67]	No	Not given	Yes	Numerical	No	Maximize within cluster similarity (implicit)	Search agent: An ant carries data points	–
[74]	No	Given	Yes	Numerical	No	Maximize within cluster similarity	Search agent: An ant searches through solution components (points)	–

Table 9 Characteristics of the hybrid algorithms for clustering

Article	Problem characteristics					Clustering objective	Particle representation	Application area
	Multi-objective	# of clusters	Arbitrary shapes	Data type	Constraints			
[21]	No	Given	No	Numerical	No	Minimize within cluster variation (sum of squared Euclidean distances from cluster centers)	Cluster representatives: Each agent (ant/particle) represents the cluster centers	Market segmentation
[31]	No	Given	No	Numerical	No	Minimize total intra-cluster distance to total inter-cluster distance ratio	Cluster representatives: Each agent (ant/particle) represents the cluster centers	–
[37]	No	Given	No	Numerical	No	Minimize within cluster distance (sum of Euclidean distances from cluster centroids)	Cluster representatives: Cluster centroids are represented by a particle	–
[51]	No	Given	No	Numerical	No	Minimize total intra-cluster variance (total mean squared quantization error)	Cluster representatives: Cluster centers are represented by a particle	Market segmentation
[69]	No	Not given	Yes	Mixed	No	Maximize within cluster similarity	Cluster representatives: A particle denotes the complete weight set of SOM	–

Ref								
[70]	No	Not given	Yes	Numerical	No	Improve cluster validity indices including CH, DB, Dunn and its variants, CS, I, and Silhouette	Cluster representatives: A particle represents the activation function values and cluster centers	–
[72]	No	Given	No	Numerical	No	Minimize total intra-cluster variance (total mean squared quantization error)	Cluster representatives: Cluster centers are represented by a particle	–
[62]	No	Not given	Yes	Numerical	No	Minimize within cluster distances	Search agent: An ant searches through solution components (points)	–
[64]	No	Not given	Yes	Numerical	No	Minimize within cluster distances	Search agent: An ant searches through solution components (points)	Credit card data

Table 10 Characteristics of the other SI-based algorithms for clustering

Article	Problem characteristics					Clustering objective	Particle representation	Application area
	Multi-objective	# of clusters	Arbitrary shapes	Data type	Constraints			
[59]	No	Given	No	Numerical	No	Minimize within cluster variation (sum of squared Euclidean distances from cluster centers)	Data point-to-cluster assignment: A particle denotes cluster assignment of a point	–
[20]	No	Given	No	Numerical	No	Minimize within cluster variation (sum of squared Euclidean distances from cluster centers)	Cluster representatives: Cluster centroids are represented by queen and drones	–
[41]	No	Given	No	Numerical	No	Minimize within cluster distance	Cluster representatives: A bee (employed and onlooker) represents cluster centers (food source)	–
[60]	No	Given	No	Numerical	No	Minimize within cluster variance (SSE)	Cluster representatives: Particles represent cluster centers	–
[68]	No	Given	No	Numerical	No	Minimize within cluster variance (SSE)	Cluster representatives: Particles represent cluster centers	–
[75]	No	Given	No	Numerical	No	Minimize sum of intra-cluster distances	Cluster representatives: A bee (employed and onlooker) represents cluster centers (food source)	–

[78]	No	Given	No	Numerical	No	Minimize sum of intra-cluster variance	Cluster representatives: A bee (employed and onlooker) represents cluster centers (food source)	–
[4]	No	Not given	Yes	Mixed	No	Build a tree with maximum similar roots and sufficiently dissimilar branches	Direct point-agent matching: An ant represents a point	Human skin analysis, web usage mining, portal site
[43]	No	Given	Yes	Numerical	No	Maximize within cluster similarity	Direct point-agent matching: Each point is represented by a particle. Position of particle in two-dimensional space denotes position of point	–
[44]	No	Given	Yes	Numerical	No	Maximize within cluster similarity	Search agent: An ant searches through data points	–

References

1. Abraham, A., Das, S., Roy, S.: Swarm Intelligence Algorithms for Data Clustering. In: Maimon, O., Rokach, L. (eds.) Soft Computing for Knowledge Discovery and Data Mining, pp. 279–313. Springer, New York (2008)
2. Alam, S., Dobbie, G., Koh, Y.S., Riddle, P., Rehman, S.U.: Research on particle swarm optimization based clustering: a systematic review of literature and techniques. Swarm Evol. Comput. **17**, 1–13 (2014)
3. Alani, H., Jones, C.B., Tudhope, D.: Voronoi-based region approximation for geographical information retrieval with gazetteers. Int. J. Geogr. Inf. Sci. **15**(4), 287–306 (2001)
4. Azzag, H., Venturini, G., Oliver, A., Guinot, C.: A hierarchical ant-based clustering algorithm and its use in three real-world applications. Eur. J. Oper. Res. **179**(3), 906–922 (2007)
5. Bandyopadhyay, S., Saha, S.: GAPS: A clustering method using a new point symmetry-based distance measure. Pattern Recogn. **40**, 3430–3451 (2007)
6. Bandyopadhyay, S., Saha, S.: A point symmetry-based clustering technique for automatic evolution of clusters. IEEE Trans. Knowl. Data Eng. **20**(11), 1441–1457 (2008)
7. Basu, S., Davidson, I.: KDD 2006 Tutorial Clustering with Constraints: Theory and Practice (2006). Available via http://www.ai.sri.com/~basu/kdd-tutorial-2006.Cited2March2015
8. Bong, C.-W., Rajeswari, M.: Multi-objective nature-inspired clustering and classification techniques for image segmentation. Appl. Soft Comput. **11**(4), 3271–3282 (2011)
9. Boryczka, U.: Finding groups in data: Cluster analysis with ants. Appl. Soft Comput. **9**(1), 61–70 (2009)
10. Celebi, M.E.: Partitional Clustering Algorithms. Springer, Switzerland (2015)
11. Chen, A., Chen, C.: A new efficient approach for data clustering in electronic library using ant colony clustering algorithm. Electron. Libr. **24**(4), 548–559 (2006)
12. Cura, T.: A particle swarm optimization approach to clustering. Expert Syst. Appl. **39**(1), 1582–1588 (2012)
13. D'Urso, P., De Giovanni, L., Disegna, M., Massari, R.: Bagged clustering and its application to tourism market segmentation. Expert Syst. Appl. **40**(12), 4944–4956 (2013)
14. Das, S., Abraham, A., Konar, A.: Automatic kernel clustering with a multi-elitist particle swarm optimization algorithm. Pattern Recogn. Lett. **29**(5), 688–699 (2008)
15. Deneubourg, J.L., Goss, S., Franks, N., Sendova-Franks, A., Detrain, C., Chrżtien, L.: The dynamics of collective sorting: robot-like ants and ant-like robots. In: Meyer, J.A., Wilson, S. (eds.) From Animals to Animats: Proceedings of the 1st International Conference on Simulation of Adaptive Behavior, pp. 356–365. Cambridge, MIT Press (2008)
16. Dorigo, M., Maniezzo, V., Colorni, A.: Positive feedback as a search strategy. Technical Report 91016 Dipartimento di Elettronica e Informatica, Politecnico di Milano, Italy (1991)
17. Dorigo, M., Maniezzo, V., Colorni, A.: Ant System: optimization by a colony of cooperating agents. IEEE Trans. Syst. Man Cybern. B Cybern. **26**(1), 29–41 (1996)
18. Elkamel, A., Gzara, M., Ben-Abdallah, H.: A bio-inspired hierarchical clustering algorithm with backtracking strategy. Appl. Intell. **42**, 174–194 (2015)
19. Ester, M., Kriegel, K.P., Sander J., Xu, X.: A density-based algorithm for discovering clusters in large spatial databases with noise. In: Proceedings of the 2nd International Conference on Knowledge Discovery and Data Mining, pp. 226–231 (1996)
20. Fathian, M., Amiri, B., Maroosi, A.: Application of honey-bee mating optimization algorithm on clustering. Appl. Math. Comput. **190**(2), 1502–1513 (2007)
21. Firouzi, B.B., Sadeghi, M.S., Niknam, T.: A new hybrid algorithm based on PSO, SA and k-means for cluster analysis. Int. J. Innovative Comput. Inf. Control **6**(7), 3177–3192 (2010)
22. Ghosh, A., Halder, A., Kothari, M., Ghosh, S.: Aggregation pheromone density based data clustering. Inf. Sci. **178**(13), 2816–2831 (2008)
23. Gower, J.: Coefficients of association and similarity, based on binary (presence-absence) data: an evaluation. Biometrics **27**, 857–871 (1971)

24. Grosan, C., Abraham, A., Chis, M.: Swarm Intelligence in Data Mining. In: Abraham, A., Grosan, C., Ramos, V. (eds.) Swarm Intelligence in Data Mining, pp. 1–20. Springer, Berlin (2006)
25. Guha, S., Rastogi, R., Shim, K.: CURE: An efficient clustering algorithm for large databases. In: Proceedings of the ACM-SIGMOD International Conference on Management of Data, pp. 73–84 (1998)
26. Gunes, O.G., Uyar, A.S.: Parallelization of an ant-based clustering approach. Kybernetes 39(4), 656–677 (2010)
27. Han, J., Kamber, M., Pei, J.: Data Mining: Concepts and Techniques, 3rd edn. Morgan Kaufman, Massachusetts (2011)
28. Handl, J., Meyer, B.: Ant-based and swarm-based clustering. Swarm Intell. 1, 95–113 (2007)
29. Handl, J., Knowles, J., Dorigo, M.: Ant-based clustering and topographic mapping. Artif. Life 12, 35–61 (2006)
30. Hruschka, E.R., Campello, R.J.G.B., de Castro, L.N.: Evolving clusters in gene-expression data. Inf. Sci. 176(13), 1898–1927 (2006)
31. Huang, C.L., Huang, W.C., Chang, H.Y., Yeh, Y.C., Tsai, C.Y.: Hybridization strategies for continuous ant colony optimization and particle swarm optimization applied to data clustering. Appl. Soft Comput. 13(9), 3864–3872 (2013)
32. Ichino, M., Yaguchi, H.: Generalized Minkowski metrics for mixed feature-type data analysis. IEEE Trans. Syst. Man Cybern. 24(4), 698–708 (1994)
33. İnkaya, T., Kayalıgil, S., Özdemirel, N.E.: Ant colony based clustering methodology. Appl. Soft Comput. 28, 301–311 (2015)
34. Jain, A.K.: Data clustering: 50 years beyond K-means. Pattern Recogn. Lett. 31, 651–666 (2010)
35. Jain, A.K., Murty, M.N., Flynn, P.J.: Data clustering: a review. ACM Comput. Surv. 31(3), 264–323 (1999)
36. Jarboui, B., Cheikh, M., Siarry, P., Rebai, A.: Combinatorial particle swarm optimization for partitional clustering problem. Appl. Math. Comput. 192, 337–345 (2007)
37. Jiang, B., Wang, N.: Cooperative bare-bone particle swarm optimization for data clustering. Soft Comput. 18, 1079–1091 (2014)
38. Jiang, L., Ding, L., Peng, Y.: An efficient clustering approach using ant colony algorithm in multidimensional search space. In: Proceedings of the 8th International Conference on Fuzzy Systems and Knowledge Discovery, pp. 1085–1089 (2011)
39. Kao, Y.T., Zahara, E., Kao, I.W.: A hybridized approach to data clustering. Expert Syst. Appl. 34(3), 1754–1762 (2008)
40. Karaboga, D.: An idea based on honey bee swarm for numerical optimization, Technical Report-TR06, Erciyes University, Engineering Faculty, Computer Engineering Department (2005)
41. Karaboga, D., Ozturk, C.: A novel clustering approach: Artificial Bee Colony (ABC) algorithm. Appl. Soft Comput. 11(1), 652–657 (2011)
42. Kennedy, J., Eberhart, R.: Particle swarm optimization. In: Proceedings of IEEE International Conference on Neural Network, vol. 4, pp. 1942–1948 (1995)
43. Khereddine, B., Gzara, M.: FDClust: a new bio-inspired divisive clustering algorithm. Advances in Swarm Intelligence. Lecture Notes in Computer Science, vol. 6729, pp. 136–145. Springer, Berlin (2011)
44. Kountche, D.A., Monmarche, N., Slimane, M.: The Pachycondyla Apicalis ants search strategy for data clustering problems. Swarm and Evolutionary Computation. Lecture Notes in Computer Science, vol. 7269, pp. 3–11. Springer, Berlin (2012)
45. Kuo, R.J., Wang, M.J., Huang, T.W.: An application of particle swarm optimization algorithm to clustering analysis. Soft Comput. 15, 533–542 (2011)
46. Lu, Y., Wang, S., Li, S., Zhou, C.: Particle swarm optimizer for variable weighting in clustering high-dimensional data. Mach. Learn. 82, 43–70 (2011)
47. Lumer, E.D., Faieta, B.: Diversity and adaptation in populations of clustering ants. In: Cliff, D., Husbands, P., Meyer, J.A., Wilson, S.W. (eds.) Proceedings of the 3rd International Conference on Simulation of Adaptive Behavior: From Animals to Animats, pp. 501–508. MIT Press/Bradford Books, Cambridge (1994)

48. Martens, D., Baesens, B., Fawcett, T.: Editorial survey: swarm intelligence for data mining. Mach. Learn. **82**, 1–42 (2011)
49. Martin, M., Chopard, B., Albquerque, P.: Formation of an ant cemetry: swarm intelligence or statistical accident? Futur. Gener. Comput. Syst. **18**(7), 951–959 (2002)
50. Nanda, S. J., Panda, G.: A survey on nature inspired metaheuristic algorithms for partitional clustering. Swarm Evol. Comput. **16**, 1–18 (2014)
51. Niknam, T., Amiri, B.: An efficient hybrid approach based on PSO, ACO and k-means for cluster analysis. Appl. Soft Comput. **10**(1), 183–197 (2010)
52. Omran, M., Salman, A., Engelbrecht, A.P.: Image classification using particle swarm optimization. In: Proceedings of the 4th Asia-Pacific Conference on Simulated Evolution and Learning, pp. 18–22 (2002)
53. Omran, M., Engelbrecht, A.P., Salman, A.: Particle swarm optimization method for image clustering. Int. J. Pattern Recognit. Artif. Intell. **19**(3), 297–321 (2005)
54. Omran, M.G.H., Salman, A., Engelbrecht, A.P.: Dynamic clustering using particle swarm optimization with application in image segmentation. Pattern. Anal. Appl. **8**, 332–344 (2006)
55. Paterlini, S., Krink, T.: Differential evolution and particle swarm optimisation in partitional clustering. Comput. Stat. Data Anal. **50**, 1220–1247 (2006)
56. Picarougne, F., Azzag, H., Venturini, G., Guinot, C.: A new approach of data clustering using a flock of agents. Evol. Comput. **15**(3), 345–367 (2007)
57. Rana, S., Jasola, S., Kumar, R.: A review on particle swarm optimization algorithms and their applications to data clustering. Artif. Intell. Rev. **35**, 211–222 (2011)
58. Runkler, T.A.: Ant colony optimization in clustering models. Int. J. Intell. Syst. **20**, 1233–1251 (2005)
59. Runkler, T.A.: Wasp swarm optimization of the c-means clustering model. Int. J. Intell. Syst. **23**, 269–285 (2008)
60. Senthilnath, J., Omkar, S.N., Mani, V.: Clustering using firefly algorithm: performance study. Swarm Evol. Comput. **1**, 164–171 (2011)
61. Shelokar, P.S., Jayaraman, V.K., Kulkarni, B.D.: An ant colony approach for clustering. Anal. Chim. Acta **509**, 187–195 (2004)
62. Sinha, A.N., Das, N., Sahoo, G.: Ant colony based hybrid optimization for data clustering. Kybernetes **36**(2), 175–191 (2007)
63. Song, W., Ma, W., Qiao, Y.: Particle swarm optimization algorithm with environmental factors for clustering analysis. Soft Comput. (2014). doi: 10.1007/s00500-014-1458-7
64. Tsai, C., Tsai, C., Wu, H., Yang, T.: ACODF: a novel data clustering approach for data mining in large databases. J. Syst. Softw. **73**, 133–145 (2004)
65. Van der Merwe, D.W., Engelbrecht, A.P.: Data clustering using particle swarm optimization. In: Proceedings of the 2003 Congress on Evolutionary Computation, pp. 215–220 (2003)
66. Veenhuis, C., Köppen, M.: Data swarm clustering. In: Abraham, A., Grosan, C., Ramos, V. (eds.) Swarm Intelligence in Data Mining, pp. 221–241. Springer, Berlin (2006)
67. Vizine, A.L., De Castro, L.N., Hruschka, E.R., Gudwin, R.R.: Towards improving clustering ants: an adaptive ant clustering algorithm. Informatica **29**, 143–154 (2005)
68. Wan, M., Wang, C., Li, L., Yang, Y.: Chaotic ant swarm approach for data clustering. Appl. Soft Comput. **12**(8), 2387–2393 (2012)
69. Xiao, X., Dow, E.R., Eberhart, R., Miled, Z.B., Oppelt, R.J.: A hybrid self-organizing maps and particle swarm optimization approach. Concurrency Comput. Pract. Exp. **16**, 895–915 (2004)
70. Xu, R., Xu, J., Wunsch, D.C.: A comparison study of validity indices on swarm-intelligence-based clustering. IEEE Trans. Syst. Man Cybern. B Cybern. **12**(4), 1243–1256 (2012)
71. Xu, X., Lu, L., He, P., Pan, Z., Chen, L.: Improving constrained clustering via swarm intelligence. Neurocomputing **116**, 317–325 (2013)
72. Yan, X., Zhu, Y., Zou, W., Wang, L.: A new approach for data clustering using hybrid artificial bee colony. Neurocomputing **97**, 241–25 (2012)
73. Yang, F., Sun, T., Zhang, C.: An efficient hybrid data clustering method based on k-harmonic means and particle swarm optimization. Expert Syst. Appl. **36**(6), 9847–9852 (2009)

74. Yang, Y., Kamel, M.S.: An aggregated clustering approach using multi-ant colonies algorithms. Pattern Recogn. **39**(7), 1278–1289 (2006)

75. Zhang, C., Ouyang, D., Ning, J.: An artificial bee colony approach for clustering. Expert Syst. Appl. **37**(7), 4761–4767 (2010)

76. Zhang, L., Cao, Q.: A novel ant-based clustering algorithm using the kernel method. Inf. Sci. **181**(20), 4658–4672 (2011)

77. Zhang, L., Cao, Q., Lee, J.: A novel ant-based clustering algorithm using Renyi entropy. Appl. Soft Comput. **13**(5), 2643–2657 (2013)

78. Zou, W., Zhu, Y., Chen, H., Sui, X.: A clustering approach using cooperative artificial bee colony algorithm. Discret. Dyn. Nat. Soc. **459796**, 1–16 (2010)

Extending Kmeans-Type Algorithms by Integrating Intra-cluster Compactness and Inter-cluster Separation

Xiaohui Huang, Yunming Ye, and Haijun Zhang

Abstract Kmeans-type clustering aims at partitioning a data set into clusters such that the objects in a cluster are compact and the objects in different clusters are well-separated. However, most of kmeans-type clustering algorithms rely on only intra-cluster compactness while overlooking inter-cluster separation. In this chapter, a series of new clustering algorithms by extending the existing kmeans-type algorithms is proposed by integrating both intra-cluster compactness and inter-cluster separation. First, a set of new objective functions for clustering is developed. Based on these objective functions, the corresponding updating rules for the algorithms are then derived analytically. The properties and performances of these algorithms are investigated on several synthetic and real-life data sets. Experimental studies demonstrate that our proposed algorithms outperform the state-of-the-art kmeans-type clustering algorithms with respects to four metrics: Accuracy, Rand Index, Fscore, and normal mutual information (NMI).

1 Introduction

Clustering is a basic operation in many applications in nature [4], such as gene analysis [38], image processing [19], text organization [33] and community detection [42], to name just a few. It is a method of partitioning a data set into clusters such that the objects in the same cluster are similar and the objects in different clusters are dissimilar according to certain pre-defined criteria [30].

There are many types of approaches [29] to solve a clustering problem: partitioning methods, hierarchical methods, density-based methods, grid-based methods and model-based methods, etc. The kmeans-type clustering algorithms are a kind

X. Huang
School of Information Engineering Departments, East China Jiaotong University, Nanchang, Jiangxi, China
e-mail: hxh016@hotmail.com

Y. Ye (✉) • H. Zhang
Shenzhen Key Laboratory of Internet Information Collaboration, Shenzhen Graduate School, Harbin Institute of Technology, Shenzhen, Guangdong, China
e-mail: yeyunming@hit.edu.cn; arrhzhang@gmail.com

© Springer International Publishing Switzerland 2016
M.E. Celebi, K. Aydin (eds.), *Unsupervised Learning Algorithms*,
DOI 10.1007/978-3-319-24211-8_13

of partitioning method [9], which has been widely used in many real-world applications. Most of the existing kmeans-type clustering algorithms consider only the similarities among the objects in a cluster by minimizing the dispersions of the cluster. The representative one of these algorithms is basic kmeans [6, 39] which addresses all the selected features equally during the process of minimizing the dispersions.

However, different features have different discriminative capabilities in real applications. For example, in the sentence "London is the first city to have hosted the modern Games of three Olympiads," the keywords "London, Olympiads" have more discriminative information than the words "city, modern" in sport news. Therefore, some researchers extended the basic kmeans algorithm by using various types of weighting ways. A weighting vector is used to weight the features in some kmeans-type algorithms, which have been reported in [14, 16–18, 30, 35]. For every feature, these algorithms compute a weighting value representing its discriminative capability in the entire data set. However, a feature may have different discriminative capabilities in different clusters in real applications. Hence, some researchers proposed matrix weighting kmeans-type algorithms in which a weighting vector is used to represent the discriminative capabilities of the features in every cluster, such as projected clustering [1], Entropy Weighting kmeans (EWkmean) [33], locally adaptive clustering [3, 22], Attributes-Weighting clustering Algorithms (AWA) [12], and feature group weighting kmeans [13], etc. Most of these methods have the same characteristic: the features must be evaluated with the large weights if the dispersions of the features in a data set are small. In essence, the discriminative capability of a feature not only relates to the dispersion, but also associates with the distances between the centroids, i.e. inter-cluster separation. As a matter of fact, inter-cluster separation plays an important role in supervised learning methods (e.g. Linear Discriminative Analysis) [8, 26]. Most of the existing kmeans-type algorithms, however, overlook the inter-cluster separation.

In this chapter, we investigate the potential framework of the kmeans-type algorithms by integrating both intra-cluster compactness and inter-cluster separation. We propose three algorithms: E-kmeans, E-Wkmeans and E-AWA, which extend basic kmeans, Wkmeans [30] and AWA [12], respectively. Moreover, the convergence theorems of our proposed algorithms are given. Extensive experiments on both synthetic and real-life data sets corroborate the effectiveness of our proposed methods. The desirable features of our algorithms can be concluded as follows:

- The generality of the extending algorithms encompasses a unified framework that considers both the intra-cluster compactness and the inter-cluster separation. Concretely, we develop a new framework for kmeans-type algorithms to include the impacts of the intra-cluster compactness and the inter-cluster separation in the clustering process.
- The proposed framework is robust because it does not introduce new parameters to balance the intra-cluster compactness and the inter-cluster separation. The proposed algorithms are also more robust for the input parameter which is used

to tune the weights of the features in comparison to the original kmeans-type algorithms.

- The extending algorithms are able to produce better clustering results in comparison to the state-of-the-art algorithms in most of cases since they can utilize more information than traditional kmeans-type algorithms such that our approaches have capability to deliver discriminative powers of different features.

The main contributions of this chapter are twofold:

- We propose a new framework of kmeans-type algorithm by combining the dispersions of the clusters which reflect the compactness of the intra-cluster and the distances between the centroids of the clusters indicating the separation between clusters.
- We give the complete proof of convergence of the extending algorithms based on the new framework.

The remaining sections of this chapter are organized as follows: a brief overview of related works on various kmeans-type algorithms is presented in Sect. 2. Section 3 introduces the extensions of the kmeans-type algorithms. Experiments on both synthetic and real data sets are presented in Sect. 4. We discuss the features of our algorithms in Sect. 5 and conclude the chapter in Sect. 6.

2   Related Work

In this section, we give a brief survey of kmeans-type clustering from three aspects: No weighting kmeans-type algorithms, Vector weighting kmeans-type algorithms and Matrix weighting kmeans-type algorithms. For detailed surveys of kmeans family algorithms, readers may refer to [31, 44].

2.1   No Weighting Kmeans-Type Algorithm

Kmeans-type clustering algorithms aim at finding a partition such that the sum of the squared distances between the empirical means of the clusters and the objects in the clusters is minimized.

2.1.1   No Weighting Kmeans-Type Algorithm Without Inter-cluster Separation

Let $X = \{X_1, X_2, \ldots, X_n\}$ be a set of n objects. Object $X_i = \{x_{i1}, x_{i2}, \ldots, x_{im}\}$ is characterized by a set of m features (dimensions). The membership matrix U is a $n \times k$ binary matrix, where $u_{ip} = 1$ indicates that object i is allocated to cluster p, otherwise, it is not allocated to cluster p. $Z = \{Z_1, Z_2, \ldots, Z_k\}$ is a set of k vectors representing the centroids of k clusters. The basic kmeans relies on minimizing an objective function [30]

$$P(U, Z) = \sum_{p=1}^{k} \sum_{i=1}^{n} \sum_{j=1}^{m} u_{ip} (x_{ij} - z_{pj})^2,$$ (1)

subject to

$$u_{ip} \in \{0, 1\}.$$ (2)

The optimization problem shown in Eq. (1) can be solved by the following two minimization steps iteratively:

1. Fix $Z = \hat{Z}$ and minimize the reduced $P(U, \hat{Z})$ to solve U with

$$u_{i,p} = \begin{cases} 1, & 1 \leq p, p' \leq K, \\ & \text{if } \sum_{j=1}^{m} (z_{pj} - x_{ij})^2 \leq \sum_{j=1}^{m} (z_{p'j} - x_{ij})^2, \\ 0, & \text{otherwise.} \end{cases}$$ (3)

2. Fix $U = \hat{U}$ and minimize the reduced $P(\hat{U}, Z)$ to solve Z with

$$z_{pj} = \frac{\sum_{i=1}^{n} u_{i,p} x_{i,j}}{\sum_{i=1}^{n} u_{i,p}}.$$ (4)

The detailed process of optimization can be found in [6, 39]. Basic kmeans algorithm has been extended in many ways. Steinbach et al. proposed a hierarchical divisive version of kmeans, called bisecting kmeans (BSkmeans) [41], which recursively partitions objects into two clusters at each step until the number of clusters is k. Bradley et al. [7] presented a fast scalable and singlepass version of kmeans that does not require all the data to be feed in the memory at the same time. Shamir and Tishby [40] studied the behavior of clustering stability using kmeans clustering framework based on an explicit characterization of its asymptotic behavior. This paper concluded that kmeans-type algorithms does not "break down" in the large sample, in the sense that even when the sample size goes to infinity and the kmeans-type algorithms becomes stable for any choice of K. Since the kmeans-type algorithms are sensitive to the choice of initial centroids and usually get stuck at local optima, many methods [5, 10, 11] are proposed to overcome this problem. For example, Arthur and Vassilvitskii proposed kmeans++ [5] which chooses the centroids according to the distances of existing centroids. Another problem of kmeans-type algorithms is to require tuning of parameter K. X-means

[37] automatically finds K by optimizing a criterion such as Akaike information criterion (AIC) or Bayesian information criterion (BIC).

2.1.2 No Weighting Kmeans-Type Algorithm with Inter-cluster Separation

To obtain the best k (the number of clusters), some validity indexes [15] which integrate both intra-cluster compactness and inter-cluster separation are used in the clustering process. Yang et al. [43, 45] proposed a fuzzy compactness and separation (FCS) algorithms which calculates the distances between the centroids of the cluster and the global centroids as the inter-cluster separation. The objective function of FCS can be formulated as

$$J_{\text{FCS}} = \sum_{p=1}^{k} \sum_{i=1}^{n} \sum_{j=1}^{m} u_{ip}^{\alpha}(x_{ij} - z_{pj})^2 - \eta_p \sum_{p=1}^{k} \sum_{i=1}^{n} \sum_{j=1}^{m} u_{ip}^{\alpha}(z_{ij} - z_{oj})^2, \tag{5}$$

subject to

$$u_{ip} \in [0, 1], \ 0 \le \eta_p \le 1, \ \alpha \ne 1, \tag{6}$$

where η_p is used to balance the importance between intra-compactness and inter-separation, α is the fuzzy index and z_{oj} is the global centroid on the jth dimension. By minimizing J_{FCS}, the following update equations can be achieved:

$$u_{ip} = \frac{\sum_{j=1}^{m} \left((x_{ij} - z_{pj})^2 - \eta_p(z_{pj} - z_{oj})^2 \right)^{\frac{-1}{\alpha-1}}}{\sum_{q=1}^{k} \sum_{j=1}^{m} \left((x_{ij} - z_{qj})^2 - \eta_q(z_{qj} - z_{oj})^2 \right)^{\frac{-1}{\alpha-1}}}, \tag{7}$$

and

$$z_{pj} = \frac{\sum_{i=1}^{n} u_{ip}^{\alpha} x_{ij} - \eta_p \sum_{i=1}^{n} u_{ip}^{\alpha} x_{oj}}{\sum_{i=1}^{n} u_{ip}^{\alpha} - \eta_p \sum_{i=1}^{n} u_{ip}^{\alpha}}. \tag{8}$$

The process of optimization is given in [43]. The promising results are achieved since FCS is more robust to noises and outliers than traditional fuzzy kmeans clustering.

2.2 Vector Weighting Kmeans-Type Algorithm

A major problem of No weighting kmeans-type algorithms lies in treating all features equally in the clustering process. In practice, an interesting clustering structure usually occurs in a subspace defined by a subset of all the features. Therefore, many studies attempt to weight features with various methods [16–18, 30, 34].

2.2.1 Vector Weighting Kmeans-Type Algorithm Without Inter-cluster Separation

De Sarbo et al. firstly introduced a feature selection method, SYNCLUS [16], which partitions features into several groups and uses weights for feature groups in the clustering process. The algorithm requires a large amount of computation [30]. It may not be applicable for large data sets. Automated variable weighting kmeans (Wkmeans) [30] is a typical vector weighting clustering algorithm, which can be formulated as

$$P(U, W, Z) = \sum_{p=1}^{k} \sum_{i=1}^{n} u_{ip} \sum_{j=1}^{m} w_j^{\beta} (x_{ij} - z_{pj})^2, \tag{9}$$

subject to

$$u_{ip} \in \{0, 1\}, \sum_{p=1}^{k} u_{ip} = 1, \sum_{j=1}^{m} w_j = 1, 0 \le w_j \le 1, \tag{10}$$

where W is a weighting vector for the features. The optimization problem shown in Eq. (9) can be solved by the following three minimization steps iteratively:

1. Fix $Z = \hat{Z}$ and $W = \hat{W}$, solve the reduced problem $P(U, \hat{W}, \hat{Z})$ with

$$u_{i,p} = \begin{cases} 1, \text{ for } 1 \le p, p' \le K, \\ \quad \text{if } \sum_{j=1}^{m} w_j^{\beta} (x_{i,j} - z_{p,j})^2 \le \sum_{j=1}^{m} w_j^{\beta} (x_{i,j} - z_{p',j})^2, \\ 0, \text{ otherwise;} \end{cases} \tag{11}$$

2. Fix $W = \hat{W}$ and $U = \hat{U}$, solve the reduced problem $P(\hat{U}, \hat{W}, Z)$ with Eq. (4);
3. Fix $U = \hat{U}$ and $Z = \hat{Z}$, solve the reduced problem $P(\hat{U}, W, \hat{Z})$ with

$$w_j = \begin{cases} 0, \text{ if } D_j = 0, \\ \dfrac{1}{\sum_{t=1}^{h} \left[\frac{D_j}{D_t}\right]^{1/(\beta-1)}}, \text{ if } D_j \ne 0, \end{cases} \tag{12}$$

where

$$D_j = \sum_{p=1}^{K} \sum_{i=1}^{n} u_{i,p}(x_{i,j} - z_{p,j})^2, \tag{13}$$

where h is the number of features when $D_j \neq 0$. The details of proof are given in [30]. Inspired by the idea of two-level weighting strategy [16], Chen et al. proposed a two-level variable weighting kmeans [14] based on Wkmeans [30].

2.2.2 Vector Weighting Kmeans-Type Algorithm with Inter-cluster Separation

De Soete [17, 18] proposed an approach to optimize feature weights for ultrametric and additive tree fitting. This approach calculates the distances between all pairs of objects and finds the optimal weight for each feature. However, this approach requires high computational cost [30] since the hierarchical clustering method used to solve the feature selection problem in this approach needs high computational cost. In order to decrease the computational cost, Makarenkov and Legendre [34] extended De Soete's approach to optimize feature weighting method for kmeans clustering. Usually, these algorithms are able to gain promising results to the data sets involving error-perturbed features or outliers.

2.3 Matrix Weighting Kmeans-Type Algorithm

Matrix weighting kmeans-type algorithms seek to group objects into clusters in different subsets of features for different clusters. It includes two tasks: identification of the subsets of features where clusters can be found and discovery of the clusters from different subsets of features.

2.3.1 Matrix Weighting Kmeans-Type Algorithm Without Inter-cluster Separation

Aggarwal et al. proposed the PROjected CLUStering (PROCLUS) [1] algorithm which is able to find a subset of features for each cluster. Using PROCLUS, a user, however, needs to specify the average number of cluster features. Different to PROCLUS, feature weighting has been studied extensively in recent years [12, 13, 19, 28, 33, 36]. Therein, AWA [12] is a typical matrix weighting clustering algorithm, which can be formulated as

$$P(U, W, Z) = \sum_{p=1}^{k} \sum_{i=1}^{n} u_{ip} \sum_{j=1}^{m} w_{pj}^{\beta}(x_{ij} - z_{pj})^2, \tag{14}$$

subject to

$$u_{ip} \in \{0, 1\}, \sum_{p=1}^{k} u_{ip} = 1, \sum_{j=1}^{m} w_{pj} = 1, 0 \le w_{pj} \le 1, \tag{15}$$

where W is a weighting matrix, each row in which denotes a weight vector of the features in a cluster. The optimization problem shown in Eq. (14) can be solved by the following three minimization steps iteratively:

1. Fix $Z = \hat{Z}$ and $W = \hat{W}$, solve the reduced problem $P(U, \hat{W}, \hat{Z})$ with

$$u_{i,p} = \begin{cases} 1, \text{ for } 1 \le p, p' \le K, \\ \quad \text{if } \sum_{j=1}^{m} w_{p,j}^{\beta}(x_{i,j} - z_{p,j}) \le \sum_{j=1}^{m} w_{p,j}^{\beta}(x_{i,j} - z_{p',j}), \\ 0, \text{ otherwise}; \end{cases} \tag{16}$$

2. Fix $U = \hat{U}$ and $W = \hat{W}$, solve the reduced problem $P(\hat{U}, \hat{W}, Z)$ with Eq. (4);
3. Fix $U = \hat{U}$ and $Z = \hat{Z}$, solve the reduced problem $P(\hat{U}, W, \hat{Z})$ with

$$w_{p,j} = \begin{cases} \frac{1}{m_j}, \text{ if } \sum_{i=1}^{n} u_{i,p}(x_{i,j} - z_{p,j})^2 = 0, \text{ and} \\ \quad m_j = \left| \left\{ t : \sum_{i=1}^{m} u_{i,p}(x_{i,t} - z_{p,t})^2 = 0 \right\} \right| \\ 0, \text{ if } \sum_{i=1}^{m} u_{i,p}(x_{i,j} - z_{p,j})^2 \neq 0, \text{ but} \\ \quad \sum_{i=1}^{m} u_{i,p}(x_{i,t} - z_{p,t})^2 = 0, \text{ for some } t, \\ \frac{1}{\sum_{t=1}^{m} \left[\frac{\sum_{i=1}^{n} u_{i,p}(x_{i,j} - z_{p,j})^2}{\sum_{i=1}^{n} u_{i,t}(x_{i,j} - z_{t,j})^2} \right]}, \text{ if } \sum_{i=1}^{n} u_{i,t}(x_{i,j} - z_{t,j})^2 \neq 0, \forall 1 \le t \le m. \end{cases} \tag{17}$$

The process of minimizing the objective function to solve U, Z and W can be found in [12]. Based on AWA [12], Jing et al. proposed an EWkmeans [33] which minimizes the intra-cluster compactness and maximizes the negative weight entropy to stimulate more features contributing to the identification of a cluster. In a later study, Chen et al. [13] proposed a two-level matrix weighting kmeans algorithm and Ahmad and Dey [2] developed a matrix kmeans-type clustering algorithm of mixed numerical and categorical data sets based on EWkmeans [33]. Domeniconi et al. [3, 22, 23] discovered clusters in subspaces spanned by different combinations of features via local weights of features. However, Jing et al. [33] pointed out that the objective functions in their methods are not differentiable while minimizing the objective functions.

2.3.2 Matrix Weighting Kmeans-Type Algorithm with Inter-cluster Separation

Friedman and Meulman [27] published the clustering objects on subsets of features algorithm for matrix weighting clustering which involves the calculation of the distances between all pairs of objects at each iterative step. This results in a high computational complexity $O(tn^2m)$ where n, m and t are the number of objects, features and iterations, respectively. Combining the FCS method [43] and EWkmeans [33], Deng et al. proposed an enhanced soft subspace clustering (ESSC) [19] algorithm that is able to use both intra-cluster compactness and inter-cluster separation. The ESSC can be formulated as the objective function:

$$
J_{\text{ESSC}}(U, W, Z) = \sum_{p=1}^{k} \sum_{i=1}^{n} u_{ip}^{\alpha} \sum_{j=1}^{m} w_{pj}(x_{ij} - z_{pj})^2 - \gamma \sum_{p=1}^{k} \sum_{j=1}^{m} w_{pj} \ln w_{pj}
$$
$$
- \eta \sum_{p=1}^{k} \left(\sum_{i=1}^{n} u_{ip}^{\alpha} \right) \sum_{j=1}^{m} w_{pj}(z_{pj} - z_{oj})^2, \tag{18}
$$

where parameters $\gamma (\gamma \geq 0)$ and $\eta (\eta \geq 0)$ are used to control the influences of entropy and the weighting inter-cluster separation, respectively. The objective function (18) can be minimized by iteratively solving the following three problems:

1. Fix $Z = \hat{Z}$ and $W = \hat{W}$, solve the reduced problem $P(U, \hat{W}, \hat{Z})$ with

$$
u_{ip} = \frac{d_{ip}^{-1/(\alpha-1)}}{\sum_{q=1}^{k} d_{iq}^{-1/(\alpha-1)}}, \quad p = 1, \ldots, k, \ i = 1, \ldots, n, \tag{19}
$$

   where

$$
d_{ip} = \sum_{j=1}^{m} w_{pj}(x_{ij} - z_{pj})^2 - \eta \sum_{j=1}^{m} w_{pj}(z_{pj} - z_{oj})^2. \tag{20}
$$

2. Fix $U = \hat{U}$ and $W = \hat{W}$, solve the reduced problem $P(\hat{U}, \hat{W}, Z)$ with

$$
z_{pj} = \frac{\sum_{i=1}^{n} u_{ip}^{\alpha}(x_{ij} - \eta z_{oj})}{\sum_{i=1}^{n} u_{ip}^{\alpha}(1 - \eta)}. \tag{21}
$$

3. Fix $U = \hat{U}$ and $Z = \hat{Z}$, solve the reduced problem $P(\hat{U}, W, \hat{Z})$ with

$$w_{pj} = \exp\left(-\frac{\sigma_{pj}}{\gamma}\right) \Big/ \sum_{j'=1}^{m} \exp\left(-\frac{\sigma_{pj'}}{\gamma}\right), \tag{22}$$

where

$$\sigma_{pj} = \sum_{i=1}^{n} u_{ip}^{\alpha}(x_{ij} - z_{pj})^2 - \eta \sum_{i=1}^{n} u_{ip}^{\alpha}(z_{pj} - z_{oj})^2. \tag{23}$$

The details of convergence proof are given in [19]. ESSC is able to effectively reduce the effect of the features on which the centroids of the clusters are close to the global centroid. However, negative values may be produced in the membership matrix if the balancing parameter is large. Moreover, ESSC has three manual input parameters. In practice, it is difficult to find a group of appropriate values for the parameters.

2.4 Summary of the Existing Kmeans-Type Algorithms

From the analysis of Sects. 2.1–2.3, the existing kmeans-type algorithms are classified to three categories: no weighting kmeans-type algorithms, vector weighting kmeans-type algorithms and matrix weighting kmeans-type algorithms. From another perspective, we can partition the kmeans-type algorithms into: the algorithms with inter-cluster separation and the algorithms without inter-cluster separation. The entire kmeans-type algorithms are summarized in Table 1.

Table 1 The summary of kmeans-type algorithms

Algorithm type	Algorithms without inter-cluster separation	Algorithms with inter-cluster separation
No weighting	kmeans [39], Bisecting kmeans [41], ISODATA [6], SinglePass [7], kmeans++ [5], X-means [37], spherical kmeans [20]	FCS [43, 45]
Vector weighting	Wkmeans [30], SYNCLUS [16], TW-kmeans [14]	OVW [17], Ovwtre [18]
Matrix weighting	AWA [12], FG-Group [13], EWkmeans [33], EWSubspace [32], PROCLUS [1], SKWIC [28], Mix-typed data kmeans [2] Ellkm [25], LAC [3, 21–23]	ESSC [19], COSA [27]

2.5 Characteristics of Our Extending Kmeans-Type Algorithms

The main feature of our proposed framework lies in the fusion of the information of inter-cluster separation in a clustering process. At present, most traditional kmeans-type algorithms (e.g. basic kmeans, Wkmeans [30] and AWA [12]) only utilize the intra-cluster compactness. On the contrary, our proposed framework synthesizes both the intra-cluster compactness and the inter-cluster separation.

On the other hand, some existing algorithms have also introduced the inter-cluster separation into their models as we mentioned in Sects. 2.1.2, 2.2.2 and 2.3.2. The schemes of using the inter-cluster separation in these algorithms can be summarized into two classes: (1) calculating the distances between all pairs of objects which belong to different clusters [17, 18, 27, 34]; (2) calculating the distance between the centroid of each cluster and the global centroid [19, 43, 45]. Both schemes are able to help the algorithms improve the clustering results. However, the objective functions involved in the algorithms using the way of class (1) are not differentiable [33]. A "subtraction" framework embedded in the objective functions, which are differentiable in class (2), is usually used to integrate the inter-cluster separation by the existing algorithms [19, 43, 45]. But a new parameter is usually required in the "subtraction" framework to balance the intra-cluster compactness and the inter-cluster separation. In practice, it is difficult to seek an appropriate value for this parameter. In our extending algorithms, to guarantee that the objective function in our proposed framework is differentiable, we calculate the distances between the centroids of the clusters with the global centroid as the inter-cluster separation. Consequently, we propose a "division" framework to integrate both the intra-cluster compactness and the inter-cluster separation.

3 The Extending Model of Kmeans-Type Algorithm

3.1 Motivation

From the analysis of the related works, most of the existing clustering methods consider only intra-cluster compactness. Take the AWA[12] as an example, the weights of features are updated according to the dispersions of the cluster. It means, in the same cluster, the features of small dispersions must be evaluated with large weights and the features of large dispersions must be evaluated with small weights. However, this does not work well in certain circumstances. For example, we have three clusters: C1 (London Olympic game), C2 (London riots) and C3 (Beijing Olympic game) as shown in Table 2, each row in which represents a document. The distribution of keyword frequencies is shown in Fig. 1. From the figure we can see that the dispersions of features are similar in all clusters. The traditional weighting kmeans will evaluate the similar weight to each feature. But we can observe, comparing C1 with C2, the features: "Olympic, game, Blackberry, riots," have

Table 2 An example of three clusters used to illustrate our motivation

Cluster	DocID	Document
C1	1	London, the capital of England, held the Olympic game in 2012
	2	2012 London Olympic game is the third Olympic game held in England
C2	5	2011 London riots in England is called "Blackberry riots"
	4	Blackberry riots is the largest riots in London, England in recent years
C3	5	Beijing, the capital of China, held the Olympic game in 2008
	6	2008 Beijing Olympic game is the first Olympic game held in China

C1

	Beijing	China	London	England	Olympic	Sport	Chaos	riots
	0	0	2	2	5	5	0	0
	0	0	3	3	6	6	0	0

W_{12}

.05	.05	.05	.05	.2	.2	.2	.2

W_{13}

.2	.2	.2	.2	.05	.05	.05	.05

C2

	Beijing	China	London	England	Olympic	Sport	Chaos	riots
	0	0	2	2	0	0	3	3
	0	0	3	3	0	0	4	4

C3

	Beijing	China	London	England	Olympic	Sport	Chaos	riots
	5	5	0	0	5	5	0	0
	5	5	0	0	5	5	0	0

Fig. 1 Term frequencies of the example in Table 2

more discriminative capabilities. Comparing C1 with C3, the features: "London, England, Beijing, China" have more discriminative capabilities. Thus, it may be ineffective to evaluate the weights of the features using only the dispersions of a data set. Under this condition, the inter-cluster separation can play an important role in distinguishing the importance of different features. In this chapter, we focus on the extending kmeans-type algorithms by integrating both the intra-cluster compactness and the inter-cluster separation. Intuitively, it is ideal to compare all the pairs of objects or centroids to utilize the inter-cluster separation. In comparison with previous weighting kmeans methods (e.g. Wkmeans and AWA), we may have a new objective function with the "subtraction" structure

$$P(U, W, Z) = \sum_{p=1}^{k} \sum_{i=1}^{n} u_{ip} \sum_{\substack{q=1 \\ q \neq p}}^{k} \sum_{i'=1}^{n} u_{i'q} \sum_{j=1}^{m} w_{pqj}^{\beta} D_{ii'pqj}, \qquad (24)$$

subject to

$$\sum_{j=1}^{m} w_{pqj} = 1 \qquad (25)$$

$$D_{ii'pqj} = (x_{ij} - z_{pj})^2 + (x_{i'j} - z_{qj})^2 - \eta(x_{ij} - x_{i'j})^2. \qquad (26)$$

This function aims to compare all pairs of objects in different clusters. In this objective function, each cluster has $k - 1$ weighting vectors, which represent the discriminative capabilities of the features while comparing to the other $k-1$ clusters. However, it is not solvable for the membership matrix U in this objective function. Instead of comparing to each pair of objects, we compare each pair of centroids to maximize the distances of different clusters in the objective function with the "subtraction" structure

$$P(U, W, Z) = \sum_{p=1}^{k} \sum_{\substack{q=1 \\ q \neq p}}^{k} \sum_{j=1}^{m} w_{pqj}^{\beta} \sum_{i=1}^{n} u_{ip} D_{pqj}, \tag{27}$$

$$D_{pqj} = (x_{ij} - z_{pj})^2 - \eta(z_{pj} - z_{qj})^2. \tag{28}$$

The membership matrix U can be solved in theory. But it is difficult to seek an appropriate value for balancing parameter η. Negative values in weight are often produced if η is large. Different to functions Eqs. (24) and (27), we may develop another objective function

$$P(U, W, Z) = \sum_{p=1}^{k} \sum_{\substack{q=1 \\ q \neq p}}^{k} \sum_{j=1}^{m} w_{pqj}^{\beta} \sum_{i=1}^{n} u_{ip} \frac{(x_{ij} - z_{pj})^2}{(z_{pj} - z_{qj})^2}. \tag{29}$$

This function employs the "division" structure similar to LDA [8, 26]. However, it is technically difficult to derive the centroid matrix Z in Eq. (29).

To make the centroid Z and membership matrix U solvable in the process of optimizing the objective functions, we can use the distances between the centroids of the clusters and the global centroid to approximate the distances of all pairs of centroids in the objective function as shown in Eq. (29). We believe that this approximation is reasonable, because making the centroid of a cluster away from the global centroid is approximately equivalent to make the cluster away from the other clusters in most of cases. For example, in Fig. 2, making cluster 1 away from the global centroid Z_0 is equal to make cluster 1 away from cluster 2 to some extent. Thus, it may attain the purpose of maximizing the distances of the clusters. It is worth noting that our extending algorithms try to maximize the distances between the centroids of the clusters and the global centroid while keeping to minimize the intra-cluster compactness which also has impact on the weight assignment. Thus, the errors produced by the approximation may be reduced. Moreover, many existing algorithms [19, 43, 45] maximize the inter-cluster separation by comparing the centroids of the clusters and the global centroid. The distances between the centroids of the clusters and the global centroid are also widely used as the inter-cluster separation in classification algorithms [8, 24, 26] (see Sect. 3.8.3). Under this framework, in the following sections we demonstrate the detailed derivative process of three extending algorithms: E-kmeans, E-Wkmeans and E-AWA.

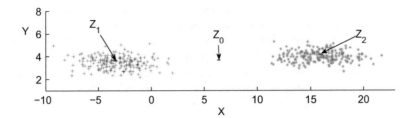

Fig. 2 An illustrative example about the effect of inter-cluster separation (Z_0 is the global centroid and Z_1, Z_2 are the centroids of cluster 1, cluster 2, respectively)

3.2 Extension of Basic Kmeans (E-kmeans)

Basic kmeans is a typical clustering algorithm which has been widely used in various data analysis. But it considers only the distances between centroids and objects, i.e. intra-cluster compactness. In order to utilize inter-cluster separation, we introduce the global centroid of a data set. Different to the basic kmeans, our proposed algorithm, E-kmeans, is expected to minimize the distances between objects and the centroid of the cluster that the objects belong to, while maximizing the distances between centroids of clusters and the global centroid.

In order to integrate intra-cluster compactness and inter-cluster separation, we modify the objective function shown in Eq. (1) into

$$P(U, Z) = \sum_{p=1}^{k} \sum_{i=1}^{n} u_{ip} \sum_{j=1}^{m} \frac{(x_{ij} - z_{pj})^2}{(z_{pj} - z_{0j})^2}, \tag{30}$$

subject to

$$u_{ip} \in \{0, 1\}, \sum_{p=1}^{k} u_{ip} = 1. \tag{31}$$

z_{0j} is the jth feature of the global centroid z_0 of a data set. We calculate z_{0j} as

$$z_{0j} = \frac{\sum_{i=1}^{n} x_{ij}}{n}. \tag{32}$$

We can minimize Eq. (30) by iteratively solving the following two problems:

1. Problem P1: Fix $Z = \hat{Z}$, and solve the reduced problem $P(U, \hat{Z})$;
2. Problem P2: Fix $U = \hat{U}$, and solve the reduced problem $P(\hat{U}, Z)$.

The problem P1 is solved by

$$u_{ip} = \begin{cases} 1, & \text{if } \sum_{j=1}^{m} \frac{(x_{ij}-z_{pj})^2}{(z_{pj}-z_{0j})^2} \leq \sum_{j=1}^{m} \frac{(x_{ij}-z_{p'j})^2}{(z_{p'j}-z_{0j})^2}, \\ 0, & \text{otherwise,} \end{cases} \tag{33}$$

where $1 \leq p' \leq k, p' \neq p$. If $(z_{pj} - z_{0j})^2 = 0$, we remove the jth feature at this iteration while calculating the membership U and the value of the objective function $P(U, Z)$. The proof procedure of minimizing objective function as shown in Eq. (30) to solve U is given in [39]. The solution to problem P2 is given by Theorem 3.1.

Theorem 3.1. *Let $U = \hat{U}$ be fixed, P2 is minimized iff*

$$z_{pj} = \begin{cases} z_{0j}, & \text{if } \sum_{i=1}^{n} u_{ip}(x_{ij} - z_{0j}) = 0, \\ \dfrac{\sum_{i=1}^{n} u_{ip}(x_{ij}-z_{0j})x_{ij}}{\sum_{i=1}^{n} u_{ip}(x_{ij}-z_{0j})}, & \text{otherwise.} \end{cases} \tag{34}$$

Proof. For minimizing objective function (30), we derive the gradient of z_{pj} as

$$\frac{\partial P(\hat{U}, Z)}{\partial z_{pj}} = \sum_{i=1}^{n} u_{ip} \frac{z_{pj}(x_{ij} - z_{0j}) - x_{ij}(x_{ij} - z_{0j})}{(z_{pj} - z_{0j})^3}. \tag{35}$$

If $z_{pj} - z_{0j} \neq 0$, we set (35) to zero, we have

$$z_{pj} = \frac{\sum_{i=1}^{n} u_{ip}(x_{ij} - z_{0j})x_{ij}}{\sum_{i=1}^{n} u_{ip}(x_{ij} - z_{0j})}. \tag{36}$$

If $\sum_{i=1}^{n} u_{ip}(x_{ij} - z_{0j}) = 0$, we set $z_{pj} = z_{0j}$. The overall procedure of E-kmeans can be described as Algorithm 1. It is noted that, since the objective function shown in Eq. (30) is strictly decreasing, when we optimize U and Z, Algorithm 1 can guarantee that the objective function converges to local minimum.

Algorithm 1: E-kmeans

Input: $X = \{X_1, X_2, \ldots, X_n\}, k$.
Output: U, Z.
Initialize: Randomly choose an initial $Z^0 = Z_1, Z_2, \ldots, Z_k$.
repeat
    Fixed Z, solve the membership matrix U with (33);
    Fixed U, solve the centroids Z with (34);
until Convergence.

3.3 Extension of Wkmeans (E-Wkmeans)

As mentioned before, basic kmeans and E-kmeans treat all the features equally. However, features may have different discriminative powers in real-world applications. Motivated by this, Huang et al. proposed the Wkmeans [30] algorithm which evaluates the importance of the features according to the dispersions of a data set. In this section, we propose the E-Wkmeans algorithm, which is able to consider the dispersions of a data set and the distances between the centroids of the clusters and the global centroid simultaneously while updating the feature weights.

Let $W = \{w_1, w_2, \ldots, w_m\}$ be the weights for m features and β be a parameter for tuning weight w_j, we extend Eq. (9) into

$$P(U, W, Z) = \sum_{p=1}^{k} \sum_{i=1}^{n} u_{ip} \left[\sum_{j=1}^{m} w_j^{\beta} \frac{(x_{ij} - z_{pj})^2}{(z_{pj} - z_{0j})^2} \right], \tag{37}$$

subject to

$$u_{ip} \in \{0, 1\}, \sum_{p=1}^{k} u_{ip} = 1, \sum_{j=1}^{m} w_j = 1, 0 \le w_j \le 1. \tag{38}$$

Similar to solve Eq. (30), we can minimize Eq. (37) by iteratively solving the following three problems:

1. Problem P1: Fix $Z = \hat{Z}$ and $W = \hat{W}$, and solve the reduced problem $P(U, \hat{Z}, \hat{W})$;
2. Problem P2: Fix $U = \hat{U}$ and $W = \hat{W}$, and solve the reduced problem $P(\hat{U}, Z, \hat{W})$;
3. Problem P3: Fix $U = \hat{U}$ and $Z = \hat{Z}$, and solve the reduced problem $P(\hat{U}, \hat{Z}, W)$;

Problem P1 is solved by

$$u_{ip} = \begin{cases} 1, & \text{if } \sum_{j=1}^{m} w_j^{\beta} \frac{(x_{ij} - z_{pj})^2}{(z_{pj} - z_{0j})^2} \le \sum_{j=1}^{m} w_j^{\beta} \frac{(x_{ij} - z_{p'j})^2}{(z_{p'j} - z_{0j})^2}, \\ 0, & \text{otherwise}, \end{cases} \tag{39}$$

where $1 \le p' \le k, p' \ne p$. Problem P2 is solved by Eq. (34) and the solution to problem P3 is given in Theorem 3.2.

Theorem 3.2. *Let $U = \hat{U}$ and $Z = \hat{Z}$ be fixed, $P(\hat{U}, \hat{Z}, W)$ is minimized iff*

$$w_j = \begin{cases} 0, & D_j = 0 \text{ or } z_{pj} = z_{0j}, \\ \dfrac{1}{\sum\limits_{i=1}^{m} \left(\frac{D_j}{D_i}\right)^{1/(\beta-1)}}, & \text{otherwise}, \end{cases} \tag{40}$$

where

$$D_j = \sum_{p=1}^{k} \sum_{i=1}^{n} u_{ip} \frac{(x_{ij} - z_{pj})^2}{(z_{pj} - z_{0j})^2}. \tag{41}$$

Proof. We consider the relaxed minimization of $P(\hat{U}, \hat{Z}, W)$ via a Lagrange multiplier obtained by ignoring the constraint $\sum_{j=1}^{m} w_j = 1$. Let α be the multiplier and $\Psi(W, \alpha)$ be the Lagrangian

$$\Psi(W, \alpha) = \sum_{j=1}^{m} w_j^\beta D_j - \alpha \left(\sum_{j=1}^{m} w_j - 1 \right). \tag{42}$$

By setting the gradient of the function Eq. (42) with respects to w_j and α to zero, we obtain the equations

$$\frac{\partial \Psi(W, \alpha)}{\partial w_j} = \beta w_j^{\beta-1} D_j - \alpha = 0, \tag{43}$$

$$\frac{\partial \Psi(W, \alpha)}{\partial \alpha} = \sum_{j=1}^{m} w_j - 1 = 0. \tag{44}$$

From Eq. (43), we obtain

$$w_j = \left(\frac{\alpha}{\beta D_j} \right)^{1/(\beta-1)}. \tag{45}$$

Substituting Eq. (45) into Eq. (44), we have

$$\alpha^{1/(\beta-1)} = 1 \Bigg/ \left[\sum_{j=1}^{m} \left(\frac{1}{\beta D_j} \right)^{1/(\beta-1)} \right]. \tag{46}$$

Substituting Eq. (46) into Eq. (45), we have

$$w_j = \frac{1}{\sum_{t=1}^{m} \left(\frac{D_j}{D_t} \right)^{1/(\beta-1)}}. \tag{47}$$

The overall procedure of E-Wkmeans can be described as Algorithm 2. Given a data partition, the goal of feature weight aims to assign a larger weight to a feature that has a smaller intra-cluster compactness and larger inter-cluster separation. The parameter β is used to control the distribution of the weight W.

When $\beta = 0$, E-Wkmeans is equivalent to E-kmeans, because $w_j^\beta = 1$ regardless of the value of w_j.

When $\beta = 1$, w_j is equal to 1 for the smallest D_j shown in Eq. (41), and the weights of other feature are 0. That means, it chooses only one feature for clustering. It is unreasonable for the high-dimensional data clustering.

Algorithm 2: E-Wkmeans

    Input: $X = \{X_1, X_2, \ldots, X_n\}, k.$
    Output: $U, Z, W.$
    Initialize: Randomly choose an initial $Z^0 = Z_1, Z_2, \ldots, Z_k$ and weight $W = \{w_1, w_2, \ldots, w_m\}.$
    repeat
        Fixed Z, W, solve the membership matrix U with (39);
        Fixed U, W, solve the centroids Z with (34);
        Fixed U, Z, solve the weight W with (40);
    until Convergence.

When $0 < \beta < 1$, the objective function Eq. (37) cannot converge to the minimization.

When $\beta < 0$, the larger the value of D_j is, the larger value of w_j we can get. This is not applicable to feature selection. The aim of feature selection is to evaluate larger weights to smaller D_j, i.e. we should evaluate larger weights to the features that have small intra-cluster compactness and large inter-cluster separation. $\beta < 0$ cannot satisfy the demand of feature selection.

When $\beta > 1$, the larger the value of D_j is, the smaller value of w_j we can get. This is able to satisfy all the demand of the algorithms and the objective function shown in Eq. (37) is strictly decreasing when we optimize U, Z and W, it is able to converge to local minimum.

3.4 Extension of AWA (E-AWA)

In Wkmeans and E-Wkmeans, the same feature in different clusters has the same weight. The same feature in different clusters, however, has different weights in most real-world applications. In this section, we focus on developing an algorithm to solve this problem under the condition of utilizing both intra-cluster compactness and inter-cluster separation.

Let $W = \{W_1, W_2, \ldots, W_k\}$ be a weight matrix for k clusters. $W_p = \{w_{p1}, w_{p2}, \ldots, w_{pm}\}$ denotes the feature weights in cluster p. We extend Eq. (14) into

$$P(U, W, Z) = \sum_{p=1}^{k} \sum_{i=1}^{n} u_{ip} \left[\sum_{j=1}^{m} w_{pj}^{\beta} \frac{(x_{ij} - z_{pj})^2}{(z_{pj} - z_{0j})^2} \right], \tag{48}$$

subject to

$$u_{ip} \in \{0, 1\}, \sum_{p=1}^{k} u_{ip} = 1, \sum_{j=1}^{m} w_{pj} = 1, 0 \le w_{pj} \le 1. \tag{49}$$

Similar to solve E-Wkmeans, we can minimize Eq. (48) by iteratively solving the following three problems:

1. Problem P1: Fix $Z = \hat{Z}$ and $W = \hat{W}$, and solve the reduced problem $P(U, \hat{Z}, \hat{W})$;
2. Problem P2: Fix $U = \hat{U}$ and $W = \hat{W}$, and solve the reduced problem $P(\hat{U}, Z, \hat{W})$;
3. Problem P3: Fix $U = \hat{U}$ and $Z = \hat{Z}$, and solve the reduced problem $P(\hat{U}, \hat{Z}, W)$.

Problem P1 is solved by

$$
u_{ip} = \begin{cases} 1, & \text{if } \sum_{j=1}^{m} w_{pj}^{\beta} \frac{(x_{ij}-z_{pj})^2}{(z_{pj}-z_{0j})^2} \leq \sum_{j=1}^{m} w_{p'j}^{\beta} \frac{(x_{ij}-z_{p'j})^2}{(z_{p'j}-z_{0j})^2}, \\ 0, & \text{otherwise,} \end{cases} \tag{50}
$$

where $1 \leq p' \leq k, p' \neq p$. Problem P2 is solved by Eq. (34) and the problem P3 is solved by

$$
w_{pj} = \begin{cases} 0, & \text{if } (z_{pj} - z_{0j})^2 = 0, \\ \frac{1}{m_i}, & \text{if } D_{pj} = 0 \text{ and } z_{pj} \neq z_{0j}, \\ m_i = |\{t : D_{pt} = 0 \text{ and } (z_{pt} - z_{0t})^2 \neq 0\}|, \\ 0, & \text{if } D_{pj} \neq 0, \text{ but } D_{pt} = 0, \text{ for some } t, \\ \frac{1}{\sum_{t=1}^{m} \left(\frac{D_{pj}}{D_{pt}}\right)^{1/(\beta-1)}}, & \forall 1 \leq t \leq m, \text{ otherwise,} \end{cases} \tag{51}
$$

where

$$
D_{pj} = \sum_{i=1}^{n} u_{ip} \frac{(x_{ij} - z_{pj})^2}{(z_{pj} - z_{0j})^2}. \tag{52}
$$

The convergence of proof process is similar to that of E-Wkmeans. The procedure of E-AWA can be described as Algorithm 3. Similar to the E-Wkmeans, the input parameter β is used to control the distribution of the weight W and β should be evaluated the value greater than 1. Since objective function is strictly decreasing in each step when optimizing U, Z and W, Algorithm 3 can assure that the objective function converges to local minimum.

Algorithm 3: E-AWA

Input: $X = \{X_1, X_2, \ldots, X_n\}, k$.
Output: U, Z, W.
Initialize: Randomly choose an initial $Z^0 = Z_1, Z_2, \ldots, Z_k$ and weight $W = \{w_{pj}\}$.
repeat
    Fixed Z, W, solve the membership matrix U with (50);
    Fixed U, W solve the centroids Z with (34);
    Fixed U, Z solve the weight W with (51);
until Convergence.

3.5 Relationship Among Algorithms

E-kmeans, E-Wkmeans and E-AWA are the extensions of basic kmeans, Wkmeans and AWA, respectively. Basic kmeans, Wkmeans and AWA employ only the intra-cluster compactness while updating the membership matrix and weights. However, the extending algorithms take the inter-cluster separation into account.

From another perspective, E-kmeans does not weight the features, i.e. all the features are treated equally. E-Wkmeans weights the features with a vector, that means, each feature has a weight representing the importance of the feature in the entire data set. E-AWA weights the features with a matrix, i.e. each cluster has a weighting vector representing the subspace of the cluster. When $\beta = 0$, E-Wkmeans and Wkmeans degenerate to E-kmeans and basic kmeans, respectively. Since $\beta = 0$, $w_j^\beta = 1$ regardless of the value of w_j. Thus, the features are treated equally while updating the membership matrix U. Likewise, when $\beta = 0$, E-AWA and AWA degenerate to the basic kmeans and E-kmeans. When $\beta = 0$, E-Wkmeans and E-AWA have the same clustering result, whilst Wkmean and AWA have the same clustering result. When the weights of the same feature in different clusters are equal, E-AWA and AWA are equivalent to E-Wkmeans and Wkmeans, respectively. But this case rarely happens in real-world data sets.

3.6 Computational Complexity

Similar to the basic kmeans, Wkmean and AWA, the extending algorithms are also iterative algorithms. The computational complexity of basic kmeans is $O(tknm)$, where t is the iterative times; $k,n,$ and m are the number of the clusters, objects and features, respectively. E-kmeans as well as basic kmeans has two computational steps: (1) updating the membership matrix; (2) updating the centroids. The complexities of updating centroids and updating membership matrix of E-kmeans are $O(knm + nm)$ and $O(knm + km)$, respectively. Therefore, the complexity of the overall E-kmeans algorithm is $O(tknm)$.

In comparison to the E-kmeans, the E-Wkmeans and E-AWA have another step: updating the weights. The complexity of updating the weights of E-Wkmeans and E-AWA is $O(knm + km)$. Therefore, the overall computational complexities of E-Wkmean and E-AWA are also $O(tknm)$. In summary, compared with the original algorithms, our extending algorithms need extra $O(km)$ computational time to calculate the distances between the centroids of the clusters and the global centroid while updating member matrix and weights, and we need extra $O(nm)$ to calculate distances between the objects and the global centroid to update the centroids of the clusters. However, it does not change the computational complexities of the algorithms in overall. Basic kmeans, E-kmeans, Wkmeans, E-kmeans, AWA and E-AWA have the same computational complexity $O(tknm)$.

4 Experiments

4.1 Experimental Setup

In experiments, the performances of proposed approaches are extensively evaluated on two synthetic data sets and nine real-life data sets. The benchmark clustering algorithms—basic kmeans (kmeans), BSkmeans [41], automated variable Weighting kmeans (Wkmeans) [30], AWA [12], EWkmeans [33] as well as ESSC [19] are chosen for the performance comparison with the proposed algorithms. Among these algorithms, kmeans, E-kmeans, and BSkmeans have no input parameter; Wkmeans, AWA, E-Wkmeans and E-AWA have a parameter β to tune the weights of the features. In our experiments, we choose $\beta = 8$ according to the empirical study of parameter β in Sects. 4.2.1 and 4.3.1. EWkmeans has parameter γ, which controls the strength of the incentive for clustering on more features. In the experiments, we set $\gamma = 5$ according to [33]. ESSC has three parameters λ, γ and η, where λ is the fuzzy index of fuzzy membership, γ and η are used to control the influences of entropy and balance the weights between intra-cluster compactness and inter-cluster separation, respectively. We have chosen the empirical values $\lambda = 1.2$, $\gamma = 5$ and $\eta = 0.1$ according to [19]. Since ESSC [19] is a fuzzy kmeans algorithms and each object corresponds to a membership vector which indicates the degree that the object belongs to the corresponding clusters, we assign the object to the cluster corresponding to the maximal value in the membership vector for simplification to compare the performance. In the experiments, we implement all the algorithms with Matlab and run the algorithms in a workstation with Intel(R)Xeon(R) 2.4 Hz CPU, 8 GB RAM.

In this chapter, four evaluation metrics including Accuracy (Acc), Rand Index (RI), Fscore and normal mutual information (NMI) are used to evaluate the results of the algorithms. Acc represents the percentage of the objects that are correctly recovered in a clustering result and RI [19, 30] considers the percentage of the pairs of objects which cluster correctly. The computational processes of Acc and RI can be found in [30]. Fscore [33] is able to leverage the information of precision and recall. NMI [19, 33] is a popular measure of clustering quality, which is more reliable to measure the imbalanced data sets (i.e. most of objects are from one cluster and only a few objects belong to other clusters) in comparison to the other three metrics. The computational processes of Fscore and NMI can be found to [33].

It is well known that the kmeans-type clustering process produces local optimal solution. The final result depends on the locations of initial cluster centroids. In weighting kmeans clustering algorithms, the initial weights also affect the final clustering result. To compare the performance between the extending algorithms and the existing algorithms, the same set of centroids which randomly generated are used to initialize the different algorithms and all the weighting algorithms are initialized with the same weights. Finally, we calculate the average Acc, RI, Fscore and NMI produced by the algorithms after running 100 times.

4.2 Synthetic Data Set

In this section, two synthetic data sets are constructed to investigate the performances of the proposed algorithms. In order to validate the effect of inter-cluster separation, different features on the two data sets are designed to different discriminative capabilities on an inter-cluster perspective. The centroids and standard deviations of synthetic data sets are given in Table 3. Each cluster contains eight features and 100 objects. The synthetic data sets are generated by three steps: (1) Generating the centroid for each cluster. For making different features have different discriminative capabilities from an inter-cluster separation perspective, we generate 3 8-dimensional vectors as the centroids of synthetic data set 1 (Synthetic1) where the first three features are well separated and the other features are very close to each other when comparing cluster 1 with cluster 2. However, comparing cluster 1 with cluster 3, the second three features are well separated and the other features are close to each other. The last two features are noisy features as they are very close to each other for all the clusters. The generating procedure of the centroids of the synthetic data set 2 (Synthetic2) is similar to that of Synthetic1. Synthetic2 has five clusters. Based on the centroids of Synthetic1, we generate another two 6-dimensional vectors as the first six features of the centroids of the last two clusters. The values of the first three, the fourth, the fifth and the sixth features of the centroids of cluster 4 are similar to the values of the corresponding features of centroids of cluster 2, 3, 3 and 2, respectively. The values of the first two, the second two, the fifth and the sixth features of the centroids of cluster 5 are similar to the values of the corresponding features of centroids of cluster 3, 2, 3 and 1, respectively. Then, we generate 5 2-dimensional vectors as the centroids of the last two features of the five clusters, (2) Generating the standard deviations for each cluster. For the Synthetic1, we generate randomly an 8-dimensional vector of which the values are between 0 and 1 as the standard deviations for each cluster. For the Synthetic2, we generate randomly an 8-dimensional vector of which the values are between 0 and 0.3 as the standard deviations for each cluster. Because we want to observe the effect of our proposed algorithms applying to the data sets which have different separability. Normally, the data sets with small standard deviations are separated easier than the data sets with large the standard deviations under the condition of similar centroids, and (3) Generating the objects of the data sets. We generate 100 points for each cluster using normal distribution with the centroids derived by the first step and the standard deviations derived the second step.

4.2.1 Parametric Study

Parameter β, which is used in the algorithms: Wkmeans, E-Wkmeans, AWA and E-AWA, is an important factor to tune the weights of the features. In this section, we give the empirical study of β for its effects on the results as in Fig. 3. The clustering results are shown from $\beta = 1.1$ until the results do not change or begin to reduce

Table 3 The centroids and the standard deviations of synthetic data sets

Data Set	Cluster centroid	Standard deviation
Synthetic1	0.503 0.325 0.728 0.103 0.814 0.613 0.510 0.893	0.225 0.115 0.020 0.132 0.294 0.257 0.064 0.715
	0.238 0.841 0.367 0.099 0.805 0.621 0.499 0.897	0.146 0.180 0.923 0.863 0.977 0.751 0.767 0.642
	0.498 0.335 0.701 0.556 0.312 0.972 0.507 0.901	0.598 0.655 0.653 0.121 0.140 0.428 0.671 0.419
Synthetic2	0.503 0.325 0.728 0.103 0.814 0.913 0.110 0.098	0.048 0.174 0.050 0.223 0.193 0.044 0.013 0.256
	0.238 0.841 0.367 0.099 0.805 0.021 0.099 0.097	0.018 0.070 0.254 0.222 0.202 0.004 0.285 0.221
	0.498 0.335 0.701 0.556 0.312 0.672 0.107 0.101	0.180 0.251 0.109 0.060 0.086 0.069 0.079 0.263
	0.240 0.838 0.365 0.560 0.310 0.074 0.109 0.103	0.246 0.046 0.137 0.004 0.288 0.192 0.238 0.211
	0.501 0.322 0.363 0.097 0.307 0.930 0.105 0.105	0.234 0.248 0.105 0.217 0.094 0.224 0.207 0.009

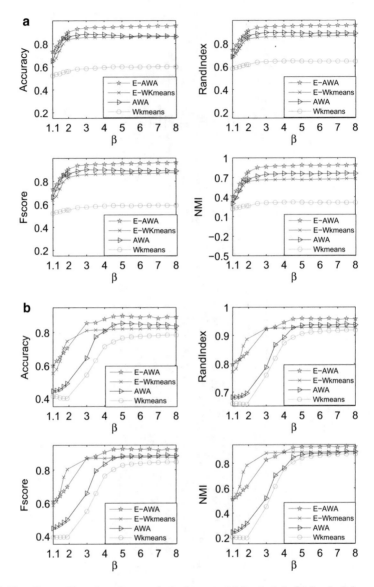

Fig. 3 The effects with various β on synthetic data sets. (**a**) Synthetic1; (**b**) Synthetic2

by increasing the value of β. We have used the increment of the value of 0.2 for β in this scope in our experiments because the performance is sensitive to the change of the values of β in the range of $(1, 2]$. From the value of $\beta = 2$, we have used the increment of the value of 0.5 for β since the performances are more stable in this range.

Figure 3 shows the changing trend of the average results produced by the compared algorithms after running 100 times on the two synthetic data sets. From the results, we can observe that the performances of the extending algorithms, E-AWA and E-Wkmeans, are consistently better than AWA and Wkmeans, respectively, across all the evaluation metrics. AWA and E-AWA perform better than Wkmeans and E-Wkmeans. E-AWA performs best in all the algorithms on both data sets. We can also observe that the performances tend to be constant when β is greater than three. Since the relatively good results can be obtained when $\beta = 8$, in the later study, we choose $\beta = 8$ in the experiments.

4.2.2 Results and Analysis

The average Acc, RI, Fscore, NMI and the standard deviations produced by the compared algorithms after running 100 times are summarized in Table 4 for the two synthetic data sets by using $\beta = 8$. In view of the overall experiment, the best performance is delivered by E-AWA with respects to all the evaluation criteria. Moreover, E-Wkmeans and E-kmeans perform better than Wkmeans and basic kmeans, respectively.

For Synthetic1, we observe from Table 4 that the performances of the E-AWA, E-Wkmeans and E-kmeans are better than that of AWA, Wkmeans and kmeans, respectively. In comparison to kmeans, Wkmeans and AWA, E-kmeans, E-Wkmeans and E-AWA are able to deliver over 25 %, 26 % and 9 % Acc improvement, respectively, on Synthetic1. In the three pairs of algorithms (basic kmeans and E-kmeans; Wkmeans and E-Wkmeans; AWA and E-AWA), AWA and E-AWA perform the best. We believe that this is caused by the same features in different clusters playing different roles in the data set. For example, feature 1 in cluster 2 plays more important than that in cluster 1 and cluster 3, because the dispersion of feature 1 in cluster 2 is small and the centroid of feature 1 in cluster 2 is far from the global centroid. ESSC is the second best algorithm for this data set. This result indicates that ESSC is able to utilize effectively the inter-cluster separation as our extending algorithms do.

In comparison to Synthetic1, Synthetic2 is a more complex data set. We can observe the results of the Synthetic2 in the Table 4 that the performances of the E-AWA, E-Wkmeans and E-kmeans also perform better than AWA, Wkmeans and kmeans, respectively. Compared with kmeans, Wkmeans and AWA, E-kmeans, E-Wkmeans and E-AWA produces 11 %, 15 % and 7 % Acc improvement, respectively. Likewise, the AWA and E-AWA perform the best in three pairs of algorithms, even better than the result of Synthetic1. The result of Synthetic2 implies that AWA and E-AWA are more applicable to the complex data set. The performance of ESSC in Synthetic2 is not as promising as that in Synthetic1. We believe that ESSC may perform worse for a complex data set due to the presence of a number of parameters which are difficult to select for a better performance. We can see the Table 4, the standard deviations of the results produced by the extending algorithms are similar to that produced by the original algorithms. In overall, the results produced by

Table 4 The results on Synthetic data sets (the Standard deviation in bracket)

Data set	Algorithm	Acc	RI	Fscore	NMI
Synthetic1	BSkmeans	0.6384(±0.08)	0.6651(±0.04)	0.6436(±0.06)	0.3064(±0.06)
	EWkmeans	0.7731(±0.15)	0.8096(±0.12)	0.8036(±0.13)	0.6563(±0.21)
	ESSC	0.9441(±0.11)	0.9457(±0.09)	0.9457(±0.10)	0.8784(±0.16)
	kmeans	0.6103(±0.06)	0.6361(±0.04)	0.6048(±0.05)	0.2955(±0.04)
	E-kmeans	0.8663(±0.13)	0.8642(±0.08)	0.8771(±0.10)	0.6643(±0.14)
	Wkmeans[a]	0.5975(±0.04)	0.6421(±0.04)	0.5884(±0.03)	0.3123(±0.06)
	E-Wkmeans[a]	0.8625(±0.14)	0.8671(±0.09)	0.8714(±0.12)	0.6782(±0.17)
	AWA[a]	0.8600(±0.17)	0.8862(±0.10)	0.8879(±0.12)	0.7617(±0.13)
	E-AWA[a]	**0.9550(±0.10)**	**0.9572(±0.07)**	**0.9613(±0.07)**	**0.8908(±0.12)**
Synthetic2	BSkmeans	0.9271(±0.02)	0.9466(±0.01)	0.9271(±0.02)	0.8452(±0.02)
	EWkmeans	0.6954(±0.16)	0.8567(±0.08)	0.7760(±0.11)	0.7713(±0.10)
	ESSC	0.7070(±0.09)	0.8685(±0.03)	0.7695(±0.06)	0.7792(±0.06)
	kmeans	0.7186(±0.16)	0.8726(±0.07)	0.7923(±0.11)	0.7536(±0.08)
	E-kmeans	0.8377(±0.15)	0.9304(±0.05)	0.8766(±0.11)	0.8446(±0.09)
	Wkmeans[a]	0.6688(±0.17)	0.8496(±0.08)	0.7516(±0.12)	0.7186(±0.11)
	E-Wkmeans[a]	0.8225(±0.16)	0.9200(±0.06)	0.8678(±0.11)	0.8398(±0.10)
	AWA[a]	0.8858(±0.14)	0.9500(±0.06)	0.9159(±0.10)	0.9137(±0.09)
	E-AWA[a]	**0.9565(±0.09)**	**0.9805(±0.03)**	**0.9659(±0.06)**	**0.9570(±0.05)**

[a]*Note*: The results in the table are produced by using $\beta = 8$

the algorithms of no input parameter have smaller standard deviations than that produced by the algorithms with one or more parameters.

In summary, our proposed algorithms, E-AWA, E-Wkmeans and E-kmeans perform better than the original kmeans-type approaches, AWA, Wkmeans and kmeans, respectively, on the synthetic data sets. We believe that this performance gain is contributed to consider the inter-cluster separation in the clustering process.

4.2.3 Feature Selection

In this section, we study the effect of feature weight with different kmeans-type algorithms. In a clustering process, both Wkmeans and E-Wkmeans produce a weight for each feature, which represents the contribution of this feature in the entire data set. Figures 4 and 5 show the comparison of feature weights of Wkmeans and E-Wkmeans on two synthetic data sets. From the Table 3, we can observe that the centroids of feature 7 and feature 8 in all the clusters are very close to each other. Therefore we can consider feature 7 and feature 8 as noisy features from the viewpoint of inter-cluster separation. From Figs. 4 and 5, we observe that E-Wkmeans can reduce the weight of noisy features (i.e. feature 7 and feature 8) and increase the weights of features which are relatively far away from each other as opposed to Wkmeans.

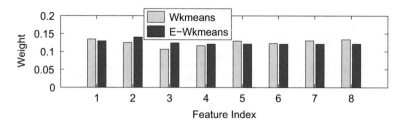

Fig. 4 The comparison of the feature weights between Wkmeans and E-Wkmeans on Synthetic1 ($\beta = 8$)

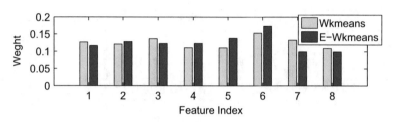

Fig. 5 The comparison of the feature weights between Wkmeans and E-Wkmeans on Synthetic2 ($\beta = 8$)

On the other hand, both E-AWA and AWA produce a weight for each feature in each cluster, which represents the contribution of a feature in a cluster. From Figs. 6 and 7, we observe that E-AWA can also reduce the weight of noisy features (i.e. feature 7 and feature 8) and increase the weights of the features which are far from the centroid of other clusters. For example, feature 3 of cluster 1 in Synthetic1, this feature has small dispersion and is far away from the centroid of other clusters. E-AWA is able to increase the weight of this feature in a significant rate comparing to AWA. And for feature 8 of cluster 5 in Synthetic2, the AWA evaluates the large weight due to the small dispersion. As a matter of fact, this feature has small discriminative capability because the mean of this feature in all the clusters are similar. E-AWA reduces the weight of this feature in cluster 5 comparing to AWA. From the analysis of the synthetic data sets, we can see that the more appropriate weights of the features can be obtained by integrating the information of the inter-cluster separation. It is worth noting that the weights of features are influenced simultaneously by parameter β, the intra-cluster compactness and the inter-cluster separation. For Wkmeans, AWA, E-Wkmeans and E-AWA, the variance of the weights of the different features will reduce with the increase of the values of β, i.e. when β is large, the weights of different features will tend to be similar.

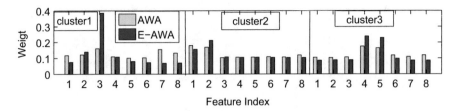

Fig. 6 The comparison of the feature weights between AWA and E-AWA on Synthetic1 ($\beta = 8$)

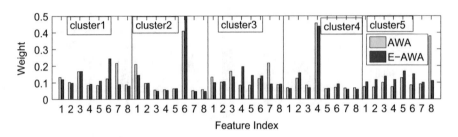

Fig. 7 The comparison of the feature weights between AWA and E-AWA on Synthetic2 ($\beta = 8$)

Table 5 The average iterations and the running time on synthetic data sets

Data set	E-AWA[a,b]	AWA[a,b]	E-Wkmeans[a,b]	Wkmeans[a,b]	E-kmeans[b]	kmeans[b]
Synthetic1	9.47(6.18)	9.30(5.20)	11.2(5.73)	13.1(5.32)	11.4(3.83)	14.6(3.83)
Synthetic2	10.5(14.6)	10.4(13.1)	9.82(11.4)	12.3(11.5)	9.56(7.81)	10.9(6.85)

[a]*Note*: The results in the table are produced by using $\beta = 8$
[b]*Note*: The values in brackets are the running time in seconds produced by running the algorithms with the same centroids in 100 runs

4.2.4 Convergence Speed

In this study, we import the inter-cluster separation into the objection functions of kmeans-type algorithms. Intuitively, the factor from inter-cluster separation may speed up the convergence process of the clustering. In this section, we investigate the convergence speed of the weighting kmeans-type algorithms with respects to the iterations and the running time. The stopping criterion of the algorithms relies on the condition that the membership matrix U is no longer changing. Table 5 lists the average iterations and the running time of the algorithms initialized by the same centroids in 100 runs with $\beta = 8$. From this table, we can observe that the iterations of the extending algorithms are similar to those of the corresponding original algorithms. However, the extending algorithms spend slightly more running time in comparison to the original algorithms. This additional time cost comes from the calculation of the distances between the centroids of the clusters and the global centroid at every iteration.

Table 6 The properties of real-life data sets

Data set	No. of features	No. of clusters	No. of objects
Wine	13	3	178
WDBC	30	2	569
Vertebral2	6	2	310
Vertebral3	6	3	310
Robot	90	4	88
Cloud	10	2	2048
LandsatSatellite	33	6	6435
Glass	9	2	214
Parkinsons	22	2	195

4.3 Real-Life Data Set

To further investigate the performance of the extending algorithms in real-life data sets, we have evaluated our algorithms in nine data sets reported in Machine Learning Repository (http://archive.ics.uci.edu/ml/). The properties of these data sets are described in Table 6.

4.3.1 Parametric Study

We show the average Acc, RI, Fscore and NMI produced by Wkmeans, E-Wkmeans, AWA and E-AWA after running 100 times from $\beta = 1.1$ until the clustering results do not change or begin to reduce by increasing the value of β as shown in Figs. 8, 9, and 10. From these figure, we can observe that E-AWA and E-Wkmeans outperform AWA and Wkmeans, respectively, for most values of β across the data sets "Vertebral2," "Vertebral3," "Robot," "Cloud," "LandsatSatellite," "Glass," and "Parkinson." The algorithms, AWA and E-AWA, perform better than Wkmeans and E-Wkmeans, respectively, on data sets "Vertebral2," "Vertebral3," "Cloud," "LandsatSatellite," and "Parkinson." On "Robot" and "Glass," E-Wkmeans outperform the other algorithms. We can observe that the results of most algorithms are unstable when the values of β is between 1 and 3 and trend to stability after the value of β is greater than 3. This observation is similar to the results of the synthetic data sets. Moreover, we can observe that the performances of our extending algorithms are more stable than the performances of the original algorithms with the change of the values of β in most of data sets, especially, when $1 < \beta \leq 3$. This indicates that the intra-cluster separation can help to improve clustering results no matter what value β is assigned when it is greater than 1 in most of cases.

However, we can see that Wkmeans performs better than E-Wkmeans on data set "Wine." Likewise, AWA achieves better results than E-AWA on data set "WDBC." We believe that performance degradation on "Wine" and "WDBC" may be caused

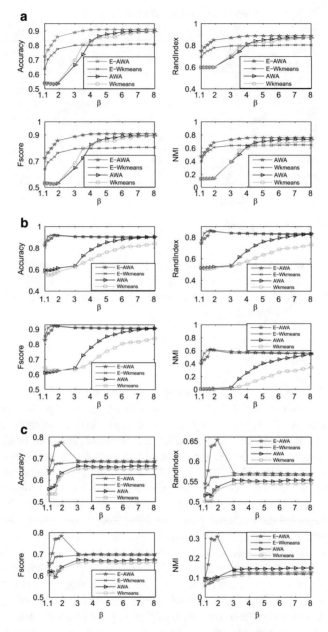

Fig. 8 The effects with various β on data sets: (**a**) "Wine," (**b**) "WDBC," (**c**) "Vertebral2"

by the resulting errors in the process of approximation when we use the distances between the centroids of the clusters and the global centroid to approximate the distances among all pairs of centroids. When the centroid of a cluster on all the

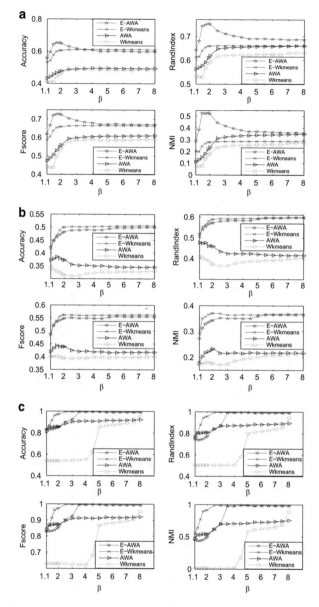

Fig. 9 The effects with various β on data sets: (**a**) "Vertebral3," (**b**) "Robot," (**c**) "Cloud"

features or some important features are very close to the corresponding features of the global centroid, but the centroids of the clusters on these features are not close to each other, the approximation may introduce errors. For example, we can observe from Table 7, the values of the centroids on the 6th, 7th, 8th and 12th features in

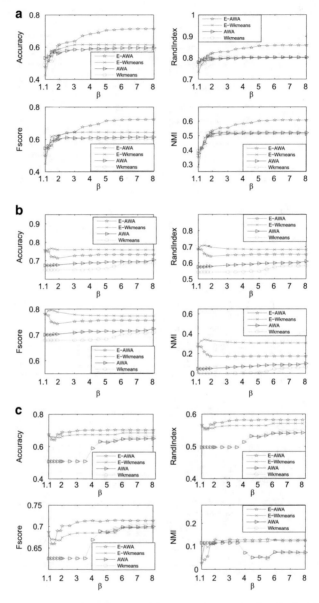

Fig. 10 The effects with various β on real data sets: (**a**) "LandsatSatellite," (**b**) "Glass," (**c**) "Parkinson"

the second clusters are very close to the values of the corresponding features in the global centroid. Thus, the small weights will be assigned to these features. However, the distances among the centroids of the clusters on these features are not very close to each other and the dispersions of the data set on these

Table 7 The characteristics of data set "Wine"

	Features 1–13
Centroid of cluster1	13.7 2.010 2.455 17.0 106. 2.840 2.982 0.290 1.899 5.528 1.062 3.157 1115.
Centroid of cluster2	12.2 1.932 2.244 20.2 94.5 **2.258 2.080 0.363** 1.630 3.086 1.056 **2.785** 519.0
Centroid of cluster3	13.1 3.333 2.437 21.4 99.3 1.678 0.781 0.447 1.153 7.396 0.682 1.683 629.8
Global centroid	13.0 2.336 2.366 19.5 99.7 **2.295 2.029 0.361** 1.590 5.058 0.957 **2.611** 746.0
Global dispersion	0.26 0.896 0.061 7.60 169. **0.179 0.247 0.011** 0.233 2.576 0.022 **0.149** 2.9E4

features are small, which indicates that these features should be assigned by large weights. Thus, it may produce the errors when assigning the objects into the second clusters. It is noteworthy that our extending algorithms have also included the intra-cluster compactness. It is able to minimize the distances between the centroid of a cluster and the objects that belong to the cluster such that the weight assignment can be accomplished in a principled manner. Thus, the problem caused by the approximation process may be relieved to some extent. From the experimental results, our proposed algorithms outperform the original algorithms in most of cases.

4.3.2 Results and Analysis

The average Acc, RI, Fscore, NMI and standard deviations produced by the compared algorithms after running 100 times are summarized in Tables 8 and 9 on nine real-life data sets by using $\beta = 8$ according to the study of Sect. 4.3.1. On the data sets "Robot" and "Cloud," the extending algorithms: E-kmeans, E-Wkmeans and E-AWA, outperform kmeans, Wkmeans and AWA, respectively, across all the evaluation metrics. For example, comparing with AWA, E-AWA obtains 15 % and 7 % Acc improvement on "Robot" and "Cloud," respectively. Compared with kmeans, Wkmeans and AWA, E-kmeans, E-Wkmeans and E-AWA achieve 6 %, 24 %, 7 % NMI improvement on data set "Glass" and 7 %, 6 %, 5 % NMI improvement on data set "Parkinson," respectively. But it is noticed that the best clustering performance as indicated by NMI, is not always consistent with that indicated by Acc, RI and Fscore. This is caused by the imbalanced properties of "Glass" and "Parkinson." Both of two data sets include two clusters. The numbers of objects in two clusters are 163 and 51, respectively, in "Glass." And the numbers of objects in two clusters are 47 and 147, respectively, in "Parkinson." Under an extreme condition, if all the objects are assigned to the same cluster, we can obtain high values of Acc, RI and Fscore, but NMI is 0. Therefore, NMI is a more reliable metric for the imbalanced data sets. For data sets "Vertebral2" and "Vertebral3," our extending algorithms achieve comparable results with the original algorithms when $\beta = 8$. The best results on the two data sets are gained by using $1 < \beta < 3$ from the study of parameter β in Figs. 8c and 9a of Sect. 4.3.1. When $\beta = 1.9$, E-AWA is able to obtain 0.7739 and 0.6539 Acc on "Vertebral2" and "Vertebral3," respectively. These results are significantly better than the results produced by other algorithms.

Table 8 The results on five real data sets (the standard deviation in bracket)

Metric	Algorithm	Wine	WDBC	Vertebral2	Vertebral3	Robot
Acc	BSkmeans	0.7182(±0.01)	0.8541(±0.01)	0.6698(±0.01)	**0.6253(±0.01)**	0.3737(±0.03)
	EWkmeans	0.4244(±0.03)	0.7882(±0.05)	0.6698(±0.01)	0.5673(±0.03)	0.3701(±0.02)
	ESSC	0.6460(±0.09)	0.8119(±0.04)	0.6426(±0.08)	0.5913(±0.09)	0.3485(±0.03)
	kmeans	0.6986(±0.02)	0.8541(±0.01)	0.6698(±0.01)	0.5672(±0.03)	0.3514(±0.03)
	E-kmeans	0.8668(±0.05)	0.8777(±0.06)	**0.6906(±0.01)**	0.6055(±0.08)	**0.5197(±0.07)**
	Wkmeans[a]	0.8998(±0.01)	0.8405(±0.01)	0.6537(±0.01)	0.4859(±0.03)	0.3299(±0.01)
	E-Wkmeans[a]	0.8069(±0.07)	0.8984(±0.06)	0.6824(±0.01)	0.6093(±0.07)	0.5059(±0.07)
	AWA[a]	0.8969(±0.02)	**0.8951(±0.07)**	0.6651(±0.01)	0.4945(±0.03)	0.3449(±0.04)
	E-AWA[a]	**0.9107(±0.04)**	0.8897(±0.08)	0.6867(±0.01)	0.5985(±0.07)	0.4994(±0.08)
RI	BSkmeans	0.7270(±0.01)	0.7504(±0.01)	0.5562(±0.01)	0.6139(±0.01)	0.5189(±0.06)
	EWkmeans	0.3694(±0.03)	0.6719(±0.05)	0.5562(±0.01)	0.6731(±0.01)	0.3884(±0.08)
	ESSC	0.6445(±0.09)	0.6974(±0.04)	0.5530(±0.04)	0.6524(±0.09)	0.4413(±0.07)
	kmeans[a]	0.7178(±0.01)	0.7504(±0.01)	0.5562(±0.01)	0.6730(±0.01)	0.4492(±0.06)
	E-kmeans	0.8478(±0.03)	0.7930(±0.06)	**0.5714(±0.01)**	0.6654(±0.03)	**0.6135(±0.10)**
	Wkmeans[a]	0.8732(±0.01)	0.7315(±0.01)	0.5459(±0.01)	0.6314(±0.01)	0.4035(±0.04)
	E-Wkmeans[a]	0.8021(±0.04)	**0.8266(±0.07)**	0.5652(±0.01)	0.6605(±0.03)	0.5983(±0.10)
	AWA[a]	0.8705(±0.02)	0.8243(±0.07)	0.5532(±0.01)	0.6599(±0.02)	0.4157(±0.07)
	E-AWA[a]	**0.8893(±0.04)**	0.8171(±0.09)	0.5686(±0.01)	**0.6861(±0.04)**	0.5962(±0.13)

Fscore	BSkmeans	0.7238(±0.01)	0.8443(±0.01)	0.6731(±0.01)	**0.6751(±0.01)**	0.4709(±0.04)
	EWkmeans	0.5142(±0.01)	0.7972(±0.04)	0.6731(±0.01)	0.6561(±0.02)	0.4367(±0.03)
	ESSC	0.6573(±0.09)	0.8136(±0.04)	0.6704(±0.06)	0.6386(±0.08)	0.4092(±0.02)
	kmeans	0.7126(±0.01)	0.8443(±0.01)	0.6731(±0.01)	0.6559(±0.02)	0.4313(±0.04)
	E-kmeans	0.8653(±0.05)	0.8822(±0.05)	**0.7032(±0.01)**	0.6607(±0.05)	**0.5602(±0.06)**
	Wkmeans[a]	0.8978(±0.01)	0.8379(±0.01)	0.6623(±0.01)	0.5896(±0.02)	0.4006(±0.01)
	E-Wkmeans[a]	0.8044(±0.06)	**0.9025(±0.05)**	0.6952(±0.01)	0.6622(±0.04)	0.5627(±0.06)
	AWA[a]	0.8936(±0.02)	0.8979(±0.06)	0.6734(±0.01)	0.6103(±0.02)	0.4173(±0.03)
	E-AWA[a]	**0.9088(±0.05)**	0.8958(±0.06)	0.7001(±0.01)	0.6701(±0.05)	0.5553(±0.08)
NMI	BSkmeans	0.3972(±0.01)	0.4672(±0.01)	**0.2542(±0.01)**	0.3865(±0.01)	0.3370(±0.06)
	EWkmeans	0.0885(±0.08)	0.3383(±0.11)	0.2542(±0.01)	0.4184(±0.01)	0.1849(±0.11)
	ESSC	0.3715(±0.12)	0.3745(±0.08)	0.1506(±0.08)	0.3482(±0.10)	0.2091(±0.05)
	kmeans	0.4285(±0.01)	0.4672(±0.01)	**0.2542(±0.01)**	**0.4186(±0.01)**	0.2650(±0.06)
	E-kmeans	0.6995(±0.06)	0.4824(±0.11)	0.1286(±0.04)	0.2970(±0.02)	**0.3565(±0.10)**
	Wkmeans[a]	0.7062(±0.01)	0.3440(±0.01)	0.1320(±0.01)	0.2734(±0.02)	0.2147(±0.04)
	E-Wkmeans[a]	0.6459(±0.07)	0.5430(±0.13)	0.1177(±0.04)	0.2873(±0.02)	0.3682(±0.11)
	AWA[a]	0.7326(±0.03)	0.5362(±0.09)	0.1496(±0.01)	0.3484(±0.04)	0.2169(±0.05)
	E-AWA[a]	**0.7593(±0.06)**	**0.5440(±0.16)**	0.1222(±0.05)	0.3582(±0.08)	0.3639(±0.15)

[a]*Note:* The results in the table are produced by using $\beta=8$

Table 9 The results on four real data sets (the standard deviation in bracket)

Metric	Algorithm	Cloud	LandsatSatellite	Glass	Parkinson
Acc	BSkmeans	0.7472(±0.01)	0.5564(±0.07)	**0.8057(±0.06)**	0.7274(±0.01)
	EWkmeans	0.8381(±0.17)	0.5926(±0.03)	0.7628(±0.02)	**0.7528(±0.01)**
	ESSC	0.7511(±0.14)	0.3919(±0.07)	0.7266(±0.12)	0.6817(±0.09)
	kmeans	0.7472(±0.01)	0.5806(±0.02)	**0.8057(±0.06)**	0.7274(±0.01)
	E-kmeans	0.9648(±0.05)	0.6296(±0.03)	0.7065(±0.10)	0.6990(±0.03)
	Wkmeans[a]	0.9529(±0.12)	0.5797(±0.02)	0.6870(±0.08)	0.6615(±0.01)
	E-Wkmeans[a]	**0.9945(±0.04)**	0.6121(±0.04)	0.7578(±0.16)	0.6846(±0.04)
	AWA[a]	0.9235(±0.13)	0.5946(±0.02)	0.7037(±0.11)	0.6491(±0.01)
	E-AWA[a]	0.9907(±0.06)	**0.7132(±0.06)**	0.7323(±0.15)	0.7016(±0.03)
RI	BSkmeans	0.6220(±0.01)	0.8147(±0.02)	**0.6934(±0.08)**	0.6015(±0.01)
	EWkmeans	0.7910(±0.17)	0.8040(±0.01)	0.6372(±0.02)	**0.6259(±0.01)**
	ESSC	0.6701(±0.17)	0.5232(±0.14)	0.6298(±0.10)	0.5834(±0.07)
	kmeans	0.6220(±0.01)	0.8007(±0.01)	**0.6934(±0.08)**	0.6015(±0.01)
	E-kmeans	0.9375(±0.06)	0.8073(±0.01)	0.6032(±0.06)	0.5798(±0.03)
	Wkmeans[a]	0.9434(±0.15)	0.8009(±0.01)	0.5820(±0.08)	0.5499(±0.01)
	E-Wkmeans[a]	**0.9940(±0.05)**	0.7999(±0.01)	0.6822(±0.14)	0.5706(±0.03)
	AWA[a]	0.8958(±0.16)	0.8020(±0.01)	0.6091(±0.10)	0.5421(±0.01)
	E-AWA[a]	0.9901(±0.06)	**0.8586(±0.02)**	0.6519(±0.11)	0.5817(±0.03)
Fscore	BSkmeans	0.7299(±0.01)	0.6002(±0.06)	**0.7893(±0.07)**	0.7371(±0.01)
	EWkmeans	0.8612(±0.13)	0.6126(±0.03)	0.7524(±0.02)	**0.7449(±0.01)**
	ESSC	0.7485(±0.14)	0.4490(±0.06)	0.7486(±0.08)	0.7021(±0.07)
	kmeans	0.7299(±0.01)	0.6003(±0.02)	**0.7893(±0.07)**	0.7371(±0.01)
	E-kmeans	0.9663(±0.04)	0.6541(±0.02)	0.7275(±0.09)	0.7105(±0.02)
	Wkmeans[a]	0.9549(±0.12)	0.5994(±0.02)	0.7086(±0.06)	0.7053(±0.01)
	E-Wkmeans[a]	**0.9962(±0.03)**	0.6421(±0.03)	0.7729(±0.14)	0.6961(±0.03)
	AWA[a]	0.9213(±0.14)	0.6130(±0.02)	0.7228(±0.09)	0.6996(±0.01)
	E-AWA[a]	0.9932(±0.04)	**0.7223(±0.05)**	0.7552(±0.10)	0.7135(±0.02)
NMI	BSkmeans	0.3416(±0.01)	0.5173(±0.05)	0.1756(±0.17)	0.0698(±0.05)
	EWkmeans	0.5092(±0.29)	0.5232(±0.02)	0.0253(±0.06)	0.0022(±0.01)
	ESSC	0.3717(±0.33)	0.3256(±0.07)	0.1589(±0.12)	0.0433(±0.09)
	kmeans	0.3416(±0.01)	0.5210(±0.01)	0.1756(±0.17)	0.0698(±0.05)
	E-kmeans	0.8427(±0.13)	0.5231(±0.02)	0.2361(±0.14)	**0.1393(±0.09)**
	Wkmeans[a]	0.8913(±0.29)	0.5213(±0.02)	0.0635(±0.14)	0.0665(±0.01)
	E-Wkmeans[a]	**0.9846(±0.09)**	0.5102(±0.03)	**0.3051(±0.23)**	0.1279(±0.08)
	AWA[a]	0.7573(±0.31)	0.5190(±0.01)	0.1000(±0.13)	0.0729(±0.01)
	E-AWA[a]	0.9813(±0.13)	**0.6070(±0.04)**	0.1727(±0.10)	0.1239(±0.08)

[a]*Note*: The results in the table are produced by using $\beta = 8$

For data set "LandsatSatellite," E-AWA achieves 12 % Acc, 5 % RI, 11 % Fscore and 9 % NMI improvement compared with AWA. For data set "Wine," E-AWA and E-kmeans obtain 1 % to 3 % Acc, RI, Fscore and NMI improvement compared with

Table 10 The frequencies that the extending algorithms produce better results than the original algorithms initialized by the same centroids in 100 runs

Metric	Accuracy							NMI	
Data set	A[a]	B[a]	C[a]	D[a]	E[a]	F[a]	G[a]	H[a]	I[a]
Fre(E-kmeans, kmeans)[b]	97	94	98	56	94	99	90	**42**	69
Fre(E-Wkmeans, Wkmeans)[b]	0	94	100	94	100	**12**	74	94	64
Fre(E-AWA, AWA)[b]	71	63	94	87	93	98	95	67	64

[a]*Note*: A, B, C, D, E, F, G, H and I represent the data sets: Wine, WDBC, Vertebral2, Vertebral3, Robot, Cloud, LandsatSatellite, Glass and Parkinson, respectively
[b]*Note*: Fre(alg1, alg2) represents the number of running times that algorithm alg1 produces better results than algorithm alg2 initialized by the same 100 controids

AWA and kmeans. Moreover, from Table 8, we can see that the results produced by our extending algorithms have slightly larger standard deviations than that produced by the original algorithms. This may suggest that the algorithms which consider twofold factors, i.e. inter-cluster compactness and inter-cluster separation, may have relatively lower stability than the algorithms which consider only one factor.

Table 10 shows the number of running times that the extending algorithms produce better results than the original algorithms initialized by the same centroids in 100 runs. Due to the imbalanced property of data sets "Glass" and "Parkinson," we show the comparative results on metric NMI for the two data sets. We can see from Table 10 that the extending algorithms can produce better results than the original algorithms in most of cases if we initialize the algorithms with the same centroids. It is worth noting that Wkmeans produces more times of better results than E-Wkmeans on "Cloud." However, from the Table 9, the results produced by Wkmeans have larger standard deviations, i.e. partial results produced by Wkmeans are inferior. Moreover, the average results produced by E-Wkmeans are better than those produced by Wkmeans on "Cloud." It is also noteworthy that Wkmeans significantly outperforms E-Wkmeans on data set "Wine." That may be caused by the resulting errors when we use the distances between the centroids of the clusters and the global centroid to approximate the distances among all pairs of centroids. We give a detailed analysis in Sect. 4.3.1. In summary, E-AWA, E-Wkmeans and E-kmeans are able to obtain performance improvement by maximizing the inter-cluster separation in contrast to the original algorithms in most of cases. Table 10 suggests that our extending algorithms have high possibilities to obtain better results in comparison to the original algorithms with the same initial centroids.

4.3.3 Convergence Speed

In this section, we study the effect of the inter-cluster separation on the convergence speed of weighting kmeans-type algorithms on real-life data sets with respects to the iterations and the running time. Table 11 lists the average iterations and the running time of the algorithms initialized by the same centroids in 100 runs with $\beta = 8$ on

Table 11 The average iterations and the running time on real-life data sets

Data set	E-AWA[a]	AWA[a]	E-Wkmeans[a]	Wkmeans[a]	E-kmeans	kmeans
Wine	9.8(2.91)	6.3(1.64)	8.1(2.97)	5.3(1.63)	7.7(1.78)	7.3(1.22)
WDBC	7.5(8.14)	8.4(7.34)	7.7(13.7)	11 (15.2)	8.0(5.02)	6.2(2.75)
Vertebral2	5.4(2.62)	6.5(2.91)	5.9(2.60)	7.7(2.66)	5.7(1.62)	9.1(1.85)
Vertebral3	10 (6.08)	10 (5.61)	8.2(4.31)	9.3(3.85)	9.2(3.21)	9.2(2.46)
Robot	4.9(9.63)	3.3(6.06)	5.5(4.63)	3.6(2.69)	4.9(1.88)	4.4(0.97)
Cloud	6.3(18.1)	6.8(16.7)	7.1(29.3)	12 (38.0)	8.1(14.1)	10 (14.1)
LandsatSatellite	28.8(499)	38.7(521)	16.6(381)	40.4(699)	17.7(245)	59 (605)
Glass	5.3(1.38)	4.8(1.03)	5.9(1.98)	5.5(1.51)	8.7(1.70)	5.3(0.78)
Parkinson	3.9(1.26)	7.7(1.81)	3.7(1.67)	11 (3.46)	2.9(0.74)	4.0(0.60)

[a]*Note*: The results in the table are produced by using $\beta = 8$
[b]*Note*: The values in brackets are the running time in seconds produced by running the algorithms with the same centroids in 100 runs

the real data sets. From this table, we can observe that the iteration of E-Wkmeans is slightly less than that of Wkmeans in most of the data sets. We can also see from the table that the extending algorithms spend slightly more running time as opposed to the original algorithms under the condition of similar or less iterations on some data sets. This is caused by the extra computational cost that the extending algorithms must spend to calculate the distances between the centroids of the clusters and the global centroid at every iteration. For "LandsatSatellite," the iterations of E-AWA, E-Wkmeans and E-kmeans reduce 26 %, 60 %, and 70 % in comparison to those of AWA, Wkmeans and kmeans, respectively. Correspondingly, the running time of E-AWA, E-Wkmeans, E-kmeans reduces 4.2 %, 45 % and 59 % as opposed to those of AWA, Wkmeans and kmeans on "LandsatSatellite," respectively.

5 Discussion

From the results in Sect. 4, E-AWA, E-Wkmeans and E-kmeans outperform AWA, Wkmeans and kmeans, respectively, in terms of various evaluation measures: Acc, RI, Fscore and NMI in most of cases. Therefore, AWA performs better than kmeans and Wkmeans, and E-AWA performs better than E-kmeans and E-Wkmeans in most of data sets. That suggests the information of the inter-cluster separation can help to improve the clustering results by maximizing the distances among the clusters. E-AWA performs the best in all the compared algorithms. Due to the parameter problem, ESSC does not perform well in comparison to our proposed algorithms in most of cases.

According to the comparison of the weights of the features, the extending algorithms can reduce the weights of the features, the centroids of which are very close to each other, and increase the weight of the features, the centroids of which are far away from each other. Therefore, our extending algorithms can effectively improve the performance of feature weighting such that they perform well for clustering.

In contrast to ESSC, the extending algorithms utilize only one parameter β as used in Wkmeans and AWA. Thus, our algorithms are more applicable for complex data sets in practice. We can observe from the experiments that the performances of E-AWA and E-Wkmeans are more smooth than those of AWA and Wkmeans, respectively, with various values of β, especially, when $1 < \beta < 3$. Therefore, our proposed algorithms are more robust than the original algorithms in overall.

The extending algorithms have the same computational complexities compared with basic kmeans algorithms. Since clustering using kmeans-type algorithm is an iterative process, the computational time also depends on the total number of iterations. From the empirical study, we can observe that the total number of iterations of the extending algorithms is similar to or fewer than that of the original algorithms. However, the extending algorithms must spend extra time to calculate the distances between the centroids of the clusters and the global centroid at each iteration. The extending algorithms may spend slightly more time in comparison to the original algorithms on some data sets.

From the experiments and analysis, we can find that our proposed algorithms have a limitation: when the centroid of certain cluster is very close to the global centroid and the centroids among the clusters are not close to each other, maximizing the distances between the centroids of clusters in place of maximizing the distances among the clusters may produce errors and the performances of our extending algorithms may decrease to some extent. However, since our extending algorithms consider both the intra-cluster compactness and the inter-cluster separation, our extending algorithms are able to obtain better results in comparison to the original algorithms in most of cases. It is also worth pointing out that the current extending algorithms are difficult to apply for categorical data sets. This is caused by the possibility that the denominators of the objective functions become zeros, i.e. division-by-zero problem.

6 Conclusion and Future Work

In this chapter, we have presented three extensions of kmeans-type algorithms by integrating both intra-cluster compactness and inter-cluster separation. This work involves the following aspects: (1) three new objective functions are proposed based on basic kmeans, Wkmeans and AWA; (2) the corresponding updating rules are derived and the convergence is proved in theory and (3) extensive experiments are carried out to evaluate the performances of E-kmeans, E-Wkmeans and E-AWA algorithms based on four evaluation metrics: Acc, RI, Fscore and NMI. The results demonstrate that the extending algorithms are more effective than the existing algorithms. In particular, E-AWA delivers the best performance in comparison to other algorithms in most of cases.

In the future work, we plan to further extend our algorithms to categorical data sets by developing new objective functions to overcome the division-by-zero problem. It will be of great importance in applying our algorithms to more real data sets. We also plan to investigate the potential of our proposed algorithms for other applications, such as gene data clustering, image clustering, and community discovery, etc.

Acknowledgements The authors are very grateful to the editors and anonymous referees for their helpful comments. This research was supported in part the National Natural Science Foundation of China (NSFC) under Grant No.61562027 and Social Science Planning Project of Jiangxi Province under Grant No.15XW12, in part by Shenzhen Strategic Emerging Industries Program under Grants No. ZDSY20120613125016389, Shenzhen Science and Technology Program under Grant No. JCYJ20140417172417128 and No. JSGG20141017150830428, National Commonweal Technology R&D Program of AQSIQ China under Grant No.201310087, National Key Technology R&D Program of MOST China under Grant No. 2014BAL05B06.

References

1. Aggarwal, C., Wolf, J., Yu, P., Procopiuc, C., Park, J.: Fast algorithms for projected clustering. ACM SIGMOD Rec. **28**(2), 61–72 (1999)
2. Ahmad, A., Dey, L.: A k-means type clustering algorithm for subspace clustering of mixed numeric and categorical datasets. Pattern Recogn. Lett. **32**(7), 1062–1069 (2011)
3. Al-Razgan, M., Domeniconi, C.: Weighted clustering ensembles. In: Proceedings of SIAM International Conference on Data Mining, pp. 258–269 (2006)
4. Anderberg, M.: Cluster Analysis for Applications. Academic, New York (1973)
5. Arthur, D., Vassilvitskii, S.: k-means++: The advantages of careful seeding. In: Proceedings of the 18th Annual ACM-SIAM Symposium on Discrete Algorithms, pp. 1027–1035. Society for Industrial and Applied Mathematics, Philadelphia (2007)
6. Bezdek, J.: A convergence theorem for the fuzzy isodata clustering algorithms. IEEE Trans. Pattern Anal. Mach. Intell. **2**(1), 1–8 (1980)
7. Bradley, P., Fayyad, U., Reina, C.: Scaling clustering algorithms to large databases. In: Proceedings of the 4th International Conference on Knowledge Discovery & Data Mining, pp. 9–15 (1998)
8. Cai, D., He, X., Han, J.: Semi-supervised discriminant analysis. In: Proceedings of IEEE 11th International Conference on Computer Vision, pp. 1–7 (2007)
9. Celebi, M.E.: Partitional Clustering Algorithms. Springer, Berlin (2014)
10. Celebi, M.E., Kingravi, H.A.: Deterministic initialization of the k-means algorithm using hierarchical clustering. Int. J. Pattern Recogn. Artif. Intell. **26**(7), 870–878 (2013)
11. Celebi, M.E., Kingravi, H.A., Vela, P.A.: A comparative study of efficient initialization methods for the k-means clustering algorithm. Expert Syst. Appl. **40**(1), 200–210 (2013)
12. Chan, E., Ching, W., Ng, M., Huang, J.: An optimization algorithm for clustering using weighted dissimilarity measures. Pattern Recogn. **37**(5), 943–952 (2004)
13. Chen, X., Ye, Y., Xu, X., Zhexue Huang, J.: A feature group weighting method for subspace clustering of high-dimensional data. Pattern Recogn. **45**(1), 434–446 (2012)
14. Chen, X., Xu, X., Huang, J., Ye, Y.: Tw-k-means: automated two-level variable weighting clustering algorithm for multi-view data. IEEE Trans. Knowl. Data Eng. **24**(4), 932–944 (2013)
15. Das, S., Abraham, A., Konar, A.: Automatic clustering using an improved differential evolution algorithm. IEEE Trans. Syst. Man Cybern. Part A Syst. Hum. **38**(1), 218–237 (2008)

16. De Sarbo, W., Carroll, J., Clark, L., Green, P.: Synthesized clustering: a method for amalgamating alternative clustering bases with differential weighting of variables. Psychometrika **49**(1), 57–78 (1984)
17. De Soete, G.: Optimal variable weighting for ultrametric and additive tree clustering. Qual. Quant. **20**(2), 169–180 (1986)
18. De Soete, G.: Ovwtre: a program for optimal variable weighting for ultrametric and additive tree fitting. J. Classif. **5**(1), 101–104 (1988)
19. Deng, Z., Choi, K., Chung, F., Wang, S.: Enhanced soft subspace clustering integrating within-cluster and between-cluster information. Pattern Recogn. **43**(3), 767–781 (2010)
20. Dhillon, I., Modha, D.: Concept decompositions for large sparse text data using clustering. Machine learning **42**(1), 143–175 (2001)
21. Domeniconi, C.: Locally adaptive techniques for pattern classification. Ph.D. thesis, University of California, Riverside (2002)
22. Domeniconi, C., Papadopoulos, D., Gunopulos, D., Ma, S.: Subspace clustering of high dimensional data. In: Proceedings of the SIAM International Conference on Data Mining, pp. 517–521 (2004)
23. Domeniconi, C., Gunopulos, D., Ma, S., Yan, B., Al-Razgan, M., Papadopoulos, D.: Locally adaptive metrics for clustering high dimensional data. Data Min. Knowl. Disc. **14**(1), 63–97 (2007)
24. Duda, R.O., Hart, P.E., Stork, D.G.: Pattern Classification. Wiley-Interscience, New York (2012)
25. Dzogang, F., Marsala, C., Lesot, M.J., Rifqi, M.: An ellipsoidal k-means for document clustering. In: Proceedings of the 12th IEEE International Conference on Data Mining, pp. 221–230 (2012)
26. Fisher, R.: The use of multiple measurements in taxonomic problems. Ann. Hum. Genet. **7**(2), 179–188 (1936)
27. Friedman, J., Meulman, J.: Clustering objects on subsets of attributes. J. R. Stat. Soc. Ser. B **66**(4), 815–849 (2004)
28. Frigui, H., Nasraoui, O.: Simultaneous clustering and dynamic keyword weighting for text documents. In: Survey of Text Mining, Springer New York, pp. 45–70 (2004)
29. Han, J., Kamber, M., Pei, J.: Data Mining: Concepts and Techniques. Morgan Kaufmann, Los Altos (2011)
30. Huang, J., Ng, M., Rong, H., Li, Z.: Automated variable weighting in k-means type clustering. IEEE Trans. Pattern Anal. Mach. Intell. **27**(5), 657–668 (2005)
31. Jain, A.: Data clustering: 50 years beyond k-means. Pattern Recogn. Lett. **31**(8), 651–666 (2010)
32. Jing, L., Ng, M., Xu, J., Huang, J.: Subspace clustering of text documents with feature weighting k-means algorithm. In: Advances in Knowledge Discovery and Data Mining, Springer Berlin Heidelberg, pp. 802–812 (2005)
33. Jing, L., Ng, M., Huang, J.: An entropy weighting k-means algorithm for subspace clustering of high-dimensional sparse data. IEEE Trans. Knowl. Data Eng. **19**(8), 1026–1041 (2007)
34. Makarenkov, V., Legendre, P.: Optimal variable weighting for ultrametric and additive trees and k-means partitioning: methods and software. J. Classif. **18**(2), 245–271 (2001)
35. Modha, D., Spangler, W.: Feature weighting in k-means clustering. Mach. Learn. **52**(3), 217–237 (2003)
36. Parsons, L., Haque, E., Liu, H.: Subspace clustering for high dimensional data: a review. ACM SIGKDD Explor. Newsl. **6**(1), 90–105 (2004)
37. Pelleg, D., Moore, A.: X-means: extending k-means with efficient estimation of the number of clusters. In: Proceedings of the 17th International Conference on Machine Learning, San Francisco, pp. 727–734 (2000)
38. Sardana, M., Agrawal, R.: A comparative study of clustering methods for relevant gene selection in microarray data. In: Advances in Computer Science, Engineering & Applications, Springer Berlin Heidelberg, pp. 789–797 (2012)

39. Selim, S., Ismail, M.: K-means-type algorithms: a generalized convergence theorem and characterization of local optimality. IEEE Trans. Pattern Anal. Mach. Intell. **6**(1), 81–87 (1984)
40. Shamir, O., Tishby, N.: Stability and model selection in k-means clustering. Mach. Learn. **80**(2), 213–243 (2010)
41. Steinbach, M., Karypis, G., Kumar, V., et al.: A comparison of document clustering techniques. In: KDD Workshop on Text Mining, vol. 400, pp. 525–526 (2000)
42. Tang, L., Liu, H., Zhang, J.: Identifying evolving groups in dynamic multi-mode networks. IEEE Trans. Knowl. Data Eng. **24**(1), 72–85 (2012)
43. Wu, K., Yu, J., Yang, M.: A novel fuzzy clustering algorithm based on a fuzzy scatter matrix with optimality tests. Pattern Recogn. Lett. **26**(5), 639–652 (2005)
44. Xu, R., Wunsch, D., et al.: Survey of clustering algorithms. IEEE Trans. Neural Netw. **16**(3), 645–678 (2005)
45. Yang, M., Wu, K., Yu, J.: A novel fuzzy clustering algorithm. In: Proceedings of the 2003 IEEE International Symposium on Computational Intelligence in Robotics and Automation, vol. 2, pp. 647–652 (2003)

A Fuzzy-Soft Competitive Learning Approach for Grayscale Image Compression

Dimitrios M. Tsolakis and George E. Tsekouras

Abstract In this chapter we develop a fuzzy-set-based vector quantization algorithm for the efficient compression of grayscale still images. In general, vector quantization can be carried out by using crisp-based and fuzzy-based methods. The motivation of the current work is to provide a systematic framework upon which the above two general methodologies can effectively cooperate. The proposed algorithm accomplishes this task through the utilization of two main steps. First, it introduces a specially designed fuzzy neighborhood function to quantify the lateral neuron interaction phenomenon and the degree of the neuron excitation of the standard self-organizing map. Second, it involves a codeword migration strategy, according to which codewords that correspond to small and underutilized clusters are moved to areas that appear high probability to contain large number of training vectors. The proposed methodology is rigorously compared to other relative approaches that exist in the literature. An interesting outcome of the simulation study is that although the proposed algorithm constitutes a fuzzy-based learning mechanism, it finally obtains computational costs that are comparable to crisp-based vector quantization schemes, an issue that can hardly be maintained by the standard fuzzy vector quantizers.

Keywords Learning vector quantization • Soft learning vector quantization • Self-organizing map • Neighborhood function • Codeword migration

1 Introduction

Image compression [1–7] is one of the most widely used and commercially successful technologies in the field of digital image processing. Web page images and high-resolution digital camera photos also are compressed routinely to save storage space and reduce transmission time. The methods developed so far to perform image compression can be classified in two categories: (a) lossless compression and

D.M. Tsolakis (✉) • G.E. Tsekouras
Laboratory of Intelligent Multimedia, Department of Cultural Technology and Communication,
University of the Aegean, 81100 Mytelene, Lesvos Island, Greece
e-mail: ctmb05018@ct.aegean.gr; gtsek@ct.aegean.gr

© Springer International Publishing Switzerland 2016   385
M.E. Celebi, K. Aydin (eds.), *Unsupervised Learning Algorithms*,
DOI 10.1007/978-3-319-24211-8_14

(b) lossy compression. Lossless compression is an error-free procedure according to which the pixel intensities of the original image are fully recovered in the compressed image representation. Although the quality of the resulting image is excellent, this approach is computationally complex and yields low compression rates. Contrary, lossy compression attempts to compromise the quality of the recovered image in exchange for higher compression rates. Thus, although the original image cannot be perfectly recovered, the computational demands are significantly reduced.

A common procedure to perform image compression is vector quantization (VQ) [4, 5, 8–14]. Vector quantization (VQ) refers to the process according to which a set of data vectors is partitioned into a number of disjoint clusters. The center elements of these clusters are referred to as the codewords, and the aggregate of all codewords as the codebook. The image is reconstructed by replacing each training vector by its closest codeword.

In general, vector quantization methods can be classified into two categories, namely crisp and fuzzy. Crisp vector quantization is usually performed in terms of the c-means algorithm and its variants [1, 5, 9, 15, 16], the self-organizing map (SOM) [17–21], and the learning vector quantization (LVQ) [9, 12, 17, 18]. On the other hand, the backbone of fuzzy techniques in designing vector quantizers is the fuzzy c-means algorithm and its variants [22–31]. The main difference between these two categories is that the former performs strong competition between the neurons by means of the winner-takes-all principle, while the latter embeds into the learning process soft conditions. The main advantages and disadvantages of the resulting behaviors can be identified as follows: A crisp-based algorithm obtains a fast learning process, yet it is very sensitive to the initialization [36, 37], and therefore it might not yield an efficient partition of the feature space. On the contrary, a fuzzy-based algorithm is less sensitive to the initialization generating effective partitions of the data set, yet it constitutes a slow process with high computational cost.

This chapter describes an algorithm that attempts to combine the merits of the above two general methodologies into a unique framework. The key idea is to alleviate the strong competitive nature of a crisp-based vector quantizer by defining a fuzzy-based novel neighborhood function able to measure the lateral neuron interaction phenomenon and the degree of the neuron excitations, as well. As a later step, a specialized codeword relocation procedure is developed in order to increase the individual contributions of the resulting codewords to the final partition.

This chapter is synthesized as follows. Section 2 reports two general methodologies to perform VQ. Section 3 presents the proposed methodology. The simulation experiments are presented in Sect. 4. Finally, the chapter's conclusions are discussed in Sect. 5.

2 Related Work

Vector quantization (VQ) is concerned with the representation of a set of training vectors by a set of codewords. The feature space is considered the p-dimensional Euclidean space. The set of training vectors is denoted as $X = \{x_1, x_2, \ldots, x_n\}$ with $x_i \in \Re^p$, while the set of codewords as $V = \{v_1, v_2, \ldots, v_c\}$ with $v_i \in \Re^p$. The corresponding set cardinalities (i.e. set sizes) are $card(X) = n$ and $card(V) = c$. Using the above nomenclature the VQ quantization is defined as a function $f : X \rightarrow V$, which yields a partition of the set X into a number of c disjoint subsets (commonly called clusters):

$$C_i = \{x_k \in X : f(x_k) = v_i, \quad 1 \leq k \leq n\} \tag{1}$$

with $i \in \{1, 2, \ldots, c\}$. The equation $f(x_k) = v_i$ in relation (1) is realized in terms of the following nearest neighbor condition:

$$\|x_k - v_i\| = \min_{1 \leq j \leq c} \{\|x_k - v_j\|\} \tag{2}$$

Note that any partition of the set X will also yield a partition of the feature space. Within the next subsections we shall describe two different approaches to VQ.

2.1 The Batch Learning Vector Quantization

The SOM constitutes a class of vector quantization algorithms based on unsupervised competitive learning. The SOM produces a similarity graph of the input data and converts the nonlinear statistical relationships between high-dimensional data into simple geometric relationships in a low-dimensional display grid, usually a regular two-dimensional grid of nodes. Therefore it can be viewed as a similarity graph and a clustering diagram, as well [17, 19, 21].

In this sense its computation procedure relies on nonparametric, recursive regression process [18]. Figure 1 shows a 16×16 neuron display grid. Each node in the grid is associated with a specific neuron weight. The neuron weights coincide with the codewords. Therefore, from now on these two concepts will be used to mean the same thing. Given the training set $X = \{x_k : 1 \leq k \leq n\}$, and providing initial values for the set of the codewords $V = \{v_i : 1 \leq i \leq c\}$, the SOM is implemented as follows. In the iteration t, for the training vector x_k we locate the nearest codeword according to the relation:

$$\|x_k - v_{i(x_k)}(t-1)\| = \min_{1 \leq j \leq c} \{\|x_k - v_j(t-1)\|\} \tag{3}$$

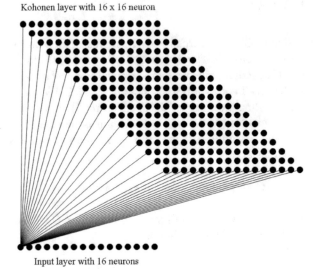

Kohonen layer with 16 x 16 neuron

Input layer with 16 neurons

Fig. 1 SOM neural network with 16 inputs

Then, the updating rule for the codewords is given in the next formula:

$$v_j(t) = v_j(t-1) + \eta(t)\, h_{i(x_k),j}\,\left(x_k - v_j(t-1)\right) \tag{4}$$

where $j = 1, 2, \ldots, c$, $\eta(t)$ is the learning parameter which is required to decrease with the number of iterations, and $h_{i(x),j}$ is the neighborhood function. The $h_{i(x),j}$ quantifies the lateral neural interaction phenomenon and the degree of excitation of the neurons. Note that the index $i(x_k) \in \{1, 2, \ldots, c\}$ denotes the codeword $v_{i(x_k)}$ that is closer to x_k, while the index $j = 1, 2, \ldots, c$ corresponds to anyone of the codewords (including $i(x_k)$). The typical form of the neighborhood function is calculated considering the corresponding nodes in the grid and reads as [17, 18, 31]:

$$h_{i(x_k),j} = \exp\left(-\left(\frac{\|r_{i(x_k)} - r_j\|}{\sqrt{2}\,\sigma(t)}\right)^2\right) \tag{5}$$

where $r_{i(x)} \in \Re^2$ and $r_j \in \Re^2$ are vectorial locations on the grid that correspond to the codewords $v_{i(x_k)}$ and v_j, and $\sigma(t)$ stands for the width, which is also required to reduce with time. A simpler way to define the neighborhood function is:

$$h_{i(x_k)j} = \begin{cases} 1, & \text{if } j = i(x_k) \\ 0, & \text{otherwise} \end{cases} \tag{6}$$

Equation (6) leads to the direct definition of the LVQ. Note that in the standard SOM the use of Eq. (5) forces all the codewords close to the winning one to move, while the LVQ updates only the position of the winning node. As a result, the very nature of the LVQ is dominated by the winner-takes-all principle. Although the LVQ is sensitive to random initialization it constitutes quite a fast procedure. The application of LVQ is accomplished in a sequential manner, meaning that the final result is an on-line procedure. In [18] Kohonen modified the on-line implementation of the LVQ and introduced a batch iterative procedure called batch LVQ, where in each iteration all the training vectors were taken into account. The basic assumption to carry out this modification relies on the observation that, for a specific training vector x_k, the on-line LVQ will converge to a stationary point, say $v_i^\oplus$. Therefore, the expectation values of $v_j(t-1)$ and $v_j(t)$ must be equal to $v_j^\oplus$ as $t \to \infty$, which implies that

$$E\left[h_{i(x_k), j}\left(x_k - v_j(t)\right)\right] = 0 \quad \text{as } t \to \infty \tag{7}$$

Applying an empirical distribution the above equation yields [18]:

$$v_j = \frac{\displaystyle\sum_{k=1}^{n} h_{i(x_k), j}\, x_k}{\displaystyle\sum_{k=1}^{n} h_{i(x_k), j}} \tag{8}$$

To this end, the batch LVQ algorithm is given by the next steps.

The Batch LVQ Algorithm

Step 1. Choose the number of neurons c, the maximum number of iterations $t_{\max}$, a small value for the parameter $\varepsilon \in (0, 1)$, and randomly initialize the matrix $V = \{[v_j] : 1 \le j \le c\}$. Set $t = 0$.

Step 2. For each x_k $(1 \le k \le n)$ calculate the nearest codeword $v_{i(x_k)}$ using the condition in Eq. (3) and calculate the quantities $h_{i(x), j}$.

Step 3. Using Eq. (8) update the codewords.

Step 4. If $\displaystyle\sum_{j=1}^{c} \|v_j(t) - v_j(t-1)\| < \varepsilon$ then stop. Else set $t = t + 1$ and go to step 2.

If we choose Eq. (5) to calculate the quantities $h_{i(x), j}$ at the step 2, then the whole approach yields a batch version of the standard SOM.

2.2 The Fuzzy Learning Vector Quantization Algorithm

The fuzzy learning vector quantization (FLVQ) algorithm was introduced in [32] as an alternative to the standard fuzzy c-means model. In later works [22, 33] certain properties of FLVQ were studied. The task of FLVQ is to design an optimal VQ scheme by minimizing the subsequent objective function:

$$J(U, V) = \sum_{k=1}^{n} \sum_{i=1}^{c} (u_{ik})^m \|x_k - v_i\|^2 \tag{9}$$

where $U = \{[u_{ik}] : 1 \leq i \leq c, 1 \leq k \leq n\}$ is the partition matrix, $V = \{[v_i] : 1 \leq i \leq c\}$ the cluster centers' matrix, and $m \in (1, \infty)$ the fuzziness parameter. The above minimization task is subjected to the constraint:

$$\sum_{i=1}^{c} u_{ik} = 1 \quad \forall k \tag{10}$$

To this end, the algorithm is summarized in the following steps.

The FLVQ Algorithm

Choose the number of clusters c, a small value for the parameter $\varepsilon \in (0, 1)$, and randomly initialize the matrix $V = \{[v_i] : 1 \leq i \leq c\}$. Also set the maximum number of iterations t_{max}, and the initial (m_0) and final (m_f) value of the fuzziness parameter.

For $t = 0, 1, 2, \ldots, t_{max}$

(a) Calculate the fuzziness parameter as:

$$m(t) = m_0 - \frac{t(m_0 - m_f)}{t_{max}} \tag{11}$$

(b) Set:

$$\eta_{ik}(t) = \left[\sum_{j=1}^{c} \left(\frac{\|x_k - v_i\|}{\|x_k - v_j\|} \right)^{\frac{2}{m(t)-1}} \right]^{-m(t)} \tag{12}$$

(c) Update the cluster centers as:

$$v_i(t) = \frac{\sum_{k=1}^{n} \eta_{ik}(t) x_k}{\sum_{k=1}^{n} \eta_{ik}(t)} \tag{13}$$

(d) If error $(t) = \sum_{i=1}^{c} \|v_i(t) - v_i(t-1)\| < \varepsilon$ then stop. Else continue.

The FLVQ consists a soft learning scheme that maintains the transition from fuzzy to crisp conditions by gradually reducing the fuzziness parameter $m(t)$ from the large value m_0 (fuzzy conditions) to the value m_f which is chosen close to unity (nearly crisp conditions). Note that in the initial stages of the learning process, all of the codewords are assigned training patterns due to the existence of fuzzy conditions. Therefore, all codewords are forced to move and to become more competitive with each another. This fact makes the FLVQ less sensitive to random initialization when compared to the batch SOM and the batch LVQ. However, the FLVQ is related to two difficulties. The first one concerns the selection of appropriate values for parameters m_0 and m_f, since different values will obtain quite different partitions of the feature space. Secondly, it requires high computational efforts due to the number of distance calculations involved in Eq. (12).

3   The Proposed Vector Quantization Approach

The motivation of the proposed vector quantization algorithm is twofold. First to alleviate the strong dependence on initialization of crisp vector quantization. Second to produce a fast fuzzy learning mechanism. To do so we propose the two-level iteration depicted in Fig. 2.

The first level concerns the implementation of a novel competitive learning scheme. This level is put into practice by developing a specialized fuzzy-set-

Fig. 2 The proposed fuzzy-soft learning vector quantizer

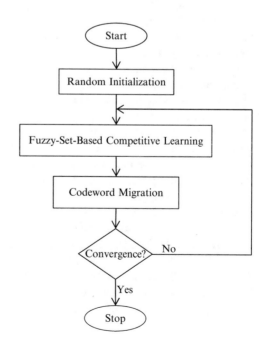

based neighborhood function able to effectively control the competition between codewords in such a way that every codeword is enabled to win as many training vectors as possible. The second level attempts to improve the partition generated by the previous level by detecting codewords that correspond to underutilized (i.e. small) clusters and relocating them to the neighborhood of large clusters. Given the nomenclature so far, the objective of the proposed vector quantizer is to minimize the distortion function:

$$D(X, V) = \frac{1}{n} \sum_{k=1}^{n} \min_{1 \le i \le c} \left\{ \|x_k - v_i\|^2 \right\} \tag{14}$$

3.1   Fuzzy-Set-Based Competitive Learning

The problem discussed in this section concerns the way the codewords must be updated so that the final learning strategy will be fast and less sensitive to initialization. Before proceeding with the mathematical formulation of our approach, let us identify how the methods described previously cope with this problem. The SOM utilizes Eq. (5) to decide which codewords should be updated, given the winner node for a specific training vector. Roughly speaking, Eq. (5) transfigures the information coded in the grid into meaningful information in the feature space. Its nature is stochastic in the sense that the existence of the Gaussian function mostly favors those codewords that correspond to nodes that are close to the winning node in the grid. However, it is highly possible for nodes that correspond to large clusters to be in the neighborhood of many winning nodes and therefore they would be continuously updated; leaving out of competition the codewords that correspond to small clusters, rendering them underutilized. Therefore, the resulting partition might not reveal the underlying structure of the data and eventually might not be efficient. This phenomenon can partially be resolved by providing the learning process with good initialization as far as the codewords are concerned. As stated in [18] such a case would be the selection of the codeword initial values from the training data set itself. Yet, even in this case, the danger of bad initial conditions is still high. The LVQ and the batch LVQ employ the condition in Eq. (6) to come up with a codeword updating decision strategy. This condition imposes an aggressive competition between the codewords. Thus, although the resulting learning is quite fast, the quality of the obtained partition is inferior. On the contrary, the FLVQ adopts Eq. (13) to update the codewords. The implementation of this equation is a direct result of the learning rate appearing in Eq. (12). Actually, this learning rate plays the role of the neighborhood function in accordance with Eq. (5) used by the SOM. As mentioned previously, we can easily observe that Eq. (12) will, indeed, force a large number of codewords to become more competitive [22, 30, 32, 33]. It thus appears that the fuzzy conditions encapsulated in this equation initiate the generation of better partitions than the SOM and the LVQ as far as the discovery of the real data structure is concerned. However, Eq. (12) constitutes a computationally complex approach.

In our opinion, the codeword updating process should take into consideration their relative locations in the feature space. To justify this claim we concentrate on the following analysis. The codeword locations are directly affected by the distribution of the clusters across this space. With its turn, the cluster distribution across the feature space should possess two main properties [22, 33, 34]: (a) clusters should be as compact as possible, and (b) clusters should be as separated as possible with each another. The former states that data belonging to the same cluster should be as similar as possible, while the latter delineates the dissimilarity between data belonging to different clusters. Both of these properties are affected by the relative locations of the codewords (i.e. cluster centers). Therefore, the first requirement the neighborhood function must fulfill is the involvement of the Euclidean distances between pairs of codewords. In a similar fashion to the SOM and the FLVQ, the second requirement states that the codewords located closer to the winning neuron must be updated with larger rates than the ones located far away. The solution to this problem comes from the area of the optimal fuzzy clustering. Specifically, the two properties mentioned previously are the basic tools to perform optimal fuzzy cluster analysis. The neighborhood function introduced here is inspired by the separation index between fuzzy clusters developed in [34] and is mathematically described next.

Assume that we have a partition of the data set $X = \{x_1, x_2, \ldots, x_n\}$ into c clusters with codewords $V = \{v_1, v_2, \ldots, v_c\}$. Define in sequence the arithmetic mean of the codewords:

$$\bar{v} = \frac{1}{c} \sum_{i=1}^{c} v_i \tag{15}$$

and the $p \times (c + 1)$ matrix:

$$Z = \begin{bmatrix} z_1 \, z_2 \cdots z_c \, z_{c+1} \end{bmatrix} = \begin{bmatrix} v_1 \, v_2 \cdots v_c \, \bar{v} \end{bmatrix} \tag{16}$$

Note that for $1 \leq i \leq c$ it holds that $z_i = v_i$ while $z_{c+1} = \bar{v}$. The need for the appearance of $\bar{v}$ in Eq. (16) will be shortly explained. Then, we view each codeword as the center of a fuzzy set, the elements of which are the rest of the codewords. Considering the codeword z_i ($1 \leq i \leq c$) the membership function of the associated fuzzy set is define as:

$$\mu_{ij} = \frac{1}{\sum_{\substack{l=1 \\ l \neq j}}^{c+1} \left(\frac{\|z_j - z_i\|}{\|z_j - z_l\|} \right)^{\frac{2}{m-1}}} \quad (1 \leq i \leq c; 1 \leq j \leq c) \tag{17}$$

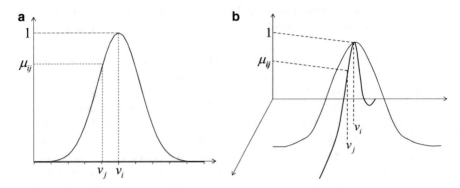

Fig. 3 The structure of the membership function μ_{ij}: (**a**) one-dimensional and (**b**) two-dimensional feature space

where $m \in (1, \infty)$ is the fuzziness parameter. Note that the indices i and j are not assigned the value $c + 1$, which corresponds to the mean $\overline{v}$. Figure 3 illustrates what the function μ_{ij} looks like in the one-dimensional and two-dimensional Euclidean space.

Now, the need for using $\overline{v}$ is obvious because if it's not going to be taken into account, when there are only two codewords (i.e. $c = 2$) Eq. (17) will give one in all cases and thus, it does not work. In addition, $\overline{v}$ provides useful information for the distribution of the codewords and its presence is necessary. For each x_k we detect the nearest codeword $v_{\ell(x_k)}$ according to the condition:

$$\left\| x_k - v_{\ell(x_k)} \right\| = \min_{1 \leq i \leq c} \left\{ \left\| x_k - v_i \right\| \right\} \tag{18}$$

To this end, the neighborhood function is defined as:

$$h_{\ell(x_k), i} = \begin{cases} 1, & \text{if } i = \ell(x_k) \\ \mu_{\ell(x_k), i}, & \text{otherwise} \end{cases} \tag{19}$$

where $\ell(x_k) \in \{1, 2, \ldots, c\}$, $1 \leq i \leq c$, and $\mu_{\ell(x_k), i}$ is calculated in (17) by changing the index i with the $\ell(x_k)$ and the index j with i.

The main properties of the function $h_{\ell(x_k), i}$ can be enumerated as follows. First, it provides the excitation degree of each codeword (with respect to the winning one) by considering the locations of the rest of codewords. Note that the decision how much each codeword is going to move is not affected only by distances between pairs but also by the relative locations of the codewords. That is, before updating a codeword, the algorithm "looks" at the overall current partition and then makes the move. Second, although the $\mu_{\ell(x_k), i}$ is a fuzzy membership degree, its calculation is fast because it includes only the codewords and not the training vectors. Specifically, the number of distance calculations performed by the FLVQ is dominated by the learning rate's estimation in Eq. (12), which calculates $(c + 1)$ c n distances per iteration. Since $(c + 1) \, c \, n \leq n(c + 1)^2$ the computational complexity of the FLVQ is $O(nc^2)$. In our case the dominant effect is provided by Eq. (17), where we can easily see that the number of distance calculations is $(c + 1)^2 \, c \leq (c + 1)^3$ meaning that the computational complexity of the proposed algorithm is $O(c^3)$. Since $c \ll n \Rightarrow c^3 \ll nc^2$, we conclude that the proposed vector quantizer is much faster that the FLVQ, something that is supported by the simulations reported later on in this chapter.

Having introduced the neighborhood function the learning rule to update the codewords is given as:

$$v_i(t) = v_i(t - 1) + \eta(t) \, h_{\ell(x_k), i} \, (x_k - v_i(t - 1)) \tag{20}$$

The above rule is an on-line LVQ scheme. As we are interested to perform a codeword migration process in the second step of the algorithm we intend to modify the above mechanism in order to produce a batch vector quantization process. In a similar way to the standard SOM model, we employ the next condition:

$$E\left[h_{\ell(x_k), i} \, (x_k - v_i(t))\right] = 0 \quad \text{as } t \to \infty \tag{21}$$

Thus, the updating rule is modified as follows:

$$v_i(t) = \frac{\sum_{k=1}^{n} h_{i(x_k), i}^{(t-1)} x_k}{\sum_{k=1}^{n} h_{i(x_k), i}^{(t-1)}} \tag{22}$$

where t stands for iteration number.

3.2   Codeword Migration Process

Although the fuzzy conditions of the above competitive learning are able to effectively deal with a bad initialization, it is highly probable the resulting partition to include small and underutilized clusters. In this step of the algorithm we intend to improve the partition by incorporating into the learning process a codeword migration mechanism. As we want all clusters to contribute as much as possible, we expect that this process will yield an optimal vector quantizer [35]. The key idea is to detect small and underutilized clusters and relocate the respective codeword to an area close to a large cluster. To perform this task, we employ the distortion utility measure developed in [16],

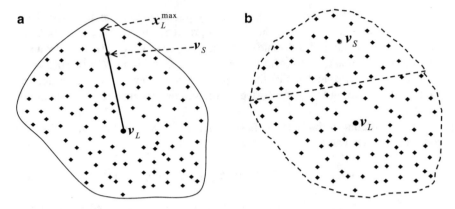

Fig. 4 (**a**) Relocation of the codeword of a small cluster to the area of a large one. (**b**) Division of the large cluster into two areas

$$\Lambda_i = \frac{D_i}{(1/c) \sum_{j=1}^{c} D_j} \tag{23}$$

where D_i is the individual distortion of the ith cluster,

$$D_i = \sum_{k=1}^{n} \psi_{ik} \|x_k - v_i\|^2 \tag{24}$$

The quantity ψ_{ik} in Eq. (24) is calculated as indicated next:

$$\psi_{ik} = \begin{cases} 1, \text{ if } \|x_k - v_i\| = \min_{1 \le j \le c} \{\|x_k - v_j\|\} \\ 0, \text{ otherwise} \end{cases} \tag{25}$$

The criterion to select a small cluster is:

$$\Lambda_i \le \gamma \tag{26}$$

with $\gamma \in (0, 1)$. Intuitively, in order to detect a big cluster (Fig. 4),

$$\Lambda_i > 1 \tag{27}$$

In each iteration we migrate the codeword of all small clusters detected by the condition (26) to the neighborhood of large clusters detected by the condition (27). In what follows, we shall concentrate to one small cluster denoted as C_s and one large denoted as C_L. The respective codewords are symbolized as v_S and v_L. As far as the cluster C_L is concerned, we detect the training vector $x_L^{\max} \in C_L$ such that

$\left\| x_L^{\max} - v_L \right\| = \max_{x \in C_L} \{ \| x - v_L \| \}$. The line segment that connects $x_L^{\max}$ and v_L is:

$$\left[v_L, x_L^{\max} \right] = v_L + \lambda \left(x_L^{\max} - v_L \right) = (1 - \lambda) v_L + \lambda x_L^{\max} \quad \forall \lambda \in [0, 1] \quad (28)$$

We select a $\lambda \in (0.5, 1)$ and remove the v_S, the location on the segment defined by this value of λ. Note that in this case the codeword v_S will be closer to the $x_L^{\max}$ than the v_L. After the migration, the training vectors of the small cluster C_S will be reassigned to their closest codewords in terms of the condition (25).

The final step of the algorithm concerns the acceptance or the rejection of the above relocation process. The relocation of a codeword is accepted if the sum of the individual distortions of the clusters C_S and C_L has been decreased [16]. To verify this, we run two iterations of the algorithm using as codebook only the codewords v_S and v_L and as training data the data belonging to the C_L [35]. Finally, all the small clusters detected in each iteration are processed in parallel.

4 Experimental Study

To test the efficiency of the proposed method we compare it with the batch LVQ and the FLVQ. The experimental data consisted of the well-known Lena, Girlface, Peppers, and Airplane grayscale images of size 512×512 pixels, which are showed in Fig. 5.

For each simulation we run all algorithms using the same initialization for each codebook size and for each image. We used ten different initial conditions for each codebook size and for each image. To carry out the experiments, each image was divided in 4×4 blocks resulting in 16,384 training vectors in the 16-dimensional feature space. The performances of the algorithms were evaluated in terms of the distortion measure given in Eq. (14) and the peak signal-to-noise ratio (PSNR):

$$\text{PSNR} = 10 \log_{10} \left(\frac{512^2 \, 255^2}{\sum_{i=1}^{512} \sum_{j=1}^{512} \left(I_{ij} - \widehat{I}_{ij} \right)^2} \right) \quad (29)$$

where 255 is the peak grayscale signal value, I_{ij} denotes the pixels of the original image, and $\widehat{I}_{ij}$ the pixels of the reconstructed image. The neighborhood function used by batch LVQ is the one calculated in Eq. (6). For the FLVQ the fuzziness parameter was gradually decreasing from $m_0 = 2$ to $m_f = 1.1$. In all simulations, the threshold in (26) was fixed to $\gamma = 0.5$ and the value λ in (28) set equal to $\lambda = 0.75$. The following five subsections present a thorough experimental investigation of the overall behavior of the algorithm.

Fig. 5 Testing images. (**a**) Airplane; (**b**) Girlface; (**c**) Peppers; (**d**) Lena

4.1 Study of the Behavior of the Distortion Measure and the PSNR

Herein, we study the distortion measure in Eq. (14) and the PSNR in (29). We choose to run the three algorithms for codebooks of sizes $c = 2^{qb}$ ($qb = 5, 6, \ldots, 10$). Since each feature vector represented a block of 16 pixels, the resulting compression rate was equal to $qb/16$ per pixel (bpp).

Tables 1, 2, 3, and 4 report the resulting mean values of the distortion measure for the four images and various codebook sizes. In all cases, our algorithm obtains the smallest distortion value.

Table 1 Distortion mean values for the Lena image

Method	$c = 32$	$c = 64$	$c = 128$	$c = 256$	$c = 512$	$c = 1024$
Batch LVQ	1548.68	1380.63	1161.19	975.22	945.53	862.75
FLVQ	1307.60	1039.02	864.71	662.24	566.09	486.51
Proposed	1260.10	977.66	770.53	602.67	479.56	415.32

Table 2 Distortion mean values for the Airplane image

Method	$c = 32$	$c = 64$	$c = 128$	$c = 256$	$c = 512$	$c = 1024$
Batch LVQ	2478.09	1888.24	1503.29	1361.53	1222.03	1098.98
FLVQ	2130.63	1654.86	1263.81	1046.38	796.68	592.37
Proposed	1968.54	1466.41	1152.49	903.61	684.63	502.49

Table 3 Distortion mean values for the Girlface image

Method	$c = 32$	$c = 64$	$c = 128$	$c = 256$	$c = 512$	$c = 1024$
Batch LVQ	1335.62	986.85	831.09	790.16	734.80	737.29
FLVQ	1224.27	971.99	819.16	662.19	543.95	470.33
Proposed	1215.94	936.52	761.67	620.03	485.47	393.55

Table 4 Distortion mean values for the Peppers image

Method	$c = 32$	$c = 64$	$c = 128$	$c = 256$	$c = 512$	$c = 1024$
Batch LVQ	1561.34	1449.77	1300.15	1170.87	1080.21	1021.44
FLVQ	1641.61	1251.23	1011.29	812.79	714.38	625.69
Proposed	1556.65	1184.17	923.01	739.87	599.65	479.78

Table 5 PSNR mean values (in dB) using various codebook sizes for the Lena image

Method	$c = 32$	$c = 64$	$c = 128$	$c = 256$	$c = 512$	$c = 1024$
Batch LVQ	28.2685	28.7669	29.5177	30.2749	30.4087	30.8068
FLVQ	29.0023	30.0011	30.7963	31.9534	32.6327	33.2892
Proposed	29.1658	30.2688	31.3061	32.3720	33.3953	34.0400

The quality of the reconstructed image was evaluated using the PSNR. Tables 5, 6, 7, and 8 summarize the mean PSNR values. The results reported in these tables are highly convincing, since in all cases the proposed algorithm outperformed the others.

Table 6 PSNR mean values (in dB) using various codebook sizes for the Airplane image

Method	$c = 32$	$c = 64$	$c = 128$	$c = 256$	$c = 512$	$c = 1024$
Batch LVQ	26.2282	27.4084	28.3985	28.8280	29.2963	29.7567
FLVQ	26.8844	27.9811	29.1497	29.9695	31.1513	32.4366
Proposed	27.2284	28.5061	29.5516	30.6073	31.8102	33.1505

Table 7 PSNR mean values (in dB) using various codebook sizes for the Girlface image

Method	$c = 32$	$c = 64$	$c = 128$	$c = 256$	$c = 512$	$c = 1024$
Batch LVQ	28.9102	30.2244	30.9678	31.1870	31.5028	31.4873
FLVQ	29.2930	30.2899	31.0299	31.9536	32.8061	33.4358
Proposed	29.3181	30.4528	31.3540	32.2484	33.3219	34.2805

Table 8 PSNR mean values (in dB) using various codebook sizes for the Peppers image

Method	$c = 32$	$c = 64$	$c = 128$	$c = 256$	$c = 512$	$c = 1024$
Batch LVQ	28.2337	28.5550	29.0272	29.4821	29.8316	30.0748
FLVQ	28.0167	29.1943	30.1177	31.0652	31.6248	32.1986
Proposed	28.2515	29.4383	30.5170	31.4821	32.3985	33.3924

4.2 Computational Demands

In this experiment we employ the Lena image. To perform the simulation, we quantify the computational demands of the batch LVQ, the FLVQ, and the proposed algorithm. We use the Lena image to generate codebooks of sizes $c = 2^{qb}$ ($qb = 5, 6, \ldots, 11$) and measure the time needed by the CPU in seconds per iteration.

All algorithms are implemented on a computer machine with dual-core 2.13GHz CPU, 4GB RAM, and the Matlab software under MS Windows 7. The results are illustrated in Fig. 6. Notice that the FLVQ exhibits an exponential growth in computational time while, as expected, the batch LVQ obtains the faster response. On the other hand the proposed algorithm maintains quite low computational efforts, which can be seen to be much better than the respective efforts of the FLVQ.

4.3 Study of the Migration Strategy

In this simulation experiment the distribution of the utility measures achieved by the three algorithms are compared by considering again the Lena image with codebook of size $c = 256$.

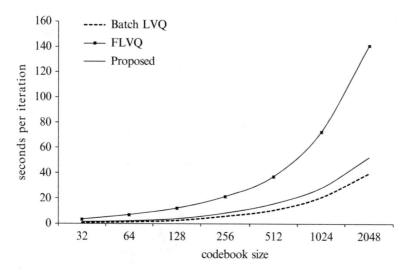

Fig. 6 Computational time in seconds per iteration as a function of the codebook size for the Lena image

Table 9 PSNR values (in dB) for the results of Fig. 7

Method	Batch LVQ	FLVQ	Proposed
PSNR	30.1920	31.6743	32.3522

Figure 7 shows the obtained distributions. The corresponding PSNR values are reported in Table 9. From the figure we can easily verify that our method provides a large percentage amount of utilities close to 1. Moreover, the distributions produced by the batch LVQ appear to have the largest deviations allowing clusters with utilities greater than 3, while the largest amount of clusters are assigned utilities much less than 1 (close to zero). On the other hand, although the FLVQ managed to reduce this variance, it creates a large number of clusters that are assigned very small utilities, something that is not desirable. The obvious conclusion of the figure is that the proposed method obtained the best distribution. To conclude, the codeword relocation approach manages to substantially increase the size of small clusters yielding an effective topology of the final partition.

4.4 Literature Comparison

In this section, we present a comparison between the best performances obtained here and the respective performances of other methods as they are reported in the corresponding references.

The comparison is based on the Lena image because it is the most used testing image in the literature.

Fig. 7 Distributions of the distortion utility measures for $c = 256$

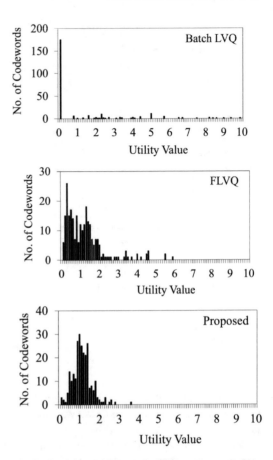

Table 10 Literature comparison results for different codebook sizes in terms of the PSNR for the Lena image

Method	$c = 32$	$c = 128$	$c = 256$	$c = 512$
Chen's model [2]	–	28.530	29.320	30.040
Hu et al. [5]	–	30.802	31.983	32.956
IFLVQ [30]	28.752	–	31.915	32.890
Proposed algorithm	29.208	31.342	32.493	33.418

Table 10 depicts the comparative results. As far as this table indicates, the proposed algorithm obtains the best results in all cases.

5 Conclusions

We have proposed a fuzzy-soft competitive LVQ for grayscale image compression. The proposed algorithm scheme combines the merits of the crisp-based and fuzzy-based methodologies into a unique framework, by utilizing two sequential steps. The first step defines a fuzzy-based neighborhood function which it is able to measure the lateral neuron interaction phenomenon and the degree of the neuron excitations. The second step involves a codeword migration strategy that is based on a cluster utility measure. Specifically, we detect clusters with small utilities and relocate the corresponding codewords to areas that appear high probability to contain large number of training vectors. This strategy increases the individual contributions of the resulting codewords to the final partition. The performance of the algorithm was evaluated in terms of the computational demands and the quality of the reconstructed image. Several simulation experiments took place, where the algorithm proved to be very fast and competitive to other related approaches.

References

1. Chang, C.C., Lin, I.C.: Fast search algorithm for vector quantization without extra look-up table using declustered subcodebooks. IEE Proc. Vis. Image Sign. Process. Lett. **7**(11), 304–306 (2005)
2. Chen, P.Y.: An efficient prediction algorithm for image quantization. IEEE Trans. Syst. Man Cybern. Part B Cybern. **34**(1), 740–746 (2004)
3. Chen, T.-H., Wu, C.-S.: Compression-unimpaired batch-image encryption combining vector quantization and index compression. Inform. Sci. **180**, 1690–1701 (2010)
4. Horng, M.-H.: Vector quantization using the firefly algorithm for image compression. Expert Syst. Appl. **39**, 1078–1091 (2012)
5. Hu, Y.-C., Su, B.-H., Tsou, C.-C.: Fast VQ codebook search algorithm for grayscale image coding. Image Vis. Comput. **26**, 657–666 (2008)
6. Poursistani, P., Nezamabadi-pour, H., Askari Moghadam, R., Saeed, M.: Image indexing and retrieval in JPEG compressed domain based on vector quantization. Math. Comput. Modell. **57**, 1005–1017 (2013)
7. Qian, S.E.: Hyper spectral data compression using a fast vector quantization algorithm. IEEE Trans. Geosci. Remote Sens. **42**(8), 1791–1798 (2004)
8. De, A., Guo, C.: An adaptive vector quantization approach for image segmentation based on SOM network. Neurocomputing **149**, 48–58 (2015)
9. Kirstein, S., Wersing, H., Gross, H.-M., Körner, E.: A life-long learning vector quantization approach for interactive learning of multiple categories. Neural Netw. **28**, 90–105 (2012)
10. Pham, T.D., Brandl, M., Beck, D.: Fuzzy declustering-based vector quantization. Pattern Recognit. **42**, 2570–2577 (2009)
11. Rizzo, F., Storer, J.A., Carpentieri, B.: Overlap and channel errors in adaptive vector quantization for image coding. Inform. Sci. **171**, 125–140 (2005)
12. Villmann, T., Haase, S., Kaden, M.: Kernelized vector quantization in gradient-descent learning. Neurocomputing **147**, 83–95 (2015)
13. Yan, S.B.: Constrained-storage multistage vector quantization based on genetic algorithms. Pattern Recognit. **41**, 689–700 (2008)
14. Zhou, S.S., Wang, W.W., Zhou, L.H.: A new technique for generalized learning vector quantization algorithm. Image Vis. Comput. **24**, 649–655 (2006)
15. Linde, Y., Buzo, A., Gray, R.M.: An algorithm for vector quantizer design. IEEE Trans. Commun. **28**(1), 84–95 (1980)
16. Patane, G., Russo, M.: The enhanced LBG. Neural Netw. **14**, 1219–1237 (2001)
17. Kohonen, T.: Self-Organizing Maps, 3rd edn. Springer, Berlin (2001)
18. Kohonen, T.: The self-organizing map. Neurocomputing **2**, 1–6 (2003)
19. Uchino, E., Yano, K., Azetsu, T.: A self-organizing map with twin units capable of describing a nonlinear input-output relation applied to speech code vector mapping. Inform. Sci. **177**, 4634–4644 (2007)

20. D'Ursoa, P., De Giovanni, L., Massari, R.: Self-organizing maps for imprecise data. Fuzzy Sets Syst. **237**, 63–89 (2014)
21. Wang, C.-H., Lee, C.-N., Hsieh, C.-H.: Classified self-organizing map with adaptive subcodebook for edge preserving vector quantization. Neurocomputing **72**, 3760–3770 (2009)
22. Bezdek, J.C., Pal, N.R.: Two soft relatives of learning vector quantization. Neural Netw. **8**, 729–743 (1995)
23. Filippi, A.M., Jensen, J.R.: Fuzzy learning vector quantization for hyperspectral coastal vegetation classification. Remote Sens. Environ. **100**, 512–530 (2006)
24. Feng, H.-M., Chen, C.-Y., Ye, F.: Evolutionary fuzzy particle swarm optimization vector quantization learning scheme in image compression. Expert Syst. Appl. **32**, 213–222 (2007)
25. Hung, W.-L., Chen, D.-H., Yang, M.-S.: Suppressed fuzzy-soft learning vector quantization for MRI segmentation. Artif. Intell. Med. **52**, 33–43 (2011)
26. Karayiannis, N.B., Randolph-Gips, M.M.: Soft learning vector quantization and clustering algorithms based on non-Euclidean norms: single-norm algorithms. IEEE Trans. Neural Netw. **16**(2), 423–435 (2005)
27. Tsekouras, G.E., Dartzentas, D., Drakoulaki, I., Niros, A.D.: Fast fuzzy vector quantization. In: Proceedings of IEEE International Conference on Fuzzy Systems, Barcelona, Spain (2010)
28. Tsolakis, D., Tsekouras, G.E., Tsimikas, J.: Fuzzy vector quantization for image compression based on competitive agglomeration and a novel codeword migration strategy. Eng. Appl. Artif. Intell. **25**, 1212–1225 (2011)
29. Tsekouras, G.E.: A fuzzy vector quantization approach to image compression. Appl. Math. Comput. **167**, 539–560 (2005)
30. Tsekouras, G.E., Mamalis, A., Anagnostopoulos, C., Gavalas, D., Economou, D.: Improved batch fuzzy learning vector quantization for image compression. Inform. Sci. **178**, 3895–3907 (2008)
31. Wu, K.-L., Yang, M.-S.: A fuzzy-soft learning vector quantization. Neurocomputing **55**, 681–697 (2003)
32. Tsao, E.C.K., Bezdek, J.C., Pal, N.R.: Fuzzy Kohonen clustering networks. Pattern Recognit. **27**(5), 757–764 (1994)
33. Pal, N.R., Bezdek, J.C., Hathaway, R.J.: Sequential competitive learning and the fuzzy c-means clustering algorithms. Neural Netw. **9**(5), 787–796 (1996)
34. Tsekouras, G.E., Sarimveis, H.: A new approach for measuring the validity of the fuzzy c-means algorithm. Adv. Eng. Software **35**, 567–575 (2004)
35. Tsolakis, D., Tsekouras, G.E., Niros, A.D., Rigos, A.: On the systematic development of fast fuzzy vector quantization for grayscale image compression. Neural Netw. **36**, 83–96 (2012)
36. Celebi, M.E., Kingravi, H., Vela, P.A.: A comparative study of efficient initialization methods for the K-means clustering algorithm. Expert Syst. Appl. **40**(1), 200–210 (2013)
37. Celebi, M.E., Kingravi, H.: Deterministic initialization of the K-means algorithm using hierarchical clustering. Int. J. Pattern Recognit. Artif. Intell. **26**(7), 1250018 (2012)

Unsupervised Learning in Genome Informatics

Ka-Chun Wong, Yue Li, and Zhaolei Zhang

Abstract With different genomes available, unsupervised learning algorithms are essential in learning genome-wide biological insights. Especially, the functional characterization of different genomes is essential for us to understand lives. In this chapter, we review the state-of-the-art unsupervised learning algorithms for genome informatics from DNA to MicroRNA.

DNA (DeoxyriboNucleic Acid) is the basic component of genomes. A significant fraction of DNA regions (transcription factor binding sites) are bound by proteins (transcription factors) to regulate gene expression at different development stages in different tissues. To fully understand genetics, it is necessary to apply unsupervised learning algorithms to learn and infer those DNA regions. Here we review several unsupervised learning methods for deciphering the genome-wide patterns of those DNA regions.

MicroRNA (miRNA), a class of small endogenous noncoding RNA (RiboNucleic acid) species, regulate gene expression post-transcriptionally by forming imperfect base-pair with the target sites primarily at the $3'$ untranslated regions of the messenger RNAs. Since the discovery of the first miRNA *let-7* in worms, a vast amount of studies have been dedicated to functionally characterizing the functional impacts of miRNA in a network context to understand complex diseases such as cancer. Here we review several representative unsupervised learning frameworks on inferring miRNA regulatory network by exploiting the static sequence-based information pertinent to the prior knowledge of miRNA targeting and the dynamic information of miRNA activities implicated by the recently available large data compendia, which interrogate genome-wide expression profiles of miRNAs and/or mRNAs across various cell conditions.

K.-C. Wong (✉)
City University of Hong Kong, Kowloon, Hong Kong
e-mail: kc.w@cityu.edu.hk

Y. Li • Z. Zhang
University of Toronto, Toronto, ON, Canada
e-mail: yueli@cs.toronto.edu; zhaolei.zhang@utoronto.ca

© Springer International Publishing Switzerland 2016   405
M.E. Celebi, K. Aydin (eds.), *Unsupervised Learning Algorithms*,
DOI 10.1007/978-3-319-24211-8_15

1   Introduction

Since the 1990s, the whole genomes of a large number of species have been sequenced by their corresponding genome sequencing projects. In 1995, the first free-living organism *Haemophilus influenzae* was sequenced by the Institute for Genomic Research [45]. In 1996, the first eukaryotic genome (*Saccharomyces cerevisiase*) was completely sequenced [58]. In 2000, the first plant genome, *Arabidopsis thaliana*, was also sequenced by Arabidopsis Genome Initiative [75]. In 2004, the human genome project (HGP) announced its completion [31]. Following the HGP, the encyclopedia of DNA elements (ENCODE) project was started, revealing massive functional putative elements on the human genome in 2011 [41]. The drastically decreasing cost of sequencing enables the 1000 Genomes Project to be carried out, resulting in an integrated map of genetic variation from 1092 human genomes published in 2012 [1]. The massive genomic data generated by those projects impose an unforeseen challenge for large-scale data analysis at the scale of gigabytes or even terabytes.

Computational methods are essential in analyzing the massive genomic data. They are collectively known as bioinformatics or computational biology; for instance, motif discovery [63] helps us distinguish real signal subsequence patterns from background sequences. Multiple sequence alignment [4] can be used to study the similarity between multiple sequences. Protein structure prediction [108, 164] can be applied to predict the 3D tertiary structure from an amino acid sequence. Gene network inference [35] are the statistical methods to infer gene networks from correlated data (e.g., microarray data). Promoter prediction [2] help us annotate the promoter regions on a genome. Phylogenetic tree inference [129] can be used to study the hierarchical evolution relationship between different species. Drug scheduling [99, 166] can help solve the clinical scheduling problems in an effective manner. Although the precision of those computational methods is usually lower than the existing wet-lab technology, they can still serve as useful preprocessing tools to significantly narrow search spaces. Thus prioritized candidates can be selected for further validation by wet-lab experiments, saving time and funding. In particular, unsupervised learning methods are essential in analyzing the massive genomic data where the ground truth is limited for model training. Therefore, we describe and review several unsupervised learning methods for genome informatics in this chapter.

2   Unsupervised Learning for DNA

In human and other eukaryotes, gene expression is primarily regulated by the DNA binding of various modulatory transcription factors (TF) onto cis-regulatory DNA elements near genes. Binding of different combinations of TFs may result in a gene being expressed in different tissues or at different developmental stages. To fully understand a gene's function, it is essential to identify the TFs that regulate the gene and the corresponding TF binding sites (TFBS). Traditionally, these regulatory sites were determined by labor-intensive experiments such as DNA footprinting or gel-shift assays. Various computational approaches have been developed to predict TFBS in *silico*. Detailed comparisons can be found in the survey by Tompa et al. [152]. TFBS are relatively short (10–20 bp) and highly degenerate sequence motifs, which makes their effective identification a computationally challenging task. A number of high-throughput experimental technologies were developed recently to determine protein-DNA binding such as protein binding microarray (PBM) [16], chromatin immunoprecipitation (ChIP) followed by microarray or sequencing (ChIP-Chip or ChIP-Seq) [79, 127], microfluidic affinity analysis [47], and protein microarray assays [71, 72].

On the other hand, it is expensive and laborious to experimentally identify TF-TFBS sequence pairs, for example, using DNA footprinting [52], gel electrophoresis [54], and SELEX [154]. The technology of ChIP [79, 127] measures the binding of a particular TF to the nucleotide sequences of co-regulated genes on a genome-wide scale *in vivo*, but at low resolution. Further processing is needed to extract precise TFBSs [102]. PBM was developed to measure the binding preference of a protein to a complete set of k-mers in vitro [16]. The PBM data resolution is unprecedentedly high, comparing with other traditional techniques. The DNA k-mer binding specificities of proteins can even be determined in a single day. It has also been shown to be largely consistent with those generated by in vivo genome-wide location analysis (ChIP-chip and ChIP-seq) [16].

To store and organize the precious data, databases have been created. TRANS-FAC is one of the largest databases for regulatory elements including TFs, TFBSs, weight matrices of the TFBSs, and regulated genes [107]. JASPAR is a comprehensive collection of TF DNA-binding preferences [123]. Other annotation databases are also available (e.g., Pfam [14], UniProbe [128], ScerTF [144], FlyTF [121], YeTFaSCo [33], and TFcat [51]). Notably, with the open-source and open-access atmosphere wide-spreads on the Internet in recent years, a database called ORegAnno appeared in 2008 [111]. It is an open-access community-driven database and literature curation system for regulatory annotation. The ENCODE consortium has also released a considerable amount of ChIP-Seq data for different DNA-binding proteins [41].

In contrast, unfortunately, it is still difficult and time-consuming to extract the high resolution 3D protein-DNA (e.g., TF-TFBS) complex structures with X-ray crystallography [142] or nuclear magnetic resonance (NMR) spectroscopic analysis [110]. As a result, there is strong motivation to have unsupervised learning methods based on existing abundant sequence data, to provide testable candidates with high confidence to guide and accelerate wet-lab experiments. Thus unsupervised learning methods are proposed to provide insights into the DNA-binding specificities of transcription factors from the existing abundant sequence data.

2.1 DNA Motif Discovery and Search

TFBSs are represented in DNA motif models to capture its sequence degeneracy [165]. They are described in the following sections.

2.1.1 Representation (DNA Motif Model)

There are several motif models proposed. For example, consensus string representation, a set of motif instance strings, count matrix, position frequency matrix (PFM), and position weight matrix (PWM). Among them, the most popular motif models are the matrix ones. They are the count matrix, PFM, and PWM. In particular, the most common motif model is the zero-order PWM which has been shown to be related to the average protein-DNA-binding energy in the experimental and statistical mechanics study [15]. Nonetheless, it assumes independence between different motif positions. A recent attempt has been made to generalize PWM but the indel operations between different nucleotide positions are still challenging [146]. Although the column dependence and indel operations could be modeled by hidden Markov model (HMM) simultaneously, the number of training parameters is increased quadratically. There is a dilemma between accuracy and model complexity.

Count Matrix

The count matrix representation is the *de facto* standard adopted in databases. In the count matrix representation, a DNA motif of width w is represented as a 4-by-w matrix C. The jth column of C corresponds to the jth position of the motif, whereas the ith row of C corresponds to the ith biological character. In the context of DNA sequence, we have four characters $\{A, C, G, T\}$. C_{ij} is the occurring frequency of the i biological character at the jth position. For example, the count matrix C_{sox9} of the SOX9 protein (JASPAR ID:MA0077.1 and UniProt ID:P48436) is tabulated in the following matrix form. The motif width is 9 so we have a 4×9 matrix here. The corresponding sequence logo is also depicted in Fig. 1.

$$
C_{\text{sox9}} = \begin{array}{c} \\ A \\ C \\ G \\ T \end{array} \begin{array}{ccccccccc} 1 & 2 & 3 & 4 & 5 & 6 & 7 & 8 & 9 \\ \left(\begin{array}{ccccccccc} 24 & 54 & 59 & 0 & 65 & 71 & 4 & 24 & 9 \\ 7 & 6 & 4 & 72 & 4 & 2 & 0 & 6 & 9 \\ 31 & 7 & 0 & 2 & 0 & 1 & 1 & 38 & 55 \\ 14 & 9 & 13 & 2 & 7 & 2 & 71 & 8 & 3 \end{array}\right) \end{array}
$$

Notably, adenine has not been found at the fourth position, resulting in a zero value. It leads to an interesting scenario. Is adenine really not found at that position or the sample size (the number of binding sites we have found so far for SOX9)

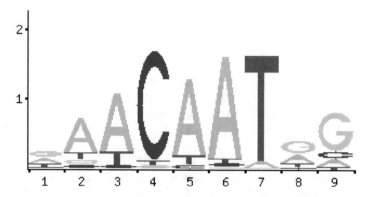

Fig. 1 DNA Motif Sequence Logo for SOX9 (JASPAR ID:MA0077.1 and UniProt ID:P48436). The vertical axis measures the information content, while the horizontal axis denotes the positions. The consensus string is DAACAATRG, following the standard IUPAC nucleotide code

is too small that the sites having adenine at that position have not been verified experimentally yet? To circumvent the problem, people usually add pseudo counts to the matrix which is justified from the use of prior probability in statistics [40, 116]. Such techniques are also found in natural language computing and machine learning.

Position Frequency Matrix

In practice, a count matrix is usually converted to PFM, and thus a zero-order PWM for scanning a long sequence. The dimension and layout of count matrix is exactly the same as those of the corresponding PFM and zero-order PWM. Their main difference is the element type. For count matrix, each element is simply a count. For PFM, each element is a maximum likelihood estimate (MLE) parameter. For zero-order PWM, each element is a weight.

To derive a PFM F from a count matrix, C, MLE is used [112]. Mathematically, we aim at maximizing the likelihood function $L(C) = P(C|F) = \prod_{i=1}^{w} \prod_{j=1}^{4} F_{ji}^{C_{ji}}$. In addition, we impose a parameter normalization constraint $\sum_{j=1}^{4} F_{ji} = 1$ for each ith position. It is added to the likelihood function with a Lagrange multiplier λ_i, resulting in a new log likelihood function:

$$\log L'(C) = \sum_{i=1}^{w} \sum_{j=1}^{4} C_{ji} \log(F_{ji}) + \sum_{i=1}^{w} \lambda_i \left(\sum_{j=1}^{4} F_{ji} - 1 \right)$$

By taking its partial derivatives to zero, it has been shown that $F_{ji} = \frac{C_{ji}}{\sum_{j=1}^{4} C_{ji}}$. The MLE parameter definition is quite intuitive. It is simply the occurring fraction of a nucleotide at the same position. For example, given the previous SOX9 count matrix C_{sox9}, we can convert it to a PFM F_{sox9} as follows:

$$
F_{\text{sox9}} = \begin{array}{c} \\ A \\ C \\ G \\ T \end{array}
\begin{array}{ccccccccc}
1 & 2 & 3 & 4 & 5 & 6 & 7 & 8 & 9 \\
\left(\begin{array}{ccccccccc}
0.31 & 0.68 & 0.75 & 0.01 & 0.83 & 0.89 & 0.06 & 0.31 & 0.13 \\
0.10 & 0.09 & 0.06 & 0.91 & 0.06 & 0.04 & 0.01 & 0.09 & 0.13 \\
0.40 & 0.10 & 0.01 & 0.04 & 0.01 & 0.03 & 0.03 & 0.49 & 0.69 \\
0.19 & 0.13 & 0.18 & 0.04 & 0.10 & 0.04 & 0.90 & 0.11 & 0.05
\end{array}\right)
\end{array}
$$

where a pseudocount $= 1$ is added to each element of C_{sox9}.

We can observe that the most invariant positions of the SOX9 motif are the 4th and 6th position. At the 4th position, cytosines have been found most of the times while guanines and thymines have just been found few times.

Position Weight Matrix

To scan a long sequence for motif matches using a PFM F, we need to derive a PWM M first so that the background distribution can be taken into account. As aforementioned, each element of PWM is a weight. Each weight can be viewed as a preference score. In practice, it is usually defined as the log likelihood ratio between the motif model and background model. Mathematically, $M_{ji} = \log(\frac{F_{ji}}{B_j})$ where B_j is the occurring fraction of the jth nucleotide in all the background sequences such that, given a subsequence a of the same width as the PWM (width w), we can compute a score $S(a)$ by summation only:

$$
S(a) = \log \frac{P(a|F)}{P(a|\text{Background})} \tag{1}
$$

$$
= \log \frac{\prod_{i=1}^{w} \prod_{j=1}^{4} F_{ji}^{[a_i=j]}}{\prod_{i=1}^{w} \prod_{j=1}^{4} B_j^{[a_i=j]}} \tag{2}
$$

$$
= \log \prod_{i=1}^{w} \prod_{j=1}^{4} \left(\frac{F_{ji}^{[a_i=j]}}{B_j^{[a_i=j]}} \right) \tag{3}
$$

$$
= \sum_{i=1}^{w} \sum_{j=1}^{4} [a_i = j] \log \left(\frac{F_{ji}}{B_j} \right) \tag{4}
$$

$$
= \sum_{i=1}^{w} \sum_{j=1}^{4} [a_i = j] M_{ji} \tag{5}
$$

where a_i is the numeric index for the nucleotide at ith position of the subsequence a. For example, given the previous SOX9 PFM F_{sox9}, we can convert it to a PWM M_{sox9} as follows:

$$
M_{\text{sox9}} = \begin{array}{c} A \\ C \\ G \\ T \end{array}
\begin{array}{ccccccccc}
1 & 2 & 3 & 4 & 5 & 6 & 7 & 8 & 9 \\
\left(\begin{array}{ccccccccc}
0.32 & 1.46 & 1.58 & -4.32 & 1.72 & 1.85 & -2.00 & 0.32 & -1.00 \\
-1.32 & -1.51 & -2.00 & 1.87 & -2.00 & -2.74 & -4.32 & -1.51 & -1.00 \\
0.68 & -1.32 & -4.32 & -2.74 & -4.32 & -3.32 & -3.32 & 0.96 & 1.49 \\
-0.42 & -1.00 & -0.51 & -2.74 & -1.32 & -2.74 & 1.85 & -1.15 & -2.32
\end{array}\right)
\end{array}
$$

where we have assumed that the background distribution is uniform (i.e., $B_j = 0.25$ for $1 \leq j \leq 4$) for illustration purposes.

2.1.2 Learning (Motif Discovery)

In general, motif discovery aims at building motif models (e.g., PFM) from related sequences. Nonetheless, there is a variety of motif discovery methods in different biological settings. From a computing perspective, they can be classified into several paradigms by its input data types:

1. A set of sequences
2. A set of sequences with quantitative measurements
3. A set of orthologous sequences

Motif Discovery for a Set of Sequences

The most classical one is *de novo* motif discovery which just takes a set of sequences as the inputs. The set of sequences is extracted such that a common transcription factor is believed to bind to them, assuming that motif models (e.g., consensus substrings) can be found from those sequences. For example, the promoter and enhancer sequences of the genes co-regulated by a common transcription factor or the sequence regions around the next generation sequencing peaks called for a common transcription factor. Theoretically, Zia and Moses have proved a theoretical upper bound on the p-value which at least one motif with a specific information content occur by chance from background distribution (false positive) for the one-occurrence per sequence motif discovery problem [179].

Chan et al. applied evolutionary computation techniques to the problem [27, 163]. Hughes et al. proposed a Gibbs sampling algorithm called AlignACE, to sample and evaluate different possible motif models using a priori log likelihood scores [74]. Workman et al. have proposed a machine learning approach using artificial neural networks (ANN-Spec) [169]. Hertz et al. utilized the maximal information content principle to greedily search for a set of candidate sequences for building motif models (Consensus) [70]. Frith et al. have adopted simulated annealing approaches to perform multiple local alignment for motif model building (GLAM) [50]. Ao et al. have used expectation maximization to determine DNA motif PWMs (Improbizer) [5]. Bailey et al. have proposed MEME to optimize the expected value of a statistic related to the information content of motif models

[9]. A parameter enumeration pipeline wrapping MEME (MUSI) was proposed for elucidating multiple specificity binding modes [86]. Eskin et al. employed a tree data structure to find composite weak motifs (MITRA) [42]. Thijs et al. have further improved the classic Gibbs sampling method and called it MotifSampler [150]. Van Helden et al. have proposed a counting algorithm to detect statistically significant motifs [155]. Regnier and Denise have proposed an exhaustive search algorithm (QuickScore) [126]. Favorov et al. have utilized Markov Chain Monte Carlo to solve the problem in a Bayesian manner (SeSiMCMC) [43]. Pavesi et al. have proposed an exhaustive enumeration and consensus-based method (Weeder) [119]. Another exhaustive search algorithm to optimize z-scores (YMF) has been proposed by Sinha and Tompa [138].

Although different statistical techniques have been developed for the motif discovery problem, most of the existing methods aim at building motif models in the form of either a set of strings or a zero-order PWM. Nonetheless, it is well known that nucleotide dependencies and indel operations exist within some TFBSs (also known as DNA motifs) [64, 151]. It is desirable to develop new methods which can capture and model such information.

Motif Discovery for a Set of Sequences with Quantitative Measurements

It has been pointed out that a fundamental bottleneck in TFBS identification is the lack of quantitative binding affinity data for a large portion of transcription factors. The advancement of new high-throughput technologies such as ChIP-Chip, ChIP-Seq, and PBM has made it possible to determine the binding affinity of these TFs (i.e., each sequence can be associated with a binding intensity value) [8]. In particular, the PBM technology can enable us to enumerate all the possible k-mers, providing an unprecedentedly high resolution binding site affinity landscape for each TF. In light of this deluge of quantitative affinity data, robust probabilistic methods were developed to take into account those quantitative affinity data. Seed and Wobble has been proposed as a seed-based approach using rank statistics [16]. RankMotif++ was proposed to maximize the log likelihood of their probabilistic model of binding preferences [28]. MatrixREDUCE was proposed to perform forward variable selections to minimize the sum of squared deviations [46]. MDScan was proposed to combine two search strategies together, namely word enumeration and position-specific weight matrix updating [102]. PREGO was proposed to maximize the Spearman rank correlation between the predicted and the actual binding intensities [148]. Notably, Wong et al. have proposed and developed a HMM approach to learn the dependence between adjacent nucleotide positions rigorously; they also show that their method (kmerHMM) can deduce multiple binding modes for a given TF [167].

Note that this paradigm is a generalization from the motif discovery for a set of sequences with binary measurements. For example, SeedSearch [11] and DME [141]. In other words, it also includes the discriminative motif discovery method in which a set of motif-containing sequences and a set of background sequences are given as the input since we can assign a value of 1 to each motif-containing sequence and 0 to each background sequence.

Motif Discovery for a Set of Orthologous Sequences

It is generally acknowledged that by comparing evolutionarily related DNA or protein sequences, functionally important sequences or motifs can be revealed by such comparison. A functional motif is assumed to be more conserved across different species than the background sequences [55]. By incorporating the evolutionary conservation with sequence-specific DNA-binding affinity, different methods have been proposed. Moses et al. have proposed an extension of MEME to take the sequence evolution into account probabilistically [113]. Kellis et al. have proposed a spaced hexamer enumeration approach to identify conserved motifs [82]. Foot-Printer has been proposed as a substring-parsimony-based approach using dynamic programming to find statistically enriched motif substrings [20]. A Gibbs sampling approach named PhyloGibbs has also been proposed [135].

2.1.3 Prediction (Motif Search)

After a motif model has been found, it is always desirable to apply it to search for motif instances over a given sequence (e.g., ChIP-Seq peak sequences). Some basic search methods have been developed to search motif instances over a sequence. Nonetheless, those methods do not have sufficient motif model complexity to distinguish false positives from true positives over a long sequence (e.g., 100 k bp) [159]. To cope with that, some improvements have been made. In general, most of them utilize the biological information beyond the motif sequence specificity to augment the motif model complexity insufficiency. In particular, multiple motif information and evolutionary conservation have been readily adopted to improve the prediction accuracy [112].

Basic Search

Likelihood Ratio

Given a sequence $b_1b_2b_3 \ldots b_l$ of length l and a PWM M of the motif x of width w, we scan $b_1b_2b_3 \ldots b_l$ with a window of width w such that the subsequences which likelihood ratio score is higher than a pre-specified threshold are considered the instances (hits) of the motif x. Mathematically, a subsequence $b_{k+1}b_{k+2} \ldots b_{k+w}$ is considered as a motif instance (hit) if and only if the following condition is satisfied:

$$S_x(b_{k+1}b_{k+2}\ldots b_{k+w}) = \sum_{i=1}^{w}\sum_{j=1}^{4} I(b_{k+i} = n_j)M_{ji} > \text{threshold}$$

where n_j is the jth nucleotide among $\{A, C, G, T\}$ and $I(\ldots)$ is the Iverson bracket. Nonetheless, different motifs may have different likelihood score distributions. It is difficult to set a single and fixed threshold which can work for all the motifs. To solve the problem, one can normalize the score to the interval [0,1] based on the maximal and minimal scores as follows:

$$S'(b_{k+1}b_{k+2}\ldots b_{k+w}) = \frac{S_x(b_k b_{k+1}b_{k+2}\ldots b_n) - \min_{\text{seq}} S_x(\text{seq})}{\max_{\text{seq}} S_x(\text{seq}) - \min_{\text{seq}} S_x(\text{seq})}$$

Posterior Ratio

If we know the prior probability of motif occurrence π, it can also be incorporated into the scoring function in a posterior manner [112]. Mathematically, given a sequence a, motif model (including PFM F and PWM M), and background distribution B, we can compute the posterior ratio as follows:

$$S''(a) = \log\frac{P(F|a)}{P(B|a)} \tag{6}$$

$$= \log\frac{\dfrac{P(a|F)P(F)}{P(a)}}{\dfrac{P(a|B)P(B)}{P(a)}} \tag{7}$$

$$= \log\frac{P(a|F)P(F)}{P(a|B)P(B)} \tag{8}$$

$$= \log\frac{\prod_{i=1}^{w}\prod_{j=1}^{4} F_{ji}^{[a_i=j]}\,\pi}{\prod_{i=1}^{w}\prod_{j=1}^{4} B_j^{[a_i=j]}(1-\pi)} \tag{9}$$

$$= \log\prod_{i=1}^{w}\prod_{j=1}^{4}(\frac{F_{ji}}{B_j})^{[a_i=j]}\left(\frac{\pi}{1-\pi}\right) \tag{10}$$

$$= S(a) + \log\left(\frac{\pi}{1-\pi}\right) \tag{11}$$

It can be observed that the posterior ratio can be computed from the likelihood ratio by simply adding the logarithm of the prior probability ratio which is related to the pseudocounts which are introduced before.

P-Value

Given the previous scoring functions, it is not easy to set a threshold since they are just ratios. For example, if $S(a) > 0$ in the above example, it just means the likelihood that the sequence a is generated by the motif model is higher than the background and vice versa. To justify it in a meaningful way, P-value distribution can be calculated from a motif model. Given a motif PWM M of width w, an exhaustive search can be applied to traverse all the possible sequences of width w. Nonetheless, it takes 4^w time complexity for the DNA alphabet $\{A, C, G, T\}$. Interestingly, if the PWM M is of zero-order, we can exploit the column independence assumption and apply dynamic programming to calculate the exact P-value distribution in $4w$ time complexity [145]. In practice, the empirical P-value distribution may also be used.

Nonetheless, the specificity of a PWM of width w is still not high if it is applied to a very long sequence of length L. Mathematically, even if we just assign the best match as the hit, $\frac{L-w+1}{4^w}$ hits are still expected (e.g., If $L = 10,000$ and $w = 6$, 2.44 hits are expected), assuming that the sequence is uniform in background nucleotide distribution. To solve the problem, people have spent efforts on incorporating additional biological information to improve the motif search.

Incorporating Multiple Motif Information

To improve motif search, multiple motif information can be incorporated. Multiple motif sites are usually clustered together, resulting in higher signal-to-noise ratios which can be easier to be detected than alone. If multiple sites of the same motif are clustered together within a short distance, it is called homotypic clustering [100]. On the other hand, if multiple sites of different motifs are clustered together within a short distance, it is called heterotypic clustering.

To exploit the additional clustering signals beyond sequence specificity, MAST was proposed to multiply the P-values of multiple motif matches (hits) together, which has demonstrated superior performance in sequence homology search than the other two methods proposed in the same study [10]. CIS-ANALYST was proposed as a sliding window approach to predict the windows which have at least *min_sites* motif matches (hits) with *p*values $<$ *site_p* [17]. Sinha et al. have proposed a probabilistic model, Stubb, to efficiently detect clusters of binding sites (i.e., cis-regulatory modules) over genomic scales using MLE [140]. To determine the window size parameter, a window size adjustment procedure has been used in ClusterBuster to find clusters of motif matches [49]. Segal et al. have also derived an expectation maximization algorithm to model the clusters of motif matches as probabilistic graphical models [132]. Recently, Hermann et al. have proposed an integrative system (*i*-cisTarget) to combine the high-throughput next generation sequencing data with motif matches to provide accurate motif cluster search [69]. Notably, Zhou and Wong have shown that it is possible to search for clusters of motifs in a *de novo* way (i.e. without any given motif model and information) [178].

Incorporating Evolutionary Conservation

Another approach to improve motif search is to incorporate evolutionary conservation. The rationale behind that is similar to that behind phylogenetic motif discovery which we have described in a previous section. Deleterious mutations will be removed from population by negative selection. To make use of that fact, we could imply that a true motif match should be more conserved across closely related species (for example, chimpanzee and mouse) than background sequences [55]. For instance, a windowing approach with several thresholds for motif matches and conservation, ConSite, was proposed by Sandelin et al. [131]. Nonetheless, it is limited to pair-wise analysis. rVISTA is a similar approach [39] using the Match program for motif matching in TRANSFAC [107]. Bayesian branch length score (BBLS) was proposed as an evolutionary conservation score without relying on any multiple sequence alignment [171]. A parsimonious method for finding statistically significant k-mers with dynamic programming was proposed (FoorPrinter) [19]. Notably, Moses et al. proposed a comprehensive probabilistic model to search for motif instances with efficient *p*-value estimation (MONKEY) [114].

To search for novel motif instances, there are programs aimed at searching for motif matches without any given motif model and information. For instance, Ovcharenko et al. have used likelihood ratio tests to distinguish conserved regions from the background [117]. Siepel et al. have demonstrated an approach in identifying conserved regions using HMMs called PhastCons [136].

Incorporating Both Approaches

Both motif clustering information and evolutionary conservation were demonstrated beneficial to motif search. Since they are independent of each other, it is straightforward to combine them. Philippakis et al. have proposed a method to combine both types of information (i.e., motif clustering and evolutionary conservation), achieving good performance on experimentally verified datasets [122]. MONKEY has been extended by Warner et al. to exploit the motif clustering information to predict motif clusters (PhylCRM) [158]. It has been reported that the misalignment errors of the input reference sequences from other species could affect the quality of phylogenetic footprinting for motif search. Thus statistical alignments have been used to assist motif search in EMMA [67]. Stub has also been extended to StubMS to take multiple species conservation into account using a HMM phylogenetic model [139]. Notably, a unified probabilistic framework which integrates multiple sequence alignment with binding site predictions, MORPH, was proposed by the same group [137]. Its effectiveness has been demonstrated and verified in an independent comparison study [147].

2.2 Genome-Wide DNA-Binding Pattern Discovery

ChIP followed by high-throughput sequencing (ChIP-Seq) measures the genome-wide occupancy of transcription factors in vivo. In a typical ChIP-Seq study, the first step is to call the peaks, i.e. determining the precise location in the genome where the TF binds. A number of peak-calling tools have been developed; for instance, model-based analysis of ChIP-Seq data (MACS) was proposed to model the shift size of ChIP-Seq tags and local biases to improve its peak-calling accuracy [176]. Spp is another method with a strong focus on background signal correction [84]. PeakSeq is a two-pass strategy method. The first pass accounts for the sequence mappability while the second pass is to filter out statistically insignificant regions comparing to controls [130]. CisGenome refines peak boundaries and uses a conditional binomial model to identify peak regions [76]. However, recent benchmark studies suggest that their predicted peaks are distinct from each other [90, 161].

Different combinations of DNA-binding protein occupancies may result in a gene being expressed in different tissues or at different developmental stages. To fully understand a gene's function, it is essential to develop unsupervised learning models on multiple ChIP-Seq profiles to decipher the combinatorial regulatory mechanisms by multiple transcription factors.

Since multiple transcription factors often work in cis-regulatory modules to confer complex gene regulatory programs, it is necessary to develop models on multiple ChIP-Seq datasets to decipher the combinatorial DNA-binding mechanism. In the following, we briefly review some of the previous works in this area. Gerstein et al. used pair-wise peak overlapping patterns to construct a human regulatory network [56]. Xie et al. proposed self organizing map methods to visualize the co-localization of DNA-binding proteins [172]. Giannopoulou et al. proposed a nonnegative matrix factorization to elucidate the clustering of DNA-binding proteins [57]. Zeng and colleagues proposed jMOSAiCS to discover histone modification patterns across multiple ChIP-Seq datasets [175]. Ferguson et al. have described a hierarchical Bayes approach to integrate multiple ChIP-Seq libraries to improve DNA-binding event predictions. In particular, they have applied the method to histone ChIP-Seq libraries and predicted the gene locations associated with the expected pathways [44]. Mahony et al. also proposed a mixture model (MultiGPS) to detect differential binding enrichment of a DNA-binding protein in different cell lines, which can improve the protein's DNA-binding location predictions (i.e., Cdx2 protein in their study) [106]. On the other hand, Chen et al. proposed a statistical framework (MM-ChIP) based on MACS to perform an integrative analysis of multiple ChIP datasets to predict ChIP-enriched regions with known motifs for a given DNA-binding protein (i.e., ER and CTCF proteins in their study) [29].

On the other hand, Ji et al. proposed a differential principal component analysis method on ChIP-Seq to perform unsupervised pattern discovery and statistical inference to identify differential protein-DNA interactions between two biological conditions [77]. Guo et al. described a generative probabilistic model (GEM) for high resolution DNA-binding site discovery from ChIP data [66]. Interestingly, that model combines ChIP signals and DNA motif discovery together to achieve precise predictions of the DNA-binding locations. The authors have further demonstrated how GEM can be applied to reveal spatially constrained TFBS pairs on a genome.

Despite the success of the methods described above, to fully understand a gene's function, it is essential to develop probabilistic models on multiple ChIP-Seq profiles to decipher the genome-wide combinatorial patterns of DNA-binding protein occupancy. Unfortunately, the majority of the previous work usually focused on large-scale clustering of called peaks, which is an intuitive and straightforward approach. However such approaches have two limitations, as (1) peak-calling ignores the contributions from weak bindings of TFs and (2) pair-wise analysis ignores the complex combinatorial binding pattern among the TFs. Thus an unsupervised learning model called SignalSpider has been proposed to directly analyze multiple normalized ChIP-Seq signal profiles on all the promoter and enhancer regions quantitatively so that weak bindings can be taken into account [30, 168]. Especially, its computational complexity has been carefully designed to scale with the increasing ChIP-Seq data (i.e., linear complexity). With such a linear complexity, the method (SignalSpider) has been successfully applied to more than one hundred ChIP-Seq profiles in an integrated way, revealing different genome-wide DNA-binding modules across the entire human genome (hg19) [168].

3 Unsupervised Learning for Inferring microRNA Regulatory Network

While transcription factors (TFs) are the major transcriptional regulator proteins, microRNA (miRNA), a small ~22 nucleotide noncoding RNA species, has been shown to play a crucial role in post-transcriptional and/or translational regulation [12]. Since the discovery of the first miRNA *let-7* in worms, a vast amount of studies have been dedicated to functionally characterizing miRNAs with a special emphasis on their roles in cancer. While TFs can serve either as a transcriptional activator or as a repressor, miRNAs are primarily known to confer mRNA degradation and/or translational repression by forming imperfect base-pair with the target sites primarily at the 3′ untranslated regions of the messenger RNAs [65]. While miRNAs are typically ~22 nt long, several experimental studies combined with computational methods [3, 23, 37, 87, 95, 96, 174] have shown that only the first six or seven consecutive nucleotides starting at the second nucleotide from the 5′ end of the miRNA are the most crucial determinants for target site recognition (Fig. 2). Accordingly, the 6mer or 7mer close to the 5′ region of the miRNA is

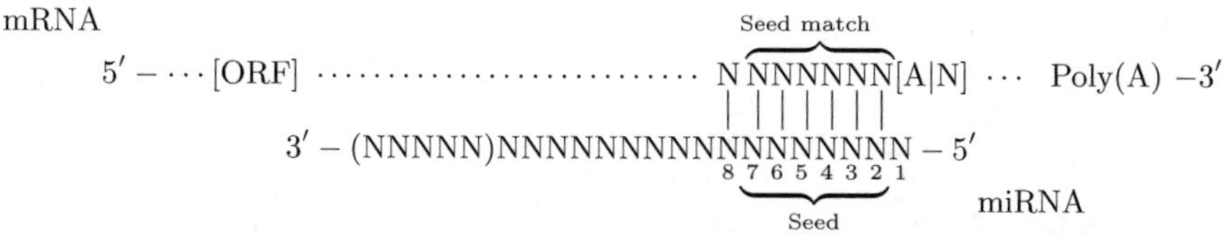

mRNA

Fig. 2 Canonical miRNA Watson-Crick base-pairing to the 3'UTR of the mRNA target site. The most critical region is a 6mer site termed as the "seed" occurs at the 2–7 position of the 5' end of the miRNA [95]. Three other variations centering at the 6mer seed are also known to be (more) conserved: 7mer-m8 site, a seed match + a Watson-Crick match to miRNA nucleotide 8; 7mer-t1A site, a seed match + a downstream A in the 3'UTR; 8mer, a seed match + both m8 and t1A. The site efficacy has also been proposed in the order of 8mer > 7mer-m8 > 7mer- A1 > 6mer [48, 61]. The abbreviations are: ORF, open reading frame; (NNNNN), the additional nucleotides to the shortest 19 nt miRNA; [A|N], A or other nucleotides; Poly(A), polyadenylated tail

termed as the "seed" region or seed. miRNAs that share common seeds belong to an miRNA family as they potentially target a vastly common set of mRNAs. Moreover, the target sites at the 3'UTR Watson-Crick (WC) pairing with the miRNA seed are preferentially more conserved within mammalian or among all of the vertebrate species [95]. In humans, more than one third of the genes harbour sites under selective pressure to maintain their pairing to the miRNA seeds [61, 95]. An important variation around this seed-target pairing scheme was discovered by Lewis et al. [96], where the target site is flanked by a conserved adenosine "A" facing to the first nucleotide of the targeting miRNA [96].

The dynamics of the miRNA regulatory network are implicated in various phenotypic changes including embryonic development and many other complex diseases [32, 143]. Although abnormal miRNA expression can sometimes be taken as a stronger indicator of carcinoma in clinical samples than aberrant mRNA expression [104, 125], the system level mechanistic effects are usually unclear. A single miRNA can potentially target ~400 distinct genes, and there are thousands of distinct endogenous miRNAs in the human genome. Thus, miRNAs are likely involved in virtually all biological processes and pathways including carcinogenesis. However, functional characterizing miRNAs hinges on the accurate identification of their mRNA targets, which has been a challenging problem due to imperfect base-pairing and condition-specific miRNA regulatory dynamics. In this section, we discuss the current state-of-art approaches in referring miRNA or miRNA-mediated transcriptional regulatory network. Table 1 summarizes these methods. As we will see, each method is established through an effective unsupervised learning model by exploiting the static sequence-based information pertinent to the prior knowledge of miRNA targeting and/or the dynamic information of miRNA activities implicated by the recently available large data compendia, which interrogate genome-wide expression profiles of miRNAs and/or mRNAs across various cell conditions.

Table 1 Unsupervised learning methods reviewed in this chapter

Method	Algorithm	Section	Ref.
PicTar	Hidden Markov model	3.1	[89]
TargetScore	Variational Bayesian mixture model	3.2	[97]
GroupMiR	Nonparametric Bayesian with Indian buffet process	3.4	[92]
SNMNMF	Constrained nonnegative matrix factorization	3.5	[177]
Mirsynergy	Deterministic overlapping neighbourhood expansion	3.6	[98]

PicTar probabilistic identification of combinations of target sites, *GroupMiR* group MiRNA target prediction, *SNMNMF* sparse network-regularized multiple nonnegative matrix factorization, *PIMiM* protein interaction-based microRNA modules

3.1   PicTar

PicTar (Probabilistic identification of combinations of Target sites) is one of the few models that rigorously considers the combinatorial miRNA regulations on the same target 3′UTR [89]. As an overview, PicTar first pre-filters target sites by their conservation across select species. However, the fundamental framework of PicTar is based on HMM with a maximum likelihood (ML) approach, which is built on the logics of several earlier works from Siggia group [24, 40, 124, 139]. Among these works, PicTar was most inspired by "Ahab," an HMM-based program developed (by the same group) to predict the combinatorial TFBS [124]. Although PicTar has been successfully applied to three studies on vertebrates [89] (where the original methodology paper was described), fly [62], and worm [91] (where some improvements were described), the description of the core HMM algorithm of PicTar is rather brief. Here we will lay out the detailed technicality of the algorithm based on the information collected from several related works [18, 24, 40, 124, 139], which will help highlight its strengths, limitations, and possible future extensions.

Let S be a 3′UTR sequence, L the length of S, and $w \in \{1, \dots, K\}$ the target sites for miRNA "word" w of length l_w, p_w the transition probability of the occurrence of miRNA w, and p_b the transition probability for the background of length $l_b = 1$, which is simply estimated from the fraction of A, U, G, C (i.e., Markov model of order 0) either in S with length >300 nt or from all query UTRs. To simplify notation, the background letters are treated as a special word w_0 so that $p_b \equiv p_{w_0}$ and $l_b \equiv l_{w_0} = 1$. Thus, S can be represented by multiple different ways of concatenating the segments corresponding to either miRNA target sites or background. The goal is to obtain at any arbitrary nucleotide position i of the 3′UTR sequence S the posterior probability $p(\pi_i = w|S, \theta)$ that i is *the last position* of the word, where θ is the model parameters controlling emission probabilities (see below).

Following Markov's assumption, $p(\pi_i = w|S, \theta)$ is proportional to the products of the probabilities before and after position i, which can be computed in time $O(L \times K)$ by Forward-Backward algorithm as described below. Formally,

$$p(\pi_i = w|S, \theta) = \frac{p(s_1, \ldots, s_L, \pi_i = w|\theta)}{p(S|\theta)} \tag{12}$$

$$= \frac{p(s_1, \ldots, s_i, \pi_i = w|\theta)p(s_{i+1}, \ldots, s_L|s_1, \ldots, s_i, \pi_i = w, \theta)}{p(S|\theta)} \tag{13}$$

$$= \frac{p(s_1, \ldots, s_i, \pi_i = w|\theta)p(s_{i+1}, \ldots, s_L|\pi_i = w, \theta)}{p(S|\theta)} \tag{14}$$

$$= \frac{Z(1, i, \pi_i = w)Z(i+1, \ldots, L|\pi_i = w)}{p(S|\theta)} \tag{15}$$

where $p(S|\theta)$ is the likelihood of sequence S or the *objective function* to be maximized in the ML framework, $Z(1, i, \pi_i = w)$ and $Z(i+1, \ldots, L|\pi_i = w)$ can be represented in recursion forms and computed via forward and backward algorithm, respectively in time $O(L \times K)$ for K words. Formally, the forward algorithm is derived as follows:

$$Z(1, i, \pi_i = w)$$

$$= p(s_1, \ldots, s_i, \pi_i = w) \tag{16}$$

$$= p(s_1, \ldots, s_i|\pi_i = w)p(\pi_i = w) \tag{17}$$

$$= p(s_{i-l_w+1}, \ldots, s_i|\pi_i = w)p(s_1, \ldots, s_{i-l_w}|\pi_i = w)p(\pi_i = w) \tag{18}$$

$$= p(s_{i-l_w+1}, \ldots, s_i|\pi_i = w)p(s_1, \ldots, s_{i-l_w}, \pi_i = w) \tag{19}$$

$$= p(s_{i-l_w+1}, \ldots, s_i|\pi_i = w) \sum_{w'} p(s_1, \ldots, s_{i-l_w}, \pi_{i-l_w} = w')p(\pi_i = w|\pi_{i-l_w} = w') \tag{20}$$

$$= e(i - l_w + 1, \ldots, i|w) \sum_{w'} Z(1, i - l_w, \pi_{i-l_w} = w')p_w \tag{21}$$

where $e(i - l_w + 1, \ldots, i|w)$ is the *emission probability* assumed known (see below) and $p_w \equiv p(\pi_i = w|\pi_{i-l_w} = w')$ is the transition probability to word w. Note that p_w is position independent. To start the recursion, $Z(1, i \leq 1) = 1$.

The backward algorithm is similarly derived:

$$Z(i+1, \ldots, L|\pi_i = w)$$

$$= p(s_{i+1}, \ldots, s_L|\pi_i = w) \tag{22}$$

$$= \sum_{w'} p(s_{i+1}, \ldots, s_L, \pi_{i+1} = w'|\pi_i = w) \tag{23}$$

$$= \sum_{w'} p(s_{i+1}, \ldots, s_L|\pi_{i+1}=w')p(\pi_{i+1}=w'|\pi_i=w) \tag{24}$$

$$= \sum_{w'} p(s_{i+2}, \ldots, s_L | \pi_{i+1} = w') p(s_{i+2-l_{w'}}, \ldots, s_{i+1} | \pi_{i+1} = w') p(\pi_{i+1} = w' | \pi_i = w)$$

$$= \sum_{w}^{'} Z(i+2, \ldots, L | \pi_{i+1} = w') e(i+2 - l_{w'}, \ldots, i+1 | w') p_{w'} \tag{25}$$

To start the backward recursion, $Z(L - l_w + 1, L) = p_0(w)$, which is simply the frequency of word w in the 3′UTR sequence S.

Given the emission probabilities, p_w's (transition probability) are the only parameters that need to be set in order to maximize $p(S|\theta)$. Following the ML solution,

$$p_w = \frac{\sum_i p(\pi_i = w | S, \theta)}{\sum_{w'} \sum_i p(\pi_i = w' | S, \theta)} \tag{26}$$

where the posterior $p(\pi_i = w | S, \theta)$ is calculated by Eq. (15), which is in turn computed by forward-backward algorithm. Finally, the likelihood (objective) function is evaluated by a simple forward pass to the end of the sequence:

$$p(S|\theta) = \sum_{w'} p(s_1, \ldots, s_L, \pi_L = w' | \theta) \tag{27}$$

$$= \sum_{w'} Z(1, L, \pi_L = w') \tag{28}$$

Together, the optimization of p_w in PicTar is performed using Baum-Welch algorithm for Expectation-Maximization (EM) as summarized in Algorithm 1. Finally, the PicTar score is defined as a log ratio ($F = -\log Z$) of the ML over background likelihood F_B:

$$\text{PicTarScore} = F_B - F \tag{29}$$

where F_B is the likelihood when only background is considered.

Algorithm 1 Baum-Welch HMM algorithm in PicTar [89]

Initialize $Z(1, i \leq 1) = 1$ and $Z(L - l_w + 1, L) = p_0(w)$
E-step:
    Forward recursion ($i = 1, \ldots, L$): compute $Z(1, i, \pi_i = w)$ by Eq (21)
    Backward recursion ($i = L - l_w, \ldots, 1$): compute $Z(i+1, \ldots, L | \pi_i = w)$ by Eq (25)
M-step:
    Update p_w by Eq (26)
Evaluate likelihood:
    Compute $p(S|\theta)$ by Eq (28)
Repeat **EM** steps until $p(S|\theta)$ increases by less than a cutoff

---

As previously mentioned, the emission probabilities $e(s|w)$ in PicTar are assumed known. In Ahab for modelling TFBS, $e(s|w)$ or $m(s|w)$ is based on the PFM: $m(s|w) = \prod_{j=1}^{l} f_j(n|w)$, where $f_j(n|w)$ is the normalized frequency of the nucleotide n at the $j-th$ position of the PFM. However, miRNAs do not have PFM. The original PicTar arbitrarily sets $e(s|w)$ to be 0.8 if there is perfect seed-match at 1–7 or 2–8 nt positions of the miRNA 5' end AND the free binding energy as estimated by RNAhybrid is no less than 33 % of the optimal binding energy of the entire miRNA sequence to the UTR [103]; otherwise $e(s|w)$ is set to 0.2 divided by M for M imperfect seed matches with only 1 mismatch allowed, provided it is above 66 % of the optimal binding energy. Thus, the setting highly disfavours imperfect seed match. The later version of PicTar changes the emission probability calculation to be the total number of occurrences in conserved 3'UTR sites divided by the total number of sites 3'UTR [91]. The setting appears to improve the sensitive/specificity of the model but makes it more dependent on the cross-species conservation, potentially prone to false negatives (for the non-conserved but functional sites).

The major advantage of PicTar over other simpler methods is that the coordinate actions of the miRNAs (synergistic in case of optimally spaced sites or antagonistic in case of overlapping binding sites) are naturally captured within the emission and transition probabilities. For instance, the *PicTarScore* as the joint ML of multiple miRNA target sites will be higher than the linear sum of individual miRNA target sites (i.e., synergistic effects). Longer 3'UTR will score less than shorter 3'UTR if both contain the same number of target sites. PicTar demonstrated a comparable signal:noise ratio relative to TargetScan and was compared favourably with some of the earlier published methods based on several surveys [3, 97, 133, 174]. When applied to vertebrates, PicTar identified roughly 200 genes per miRNA, which is a rather conservative estimate compared to the recent findings by TargetScan with a conserved targeting scoring approach [48]. When applied to *C. elegans* (worm), PicTar identified 10 % of the *C. elegans* genes that are under conserved miRNA regulation.

Nonetheless, PicTar has three important limitations. First, PicTar does not consider the correlation between miRNA target sites since p_w is essentially position independent. This is perhaps largely due to the increased model complexity when considering all pairwise transition probabilities between K miRNAs and background since there will be $(K + 1) \times K + 1$ parameters to model (as apposed to only $K + 1$). Supported by Grimson et al. [61], however, the specific spatial arrangement of the target sites may be functionally important. In particular, the optimal distance between miRNA sites was estimated as 8 to ~40 nt based on transfection followed by regression analysis [61]. Although unsupported by experimental evidence, the ordering of some specific target sites may be also important. For instance, target site x must be located before target site y (for the same or different miRNA) to achieve optimal synergistic repression. The model that takes into account the spatial correlation between motifs is called the hcHMM (history conscious HMM) implemented in a program called Stubb for detecting TFBS rather than miRNA target predictions [140].

Second, the ML approach is prone to local optimal especially for long UTRs or many coordinated miRNA actions considered simultaneously (i.e., many p_w's). An alternative HMM formulation is to impose Bayesian priors on the HMM parameters [105]. In particular, [170] demonstrated such Bayesian formalism of HMM in modelling combinatorial TFBS. In their model so-called module HMM, the transition probabilities are assumed to follow a Dirichlet distribution with hyperparameters $\boldsymbol{\alpha} = \{\alpha_0, \alpha_1, \ldots, \alpha_K, \alpha_{K+1}\}$, where α_0 corresponds to the background, $\{\alpha_1 \ldots \alpha_K\}$ to the K TFs, and α_{K+1} to the background inside of *cis*-regulatory module. Inference is performed via Markov Chain Monte Carlo (MCMC) procedure or Gibbs sampling in particular. Briefly, a forward pass and backward pass are run to generate marginal probabilities at each nucleotide position. Starting at the end of the sequence, hidden states are sampled at each position based on the marginals until reaching the front of the sequence. Given the hidden states, the hyperparameters for Dirichlet distribution of the transition probabilities are then updated by simply counting the occurrences of each state. The posteriors of the hidden states at each nucleotide position are inferred as the averaged number of times the states are sampled in 1000 samplings, and the states with the *maximum a posteriori* (MAP) are chosen. The model is demonstrated to perform better than Ahab and Stubb (which are the basis for PicTar) in TFBS predictions but have not yet been adapted to miRNA target predictions.

Third, since data for expression profiling of mRNAs and miRNAs by microarrays or RNA-seq is now rather abundant (e.g., [7], GSE1738; GSE31568, [81]; GSE40499, [109] from GEO) or ENCODE (GSE24565, [36]) or TCGA [25, 160], the combinatorial regulation needs to be revisited by taking into account whether or not the co-operative miRNAs are indeed expressed in vivo and/or the expression correlation between mRNA and miRNA. In particular, the emission probabilities $e(s|w)$ need to be redefined to integrate both sequence-based and expression-based information.

3.2 A Probabilistic Approach to Explore Human miRNA Target Repertoire by Integrating miRNA-Overexpression Data and Sequence Information

One of the most direct way to query the targets of a given miRNA is by transfecting the miRNA into a cell and examine the expression changes of the cognate target genes [101]. Presumably, a *bonafide* target will exhibit decreased expression upon the miRNA transfection. In particular, overexpression of miRNA coupled with expression profiling of mRNA by either microarray or RNA-seq has proved to be a promising approach [6, 101]. Consequently, genome-wide comparison of differential gene expression holds a new promise to elucidate the global impact of a specific miRNA regulation without solely relying on evolutionary conservation. However, miRNA transfection is prone to off-target effects. For instance, overexpressing a miRNA may not only repress the expression of its direct targets but also cause a cascading repression of in-direct targets of the affected transcription activators. To improve prediction accuracy for direct miRNA targets, this chapter describes a novel model called *TargetScore* that integrates expression change due to miRNA overexpression and sequence information such as context score [53, 61] and other orthogonal sequence-based features such as conservation [48] into a probabilistic score.

In one of our recent papers, we described a novel probabilistic method for miRNA target prediction problem by integrating miRNA-overexpression data and sequence-based scores from other prediction methods [97]. Briefly, each score feature is considered as an independent observed variable, which is the input to a Variational Bayesian-Gaussian Mixture Model (VB-GMM). We chose a Bayesian over a maximum likelihood approach to avoid overfitting. Specifically, given expression fold-change (due to miRNA transfection), we use a three-component VB-GMM to infer down-regulated targets accounting for genes with little or positive fold-change (due to off-target effects [83]). Otherwise, two-component VB-GMM is applied to unsigned sequence scores. The parameters for the VB-GMM are optimized using variational Bayesian expectation-maximization (VB-EM) algorithm. The mixture component with the largest absolute means of observed negative fold-change or sequence score is associated with miRNA targets and denoted as "target component." The other components correspond to the "background component." It follows that inferring miRNA-mRNA interactions is equivalent to inferring the posterior distribution of the target component given the observed variables. The targetScore is computed as the sigmoid-transformed fold-change weighted by the averaged posteriors of target components over all of the features.

3.2.1   Bayesian Mixture Model

Assuming there are N genes, we denote $\mathbf{x} = (x_1, \ldots, x_N)^T$ as the log expression fold-change ($\mathbf{x}_f$) or sequence scores ($\mathbf{x}_l, l \in \{1, \ldots, L\}$). Thus, for L sets of sequence scores, $\mathbf{x} \in \{\mathbf{x}_f, \mathbf{x}_1, \ldots, \mathbf{x}_L\}$. To simplify the following equations, we use $\mathbf{x}$ to represent one of the independent variables without loss of generality. To infer target genes for a miRNA given $\mathbf{x}$, we need to obtain the posterior distribution $p(\mathbf{z}|\mathbf{x})$ of the latent variable $\mathbf{z} \in \{z_1, \ldots, z_K\}$, where $K = 3$ ($K = 2$) for modelling signed (unsigned) scores such as logarithmic fold-changes (sequence scores).

We follow the standard Bayesian GMM based on [18] (pp. 474–482) with only minor modifications. Although univariate GMM ($D = 1$) is applied to each variable separately, we implemented and describe the following formalism as a more general multivariate GMM, allowing modeling the covariance matrices. Briefly, the latent variables $\mathbf{z}$ are sampled at probabilities $\boldsymbol{\pi}$ (mixing coefficient), that follow a Dirichlet prior $\text{Dir}(\boldsymbol{\pi}|\boldsymbol{\alpha}_0)$ with hyperparameters $\boldsymbol{\alpha}_0 = (\alpha_{0,1}, \ldots, \alpha_{0,K})$. To account for the relative frequency of targets and non-targets for any miRNA, we set the $\alpha_{0,1}$ (associated with the target component) to aN and other $\alpha_{0,k} = (1 - a) \times N/(K-1)$, where $a = 0.01$ (by default). Assuming $\mathbf{x}$ follows a Gaussian distribution $\mathcal{N}(\mathbf{x}|\boldsymbol{\mu}, \boldsymbol{\Lambda}^{-1})$, where $\boldsymbol{\Lambda}$ (precision matrix) is the inverse covariance matrix, $p(\boldsymbol{\mu}, \boldsymbol{\Lambda})$

together follow a Gaussian–Wishart prior $\prod_k^K \mathcal{N}(\mu_k|\mathbf{m}_0, (\beta_0 \Lambda)^{-1})\mathcal{W}(\Lambda_k|\mathbf{W}_0, \nu_0)$, where the hyperparameters $\{\mathbf{m}_0, \beta_0, \mathbf{W}_0, \nu_0\} = \{\hat{\mu}, 1, \mathbf{I}_{D \times D}, D + 1\}$.

3.2.2 Variational Bayesian Expectation Maximization

Let $\boldsymbol{\theta} = \{\mathbf{z}, \boldsymbol{\pi}, \boldsymbol{\mu}, \boldsymbol{\Lambda}\}$. The marginal log likelihood can be written in terms of lower bound $\mathcal{L}(q)$ (first term) and Kullback-Leibler divergence $\mathcal{KL}(q||p)$ (second term):

$$\ln p(\mathbf{x}) = \int q(\boldsymbol{\theta}) \ln \frac{p(\mathbf{x}, \boldsymbol{\theta})}{q(\boldsymbol{\theta})} d\boldsymbol{\theta} + \int q(\boldsymbol{\theta}) \ln \frac{q(\boldsymbol{\theta})}{p(\boldsymbol{\theta}|\mathbf{x})} d\boldsymbol{\theta} \tag{30}$$

where $q(\boldsymbol{\theta})$ is a proposed distribution for $p(\boldsymbol{\theta}|\mathbf{x})$, which does not have a closed-form distribution. Because $\ln p(\mathbf{x})$ is a constant, maximizing $\mathcal{L}(q)$ implies minimizing $\mathcal{KL}(q||p)$. The general optimal solution $\ln q_j^*(\theta_j)$ is the expectation of variable j w.r.t other variables, $\mathbb{E}_{i \neq j}[\ln p(\mathbf{x}, \boldsymbol{\theta})]$. In particular, we define $q(\mathbf{z}, \boldsymbol{\pi}, \boldsymbol{\mu}, \boldsymbol{\Lambda}) = q(\mathbf{z})q(\boldsymbol{\pi})q(\boldsymbol{\mu}, \boldsymbol{\Lambda})$. The expectations for the three terms (at log scale), namely $\ln q^*(\mathbf{z}), \ln q^*(\boldsymbol{\pi}), \ln q^*(\boldsymbol{\mu}, \boldsymbol{\Lambda})$, have the same forms as the initial distributions due to the conjugacy of the priors. However, they require evaluation of the parameters $\{\mathbf{z}, \boldsymbol{\pi}, \boldsymbol{\mu}, \boldsymbol{\Lambda}\}$, which in turn all depend on the expectations of $\mathbf{z}$ or the posterior of interest:

$$p(z_{nk}|\mathbf{x}_n, \boldsymbol{\theta}) \equiv \mathbb{E}[z_{nk}] = \frac{\rho_{nk}}{\sum_{j=1}^K \rho_{nj}} \tag{31}$$

where $\ln \rho_{nk} = \mathbb{E}[\ln \pi_k] + \frac{1}{2}\mathbb{E}[\ln |\Lambda_k|] - \frac{D}{2}\ln(2\pi) - \frac{1}{2}\mathbb{E}_{\mu_k, \Lambda_k}[(\mathbf{x}_n - \boldsymbol{\mu}_k)^T \Lambda_k(\mathbf{x}_n - \boldsymbol{\mu}_k)]$. The inter-dependence between the expectations and model parameters falls naturally into an EM framework, namely VB-EM. Briefly, we first initialize the model parameters based on priors and randomly sample K data points $\boldsymbol{\mu}$. At the ith iteration, we evaluate (31) using the model parameters (VB-E step) and update the model parameters using (31) (VB-M step). The EM iteration terminates when $\mathcal{L}(q)$ improves by less than 10^{-20} (default). Please refer to [18] for more details.

3.2.3 TargetScore

We define the targetScore as an integrative probabilistic score of a gene being a target t (meaning that $z_{nk} = 1$ for the target component k) of a miRNA:

$$\text{targetScore} = \sigma(-\log FC)\left(\frac{1}{L+1}\sum_{\mathbf{x} \in \{\mathbf{x}_f, \mathbf{x}_1, \dots, \mathbf{x}_L\}} p(t|\mathbf{x})\right) \tag{32}$$

where $\sigma(-\log FC) = \frac{1}{1+\exp(\log FC)}$, $p(t|\mathbf{x})$ is the posterior in (31).

TargetScore demonstrates superior statistical power compared to existing methods in predicting validated miRNA targets in various human cell lines. Moreover, the confidence targets from TargetScore exhibit comparable protein downregulation and are more significantly enriched for Gene Ontology terms. Using TargetScore, we explored oncomir-oncogenes network and predicted several potential cancer-related miRNA-messenger RNA interactions. TargetScore is available at Bioconductor http://www.bioconductor.org/packages/devel/bioc/html/TargetScore.html.

3.3 Network-Based Methods to Detect miRNA Regulatory Modules

Although targets of individual miRNAs are significantly enriched for certain biological processes [118, 153], it is also likely that multiple miRNAs are coordinated together to synergistically regulate one or more pathways [21, 89, 173]. Indeed, despite their limited number (2578 mature miRNAs in human genome, miRBase V20, [88]), miRNAs may be in charge of more evolutionarily robust and potent regulatory effects through coordinated collective actions. The hypothesis of miRNA synergism is also parsimonious or biologically plausible because the number of possible combinations of the 2578 human miRNAs is extremely large, enough to potentially react to virtually countless environmental changes. Intuitively, if a group of (miRNA) workers perform similar tasks together, then removing a single worker will not be as detrimental as assigning each worker a unique task [21].

Several related methods have been developed to study miRNA synergism. Some early methods were based on pairwise overlaps [134] or score-specific correlation [173] between predicted target sites of any given two (co-expressed) miRNAs. For instance, Shalgi et al. [134] devised an overlapping scoring scheme to account for differential 3′UTR lengths of the miRNA targets, which may otherwise bias the results if standard hypergeometric test was used [134]. Methods beyond pairwise overlaps have also been described. These methods considered not only the sequence-based miRNA-target site information but also the respective miRNA-mRNA expression correlation (MiMEC) across various conditions to detect miRNA regulatory modules (MiRMs).

For instance, Joung et al. [80] developed a probabilistic search procedure to separately sample from the mRNA and miRNA pools candidate module members with probabilities proportional to their overall frequency of being chosen as the "fittest," which was determined by their target sites and MiMEC relative to the counterparts [80]. The algorithm found only the single best MiRM, which varied depending on the initial mRNA and miRNA set. Other network-based methods using either the sequence information only or using m/miRNA expression profiles only as a filter for a more disease-focused network construction on only the differentially expressed (DE) m/miRNAs. For instance, Peng et al. (2006) employed an enumeration approach to search for maximal bi-clique on DE m/miRNAs to discover complete bipartite subgraphs, where every miRNA is connected with every mRNA [120]. The approach operated on unweighted edges only, which required discretizing miRNA-mRNA expression correlation. Also, maximal bi-clique does not necessarily imply functional MiRMs and vice versa.

The following subsections review in details three recently developed network methods (Table 1) to detect MiRMs. Despite distinct unsupervised learning frameworks, all three methods exploit the widely available paired m/miRNA expression profiles to improve upon the accuracy of earlier developed (sequence-based) network approaches.

3.4 GroupMiR: Inferring miRNA and mRNA Group Memberships with Indian Buffet Process

The expression-based methods reviewed elsewhere [174] were essentially designed to explain the expression of each mRNA in isolation using a subset of the miRNA expression in a linear model with a fixed set of parameters. However, the same mRNAs (miRNAs) may interact with different sets of miRNAs (mRNAs) in different pathways. The exact number of pathways is unknown and may grow with an increase of size or quality of the training data. Thus, it is more natural to *infer* the number of common features shared among different *groups* of miRNAs and mRNAs. Accordingly, Le and Bar-Joseph [92] proposed a powerful alternative model called GroupMiR (Group MiRNA target prediction) [92]. As an overview, GroupMiR first explored the latent binary features or memberships possessed within mRNAs, miRNAs, or shared between mRNAs and miRNAs on a potentially infinite binary feature space empowered by a *nonparametric Bayesian* (NBP) formalism. Thus, the number of features was inferred rather than determined arbitrarily. Importantly, the feature assignment took into account the prior information for miRNA and mRNA targeting relationships, obtained from sequence-based target prediction tools such as TargetScan or PicTar. Based on the shared memberships, mRNAs and/or miRNAs formed groups (or clubs). The same miRNAs (mRNAs) could possess multiple memberships and thus belong to multiple groups each corresponding to a latent feature. This was also biologically plausible since a miRNA (mRNA) may participate in several biological processes. Similar to GenMiR++ [73], GroupMiR then performed a Bayesian linear regression on each mRNA expression using *all miRNA expression* but placing more weight on the expression of miRNAs that shared one or more common features with that mRNA.

Specifically, the framework of GroupMiR was based on a recently developed general nonparametric Bayesian prior called the Indian buffet process (IBP) [59] (which was later on proved to be equivalent to Beta process [149]). As the name suggests, IBP can be understood from an analogy of a type of an "Indian buffet" as follows. A finite number of N customers or objects form a line to enter one after another a buffet comprised of K dishes or features. Each customer i samples

$\sum_k \frac{m_k}{i}$ dishes selected by m_k previous customers, and Poisson($\frac{\alpha}{i}$) new dishes, where α is a model parameter. The choices of N customers on the K dishes are expressed in an $N \times K$ binary matrix $\mathbf{Z}$. A left-order function lof($\cdot$) maps a binary matrix $\mathbf{Z}$ to a left-ordered binary matrix with columns (i.e., dishes) sorted from left to right by decreasing order of m_k and breaking ties in favour of customers who enter the buffet earlier. This process defines an exchangeable distribution on *equivalence class* [$\mathbf{Z}$] comprising all of the $\mathbf{Z}$ that have the same left-ordered binary matrix lof($\mathbf{Z}$) regardless of the order the customers enter the buffet (i.e., row order) or the dish order (i.e., column order).

Before reviewing the IBP derivation, we need to establish some notations [59]. $(z_{1k}, \ldots, z_{(i-1)k})$ ($i \in \{1, \ldots, N\}$) denotes the history h of feature k at object i, which is encoded by a single decimal number. At object $i = 3$, for instance, a feature k has one of four histories encoded by $0, 1, 2, 3$ corresponding to all of the four possible permutations of choices for objects 1 and 2: $(0, 0), (0, 1), (1, 0), (1, 1)$. Accordingly, for N objects, there are 2^N histories for each feature k and $2^N - 1$ histories excluding the history of all zeros (i.e., $(0)_{1 \times N}$). Additionally, K_h denotes the number of features possessing the same history h, K_0 for all features with $m_k = 0$, and $K_+ = \sum_{h=1}^{2^N - 1} K_h$ for all features for which $m_k > 0$. Thus, $K = K_0 + K_+$. It is easy to see that binary matrices belong to an equivalence class if and only if they have the same history profile h for each feature k. The cardinality (*card*) of an equivalence class [$\mathbf{Z}$] is the number of all of the binary matrices with the same history profile:

$$\text{card}([\mathbf{Z}]) = \binom{K}{K_0 \ldots K_{2^N-1}} = \frac{K!}{\prod_{h=0}^{2^N-1} K_h!} \tag{33}$$

As shown below, Eq. (33) is essential in order to establish the close-formed solution of IBP prior when $K \to \infty$ leads to an infinite feature space or infinite number of columns in $\mathbf{Z}$. After establishing the above properties, the central steps in deriving the IBP prior used in GroupMiR is reviewed below. I will focus on some steps neglected from the original work and refer the reader to the full derivation when appropriate. As in the original papers, we first derive IBP on a finite number of latent features K and then take the limit making use of Eq. (33).

Let $\mathbf{Z}$ be an $N \times K$ binary matrix, where $N = M + R$ for M mRNAs and R miRNAs, and K is the number of latent features. Assuming the binary value z_k in $\mathbf{Z}$ for each feature k is sampled from *Bernoulli*(π_k) and are conditionally independent given π_k, the joint distribution of z_k is then:

$$p(z_k|\pi_k) = \prod_i (1 - \pi_k)^{1-z_{ik}} \pi_k^{z_{ik}}$$

$$= \exp\left(\sum_i (1 - z_{ik}) \log(1 - \pi_k) + z_{ik} \log \pi_k\right) \tag{34}$$

where π_k follows a Beta prior $\pi_k|\alpha \sim \text{Beta}(r, s)$ with $r = \frac{\alpha}{K}, s = 1$:

$$p(\pi_k | \alpha) = \frac{\pi_k^{r-1}(1 - \pi_k)^{s-1}}{B(r, s)} = \frac{\pi_k^{\frac{\alpha}{k}-1}}{B(\frac{\alpha}{K}, 1)} \tag{35}$$

where $B(\cdot)$ is a Beta function. To take into account the prior information between miRNA and mRNA targeting from sequence-based predictions, GroupMiR incorporated in Eq. (35) an $N \times N$ weight matrix $\mathbf{W}$:

$$\mathbf{W} = \begin{pmatrix} \mathbf{0} & \mathbf{C} \\ \mathbf{C}^T & \mathbf{0} \end{pmatrix} \tag{36}$$

where interaction within mRNAs and within miRNAs were set to zeros and interaction between mRNA and miRNA followed the $R \times M$ scoring matrix $\mathbf{C}$ obtained from a quantitative sequence-based predictions. In particular, MiRanda scores were used in their paper. Thus, w_{ij} is either 0 or defined as a pairwise potential of interactions between mRNA i and miRNA j. The modified $p^*(\pi_k | \alpha)$ was then defined as:

$$p^*(\pi_k | \alpha) = \frac{\pi_k^{\frac{\alpha}{K}-1}}{\mathbf{Z}'} \boldsymbol{\Phi}_{z_k} \tag{37}$$

where $\boldsymbol{\Phi}_{z_k}$ and the partition function $\mathbf{Z}'$ were defined as:

$$\boldsymbol{\Phi}_{z_k} = \exp\left(\sum_{i<j} w_{ij} z_{ik} z_{jk}\right) \tag{38}$$

$$\mathbf{Z}' = \sum_{h=0}^{2^N-1} \boldsymbol{\Phi}_h B\left(\frac{\alpha}{K} + m_h, N - m_h + 1\right) \tag{39}$$

The marginal probability of $P(\mathbf{Z})$ is derived by integrating out π_k as follows:

$P(\mathbf{Z})$

$$= \prod_{k=1}^{K} \int_0^1 P(z_k | \pi_k) P(\pi_k | \alpha) d\pi_k \tag{40}$$

$$= \prod_{k=1}^{K} \int_0^1 \exp\left(\sum_i (1 - z_{ik}) \log(1 - \pi_k) + z_{ik} \log \pi_k\right) \left(\frac{\pi_k^{\frac{\alpha}{K}-1}}{\mathbf{Z}'/\boldsymbol{\Phi}_{z_k}}\right) d\pi_k \tag{41}$$

$$= \prod_{k=1}^{K} \int_0^1 \frac{\boldsymbol{\Phi}_{z_k}}{\mathbf{Z}'} \exp\left(\sum_i (1-z_{ik}) \log(1-\pi_k) + z_{ik} \log \pi_k\right) \exp\left[\left(\frac{\alpha}{K}-1\right) \log \pi_k\right] d\pi_k \tag{42}$$

$$= \prod_{k=1}^{K} \frac{\boldsymbol{\Phi}_{z_k}}{\mathbf{Z}'} \int_0^1 \exp\left((N - m_k) \log(1 - \pi_k) + m_k \log \pi_k + \left(\frac{\alpha}{K} - 1 \right) \log \pi_k \right) d\pi_k$$

(43)

$$= \prod_{k=1}^{K} \frac{\boldsymbol{\Phi}_{z_k}}{\mathbf{Z}'} \int_0^1 \exp\left((N - m_k) \log(1 - \pi_k) + \left(\frac{\alpha}{K} + m_k - 1 \right) \log \pi_k \right) d\pi_k \qquad (44)$$

$$= \prod_{k=1}^{K} \frac{\boldsymbol{\Phi}_{z_k}}{\mathbf{Z}'} B\left(\frac{\alpha}{K} + m_k, N - m_k + 1 \right)$$

(45)

where m_k in Eq. (43) is the sum over all $z_{ik} = 1$, and (45) directly follows the definition of Beta function. However, when $\lim_{K \to \infty} P(\mathbf{Z}) = 0$ since the probability of sampling a specific binary matrix from an infinite number of matrices is 0. Instead, the inference was performed over the equivalence class $[\mathbf{Z}]$ with the number of lof-equivalent matrices defined above:

$$P([\mathbf{Z}]) = \sum_{\mathbf{Z} \in [\mathbf{Z}]} P(\mathbf{Z})$$

(46)

$$= \frac{K!}{\prod_{h=0}^{2^N - 1} K_h!} \prod_{k=1}^{K} \frac{\boldsymbol{\Phi}_{z_k}}{\mathbf{Z}'} B\left(\frac{\alpha}{K} + m_k, N - m_k + 1 \right) \qquad \left[\text{Eq. (33), (45)} \right]$$

(47)

$$\lim_{K \to \infty} P([\mathbf{Z}]) = \frac{\alpha^{K_+}}{\prod_{h=0}^{2^N - 1} K_h!} \prod_{k=1}^{K_+} \boldsymbol{\Phi}_{z_k} \frac{(N - m_k)!(m_k - 1)!}{N!} \exp(-\alpha \boldsymbol{\Psi})$$

(48)

where

$$\boldsymbol{\Psi} = \sum_{h=0}^{2^N - 1} \boldsymbol{\Phi}_h \frac{(N - m_k)!(m_k - 1)!}{N!}$$

(49)

A more elaborate derivation of (48) was described in the Appendix from [92] and omitted here. Additionally, the authors also showed that when $\mathbf{W} = 0$ or equivalently $\boldsymbol{\Phi}_h = 1$ for all histories h, then Eq. (48) reduces to the original IBP introduced in [59], which is thus a special case of the weighted IBP in GroupMiR.

Given the IBP prior Eq. (48), the generative process for z_{ik} corresponding to an existing feature k (where $m_k > 0$) was derived as follows:

$$P(z_{ik} = 1 | \mathbf{Z}_{-ik})$$

$$= \frac{P(z_{ik} = 1, \mathbf{Z}_{-ik})}{P(z_{ik} = 0, \mathbf{Z}_{-ik}) + P(z_{ik} = 1, \mathbf{Z}_{-ik})}$$

(50)

$$= \frac{\boldsymbol{\Phi}_{z_k, z_{ik}=1}(N - m_{-ik} - 1)!(m_{-ik} + 1 - 1)!}{\boldsymbol{\Phi}_{z_k, z_{ik}=0}(N - m_{-ik})!(m_{-ik} - 1)! + \boldsymbol{\Phi}_{z_k, z_{ik}=1}(N - m_{-ik} - 1)!(m_{-ik} + 1 - 1)!} \tag{51}$$

$$= \frac{\exp(\sum_{j \neq i} w_{ij} \cdot 1 \cdot z_{jk})(N - m_{-ik} - 1)!(m_{-ik} + 1 - 1)!}{\exp(\sum_{j \neq i} w_{ij} \cdot 0 \cdot z_{jk})(N - m_{-ik})!(m_{-ik} - 1)! +}$$

$$\exp\left(\sum_{j \neq i} w_{ij} \cdot 1 \cdot z_{jk}\right)(N - m_{-ik} - 1)!(m_{-ik} + 1 - 1)!$$

$$= \frac{\exp(\sum_{j \neq i} w_{ij} z_{jk}) m_{-ik}}{(N - m_{-ik}) + \exp(\sum_{j \neq i} w_{ij} z_{jk}) m_{-ik}} \tag{52}$$

where the subscript $-ik$ (e.g., $\mathbf{Z}_{-ik}$ or m_{-ik}) denotes all objects for k except for i, Eq. (51) arose from cancellations of the common terms in (48), and similar for (52). The number of new feature k^* (where $m_{k^*} = 0$) are sampled from Poisson$(\frac{\alpha}{i})$.

Notably, $\mathbf{Z}$ can be expressed as $\mathbf{Z} = (\mathbf{U}^T, \mathbf{V}^T)^T$, where $\mathbf{U}$ is a $M \times K$ binary matrix for mRNA and $\mathbf{V}$ is a $R \times K$ binary matrix for miRNA. Thus, mRNA i and miRNA j are in the same group k if $u_{ik} v_{jk} = 1$. Given $\mathbf{U}$ and $\mathbf{V}$, the regression model in GroupMiR was defined as:

$$x_i \sim \mathcal{N}\left(\mu - \sum_j \left(r_j + \sum_{k: u_{ik} v_{jk} = 1} s_k\right) y_j, \sigma^2 I\right) \tag{53}$$

where x_i is the expression of mRNA i, μ is the baseline expression for i, r_j is the regulatory weight of miRNA j, s_k is the group-specific coefficient for group k, and y_j is the expression of miRNA j. With the Gaussian distribution assumption for x_i, the data likelihood then follows as:

$$P(\mathbf{X}, \mathbf{Y} | \mathbf{Z}, \Theta) \propto \exp\left(-\frac{1}{2\sigma^2} \sum_i (x_i - \bar{x}_i)^T (x_i - \bar{x}_i)\right) \tag{54}$$

where $\Theta = (\mu, \sigma^2, \mathbf{s}, \mathbf{r})$ and $\bar{x}_i = \mu - \sum_j (r_j + \sum_{k: u_{ik} v_{jk} = 1} s_k) y_j$. The (conjugate) priors over the parameters in Θ were defined and omitted here.

Finally, the marginal posterior of $\mathbf{Z}$ are defined as:

$$P(z_{ik} | \mathbf{X}, \mathbf{Y}, \mathbf{Z}_{-(ik)}) \propto P(\mathbf{X}, \mathbf{Y} | \mathbf{Z}_{-(ik)}, z_{ik}) P(z_{ik} | z_{-ik}) \tag{55}$$

where $P(\mathbf{X}, \mathbf{Y} | \mathbf{Z}_{-(ik)}, z_{ik})$ was obtained by integrating the likelihood [Eq. (54)] over all parameters in Θ and $P(z_{ik} | z_{-ik})$ from Eq. (50).

Due to the integral involved above, analytical solution for posterior in (55) is difficult to obtain. Accordingly, the inference in GroupMiR was performed via MCMC:

1. Sample an existing column z_{ik} from Eq. (50);
2. Assuming object i is the last customer in line (i.e., $i = N$), sample Poisson($\frac{\alpha}{N}$) new columns and sample s_k from its prior (Gamma distribution) for each new column;
3. Sample the remaining parameters by Gibbs sampler if closed-form posterior of the parameter exists (due to conjugacy) or by Metropolis-Hasting using likelihood Eq. (54) to determine the acceptance ratio;
4. Repeat 1–3 until convergence.

Finally, the posteriors of $\mathbf{Z}$ in Eq. (55) for all feature column with at least one nonzero entry serve as the target prediction of GroupMiR.

GroupMiR was applied to simulated data generated from Eq. (53) with $K = 5$ (i.e., 5 latent features shared among miRNA and mRNA) and increasing noise level (0.1, 0.2, 0.4, 0.8) to the prior scoring matrix $\mathbf{C}$, mimicking the high false positive and negative rates from the sequence-based predictors. At the low noise levels of 0.1, 0.2, or 0.4, GroupMiR was able to identify exactly 5 latent features and above 90 % accuracy in predicting the correct memberships between miRNA and mRNA. At the high noise level of 0.8, on the other hand, GroupMiR started to identify > 5 latent features but the accuracy remained above 90 %, demonstrating its robustness. In contrast, GenMiR++ had much a lower performance than GroupMiR on the same data, scoring lower than 60 % accuracy at the high noise level. GroupMiR was also applied to the real microarray data with 7 time points profiling the expression of miRNA and mRNA in mouse lung development. However, only the top 10 % of the genes with highest variance were chosen leading to 219 miRNAs and 1498 mRNAs. Although the authors did not justify using such a small subset of the data, it is likely due to the model complexity that prohibited the full exploration of the data. Nonetheless, GroupMiR identified higher network connectivity and higher GO enrichment for the predicted targets than GenMiR++ on the same dataset. It would have been even more convincing, however, if GroupMiR was also tested on the same datasets of 88 human tissues, which were used in the GenMiR++ study [73].

Although the time complexity was not analyzed for GroupMiR, it appears similar to, if not higher, than the general IBP, which has a time complexity of $O(N^3)$ per iteration for N objects [38]. The slow mixing rate is the main issue for IBP-based framework due to the intensive Gibbs samplings required to perform the inference, which prevented GroupMiR from fully exploring the data space at a genome scale offered by microarray or RNA-seq platforms, and consequently compromised the model accuracy. Adaptations of efficient inference algorithms that were recently developed for IBP are crucial to unleash the full power of the NPB framework [38]. Additionally, GroupMiR did not consider the spatial relationships between adjacent target sites of the same mRNA 3'UTR for the same or different miRNAs as in PicTar (Sect. 3.1). A more biologically meaningful (IBP) prior may improve the accuracy and/or the model efficiency by restricting the possible connections in $\mathbf{Z}$. Finally, it would be interesting to further examine the biological revelation of the groupings from $\mathbf{Z}$ on various expression consortia. In particular, miRNAs participating in many groups or having higher out-degrees in network context are likely to be more functionally important than others. Moreover, the groupings may not only reveal miRNA and mRNA targeting relationships but also the regulatory roles of mRNAs as TFs on miRNA when a modified IBP prior is used. Taken together, many directions remained unexplored with the powerful NBP framework.

3.5 SNMNMF: Sparse Network-Regularized Multiple Nonnegative Matrix Factorization

The nonnegative matrix factorization (NMF) algorithm was originally developed to extract latent features from images [94, 157]. NMF serves as an attractive alternative to conventional dimensionality reduction techniques such as principle component analysis (PCA) because it factorizes the original matrix $V_{s \times n}$ (for s images and n pixels) into two nonnegative matrices $V_{s \times n} = W_{s \times k} H_{k \times n}$, where $W_{s \times k}$ and $H_{k \times n}$ are the "image encoding" and "basis image" matrices, respectively.[4] Notably, k needs to be known beforehand. The nonnegativity of the two factorized matrices enforced by the NMF algorithm provides the ground for intuitive interpretations of the latent features because the factorized matrices tend to be sparse and reflective to certain distinct local features of the original image matrix. Kim and Tidor (2003) were among the very first groups that introduced NMF into the world of computational biology [85]. In particular, the authors used NMF to assign memberships to genes based on the "image encoding" matrix $H_{k \times n}$ in order to decipher yeast expression network using gene expression data measured by microarray. Since then, many NMF-based frameworks were developed [34].

In particular, Zhang et al. [177] extended the NMF algorithm to detecting miRNA regulatory modules (MiRMs) [177]. Specifically, the authors proposed a sparse network-regularized multiple NMF (SNMNMF) technique to minimize the following objective function:

$$W, H_1, H_2 \leftarrow \underset{W, H_1, H_2}{\arg \min} \sum_{l=1,2} ||X_l - WH_l||^2 - \lambda_1 Tr(H_2 A H_2^T) - \lambda_2 Tr(H_1 B H_2^T)$$

$$+ \gamma_1 ||W||^2 + \gamma_2 \left(\sum_j ||h_j||^2 + \sum_{j'} ||h_{j'}||^2 \right) \tag{56}$$

where

- X_1 and X_2 are the $s \times n$ mRNA and $s \times m$ miRNA expression matrices, respectively, for s samples, n mRNAs, and m miRNAs;

---

[4]In the original paper [94], the image matrix $V_{s \times n}$ was transposed, where the rows and columns represent the pixel and image, respectively. The representation used here is to be consistent with the one used by Zhang et al. [177] reviewed below.

- W is the $s \times k$ encoding matrix using $k = 50$ latent features (chosen based on the number of spatially separable miRNA clusters in the human genome);
- H_1 and H_2 are the $k \times n$ and $k \times m$ "image basis" matrices for genes and miRNAs, respectively;
- A is the $n \times n$ binary gene-gene interactions (GGI) matrices as a union of the TFBS from TRANSFAC [162] and the protein-protein interactions from [22];
- B is the $n \times m$ binary miRNA-mRNA interaction matrix obtained from Micro-Cosm [60] database that hosts the target predictions from MiRanda [78];
- $\gamma_1 ||W||^2 + \gamma_2(\sum_j ||h_j||^2 + \sum_{j'} ||h_{j'}||^2)$ are regularization terms that prevent the parameter estimates from growing too large;
- the weights $\lambda_{1,2}$ and regularization parameters $\gamma_{1,2}$ were selected *post hoc*.

The original optimization algorithm of NMF was based on a simple gradient decent procedure, which operated on only a single input matrix. Here, however, the partial derivative of Eq. (56) with respect to each matrix (W, H_1, H_2) depends on the optimal solution from the other two matrices. Accordingly, the authors developed a two-stage heuristic approach, which nonetheless guarantees to converge to a local optimal: (1) update W fixing H_1, H_2 (which are initialized randomly); (2) update H_1, H_2 fixing W, repeat 1 & 2 until convergence. To cluster m/miRNAs, the authors exploited the encoding matrices H_1 $(k \times n)$ and H_2 $(k \times m)$ by transforming each entry into z-score: $z_{ij} = (h_{ij} - \bar{h}_{.j})/\sigma_j$, where $\bar{h}_{.j}$ and σ_j are the mean and standard deviation of column j of H_1 (or H_2) for mRNA j (or miRNA j'), respectively. Thus, each m/miRNA was assigned to zero or more features if their corresponding z-score is above a threshold.

SNMNMF was applied to ovarian cancer dataset containing paired m/miRNA expression profiles from TCGA measuring 559 miRNAs and 12456 genes for each of the 385 patient samples. The authors found that more than half of the 49 modules (1 module was empty) identified by SNMNMF were enriched for at least one GO terms or KEGG pathway. Also, miRNAs involved in the SNMNMF-MiRMs were enriched for cancer-related miRNAs. Moreover, Kaplan-Meier survival analysis revealed that some of the k latent features from the basis matrix $W_{s \times k}$ offered promising prognostic power. Finally, SNMNMF compared favourably with the enumeration of bi-clique (EBC) algorithm proposed by Peng et al. [120] in terms of the number of miRNAs involved in the modules and GO/pathway enrichments [120]. Specifically, the authors found that EBC tended to produce modules involving only a single miRNA and multiple mRNAs. The star-shape modules are instances of a trivial case that can be derived directly from miRNA-mRNA interaction scores rather than network analysis.

Despite the statistical rigor, there are several limitations of the SNMNMF algorithm. First, the NMF approach requires a predefined number of modules in order to perform the matrix factorization, which may be data-dependent and difficult to determine beforehand. Additionally, the NMF solution is often not unique, and the identified modules do not necessarily include both miRNAs and mRNAs, which makes reproducing and interpreting the results difficult. Moreover, the SNMNMF does not enforce negative MMEC (miRNA-mRNA expression correlation), whereas the negative MMEC is necessary to ascertain the repressive function of the miRNAs on the mRNAs within the MiRMs. Finally, SNMNMF incurs a high time complexity of $O(tk(s+m+n)^2)$ for t iterations, k modules, s samples, m miRNAs, and n mRNAs. Because n is usually large (e.g., 12,456 genes in the ovarian cancer dataset), the computation is expensive even for a small number of iterations or modules. Thus, an intuitively simple and efficient deterministic framework may serve as an attractive alternative, which we describe next.

3.6 Mirsynergy: Detecting Synergistic miRNA Regulatory Modules by Overlapping Neighborhood Expansion

In one of our recent works, we described a novel model called *Mirsynergy* that integrates m/miRNA expression profiles, target site information, and GGIs to form MiRMs, where an m/miRNA may participate in multiple MiRMs, and the module number is systematically determined given the predefined model parameters [98]. The clustering algorithm of Mirsynergy adapts from ClusterONE [115], which was intended to identify protein complex from PPI data. The ultimate goal here however is to construct a priori the MiRMs and exploit them to better explain clinical outcomes such as patient survival rate.

We formulate the construction of synergistic miRNA regulatory modules (MiRMs) as an overlapping clustering problem with two main stages. Prior to the two clustering stages, we first inferred miRNA-mRNA interaction weights (MMIW) (**W**) using m/miRNA expression data and target site information. At stage 1, we only cluster miRNAs to greedily maximize miRNA-miRNA synergy, which is proportional to the correlation between miRNAs in terms of their MMIW. At stage 2, we fix the MiRM assignments and greedily add (remove) genes to (from) each MiRM to maximize the synergy score, which is defined as a function of the MMIW matrix and the GGI weight (GGIW) matrix (**H**).

3.6.1 Two-Stage Clustering

Let **W** denote the expression-based $N \times M$ MMIW matrix obtained from the coefficients of a linear regression model such as LASSO, determined as the best performing target prediction model on our data, where $w_{i,k}$ is the scoring weight for miRNA k targeting mRNA i. Similar to the "Meet/Min" score defined by Shalgi et al. [134] for binary interactions of co-occurring targets of miRNA pairs, we define an $M \times M$ scoring matrix denoted as **S**, indicating miRNA-miRNA synergistic scores between miRNA j and k ($j \neq k$):

$$s_{j,k} = \frac{\sum_{i=1}^{N} w_{i,j} w_{i,k}}{\min[\sum_i w_{i,j}, \sum_i w_{i,k}]} \tag{57}$$

Notably, if $\mathbf{W}$ were a binary matrix, Eq. (57) became the ratio of number of targets shared between miRNA j and k over the minimum number of targets possessed by j or k, which is essentially the original "Meet/Min" score. We chose such scoring system to strictly reflect the overlapping between the two miRNA target repertoires rather than merely correlated trends as usually intended by alternative approaches such as Pearson correlation.

Similar to the cohesiveness defined by Nepusz et al. [115], we define *synergy* score $s(V_c)$ for any given MiRM V_c as follows. Let $w^{\mathrm{in}}(V_c)$ denote the total weights of the internal edges within the miRNA cluster, $w^{\mathrm{bound}}(V_c)$ the total weights of the boundary edges connecting the miRNAs within V_c to the miRNAs outside V_c, and $\alpha(V_c)$ the penalty scores for forming cluster V_c. The synergy of V_c (i.e., the objective function) is:

$$s(V_c) = \frac{w^{\mathrm{in}}(V_c)}{w^{\mathrm{in}}(V_c) + w^{\mathrm{bound}}(V_c) + \alpha(V_c)} \tag{58}$$

where $\alpha(V_c)$ reflects our limited knowledge on potential unknown targets of the added miRNA as well as the false positive targets within the cluster. Presumably, these unknown factors will affect our decision on whether miRNA k belong to cluster V_c. For instance, miRNA may target noncoding RNAs and seedless targets, which are the mRNAs with no perfect seed-match [68]. We considered only mRNA targets with seed-match to minimize the number of false positives. By default, we set $\alpha(V_c) = 2|V_c|$, where $|V_c|$ is the cardinality of V_c. Additionally, we define two scoring functions to assess the overlap $\omega(V_c, V_{c'})$ between V_c and $V_{c'}$ for $c \neq c'$ and the density $d_1(V_c)$ of any given V_c:

$$\omega(V_c, V_{c'}) = \frac{|V_c \cap V_{c'}|^2}{|V_c||V_{c'}|} \tag{59}$$

$$d_1(V_c) = \frac{2w^{\mathrm{in}}(V_c)}{m(m-1)} \tag{60}$$

where $|V_c \cap V_{c'}|$ is the total number of common elements in V_c and $V_{c'}$, and m is the number of miRNAs in V_c.

The general solution for solving an overlapping clustering problems is NP-hard [13]. Thus, we adapt a greedy-based approach [115]. The algorithm can be divided into two major steps. In step 1, we select as an initial *seed* miRNA k with the highest total weights. We then grow an MiRM V_t from seed k by iteratively including boundary or excluding internal miRNAs to maximize the synergy $s(V_t)$ [Eq. (58)] until no more node can be added or removed to improve $s(V_t)$. We then pick another miRNA that has neither been considered as seed nor included in any previously expanded V_t to form V_{t+1}. The entire process terminates when all of the miRNAs are considered. In step 2, we treat the clusters as a graph with V_c as nodes and $\omega(V_c, V_{c'}) \geq \tau$ as edges. Here τ is a free parameter. Empirically, we observed that most MiRMs are quite distinct from one another in terms of $\omega(V_c, V_{c'})$ (before the merging). Accordingly, we set τ to 0.8 to ensure merging only very similar MiRMs, which avoids producing very large MiRMs (when τ is too small). We then perform a breath-first search to find all of the weakly connected components (CC), each containing clusters that can reach directly/indirectly to one another within the CC. We merge all of the clusters in the same CC and update the synergy score accordingly.

After forming MiRMs at stage 1, we perform a similar clustering procedure by adding (removing) *only the mRNAs* to (from) each MiRM. Different from stage 1, however, we grow each existing MiRM separately with no prioritized seed selection or cluster merging, which allows us to implement a parallel computation by taking advantage of the multicore processors in the modern computers. In growing/contracting each MiRM, we maximize the same synergy function [Eq. (58)] but changing the edge weight matrix from $\mathbf{S}$ to a $(N+M) \times (N+M)$ matrix by combining $\mathbf{W}$ (the $N \times M$ MMIW matrix) and $\mathbf{H}$ (the $N \times N$ GGIW matrix). Notably, here we assume miRNA-miRNA edges to be zero. Additionally, we do not add/remove miRNAs to/from the MiRM at each greedy step at this stage. Finally, we define a new density function due to the connectivity change at stage 2:

$$d_2(V_c) = \frac{w^{\text{in}}(V_c)}{n(m + n - 1)} \tag{61}$$

where n (m) are the number of mRNAs (miRNAs) in the V_c. By default, we filter out MiRMs with $d_1(V_i) < 1e\text{-}2$ and $d_2(V_j) < 5e\text{-}3$ at stage 1 and 2, respectively. Both density thresholds were chosen based on our empirical analyses. For some datasets, in particular, we found that our greedy approach tends to produce a very large cluster involving several hundred miRNAs or several thousand mRNAs at stage 1 or 2, respectively, which are unlikely to be biologically meaningful. Despite the ever increasing synergy (by definition), however, the anomaly modules all have very low density scores, which allows us to filter them out using the above-chosen thresholds.

Notably, standard clustering methods such as k-means [26] or hierarchical clustering are not suitable for constructing MiRMs since these methods assign each data point to a unique cluster [156]. A recently developed greedy-based clustering method ClusterONE is more realistic because it allows overlap between clusters [115]. However, ClusterONE was developed with physical PPI in mind. Mirsynergy extends from ClusterONE to detecting MiRMs. The novelty of our approach resides in a two-stage clustering strategy with each stage maximizing a synergy score as a function of either the miRNA-miRNA synergistic co-regulation or miRNA-mRNA/GGI. Several methods have incorporated GGI as PPI and TFBS into predicting MiRMs [93, 177], which proved to be a more accurate approach than using miRNA-mRNA alone. Comparing with recent methods such as SNMNMF [177] and PIMiM [93], however, an advantage of our deterministic formalism is the automatic determination of module number (given the predefined thresholds to merge and filter low quality clusters) and efficient computation with the theoretical bound reduced from $O(K(T + N + M)^2)$ per iteration to only $O(M(N + M))$

for N (M) mRNA (miRNA) across T samples. Because N is usually much larger than M and T, our algorithm runs orders faster. Based on our tests on a linux server, Mirsynergy took about 2 h including the run time for LASSO to compute OV (N = 12456; M = 559; T = 385), BRCA or THCA (N = 13306; M = 710; T = 331 or 543, respectively), whereas SNMNMF took more than a day for each dataset. Using expression data for ovarian, breast, and thyroid cancer from TCGA, we compared Mirsynergy with internal controls and existing methods including SNMNMF reviewed above. Mirsynergy-MiRMs exhibit significantly higher functional enrichment and more coherent miRNA-mRNA expression anti-correlation. Based on the Kaplan-Meier survival analysis, we proposed several prognostically promising MiRMs and envisioned their utility in cancer research. Mirsynergy is available as an R/Bioconductor package at http://www.bioconductor. org/packages/release/bioc/html/Mirsynergy.html.

The success of our model is likely attributable to its ability to explicitly leverage two types of information at each clustering stage: (1) the miRNA-miRNA synergism based on the correlation of the inferred miRNA target score profiles from MMIW matrix; (2) the combinatorial miRNA regulatory effects on existing genetic network, implicated in the combined MMIW and GGIW matrices. We also explored other model formulations such as clustering m/miRNAs in a single clustering stage or using different MMIW matrices other than the one produced from LASSO, which tends to produce MiRMs each containing only one or a few miRNAs or several very large low quality MiRMs, which were then filtered out by the density threshold in either clustering stage. Notably, an MiRM containing only a single miRNA can be directly derived from the MMIW without any clustering approach. Moreover, Mirsynergy considers only neighbor nodes with nonzero edges. Thus, our model works the best on a sparse MMIW matrix such as the outputs from LASSO, which is the best performing expression-based methods based on our comparison with other alternatives. Nonetheless, the performance of Mirsynergy is sensitive to the quality of MMIW and GGIW. In this regard, other MMIW or GGIW matrices (generated from improved methods) can be easily incorporated into Mirsynergy as the function parameters by the users of the Bioconductor package (please refer to the package vignette for more details). In conclusion, with large amount of m/miRNA expression data becoming available, we believe that Mirsynergy will serve as a powerful tool for analyzing condition-specific miRNA regulatory networks.

References

1. Abecasis, G.R., Auton, A., Brooks, L.D., DePristo, M.A., Durbin, R.M., et al.: An integrated map of genetic variation from 1,092 human genomes. Nature **491**(7422), 56–65 (2012)
2. Abeel, T., Van de Peer, Y., Saeys, Y.: Toward a gold standard for promoter prediction evaluation. Bioinformatics **25**(12), i313–i320 (2009). http://www.dx.doi.org/10.1093/bioinformatics/btp191
3. Alexiou, P., Maragkakis, M., Papadopoulos, G.L., Reczko, M., Hatzigeorgiou, A.G.: Lost in translation: an assessment and perspective for computational microRNA target identification. Bioinformatics (Oxford, England) **25**(23), 3049–3055 (2009)

4. Altschul, S.F., Gish, W., Miller, W., Myers, E.W., Lipman, D.J.: Basic local alignment search tool. J. Mol. Biol. **215**(3), 403–410 (1990). doi:10.1006/jmbi.1990.9999. http://www.dx.doi.org/10.1006/jmbi.1990.9999

5. Ao, W., Gaudet, J., Kent, W.J., Muttumu, S., Mango, S.E.: Environmentally induced foregut remodeling by PHA-4/FoxA and DAF-12/NHR. Science **305**, 1743–1746 (2004)

6. Arvey, A., Larsson, E., Sander, C., Leslie, C.S., Marks, D.S.: Target mRNA abundance dilutes microRNA and siRNA activity. Mol. Syst. Biol. **6**, 1–7 (2010)

7. Babak, T., Zhang, W., Morris, Q., Blencowe, B.J., Hughes, T.R.: Probing microRNAs with microarrays: tissue specificity and functional inference. RNA (New York, NY) **10**(11), 1813–1819 (2004)

8. Badis, G., Berger, M.F., Philippakis, A.A., Talukder, S., Gehrke, A.R., Jaeger, S.A., Chan, E.T., Metzler, G., Vedenko, A., Chen, X., Kuznetsov, H., Wang, C.F., Coburn, D., Newburger, D.E., Morris, Q., Hughes, T.R., Bulyk, M.L.: Diversity and complexity in DNA recognition by transcription factors. Science **324**(5935), 1720–1723 (2009)

9. Bailey, T.L., Elkan, C.: The value of prior knowledge in discovering motifs with MEME. Proc. Int. Conf. Intell. Syst. Mol. Biol. **3**, 21–29 (1995)

10. Bailey, T.L., Gribskov, M.: Methods and statistics for combining motif match scores. J. Comput. Biol. **5**(2), 211–221 (1998)

11. Barash, Y., Bejerano, G., Friedman, N.: A simple hyper-geometric approach for discovering putative transcription factor binding sites. In: Proceedings of the First International Workshop on Algorithms in Bioinformatics, WABI '01, pp. 278–293. Springer, London (2001). http://www.dl.acm.org/citation.cfm?id=645906.673098

12. Bartel, D.P.: MicroRNAs: target recognition and regulatory functions. Cell **136**(2), 215–233 (2009)

13. Barthélemy, J.P., Brucker, F.: Np-hard approximation problems in overlapping clustering. J. Classif. **18**(2), 159–183 (2001)

14. Bateman, A., Coin, L., Durbin, R., Finn, R.D., Hollich, V., GrifRths-Jones, S., Khanna, A., Marshall, M., Moxon, S., Sonnhammer, E.L.L., Studholme, D.J., Yeats, C., Eddy, S.R.: The pfam protein families database. Nucleic Acids Res. **32**, D138–D141 (2004)

15. Berg, O.G., von Hippel, P.H.: Selection of DNA binding sites by regulatory proteins. Statistical-mechanical theory and application to operators and promoters. J. Mol. Biol. **193**(4), 723–750 (1987)

16. Berger, M.F., Philippakis, A.A., Qureshi, A.M., He, F.S., Estep, P.W., Bulyk, M.L.: Compact, universal DNA microarrays to comprehensively determine transcription-factor binding site specificities. Nat. Biotechnol. **24**, 1429–1435 (2006)

17. Berman, B.P., Nibu, Y., Pfeiffer, B.D., Tomancak, P., Celniker, S.E., Levine, M., Rubin, G.M., Eisen, M.B.: Exploiting transcription factor binding site clustering to identify cis-regulatory modules involved in pattern formation in the Drosophila genome. Proc. Natl. Acad. Sci. USA **99**(2), 757–762 (2002)

18. Bishop, C.: Pattern Recognition and Machine Learning. Information Science and Statistics, No. 605–631. Springer Science, New York, NY (2006)

19. Blanchette, M., Tompa, M.: Discovery of regulatory elements by a computational method for phylogenetic footprinting. Genome Res. **12**(5), 739–748 (2002)

20. Blanchette, M., Schwikowski, B., Tompa, M.: Algorithms for phylogenetic footprinting. J. Comput. Biol. **9**(2), 211–223 (2002)

21. Boross, G., Orosz, K., Farkas, I.J.: Human microRNAs co-silence in well-separated groups and have different predicted essentialities. Bioinformatics (Oxford, England) **25**(8), 1063–1069 (2009)

22. Bossi, A., Lehner, B.: Tissue specificity and the human protein interaction network. Mol. Syst. Biol. **5**, 260 (2009)

23. Burgler, C., Macdonald, P.M.: Prediction and verification of microRNA targets by Moving-Targets, a highly adaptable prediction method. BMC Genomics **6**, 88 (2005)

24. Bussemaker, H.J., Li, H., Siggia, E.D.: Building a dictionary for genomes: identification of presumptive regulatory sites by statistical analysis. Proc. Natl. Acad. Sci. USA **97**(18), 10,096–10,100 (2000)

25. Cancer Genome Atlas Research Network: Comprehensive genomic characterization defines human glioblastoma genes and core pathways. Nature **455**(7216), 1061–1068 (2008)
26. Celebi, M.E. (ed.): Partitional Clustering Algorithms. Springer, Berlin (2014)
27. Chan, T.M., Leung, K.S., Lee, K.H.: TFBS identification based on genetic algorithm with combined representations and adaptive post-processing. Bioinformatics **24**, 341–349 (2008)
28. Chen, X., Hughes, T.R., Morris, Q.: RankMotif++: a motif-search algorithm that accounts for relative ranks of K-mers in binding transcription factors. Bioinformatics **23**, i72–i79 (2007)
29. Chen, Y., Meyer, C.A., Liu, T., Li, W., Liu, J.S., Liu, X.S.: Mm-chip enables integrative analysis of cross-platform and between-laboratory chip-chip or chip-seq data. Genome Biol. **12**(2), R11 (2011)
30. Cheng, C., Alexander, R., Min, R., Leng, J., Yip, K.Y., Rozowsky, J., Yan, K.K., Dong, X., Djebali, S., Ruan, Y., Davis, C.A., Carninci, P., Lassman, T., Gingeras, T.R., Guigo, R., Birney, E., Weng, Z., Snyder, M., Gerstein, M.: Understanding transcriptional regulation by integrative analysis of transcription factor binding data. Genome Res. **22**(9), 1658–1667 (2012)
31. Human Genome Project: Finishing the euchromatic sequence of the human genome. Nature **431**(7011), 931–945 (2004)
32. Croce, C.M.: Causes and consequences of microRNA dysregulation in cancer. Nat. Rev. Genet. **10**(10), 704–714 (2009)
33. de Boer, C.G., Hughes, T.R.: YeTFaSCo: a database of evaluated yeast transcription factor sequence specificities. Nucleic Acids Res. **40**(Database Issue), D169–D179 (2012)
34. Devarajan, K.: Nonnegative matrix factorization: an analytical and interpretive tool in computational biology. PLoS Comput. Biol. **4**(7), e1000,029 (2008)
35. D'Haeseleer, P., Liang, S., Somogyi, R.: Genetic network inference: from co-expression clustering to reverse engineering. Bioinformatics (Oxford, England) **16**(8), 707–726 (2000). http://www.dx.doi.org/10.1093/bioinformatics/16.8.707
36. Djebali, S., Davis, C.A., Merkel, A., Dobin, A., Lassmann, T., Mortazavi, A., Tanzer, A., Lagarde, J., Lin, W., Schlesinger, F., et al.: Landscape of transcription in human cells. Nature **488**(7414), 101–108 (2013)
37. Doench, J.G., Sharp, P.A.: Specificity of microRNA target selection in translational repression. Genes Dev. **18**(5), 504–511 (2004)
38. Doshi-Velez, F., Ghahramani, Z.: Accelerated sampling for the Indian buffet process. In: Proceedings of the 26th Annual International Conference on Machine Learning, ICML '09, pp. 273–280. ACM, New York, NY (2009). doi:10.1145/1553374.1553409
39. Dubchak, I., Ryaboy, D.V.: VISTA family of computational tools for comparative analysis of DNA sequences and whole genomes. Methods Mol. Biol. **338**, 69–89 (2006)
40. Durbin, R., Eddy, S.R., Krogh, A., Mitchison, G.: Biological Sequence Analysis: Probabilistic Models of Proteins and Nucleic Acids. Cambridge University Press, Cambridge (1998). http://www.amazon.ca/exec/obidos/redirect?tag=citeulike09-20&path=ASIN/0521629713
41. ENCODE: An integrated encyclopedia of DNA elements in the human genome. Nature **489**(7414), 57–74 (2012)
42. Eskin, E., Pevzner, P.A.: Finding composite regulatory patterns in DNA sequences. Bioinformatics **18** (Suppl 1), S354–S363 (2002)
43. Favorov, A.V., Gelfand, M.S., Gerasimova, A.V., Ravcheev, D.A., Mironov, A.A., Makeev, V.J.: A Gibbs sampler for identification of symmetrically structured, spaced DNA motifs with improved estimation of the signal length. Bioinformatics **21**, 2240–2245 (2005)
44. Ferguson, J.P., Cho, J.H., Zhao, H.: A new approach for the joint analysis of multiple chip-seq libraries with application to histone modification. Stat. Appl. Genet. Mol. Biol. **11**(3), 5–25 (2012)
45. Fleischmann, R., Adams, M., White, O., Clayton, R., Kirkness, E., Kerlavage, A., Bult, C., Tomb, J., Dougherty, B., Merrick, J.: Whole-genome random sequencing and assembly of Haemophilus influenzae Rd. Science **269**, 496–512 (1995)

46. Foat, B.C., Houshmandi, S.S., Olivas, W.M., Bussemaker, H.J.: Profiling condition-specific, genome-wide regulation of mRNA stability in yeast. Proc. Natl. Acad. Sci. USA **102**, 17, 675–17,680 (2005)
47. Fordyce, P.M., Gerber, D., Tran, D., Zheng, J., Li, H., DeRisi, J.L., Quake, S.R.: De novo identification and biophysical characterization of transcription-factor binding sites with microfluidic affinity analysis. Nat. Biotechnol. **28**(9), 970–975 (2010)
48. Friedman, R.C., Farh, K.K.H., Burge, C.B., Bartel, D.P.: Most mammalian mRNAs are conserved targets of microRNAs. Genome Res. **19**(1), 92–105 (2009)
49. Frith, M.C., Li, M.C., Weng, Z.: Cluster-Buster: Finding dense clusters of motifs in DNA sequences. Nucleic Acids Res. **31**(13), 3666–3668 (2003)
50. Frith, M.C., Hansen, U., Spouge, J.L., Weng, Z.: Finding functional sequence elements by multiple local alignment. Nucleic Acids Res. **32**, 189–200 (2004)
51. Fulton, D.L., Sundararajan, S., Badis, G., Hughes, T.R., Wasserman, W.W., Roach, J.C., Sladek, R.: TFCat: the curated catalog of mouse and human transcription factors. Genome Biol. **10**(3), R29 (2009)
52. Galas, D.J., Schmitz, A.: DNAse footprinting: a simple method for the detection of protein-DNA binding specificity. Nucleic Acids Res. **5**(9), 3157–3170 (1987)
53. Garcia, D.M., Baek, D., Shin, C., Bell, G.W., Grimson, A., Bartel, D.P.: Weak seed-pairing stability and high target-site abundance decrease the proficiency of lsy-6 and other microRNAs. Nat. Struct. Mol. Biol. **18**(10), 1139–1146 (2011)
54. Garner, M.M., Revzin, A.: A gel electrophoresis method for quantifying the binding of proteins to specific DNA regions: application to components of the escherichia coli lactose operon regulatory system. Nucleic Acids Res. **9**(13), 3047–3060 (1981)
55. Gasch, A.P., Moses, A.M., Chiang, D.Y., Fraser, H.B., Berardini, M., Eisen, M.B.: Conservation and evolution of cis-regulatory systems in ascomycete fungi. PLoS Biol. **2**(5), e398 (2004). doi:10.1371/journal.pbio.0020398
56. Gerstein, M.B., Kundaje, A., Hariharan, M., Landt, S.G., Yan, K.K., Cheng, C., Mu, X.J., Khurana, E., Rozowsky, J., Alexander, R., Min, R., Alves, P., Abyzov, A., Addleman, N., Bhardwaj, N., Boyle, A.P., Cayting, P., Charos, A., Chen, D.Z., Cheng, Y., Clarke, D., Eastman, C., Euskirchen, G., Frietze, S., Fu, Y., Gertz, J., Grubert, F., Harmanci, A., Jain, P., Kasowski, M., Lacroute, P., Leng, J., Lian, J., Monahan, H., O'Geen, H., Ouyang, Z., Partridge, E.C., Patacsil, D., Pauli, F., Raha, D., Ramirez, L., Reddy, T.E., Reed, B., Shi, M., Slifer, T., Wang, J., Wu, L., Yang, X., Yip, K.Y., Zilberman-Schapira, G., Batzoglou, S., Sidow, A., Farnham, P.J., Myers, R.M., Weissman, S.M., Snyder, M.: Architecture of the human regulatory network derived from ENCODE data. Nature **489**(7414), 91–100 (2012)
57. Giannopoulou, E.G., Elemento, O.: Inferring chromatin-bound protein complexes from genome-wide binding assays. Genome Res. **23**(8), 1295–1306 (2013)
58. Goffeau, A., Barrell, B., Bussey, H., Davis, R., Dujon, B., Feldmann, H., Galibert, F., Hoheisel, J., Jacq, C., Johnston, M., Louis, E., Mewes, H., Murakami, Y., Philippsen, P., Tettelin, H., Oliver, S.: Life with 6000 genes. Science **274**, 563–567 (1996)
59. Griffiths, T., Ghahramani, Z.: Infinite latent feature models and the Indian buffet process. In: Neural Information Processing Systems, pp. 475–482. MIT Press, Cambridge (2005)
60. Griffiths-Jones, S., Saini, H.K., van Dongen, S., Enright, A.J.: miRBase: tools for microRNA genomics. Nucleic Acids Res. **36**(Database Issue), D154–D158 (2008)
61. Grimson, A., Farh, K.K.H., Johnston, W.K., Garrett-Engele, P., Lim, L.P., Bartel, D.P.: MicroRNA targeting specificity in mammals: determinants beyond seed pairing. Mol. Cell **27**(1), 91–105 (2007)
62. Grün, D., Wang, Y.L., Langenberger, D., Gunsalus, K.C., Rajewsky, N.: microRNA target predictions across seven Drosophila species and comparison to mammalian targets. PLoS Computational Biology **1**(1), e13 (2005)
63. GuhaThakurta, D.: Computational identification of transcriptional regulatory elements in DNA sequence. Nucleic Acids Res. **34**, 3585–3598 (2006)
64. Gunewardena, S., Zhang, Z.: A hybrid model for robust detection of transcription factor binding sites. Bioinformatics **24**(4), 484–491 (2008)

65. Guo, H., Ingolia, N.T., Weissman, J.S., Bartel, D.P.: Mammalian microRNAs predominantly act to decrease target mRNA levels. Nature **466**(7308), 835–840 (2010)
66. Guo, Y., Mahony, S., Gifford, D.K.: High resolution genome wide binding event finding and motif discovery reveals transcription factor spatial binding constraints. PLoS Comput. Biol. **8**(8), e1002,638 (2012)
67. He, X., Ling, X., Sinha, S.: Alignment and prediction of cis-regulatory modules based on a probabilistic model of evolution. PLoS Comput. Biol. **5**(3), e1000,299 (2009)
68. Helwak, A., Kudla, G., Dudnakova, T., Tollervey, D.: Mapping the human miRNA interactome by CLASH reveals frequent noncanonical binding. Cell **153**(3), 654–665 (2013)
69. Herrmann, C., Van de Sande, B., Potier, D., Aerts, S.: i-cisTarget: an integrative genomics method for the prediction of regulatory features and cis-regulatory modules. Nucleic Acids Res. **40**(15), e114 (2012)
70. Hertz, G.Z., Stormo, G.D.: Identifying DNA and protein patterns with statistically significant alignments of multiple sequences. Bioinformatics **15**, 563–577 (1999)
71. Ho, S.W., Jona, G., Chen, C.T., Johnston, M., Snyder, M.: Linking DNA-binding proteins to their recognition sequences by using protein microarrays. Proc. Natl. Acad. Sci. USA **103**(26), 9940–9945 (2006)
72. Hu, S., Xie, Z., Onishi, A., Yu, X., Jiang, L., Lin, J., Rho, H.S., Woodard, C., Wang, H., Jeong, J.S., Long, S., He, X., Wade, H., Blackshaw, S., Qian, J., Zhu, H.: Profiling the human protein-DNA interactome reveals ERK2 as a transcriptional repressor of interferon signaling. Cell **139**(3), 610–622 (2009)
73. Huang, J.C., Babak, T., Corson, T.W., Chua, G., Khan, S., Gallie, B.L., Hughes, T.R., Blencowe, B.J., Frey, B.J., Morris, Q.D.: Using expression profiling data to identify human microRNA targets. Nat. Methods **4**(12), 1045–1049 (2007)
74. Hughes, J.D., Estep, P.W., Tavazoie, S., Church, G.M.: Computational identification of cis-regulatory elements associated with groups of functionally related genes in Saccharomyces cerevisiae. J. Mol. Biol. **296**, 1205–1214 (2000)
75. Initiative, A.G.: Analysis of the genome sequence of the flowering plant Arabidopsis thaliana. Nature **408**, 796–815 (2000)
76. Ji, H., Jiang, H., Ma, W., Johnson, D.S., Myers, R.M., Wong, W.H.: An integrated software system for analyzing ChIP-chip and ChIP-seq data. Nat. Biotechnol. **26**(11), 1293–1300 (2008)
77. Ji, H., Li, X., Wang, Q.F., Ning, Y.: Differential principal component analysis of chip-seq. Proc. Natl. Acad. Sci. **110**(17), 6789–6794 (2013)
78. John, B., Enright, A.J., Aravin, A., Tuschl, T., Sander, C., Marks, D.S.: Human MicroRNA targets. PLoS Biol. **2**(11), e363 (2004)
79. Johnson, D.S., Mortazavi, A., Myers, R.M., Wold, B.: Genome-wide mapping of in vivo protein-DNA interactions. Science **316**(5830), 1497–1502 (2007)
80. Joung, J.G., Hwang, K.B., Nam, J.W., Kim, S.J., Zhang, B.T.: Discovery of microRNA-mRNA modules via population-based probabilistic learning. Bioinformatics (Oxford, England) **23**(9), 1141–1147 (2007)
81. Keller, A., Leidinger, P., Bauer, A., ElSharawy, A., Haas, J., Backes, C., Wendschlag, A., Giese, N., Tjaden, C., Ott, K., Werner, J., Hackert, T., Ruprecht, K., Huwer, H., Huebers, J., Jacobs, G., Rosenstiel, P., Dommisch, H., Schaefer, A., Müller-Quernheim, J., Wullich, B., Keck, B., Graf, N., Reichrath, J., Vogel, B., Nebel, A., Jager, S.U., Staehler, P., Amarantos, I., Boisguerin, V., Staehler, C., Beier, M., Scheffler, M., Büchler, M.W., Wischhusen, J., Haeusler, S.F.M., Dietl, J., Hofmann, S., Lenhof, H.P., Schreiber, S., Katus, H.A., Rottbauer, W., Meder, B., Hoheisel, J.D., Franke, A., Meese, E.: Toward the blood-borne miRNome of human diseases. Nat. Methods **8**(10), 841–843 (2011)
82. Kellis, M., Patterson, N., Birren, B., Berger, B., Lander, E.S.: Methods in comparative genomics: genome correspondence, gene identification and regulatory motif discovery. J. Comput. Biol. **11**(2–3), 319–355 (2004)
83. Khan, A.A., Betel, D., Miller, M.L., Sander, C., Leslie, C.S., Marks, D.S.: Transfection of small RNAs globally perturbs gene regulation by endogenous microRNAs. Nat. Biotechnol. **27**(6), 549–555 (2009)

84. Kharchenko, P.V., Tolstorukov, M.Y., Park, P.J.: Design and analysis of ChIP-seq experiments for DNA-binding proteins. Nat. Biotechnol. **26**(12), 1351–1359 (2008)

85. Kim, P.M., Tidor, B.: Subsystem identification through dimensionality reduction of large-scale gene expression data. Genome Res. **13**(7), 1706–1718 (2003)

86. Kim, T., Tyndel, M.S., Huang, H., Sidhu, S.S., Bader, G.D., Gfeller, D., Kim, P.M.: MUSI: an integrated system for identifying multiple specificity from very large peptide or nucleic acid data sets. Nucleic Acids Res. **40**(6), e47 (2012)

87. Kiriakidou, M., Nelson, P.T., Kouranov, A., Fitziev, P., Bouyioukos, C., Mourelatos, Z., Hatzigeorgiou, A.: A combined computational-experimental approach predicts human microRNA targets. Genes Dev. **18**(10), 1165–1178 (2004)

88. Kozomara, A., Griffiths-Jones, S.: miRBase: annotating high confidence microRNAs using deep sequencing data. Nucleic Acids Res. **42**(1), D68–D73 (2014)

89. Krek, A., Grün, D., Poy, M.N., Wolf, R., Rosenberg, L., Epstein, E.J., MacMenamin, P., da Piedade, I., Gunsalus, K.C., Stoffel, M., Rajewsky, N.: Combinatorial microRNA target predictions. Nat. Genet. **37**(5), 495–500 (2005)

90. Laajala, T.D., Raghav, S., Tuomela, S., Lahesmaa, R., Aittokallio, T., Elo, L.L.: A practical comparison of methods for detecting transcription factor binding sites in ChIP-seq experiments. BMC Genomics **10**, 618 (2009)

91. Lall, S., Grün, D., Krek, A., Chen, K., Wang, Y.L., Dewey, C.N., Sood, P., Colombo, T., Bray, N., MacMenamin, P., Kao, H.L., Gunsalus, K.C., Pachter, L., Piano, F., Rajewsky, N.: A genome-wide map of conserved MicroRNA targets in C. elegans. Curr. Biol. **16**(5), 460–471 (2006)

92. Le, H.S., Bar-Joseph, Z.: Inferring interaction networks using the IBP applied to microRNA target prediction. In: Advances in Neural Information Processing Systems, pp. 235–243 (2011)

93. Le, H.S., Bar-Joseph, Z.: Integrating sequence, expression and interaction data to determine condition-specific miRNA regulation. Bioinformatics (Oxford, England) **29**(13), i89–i97 (2013)

94. Lee, D.D., Seung, H.S.: Learning the parts of objects by non-negative matrix factorization. Nature **401**(6755), 788–791 (1999)

95. Lewis, B.P., Shih, I.H., Jones-Rhoades, M.W., Bartel, D.P., Burge, C.B.: Prediction of mammalian microRNA targets. Cell **115**(7), 787–798 (2003)

96. Lewis, B.P., Burge, C.B., Bartel, D.P.: Conserved seed pairing, often flanked by Adenosines, indicates that thousands of human genes are microRNA targets. Cell **120**(1), 15–20 (2005)

97. Li, Y., Goldenberg, A., Wong, K.C., Zhang, Z.: A probabilistic approach to explore human miRNA targetome by integrating miRNA-overexpression data and sequence information. Bioinformatics (Oxford, England) **30**(5), 621–628 (2014)

98. Li, Y., Liang, C., Wong, K.C., Luo, J., Zhang, Z.: Mirsynergy: detecting synergistic miRNA regulatory modules by overlapping neighbourhood expansion. Bioinformatics (Oxford, England) **30**(18), 2627–2635 (2014)

99. Liang, Y., Leung, K.S., Mok, T.S.K.: Evolutionary drug scheduling models with different toxicity metabolism in cancer chemotherapy. Appl. Soft Comput. **8**(1), 140–149 (2008). http://www.dx.doi.org/10.1016/j.asoc.2006.12.002

100. Lifanov, A.P., Makeev, V.J., Nazina, A.G., Papatsenko, D.A.: Homotypic regulatory clusters in Drosophila. Genome Res. **13**(4), 579–588 (2003)

101. Lim, L.P., Lau, N.C., Garrett-Engele, P., Grimson, A., Schelter, J.M., Castle, J., Bartel, D.P., Linsley, P.S., Johnson, J.M.: Microarray analysis shows that some microRNAs downregulate large numbers of target mRNAs. Nature **433**(7027), 769–773 (2005)

102. Liu, X.S., Brutlag, D.L., Liu, J.S.: An algorithm for finding protein-DNA binding sites with applications to chromatin-immunoprecipitation microarray experiments. Nat. Biotechnol. **20**, 835–839 (2002)

103. Lorenz, R., Bernhart, S.H., Höner Zu Siederdissen, C., Tafer, H., Flamm, C., Stadler, P.F., Hofacker, I.L.: ViennaRNA package 2.0. Algorithms Mol. Biol. **6**, 26 (2011)

104. Lu, J., Getz, G., Miska, E.A., Alvarez-Saavedra, E., Lamb, J., Peck, D., Sweet-Cordero, A., Ebert, B.L., Mak, R.H., Ferrando, A.A., Downing, J.R., Jacks, T., Horvitz, H.R., Golub, T.R.: MicroRNA expression profiles classify human cancers. Nature 435(7043), 834–838 (2005)
105. MacKay, D.J.: Ensemble learning for hidden markov models. Tech. Rep., Cavendish Laboratory, Cambridge (1997)
106. Mahony, S., Edwards, M.D., Mazzoni, E.O., Sherwood, R.I., Kakumanu, A., Morrison, C.A., Wichterle, H., Gifford, D.K.: An integrated model of multiple-condition chip-seq data reveals predeterminants of cdx2 binding. PLoS Comput. Biol. 10(3), e1003,501 (2014)
107. Matys, V., Kel-Margoulis, O.V., Fricke, E., Liebich, I., Land, S., Barre-Dirrie, A., Reuter, I., Chekmenev, D., Krull, M., Hornischer, K., Voss, N., Stegmaier, P., Lewicki-Potapov, B., Saxel, H., Kel, A.E., Wingender, E.: Transfac and its module transcompel: transcriptional gene regulation in eukaryotes. Nucleic Acids Res. 34, 108–110 (2006)
108. McGuffin, L.J., Bryson, K., Jones, D.T.: The psipred protein structure prediction server. Bioinformatics (Oxford, England) 16(4), 404–405 (2000). doi:10.1093/bioinformatics/16.4.404. http://www.dx.doi.org/10.1093/bioinformatics/16.4.404
109. Meunier, J., Lemoine, F., Soumillon, M., Liechti, A., Weier, M., Guschanski, K., Hu, H., Khaitovich, P., Kaessmann, H.: Birth and expression evolution of mammalian microRNA genes. Genome Res. (2012)
110. Mohan, P.M., Hosur, R.V.: Structure-function-folding relationships and native energy landscape of dynein light chain protein: nuclear magnetic resonance insights. J. Biosci. 34, 465–479 (2009)
111. Montgomery, S., Griffith, O., Sleumer, M., Bergman, C., Bilenky, M., Pleasance, E., Prychyna, Y., Zhang, X., Jones, S.: ORegAnno: an open access database and curation system for literature-derived promoters, transcription factor binding sites and regulatory variation. Bioinformatics 22, 637–640 (2006)
112. Moses, A.M., Sinha, S.: Regulatory motif analysis. In: Edwards, D., Stajich, J., Hansen, D. (eds.) Bioinformatics: Tools and Applications. Springer Biomedical and Life Sciences Collection, pp. 137–163. Springer, New York (2009)
113. Moses, A.M., Chiang, D.Y., Eisen, M.B.: Phylogenetic motif detection by expectation-maximization on evolutionary mixtures. Pac. Symp. Biocomput., 324–335 (2004)
114. Moses, A.M., Chiang, D.Y., Pollard, D.A., Iyer, V.N., Eisen, M.B.: MONKEY: identifying conserved transcription-factor binding sites in multiple alignments using a binding site-specific evolutionary model. Genome Biol. 5(12), R98 (2004)
115. Nepusz, T., Yu, H., Paccanaro, A.: Detecting overlapping protein complexes in protein-protein interaction networks. Nat. Methods 9(5), 471–472 (2012)
116. Nishida, K., Frith, M.C., Nakai, K.: Pseudocounts for transcription factor binding sites. Nucleic Acids Res. 37(3), 939–944 (2009)
117. Ovcharenko, I., Boffelli, D., Loots, G.G.: eShadow: a tool for comparing closely related sequences. Genome Res. 14(6), 1191–1198 (2004)
118. Papadopoulos, G.L., Alexiou, P., Maragkakis, M., Reczko, M., Hatzigeorgiou, A.G.: DIANA-mirPath: integrating human and mouse microRNAs in pathways. Bioinformatics (Oxford, England) 25(15), 1991–1993 (2009)
119. Pavesi, G., Mereghetti, P., Mauri, G., Pesole, G.: Weeder Web: discovery of transcription factor binding sites in a set of sequences from co-regulated genes. Nucleic Acids Res. 32, 199–203 (2004)
120. Peng, X., Li, Y., Walters, K.A., Rosenzweig, E.R., Lederer, S.L., Aicher, L.D., Proll, S., Katze, M.G.: Computational identification of hepatitis C virus associated microRNA-mRNA regulatory modules in human livers. BMC Genomics 10, 373 (2009)
121. Pfreundt, U., James, D.P., Tweedie, S., Wilson, D., Teichmann, S.A., Adryan, B.: FlyTF: improved annotation and enhanced functionality of the Drosophila transcription factor database. Nucleic Acids Res. 38(Database Issue), D443–D447 (2010)
122. Philippakis, A.A., He, F.S., Bulyk, M.L.: Modulefinder: a tool for computational discovery of cis regulatory modules. Pac. Symp. Biocomput. pp. 519–530 (2005)

123. Portales-Casamar, E., Thongjuea, S., Kwon, A.T., Arenillas, D., Zhao, X., Valen, E., Yusuf, D., Lenhard, B., Wasserman, W.W., Sandelin, A.: JASPAR 2010: the greatly expanded open-access database of transcription factor binding profiles. Nucleic Acids Res. **38**(Database Issue), D105–D110 (2010)

124. Rajewsky, N., Vergassola, M., Gaul, U., Siggia, E.D.: Computational detection of genomic cis-regulatory modules applied to body patterning in the early Drosophila embryo. BMC Bioinf. **3**, 30 (2002)

125. Ramaswamy, S., Tamayo, P., Rifkin, R., Mukherjee, S., Yeang, C.H., Angelo, M., Ladd, C., Reich, M., Latulippe, E., Mesirov, J.P., Poggio, T., Gerald, W., Loda, M., Lander, E.S., Golub, T.R.: Multiclass cancer diagnosis using tumor gene expression signatures. Proc. Natl. Acad. Sci. USA **98**(26), 15,149–15,154 (2001)

126. Régnier, M., Denise, A.: Rare events and conditional events on random strings. Discret. Math. Theor. Comput. Sci. **6**(2), 191–214 (2004). http://www.dmtcs.loria.fr/volumes/abstracts/dm060203.abs.html

127. Ren, B., Robert, F., Wyrick, J.J., Aparicio, O., Jennings, E.G., Simon, I., Zeitlinger, J., Schreiber, J., Hannett, N., Kanin, E., Volkert, T.L., Wilson, C.J., Bell, S.P., Young, R.A.: Genome-wide location and function of DNA binding proteins. Science **290**(5500), 2306–2309 (2000)

128. Robasky, K., Bulyk, M.L.: UniPROBE, update 2011: expanded content and search tools in the online database of protein-binding microarray data on protein-DNA interactions. Nucleic Acids Res. **39**, D124–D128 (2011)

129. Ronquist, F., Huelsenbeck, J.P.: Mrbayes 3: Bayesian phylogenetic inference under mixed models. Bioinformatics **19**(12), 1572–1574 (2003). doi:10.1093/bioinformatics/btg180. http://www.dx.doi.org/10.1093/bioinformatics/btg180

130. Rozowsky, J., Euskirchen, G., Auerbach, R.K., Zhang, Z.D., Gibson, T., Bjornson, R., Carriero, N., Snyder, M., Gerstein, M.B.: PeakSeq enables systematic scoring of ChIP-seq experiments relative to controls. Nat. Biotechnol. **27**(1), 66–75 (2009)

131. Sandelin, A., Wasserman, W.W., Lenhard, B.: ConSite: web-based prediction of regulatory elements using cross-species comparison. Nucleic Acids Res. **32**(Web Server Issue), W249–W252 (2004)

132. Segal, E., Yelensky, R., Koller, D.: Genome-wide discovery of transcriptional modules from DNA sequence and gene expression. Bioinformatics **19**(Suppl 1), i273–i282 (2003)

133. Sethupathy, P., Megraw, M., Hatzigeorgiou, A.G.: A guide through present computational approaches for the identification of mammalian microRNA targets. Nat. Methods **3**(11), 881–886 (2006)

134. Shalgi, R., Lieber, D., Oren, M., Pilpel, Y.: Global and local architecture of the mammalian microRNA-transcription factor regulatory network. PLoS Comput. Biol. **3**(7), e131 (2007)

135. Siddharthan, R., Siggia, E.D., van Nimwegen, E.: PhyloGibbs: a Gibbs sampling motif finder that incorporates phylogeny. PLoS Comput. Biol. **1**(7), e67 (2005)

136. Siepel, A., Bejerano, G., Pedersen, J.S., Hinrichs, A.S., Hou, M., Rosenbloom, K., Clawson, H., Spieth, J., Hillier, L.W., Richards, S., Weinstock, G.M., Wilson, R.K., Gibbs, R.A., Kent, W.J., Miller, W., Haussler, D.: Evolutionarily conserved elements in vertebrate, insect, worm, and yeast genomes. Genome Res. **15**(8), 1034–1050 (2005)

137. Sinha, S., He, X.: MORPH: probabilistic alignment combined with hidden Markov models of cis-regulatory modules. PLoS Comput. Biol. **3**(11), e216 (2007)

138. Sinha, S., Tompa, M.: YMF: A program for discovery of novel transcription factor binding sites by statistical overrepresentation. Nucleic Acids Res. **31**, 3586–3588 (2003)

139. Sinha, S., van Nimwegen, E., Siggia, E.D.: A probabilistic method to detect regulatory modules. Bioinformatics (Oxford, England) **19**(Suppl 1), i292–i301 (2003)

140. Sinha, S., Liang, Y., Siggia, E.: Stubb: a program for discovery and analysis of cis-regulatory modules. Nucleic Acids Res. **34**(Web Server Issue), W555–W559 (2006)

141. Smith, A.D., Sumazin, P., Zhang, M.Q.: Identifying tissue-selective transcription factor binding sites in vertebrate promoters. Proc. Natl. Acad. Sci. USA **102**(5), 1560–1565 (2005)

142. Smyth, M.S., Martin, J.H.: X ray crystallography. Mol. Pathol. **53**(1), 8–14 (2000). http://www.view.ncbi.nlm.nih.gov/pubmed/10884915

143. Song, L., Tuan, R.S.: MicroRNAs and cell differentiation in mammalian development. Birth Defects Res. C Embryo Today **78**(2), 140–149 (2006)

144. Spivak, A.T., Stormo, G.D.: ScerTF: a comprehensive database of benchmarked position weight matrices for Saccharomyces species. Nucleic Acids Res. **40**(Database Issue), D162–D168 (2012)

145. Staden, R.: Methods for calculating the probabilities of finding patterns in sequences. Comput. Appl. Biosci. **5**(2), 89–96 (1989)

146. Stormo, G.D.: Maximally efficient modeling of dna sequence motifs at all levels of complexity. Genetics **187**(4), 1219–1224 (2011). http://www.dx.doi.org/10.1534/genetics.110.126052

147. Su, J., Teichmann, S.A., Down, T.A.: Assessing computational methods of cis-regulatory module prediction. PLoS Comput. Biol. **6**(12), e1001,020 (2010)

148. Tanay, A.: Extensive low-affinity transcriptional interactions in the yeast genome. Genome Res. **16**, 962–972 (2006)

149. Thibaux, R., Jordan, M.I.: Hierarchical beta processes and the Indian buffet process. Int. Conf. Artif. Intell. Stat. **11**, 564–571 (2007)

150. Thijs, G., Lescot, M., Marchal, K., Rombauts, S., De Moor, B., Rouze, P., Moreau, Y.: A higher-order background model improves the detection of promoter regulatory elements by Gibbs sampling. Bioinformatics **17**, 1113–1122 (2001)

151. Tomovic, A., Oakeley, E.J.: Position dependencies in transcription factor binding sites. Bioinformatics **23**(8), 933–941 (2007)

152. Tompa, M., Li, N., Bailey, T.L., Church, G.M., Moor, B.D., Eskin, E., Favorov, A.V., Frith, M.C., Fu, Y., Kent, W.J., Makeev, V.J., Mironov, A.A., Noble, W.S., Pavesi, G., Pesole, G., Regnier, M., Simonis, N., Sinha, S., Thijs, G., van Helden, J., Vandenbogaert, M., Weng, Z., Workman, C., Ye, C., , Zhu, Z.: Assessing computational tools for the discovery of transcription factor binding sites. Nat. Biotechnol. **23**(1), 137–144 (2005)

153. Tsang, J.S., Ebert, M.S., van Oudenaarden, A.: Genome-wide dissection of microRNA functionsand cotargeting networks using gene set signatures. Mol. Cell **38**(1), 140–153 (2010)

154. Tuerk, C., Gold, L.: Systematic evolution of ligands by exponential enrichment: RNA ligands to bacteriophage T4 DNA polymerase. Science **249**(4968), 505–510 (1990)

155. van Helden, J., Andre, B., Collado-Vides, J.: Extracting regulatory sites from the upstream region of yeast genes by computational analysis of oligonucleotide frequencies. J. Mol. Biol. **281**, 827–842 (1998)

156. Wang, J.J., Bensmail, H., Gao, X.: Multiple graph regularized protein domain ranking. BMC Bioinf. **13**, 307 (2012)

157. Wang, J.J.Y., Bensmail, H., Gao, X.: Multiple graph regularized nonnegative matrix factorization. Pattern Recogn. **46**(10), 2840–2847 (2013)

158. Warner, J.B., Philippakis, A.A., Jaeger, S.A., He, F.S., Lin, J., Bulyk, M.L.: Systematic identification of mammalian regulatory motifs' target genes and functions. Nat. Methods **5**(4), 347–353 (2008)

159. Wasserman, W.W., Sandelin, A.: Applied bioinformatics for the identification of regulatory elements. Nat. Rev. Genet. **5**(4), 276–287 (2004)

160. Weinstein, J.N., Collisson, E.A., Mills, G.B., Shaw, K.R.M., Ozenberger, B.A., Ellrott, K., Shmulevich, I., Sander, C., Stuart, J.M.: The cancer genome atlas pan-cancer analysis project. Nat. Genet. **45**(10), 1113–1120 (2013)

161. Wilbanks, E.G., Facciotti, M.T.: Evaluation of algorithm performance in ChIP-seq peak detection. PLoS ONE **5**(7), e11,471 (2010)

162. Wingender, E., Chen, X., Hehl, R., Karas, H., Liebich, I., Matys, V., Meinhardt, T., Prüss, M., Reuter, I., Schacherer, F.: TRANSFAC: an integrated system for gene expression regulation. Nucleic Acids Res. **28**(1), 316–319 (2000)

163. Wong, K.C., Leung, K.S., Wong, M.H.: An evolutionary algorithm with species-specific explosion for multimodal optimization. In: GECCO '09: Proceedings of the 11th Annual Conference on Genetic and Evolutionary Computation, pp. 923–930. ACM, New York, NY (2009). http://www.doi.acm.org/10.1145/1569901.1570027

164. Wong, K.C., Leung, K.S., Wong, M.H.: Protein structure prediction on a lattice model via multimodal optimization techniques. In: Proceedings of the 12th Annual Conference on Genetic and Evolutionary Computation, pp. 155–162. ACM (2010)

165. Wong, K.C., Peng, C., Wong, M.H., Leung, K.S.: Generalizing and learning protein-dna binding sequence representations by an evolutionary algorithm. Soft Comput. **15**(8), 1631–1642 (2011). doi:10.1007/s00500-011-0692-5. http://www.dx.doi.org/10.1007/s00500-011-0692-5

166. Wong, K.C., Wu, C.H., Mok, R.K.P., Peng, C., Zhang, Z.: Evolutionary multimodal optimization using the principle of locality. Inf. Sci. **194**, 138–170 (2012)

167. Wong, K.C., Chan, T.M., Peng, C., Li, Y., Zhang, Z.: DNA motif elucidation using belief propagation. Nucleic Acids Res. **41**(16), e153 (2013)

168. Wong, K.C., Li, Y., Peng, C., Zhang, Z.: Signalspider: probabilistic pattern discovery on multiple normalized chip-seq signal profiles. Bioinformatics **31**(1), 17–24 (2014)

169. Workman, C.T., Stormo, G.D.: ANN-Spec: a method for discovering transcription factor binding sites with improved specificity. Pac. Symp. Biocomput., 467–478 (2000)

170. Wu, J., Xie, J.: Computation-based discovery of cis-regulatory modules by hidden Markov model. J. Comput. Biol.: J. Comput. Mol. Cell Biol. **15**(3), 279–290 (2008)

171. Xie, X., Rigor, P., Baldi, P.: MotifMap: a human genome-wide map of candidate regulatory motif sites. Bioinformatics **25**(2), 167–174 (2009)

172. Xie, D., Boyle, A.P., Wu, L., Zhai, J., Kawli, T., Snyder, M.: Dynamic trans-acting factor colocalization in human cells. Cell **155**(3), 713–724 (2013)

173. Xu, J., Li, C.X., Li, Y.S., Lv, J.Y., Ma, Y., Shao, T.T., Xu, L.D., Wang, Y.Y., Du, L., Zhang, Y.P., Jiang, W., Li, C.Q., Xiao, Y., Li, X.: MiRNA-miRNA synergistic network: construction via co-regulating functional modules and disease miRNA topological features. Nucleic Acids Res. **39**(3), 825–836 (2011)

174. Yue, D., Liu, H., Huang, Y.: Survey of Computational Algorithms for MicroRNA Target Prediction. Curr. Genomics **10**(7), 478–492 (2009)

175. Zeng, X., Sanalkumar, R., Bresnick, E.H., Li, H., Chang, Q., Kele, S.: jMOSAiCS: joint analysis of multiple ChIP-seq datasets. Genome Biol. **14**(4), R38 (2013)

176. Zhang, Y., Liu, T., Meyer, C.A., Eeckhoute, J., Johnson, D.S., Bernstein, B.E., Nusbaum, C., Myers, R.M., Brown, M., Li, W., Liu, X.S.: Model-based analysis of ChIP-Seq (MACS). Genome Biol. **9**(9), R137 (2008)

177. Zhang, S., Li, Q., Liu, J., Zhou, X.J.: A novel computational framework for simultaneous integration of multiple types of genomic data to identify microRNA-gene regulatory modules. Bioinformatics (Oxford, England) **27**(13), i401–i409 (2011)

178. Zhou, Q., Wong, W.H.: CisModule: de novo discovery of cis-regulatory modules by hierarchical mixture modeling. Proc. Natl. Acad. Sci. USA **101**(33), 12,114–12,119 (2004)

179. Zia, A., Moses, A.M.: Towards a theoretical understanding of false positives in DNA motif finding. BMC Bioinformatics **13**, 151 (2012)

The Application of LSA to the Evaluation of Questionnaire Responses

Dian I. Martin, John C. Martin and Michael W. Berry

Abstract This chapter will discuss research surrounding the development of applications for automated evaluation and generation of instructional feedback based on the analysis of responses to open-ended essay questions. While an effective means of evaluation, examining responses to essay questions requires natural language processing that often demands human input and guidance which can be labor intensive, time consuming, and language dependent. In order to provide this means of evaluation on a larger scale, an automated unsupervised learning approach is necessary that overcomes all these limitations. Latent Semantic Analysis (LSA) is an unsupervised learning system used for deriving and representing the semantic relationships between items in a body of natural language content. It mimics the representation of meaning that is formed by a human who learns linguistic constructs through exposure to natural language over time. The applications described in this chapter leverage LSA as an unsupervised system to learn language and provide a semantic framework that can be used for mapping natural language responses, evaluating the quality of those responses, and identifying relevant instructional feedback based on their semantic content. We will discuss the learning algorithms used to construct the LSA framework of meaning as well as methods to apply that framework for the evaluation and generation of feedback.

Keywords Latent Semantic Analysis (LSA) • Essay grading • Feedback generation • Simulating human raters • Cognitive model • Unsupervised learning • Natural language responses • Open-ended essay questions

D.I. Martin (✉) • J.C. Martin
Small Bear Technologies, Inc., Thorn Hill, TN, USA
e-mail: Dian.Martin@SmallBearTechnologies.com

M.W. Berry
University of Tennessee, Knoxville, TN, USA

© Springer International Publishing Switzerland 2016
M.E. Celebi, K. Aydin (eds.), *Unsupervised Learning Algorithms*,
DOI 10.1007/978-3-319-24211-8_16

449

1 Introduction

The use of computerized systems for evaluating and understanding open-ended responses to test and survey questions has been researched and discussed for over 50 years in an effort to overcome the fundamental problem of a need for human effort to score them. While closed response questions where a subject picks from a set of pre-defined answers are readily amenable to automated scoring and have shown themselves to be very useful in standardized testing and information surveys, they cannot be used in all situations and do not permit the sort of assessment options that open-ended prompts can provide. The use of essay and short answer questions in testing and surveys persists because of the unique insight they can provide. Advances in computer technology and our understanding of language, thought, and learning have made it possible to produce automated systems for evaluating the unstructured text of open-ended responses that could only have been imagined as little as 20 years ago.

In this chapter we will present one particularly promising machine learning technology used to address open-ended responses, Latent Semantic Analysis (LSA), and its application as a theory of learning, a computational model of human thought, and a powerful text analytics tool. First, we will discuss the nature of open-ended response items or essay questions themselves, covering both their application advantages and the complications involved in their use. We will describe the background of LSA, exploring its application beyond its initial use in information retrieval and its development into a theory and model of learning. Looking toward the subject at hand, we will elaborate on the application of LSA toward essay questions and in the realm of scoring and feedback generation. As an example, we will present a specific case study describing the application of LSA to form an unsupervised learning system used to evaluate open-ended questionnaire responses and generate constructive feedback for the purpose of driver training in a NIH-funded research project. Finally, we will touch on other projects involving LSA for scoring, evaluation, and content analysis, and discuss ongoing research work with LSA.

2 Essays for Evaluation

Humans think, speak, learn, and reason in natural language. The ability to express an answer to a question in one's own words without being prompted by a set of predefined choices has been a measure of understanding and achieved learning probably since the invention of students. While closed response questions in the form of multiple-choice tests have proven their worth in many testing arenas, the essay question still remains as a primary, though problematic, evaluation tool in many areas of learning.

2.1 Open-Ended Responses Provide Unique Evaluation Leverage

One of the key advantages of open-ended response items is that they demonstrate the ability to recall relevant information as well as the synthesis and understanding of concepts rather than the mere recognition of facts [1]. Open-ended responses also allow for creative and divergent thoughts to be expressed as they provide an opportunity for respondents to express ideas or analyze information using their own words. In educational contexts, essays in many ways exhibit a more useful level of knowledge from a student, while providing cues to educators guiding additional instruction. Feedback from essay evaluation can help the students learn certain content as well as the skill of thinking [2]. Additionally, there is a universal need for people to be able to communicate what they know, not just pick it off of a list, and educators need to promote the expressing of knowledge in self-generated prose. Outside of a testing environment, communication is at times perhaps more critical than factual knowledge.

Another consideration favoring open response items is that some types of questions do not lend themselves to closed response prompts. The formation of high-quality closed response questions when they are suitable is still time consuming and expensive. The ability to creatively combine multiple pieces of information and then to recognize and describe their interconnectedness demonstrates a higher order of thinking than can be captured in a closed response evaluation item. We desire to know what people can do, not what they can choose [3].

In short, essay and open-ended response questions provide several advantages for assessment:

- Demonstrate the ability to recall relevant information
- Demonstrate the synthesis and understanding of information
- Allows creative and divergent thoughts to be expressed
- Allows for different type of questions to be posed
- Avoids leading the respondent with pre-formed answers

2.2 The Problem with Essays: Human Raters Don't Scale

Essay questions are not the problem. The problem is essay assessment. Measurement tools for gauging the quality of an open response item need to reliable, valid, and generalizable. Ideally, essay scoring should meet the same level of objectivity as a well-crafted multiple choice test. The standard approach to evaluating open-ended responses has been to use human raters to read them and assign scores. Even though individual human scoring can be unreliable as it is subject to a rater's variability, biases, and inadequacies, human judgment is still considered the standard of quality in the evaluation of essay responses [4]. Human essay assessment however is fraught with issues. It is both expensive and time consuming. Language dependency makes it difficult or impossible to generalize human assessment for large collections of responses. Finally, such assessment is low in reliability even for relatively small sets of items. Consistency issues increase as the volume of items to be evaluated grows.

2.2.1 Expense

The process of human assessment is more complicated than would appear on its face. It is both time consuming and labor intensive. A human rater can read for comprehension at a certain maximum rate. While repetition may result in some speedup, the rate cannot be increased beyond a certain point. Large volumes of items to be rated will overwhelm the abilities of a human rater and delay the availability of assessment scores. Additionally, using a single human rater can cloud the assessment ratings with questions of subjective bias.

In order to deal both with problems of scale and questions of bias, it is apparent that multiple human raters must be employed. In order to ensure reliable assessment and eliminate questions of bias, more than one expert human needs to grade each essay [4]. This redundancy is expensive both in time and money while doing nothing to increase the throughput of scored items. Even more additional raters are required to deal with larger numbers of items. Using multiple raters at any level introduces both additional costs and problems. First, to facilitate consistent scoring a set of evaluation criteria, a rubric, must be established. The essay items to be evaluated must then be read by the human raters in the light of this rubric and scores assigned. This group of raters must be trained to correctly apply the scoring rubric in a uniform manner, and their performance must be monitored to ensure some level of inter-rater agreement. Human assessment requires redundant scoring of items, perhaps at multiple levels, to both moderate bias and provide comparability between rater groups. If the measured inter-rater reliability is not at the desired level, additional rater training is often prescribed or perhaps the rubric is adjusted and the raters re-trained. In either case items are generally rescored yet again.

2.2.2 Language Dependencies

Aside from the issues of volume, human raters are also limited in the type of items they can read and process. A human rater will typically only be able to evaluate response items within a single language, so generalizing the evaluation of essay responses across multiple languages presents an obvious difficulty. Even within a single language, however, multiple raters have different understanding and vocabulary dependent on several factors (regional idioms, fluency, use of technical jargon, age, etc.). While training of the raters attempts to mitigate these inconsistencies, they still remain at some level. These issues manifest themselves in the form of poor inter-rater agreement or inconsistency in rating reliability over large groups of essays or long periods of time.

2.2.3 Consistency Issues

Key to large-scale generalizable evaluation of open response items is consistency. This consistency must exist at the individual rater level, across teams of multiple raters, and over time as responses are collected, perhaps over a span of years.

Consistent objective analysis by human raters is difficult to achieve for both individual and multiple raters. In many cases, an individual rater will not agree with their own scores when repeating their assessment of the same essay some time later after they have forgotten the first score. The correlation of repeat ratings of the same items has been shown to be only about 0.70 [3]. One factor that certainly contributes to this is rater fatigue. Faced with a large collection of items to be assessed, it is no surprise that the time and attention given to the evaluation of item number 1 will differ from that given to item number 300. The other potential is that after examining the same sort of item repeatedly the evaluation will devolve from a holistic analysis to a simple check-off of certain prominent attributes.

Multiple raters used to increase the throughput of open item scoring present another set of problems impacting consistency. It has been noted that "with human raters, it is seldom clear that any two are grading an essay on exactly the same criteria, even when trained" [4]. Individual raters agree with each other weakly, generally correlating about 0.50–0.60 of the time. The average grade given to an essay by a group of graders is more dependable. For example, if a panel of four experts scores an essay, each expert will correlate about 0.80 with the average grade of the four experts [1]. Many tests are high stakes and need humans to grade in the capacity of objective testing, requiring better than a 0.65 correlation with a reliability of 0.78 [3]. Achieving this goal requires more raters, more training, and consequently more expense.

Consistency Over Time

Many assessments are gathered over a long period of time, perhaps even years. In such cases rater consistency becomes a critical issue as raters change and even those that do not may be subject to greater drifting in their scoring. The quality and consistency of different groups of human raters is difficult to measure and requires repeated rating of the same items to establish a basis for comparison. This results in some amount of redundant work if a quality comparison is to be performed. Human judgments do not meet the need of group descriptions or group trends due to the fact that raters usually change from year to year or for each time a test is given and assessed. Human raters tend to adapt their grading or rating to the essays in hand or at the time. Therefore, it is hard to draw conclusions on trends over time. How can current tests indicate how a group of students are doing compared to last year or another state or other demographic groups [3]? This is important for improvement in education of any kind.

2.3 Automated Scoring Is Needed

The use of human raters for essay evaluation is limited in both consistent repeatability and scalability. The only way to scale up to handle large volumes of responses while providing reliable and generalizable ratings is through automated systems. There has been much work and research on using computers to grade essays and perform evaluation with the goal of producing "pedagogically adequate, psychometrically sound, and socially acceptable machine assessment and tutorial feedback for expository essays" [5]. Several such systems have been developed, demonstrating that computerized evaluation of open response items and essays is feasible [5–8].

Just as with human raters, the performance of automated scoring systems must meet standards for reliability and quality. Research has shown that computerized essay scoring systems correlate to expert human ratings as well as or more highly than the human experts do with each other [3, 8]. Of course automated systems have no issue with consistency and are not subject to fatigue. When properly developed they can eliminate the issue of rater bias [5]. The use of computerized systems also provides capabilities that are not possible using human raters, such as allowing an evaluation item to be compared with every other evaluation item. Essay items being evaluated can then be described in different aspects of the full collection of items as groups exhibiting certain similarities, differences, trends, etc. All of this can be performed at large volumes, with shorter turnaround times, and at lower cost.

2.3.1 Creating Automated Methods Requires Learning Systems

The objective of all this is not to have computers replace human judgments or grading of essay items completely. Rather, the computer is used to replicate human judgments by learning the process of grading, reaching an acceptable agreement with humans, and then is deployed for large-scale assessment [3]. Due to the need to handle varying prompts across many different subject areas and multiple languages, an automated evaluation system must be easily adaptable to different applications and different contexts. One such technology that supports this adaptable learning approach is LSA.

3 LSA as an Unsupervised Learning System

LSA has evolved into a theory of learning, a computational model of human thought, and a powerful text analytics tool. It takes its name from the fact that it presumes the existence of an underlying or "latent" structure relating the meanings (semantic value) of words within a body of text [9]. This section discusses the historical development of LSA, briefly describes the mathematics that make up its foundations, and explores the application of LSA as a learning system to analyze language and meaning.

3.1   Brief History of LSA

LSA has its historical roots in work performed during the 1980s at Bell Communications Research (Bellcore) aimed at what is known as the Vocabulary Problem. In computerized data retrieval systems of the time, the ability to access a relevant object or item was dependent upon lexical matching which required referencing the correct word tied to the information of interest. It was noted that users of such a retrieval system often used the "wrong" words to indicate the object of their request, and were therefore unable to retrieve the items they desired. A research study was conducted collecting data from human language usage in five different domains. For each of the five experiments, subjects were asked to provide a word to identify a given object or item. One of the key results from the study was the finding that people use the same term to describe a given object less than 20 % of the time. This fact leads to the Vocabulary Problem [10].

The Vocabulary Problem has come to be known in the information retrieval (IR) field as the problem of synonymy. It refers to the diversity in word usage by people referencing the same object [11]. Further complicating matters, there is also the case where people use the same word to reference different objects, a condition known as polysemy. Both synonymy and polysemy complicate the task of information retrieval. The key to the performance of an information retrieval system is addressing these issues [11–13].

Following up their earlier work, the group at Bellcore proceeded to develop an automated technology to overcome the lexical keyword matching problems that were being encountered in attempts to access their document databases. This approach, LSA, provided new leverage for indexing and retrieval and was applied to a number of internal projects to improve success of retrieval. LSA showed remarkable improvement in information retrieval because objects in a database were retrieved for a given query based on semantics and were no longer restricted to lexical keyword matching [9].

Additional work at Bellcore showed that LSA improved recall (the finding of relevant documents for a query), solving the synonym problem, but was less successful improving precision (the number of relevant documents in the top k spots retrieved for a query), which would address the polysemy problem [14]. It is important to note that all of this early Bellcore work was performed on relatively small data sets (a couple thousand documents or less) due in part to the limited computational power available at the time. The issue of polysemy has been addressed more recently as it has come to be recognized that a great many more documents are needed to support the learning of the linguistic constructs necessary for addressing the problem, and the computational abilities of modern processors have grown to allow these larger sets to be processed (to be elaborated on later in this chapter). In this early work LSA showed potential for being a viable option, better than lexical term matching, for use as an effective retrieval method. The LSA approach was able to overcome the Vocabulary Problem, was widely applicable, fully automated (not requiring human intervention), and gave a compact representation of the original textual data with noise removed [9].

Using the performance measures of precision and recall, initial experiments with LSA showed 20 % better performance than lexical matching techniques and a 30 % improvement over standard vector methods. In the early 1990s much progress was made applying LSA in the field of information retrieval. One of the first adjustments explored was the application of different term weighting schemes. Significantly improved retrieval performance was noted when using the log-entropy weighting scheme. This yielded a 40 % improvement in performance over the straight term frequency weighting that had been used up to that point [12]. Over time, it was realized that performance improved as increasing numbers of documents were added to the data collections being processed. Growth in the size of the available corpora for analysis and the continued improvement in results led to further development of LSA as a theory of meaning and its recognition as a learning system.

3.2   Mathematical Background

Before the discussion of how LSA can be used as an unsupervised learning system, the mathematical basis for the LSA needs to be presented. After all, it is this mathematical representation of the original textual database and its subsequent processing that give LSA its power. A brief overview of the concepts is presented here, but for a detailed description one should refer to Martin and Berry 2007 or Martin and Berry 2010 [11, 15].

3.2.1   Parsing: Turning Words into Numbers

The first step in the process is the conversion of strings of text into a suitable numeric representation. This is a fairly straightforward process. Given a textual body of information, a matrix (conceived of as a simple table) is constructed where the rows are unique token types and the columns are contexts in which the token types are used. Token types are typically formed from the single word items in the text. Contexts, also referred to as documents or passages, can be phrases, sentences, paragraphs, or multiple paragraphs, but are usually selected to be single paragraphs [16]. A number of policy rules must be established for how the text stream will be broken down into tokens in the parsing process. Some of these rules are fairly standard, such as using a single word per token, treating whitespace characters as boundary markers for tokens, or ignoring character case distinctions. Other rules are decided based on the application and the nature of the text being processed. These might include rules on how to handle punctuation characters, numeric values, or even designating certain words to be ignored altogether. In general, these policy rules are meant simply to establish a uniform parsing scheme, not to influence the interpretation of meaning in the text.

Once the parsing process has been completed, every cell in this matrix will have been assigned a value indicating the number of times each token type appears in each individual context. The result is a large sparse matrix, often referred to as the term-by-document matrix.

To this sparse matrix a weighting function is applied. The use of a weighting function is common practice in vector space models for information retrieval. The weighting function serves to somewhat normalize the importance of token types (terms) within contexts and across the collection of contexts (documents). The value of each element is adjusted by a global weight, the overall importance of the term in the collection, and local weight, the importance of a term within a document. Local weighting schemes are generally intended to dampen the effect of terms that occur frequently within a single document. Local weighting functions that might be applied include the use of straight unadjusted term frequency, a binary weighting indicating simply the absence or presence of a term, and finally the natural log of term frequency $+1$.

Similar to the local weight, the global weighting function is intended to dampen the effect of terms appearing frequently across the documents in a collection. Common global weighting functions include:

Normal

$$\sqrt{\frac{1}{tf_{ij}^{2}}},$$

where tf_{ij} is the number of times term i appears in document j,
  Gfldf

$$\frac{gf_{i}}{df_{i}},$$

where gf_{i} is the number of times term i appears in the collection (global frequency of term i) and df_{i} is the number of documents in which term i appears (document frequency of term i),
  Idf

$$\log_{2}\left(\frac{n}{df_{i}}\right)+1,$$

where n is the total number of documents in the collection, and
  Entropy[1]

$$1+\sum\frac{p_{ij}\log_{2}\left(p_{ij}\right)}{\log_{2}(n)}, \quad \text{where } p_{ij}=\frac{tf_{ij}}{gf_{i}}.$$

---

[1]There was an error in the entropy formula that was published in Dumais [12]. The formula as it appears here has been corrected. This correction also appeared in Martin and Berry [15].

Entropy is the most sophisticated global weighting scheme, taking into account the distribution of terms over documents in the collection. In tests of various combinations of these local and global weighting schemes, log-entropy yielded the best retrieval performance, performing 40 % better than just straight term frequency alone [12].

These weighting transformations are akin to classical conditioning in psychology because the transformations depend on co-occurrence of words in context weighted first by local co-occurrence frequency (local weighting) and then inversely by the extent to which its occurrences are spread evenly over the collection (global weighting). Using log as the local weighting factor approximates the standard empirical growth function of simple learning and using entropy as the global factor indicates the degree to which seeing the word indicates its context [17].

3.2.2 Singular Value Decomposition

The data represented in the term by document matrix at this point describes the content as a co-occurrence of words (terms) and contexts (documents). A document derives its meaning from the terms it contains. Each term contributes something to its meaning. At the same time, two documents can be similar in meaning and not contain the same terms. Similarly, terms that appear together in one document do not necessarily have similar meaning in different document contexts. This leads to the definition of the Compositionality Constraint, that the meaning of a document is the sum of the meaning of its terms [16]. Recognizing that this matrix reflects a coefficient matrix for a system of simultaneous linear equations with each equation representing a passage, the basis of LSA is the computation of a solution to this system in order to infer the meaning for each term based on the contexts in which they do and do not appear. Singular Value Decomposition (SVD) is used to solve this system of equations, yielding as the output a set of vectors in a high-dimensional "semantic space" with each passage and word represented by a vector. A document, or context, vector can then be considered as the vector sum of all the term vectors corresponding to the terms it contains. Similarly, a term vector represents a term in all its different senses [18].

The SVD produces a factorization of the original term by document matrix A in three parts:

$$A = U\Sigma V^{\mathrm{T}}$$

The rows of matrix U correspond to the term vectors, and the rows of matrix V correspond to the document vectors. The nonzero diagonal elements of Σ, the scaling factors, are known as the singular values.

From a full SVD, it is possible to reconstruct the original matrix A from the three matrices U, Σ, and V; however, the truncated SVD that is computed for LSA will equate to the best k-rank approximation of A. This dimensional reduction is desirable because it has the effect of removing noise from the original representation

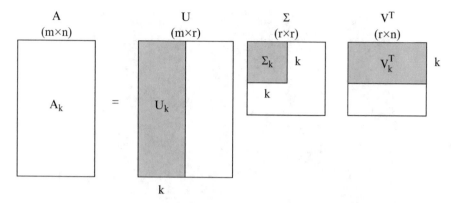

Fig. 1 Pictorial representation of the truncated SVD

of A (countering the dissimilarity of related documents that use synonymous terms and separating those that contain polysemic terms).

The truncated SVD (see Fig. 1) is defined as

$$A_k = U_k \textstyle\sum_k V_k^T .$$

Use of the truncated SVD is a hallmark of LSA. The truncated SVD is used to produce a set of dimensions based on the k extremal singular values of the term-by-document matrix A. This results in the most significant dimensions of the space being used to define the truncated vector space used for the LSA representation. The importance of this has become more apparent over time as the study of LSA has continued. The truncated vector space can be considered as a multi-dimensional hyperspace where each item is represented by a vector projecting into this space. This concept can be roughly pictured with a simple 3-dimensional representation where the vectors point out into the 3D space, see Fig. 2.

This illustration is extremely simplified to facilitate visualization. In practice, k is typically selected to be anywhere from 300 to 500 dimensions. Empirical testing has shown dimensionalities in this range to be most effective for synonym recognition applications [17, 20]. Within this hyper-spatial representation, information items are left clustered together based on the latent semantic relationships between them. The result of this clustering is that items which are similar in meaning are clustered close to each other in the space and dissimilar items are distant from each other. In many ways, this is how a human brain organizes the information an individual accumulates over a lifetime [19].

Calculation of the SVD is mathematically nontrivial and computationally intensive. Early research in LSA was certainly restricted by the available computing power of the era. What would today be considered very small data sets (2000 documents or less), were used in much of this initial testing. Advances in available computing power have made the processing of much larger bodies of content

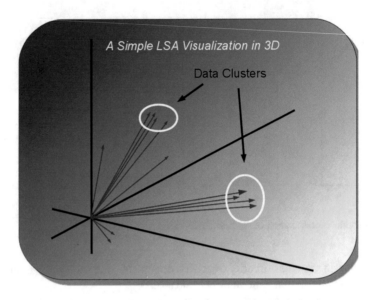

Fig. 2 A simplified visualization of a LSA semantic space [19]

feasible, as have advances in the algorithms and software for performing this work. Our own efforts have contributed to the processing of larger data sets through the optimization of resource usage [21] and implementation of large-scale parallelized processing software [22]. For our current research, we use the LSA_Toolkit™, a commercially available library for both efficiently constructing large LSA spaces and performing many additional analysis functions [23].

3.2.3 Query and Analysis Processing

Once the truncated SVD has been computed for a text collection, two forms of analysis can be performed. Items within the space can be compared for semantic similarity, and new document items can be constructed and projected into the space mapped according to their semantic content.

Comparison of items within the space can be used to analyze the content of a document collection, looking for items that are closely related or perhaps clustered together by their similar semantic mapping. By evaluating term proximities, it is possible to identify synonyms and examine term usage across the collection. An item within the LSA space, whether a document or an individual term, is represented by a k-dimensional vector. Vectors for two such items may be compared by computing a distance measure or a similarity measure that quantifies their semantic proximity or separation. The similarity measure that is typically used for LSA is the cosine similarity which has been shown to be a reliable measure of semantic relatedness within the LSA space [24].

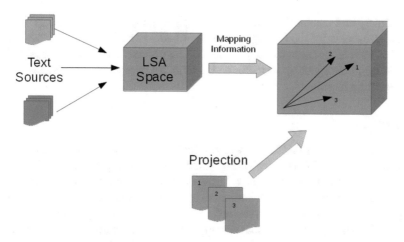

Fig. 3 Conceptual view of items projected into a LSA space

Cosine similarity is defined as:

$$\frac{u \cdot v}{\|u\| \, \|v\|},$$

that is, the dot product of vectors u and v divided by the product of their vector lengths. Other options for similarity measures include Euclidean distance and the dot product itself.

Projection of New Content: Subject Items for Evaluation

Projection of items (see Fig. 3) exploits the term definitions provided by the space to map new document items that were not included in the original content within the semantic context of the space. This can be used to probe the space as in the case of an information retrieval query, or for simply establishing semantic relationships between items of interest within the context of meaning represented by the LSA space. In mathematical terms, a new document projection is computed as the weighted sum of the term vectors corresponding to the terms in the item being projected scaled by the inverse of the singular values.

Parsing

Projection begins by parsing the text of the item to be evaluated. This is performed in the same manner as the parsing of the content used to form the base space. Policy for handling punctuation, casing, numeric values, etc. are all followed to break the evaluation item down into a set of terms. Unlike the formation of the base space, however, new terms that are not present in the base space cannot be added to the vocabulary and cannot be used in the projection mapping. The terms that do exist in the vocabulary of the base space are identified so that their term vectors may be used in the computation of the projection. The occurrence of each mappable term in the evaluation item is counted, and a term frequency vector (TFV) is formed representing its content.

Weighting

Just as when the original LSA space was computed, a weighting is applied to the evaluation item TFV. This weighting scheme must correspond to the one used in the construction of the LSA space. If the typical log entropy weighting method is used, for each term in the evaluation item compute the local weighting for the TFV: $\ln\left(tf_{ij} + 1\right)$. This value is then multiplied by the global entropy value for that term within the base space. The global entropy value was computed for each term in the original sparse matrix when it was weighted in preparation for formation of the LSA space.

Composition

Finally, to compose the projection vector, z, the weighted term frequency values q for the terms in the evaluation item are multiplied by their respective term vectors U_k from the base space. The sum of these vector products is then the new projection vector for the evaluation item being considered.

The projection vector computation is

$$z = \frac{q^{\mathsf{T}} U_k}{\Sigma_k}.$$

Once computed, the projection of an item can be compared to items in the space or to other projection items using the same distance measurement functions described previously.

3.3 LSA Learns Language

The pioneer research presenting LSA as an unsupervised learning system was presented in the 1997 paper by Landauer and Dumais describing LSA as "a solution to Plato's problem" [17]. Plato's problem is the name of a persistent problem in cognitive science. It is basically the question of how humans develop the knowledge they have based on the relatively limited amount of information they have received. Peoples' knowledge exceeds simply the facts and experiences that they have been exposed to over time. In their paper, Landauer and Dumais put forward LSA as a model of human learning as an answer to this question, marking a change from viewing LSA as retrieval method to that of a powerful learning system.

Landauer and Dumais proposed that underlying domain knowledge is a vast number of interrelations. When properly exploited, these interrelations can increase learning by the process of inference. This mechanism, under the right choice of dimensionality, can represent the meaning of words and contexts and their similarities in such a way that the acquisition of knowledge is enhanced, bridging the gap between "information in local contiguity and what people know after large amounts of experience" [17]. More succinctly, people learn things that they were not directly taught. LSA is a mathematical model of the meaning represented in human language and the acquisition of knowledge. While it lacks certain capabilities present in our human cognitive model such as word order, syntax, morphology, etc., the representation produced by the LSA process is highly similar to that of humans. It is based on concept and semantics not keywords or syntactic constructs [17, 25].

The LSA representation of words and contexts as vectors in a vector space where the meaning of each is represented as a point is a model of how people acquiring meaning from experience. The assumption is that people learn by associating words and contexts that are experienced together over time. This provides a dynamic framework for predicting new experiences. Human cognition takes in all these experiences, word and context co-occurrences, and fits them into a semantic map that represents how each word, object, and context is related to each other. The mathematical model of LSA does much the same thing. It constructs a semantic space by first digesting a large body of textual information, and then it combines all these links to form a common semantic space. This second step is what constitutes LSA's understanding of meaning. LSA will bring those words that never co-occur together, like synonyms, together and separate those contexts that share common terms apart if they do not mean the same thing [26].

3.3.1 Unsupervised Learning

Using LSA a machine can learn in much the same way that a human does in the fact that it can be fed much of the same empirical associational data that literate humans use in learning. While the input is electronic text and a human has multiple learning sources, LSA does remarkably well as a learning system. Even without prior specific, or any, knowledge, LSA can produce a linguistic and cognitively effective representation of word meanings [18]. LSA is different from traditional natural language processing (NLP) methods because it requires no prior human intervention: no dictionaries, no grammars, no syntactic parsers, no morphologies, and no knowledge base. LSA simply takes raw text as input, separated into passages, and parsed into a collection of distinguishable words [20]. Traditional NLP uses cues from the structure of a language to infer meaning. It is syntax based. The rules for such systems must be developed with expert human input, are language specific, perhaps domain specific (limited to a certain form of usage for a particular language), and are subject to aging. NLP is not an unsupervised learning system; it may not even be considered a learning system in many cases.

3.3.2   The LSA Model of Learning

As previously mentioned, LSA is based on what has come to be known as the Compositionality Constraint: The meaning of a document is a sum of the meaning of its words, and the meaning of a word is defined by all the contexts in which it appears (and does not appear). By using the SVD to solve a set of simultaneous linear equations representing the content of a collection, a hyper-dimensional semantic space is obtained where each term, and each document, is mapped by a vector. This representational technique exploits mutual constraints on the occurrence of many words in many contexts, and the resulting representation allows for similarities to be observed between the words and contexts based on their position in the mapping space. Word association is extremely important to human cognition, and that association is what LSA models [18]. The resulting system can be viewed as both a way to automatically learn the meaning of words and contexts as well as a computational model for the process of human learning itself [27].

It is important to note that since LSA only uses the supplied set of input text, it does not always capture all human knowledge. Some of this shortcoming is due to the insufficiency of the training corpora to represent the language experience of a typical person [27]. LSA only induces the meaning of words and contexts through the analysis of text alone, not from perceptual information, instinct, feelings, etc. However, LSA can access knowledge about those processes through the written word and produce a close enough approximation to peoples' knowledge to represent those concepts. Landauer describes LSA's capabilities "to be analogous to a well-read nun's knowledge of sex, a level of knowledge often deemed a sufficient basis for advising the young" [20].

Dimensionality

An important factor in the formation of the LSA space is the choice of dimensionality. LSA does not use the complete SVD, but rather a truncated SVD that will equate to a k-rank approximation of the term-by-document matrix A. This dimensional reduction is desirable because it has the effect of removing noise from the original representation of A. Selection of k, the number of dimensions, is crucial. Including too few dimensions will leave out important underlying concepts in the data, but using too many dimensions will introduce excessive differentiation between the items making it difficult to find the underlying important semantic relationships [8, 14, 16]. It is theorized that reducing the number of dimensions to be smaller than the initial number of contexts or words produces the same approximate relations as occur in human cognition. Ideally, finding the right number of LSA dimensions produces a representation that is analogous to the same dimensionality as the source that generates the semantic space of a human [25]. The high dimensionality of LSA's semantic space is key and mimics the structure of the brain along with the statistical structure of experience [26].

Orthogonal Axes

One of the products of the SVD computation is a set of orthogonal axes that are the mapping dimensions of the semantic space. Terms and documents derive their meaning from the mapping on these axes, but they do not define the axes [8, 16]. Dimensional axes in the semantic space are an abstract feature. They are the foundation or framework for the vector space, not the words or any nameable characteristic, and they should not be interpreted as such [18, 50]. Attempts to force a labeling of these axes in humanly recognizable concepts tend to be contrived. There is no reason to require that these axes be more than what they are, namely a coordinate system for locating items within the semantic space. Identifying concepts and features represented in the LSA space is best achieved by identifying semantic clusters within the space and other analysis techniques for investigating the relationships it contains.

Meaning

Success in inferring meaning from this system of linear equations is critically dependent on the availability of a sufficient amount and adequate quality (garbage in = garbage out) of input. In a LSA learning system, the meaning of a word is learned not just by the number of times it has been seen in a context, but also by the tens of thousands of times it has been observed absent from a context. "Greater amounts of text help define the space by providing more contexts in which words can co-occur with other words" [28]. LSA must be trained on enough passages before the meaning of any word can be distinguishable from other words. LSA cannot learn word meanings used in contexts on which it was not trained. Small or domain specific corpora are not sufficient to train the LSA learning system [18].

The SVD uses all the linear relations in the given dimensionality to determine the word vector that best predicts the meaning of all of the contexts (pieces of text) in which the word occurs. Additionally, an LSA learning system will not recall a passage verbatim, but it will convey a meaning in the semantic space [18]. "This expresses the belief that a representation that captures much of how words are used in natural context captures much of what we mean by meaning" [17].

LSA mimics thinkable meaning. Word order is not considered, but it is not the most important thing in learning. Inductive association from the word usage in contexts is sufficient for learning the meaning of words and performing the way humans do. It is true that LSA will not represent the meaning of passages that depend strongly on word order syntax or perceptual knowledge not captured in the textual corpus from which LSA learns. The comprehension and representation of meaning is modeled by LSA, not how the passages are constructed or produced [18].

3.3.3   Evidence of the Model

The semantic representation employed by LSA has been shown to induce meaning of words and passages and their similarities in a way that simulates human like similarity judgments and behavior. LSA is not necessarily the way the brain actually acquires knowledge, but is a model, though not a complete one. The pioneering study described in detail in Landauer and Dumais [17] demonstrated this through an experiment using an LSA simulation of an older grade-school child which learned on average 10 or more new words after reading a modest amount of text (approximately 50 paragraphs). Notably, the learned words were not necessarily included in the text that was read. This excessive learning rate observed in the experiment was analogous to the vocabulary acquisition of a school-age child suggesting that the LSA model was an answer to "Plato's problem." Evidence that LSA is mimicking how a human learns language and acquires knowledge comes from this and other simulation of human performance. The performance of LSA on multiple choice and domain knowledge tests as well as essay grading applications show this is true [18].

Text Comprehension and Coherence

One of the first research works to indicate that LSA could be used for much more than information retrieval was in text comprehension research. The semantic structure revealed in the vector representation of LSA allowed for semantic similarity between textual information to be measured quantitatively. In the study of text comprehension, it is essential to understand what components affect the ability of a reader to comprehend a passage. Common practice for gauging reading comprehension is to have a student read a given text and then write a summary for that text. The summarization is examined to determine how much information the student has learned from the text [28].

One method for studying text comprehension is to develop a human coded cognitive model of student's representation of the text read. This can be done using propositions, which are sets of semantic components drawn from clauses in the text, to represent the semantic information in a passage. A linking of the propositions within the textual information is constructed to represent the semantic structure of both the original text and the summary. Then, a semantic comparison is performed between the two propositional analyses [29]. LSA provides a fully automated way to implement this cognitive modeling process based on propositions without the need for human intervention. The cosine similarity between two texts represented in a LSA semantic space provides an estimated semantic similarity.

First, LSA was used to look at a subject's summary to determine how much information he/she had learned by examining what textual documents had the most influence on the subject's recall. Sentences from the subject's summary were compared to the sentences in the original text to determine which ones had the most influence on the subject's understanding. The results were promising, performing within 7 % agreement of human raters. A second experiment involved characterizing the content quality of each summary as a whole. The results indicated that LSA performs as well as human raters at judging the quality of the summary. Finally, a third experiment was performed to measure text coherence in the summaries. LSA was used to examine the semantic similarity or overlap between sections of a summary. The LSA predictions of coherence were generated by calculating the semantic overlap between adjoining sentences in a text and then computing a mean for all the cosines between the sentences. This single number or score was used to represent the mean coherence of the text and was compared to scoring based on straight word overlap between sentences. The results indicated that LSA did capture coherence based on semantics not shared words between adjoining sentences [28]. As demonstrated in these experiments and in extensions of them, LSA provides an accurate model of the coherence exhibited in a text that is similar to the propositional modeling that can be done by humans. The representation for a given text by LSA corresponds to the subject's semantic interpretation of the same text, whether written or read. It was also shown that the LSA coherence measurement was, in many cases, an accurate predictor of a reader's comprehension of the text [25].

Synonym Test

In one early study assessing how LSA mimics humans in synonym recognition, an LSA simulated student was given a Educational Testing Services (ETS) multiple-choice synonym test for students where "English is a Foreign Language" (TOEFL). The overall test score achieved by the LSA-simulated student was the same as the average score for students from non-English speaking countries successfully seeking entrance into a college in the USA. From the performance on this test, the LSA model could be considered to be similar to the behavior of a group of moderately proficient English readers, with respect to judging the meaning of similarity between two words [17].

Semantic Priming

Semantic priming is an experiment in which a person reads short paragraphs word by word, and then at selected time points he/she is presented with a lexical decision. This lexical decision involves either choosing a word that is related to one or another sense of an ambiguous word in the paragraph that was just presented, or picking words related to words that did not appear in the paragraph but which are related inferentially to the text the subject should have comprehended up to the stopping point.

In this experiment a Latent Semantic Analysis simulation LSA simulation of semantic priming was performed. In the cases of the second lexical decision, the average cosine between the vector for the ambiguous word presented in the paragraphs and the vectors of the related words presented were significantly higher than the average vector cosine with the unrelated word vectors. For the first type of lexical decision, it was observed that the cosines of the related words to the homograph in the paragraphs and passage vector representation were significantly higher than the cosines of the unrelated, different sense, word vectors, and the passage vector. The LSA representation of the passage provided the correct meaning to select the "homograph's contextually appropriate associate" [17]. The similarity relations examined in the study support the LSA simulation of the construction-integration theory in cognition [29].

Learning from Text

Learning is a constructive process. Just as when building a house one cannot start with the roof, learning advanced knowledge requires a foundation of primary learning. Much of what a person learns, they learn from reading. In order to learn effectively from a particular piece of text, that text cannot be too difficult or too easy to comprehend based on the background knowledge a person already possesses. New information being processed must integrate with prior knowledge, both in terms of understanding the text and in remembering it for use later in other situations. This information must be somewhat relevant to prior knowledge for this learning to occur; otherwise, a person has no relevant contextual basis with which to integrate the new information.

This study was performed to test whether LSA could be used to judge prior knowledge and then to determine if learning had occurred by analyzing a series of essays written by humans. This involved presenting a group of students with varied amounts of prior knowledge texts of different difficulty level on the same subject. Students provided answers to open-ended prompts on the subject before reading a selected text and again after reading the text. LSA was used to compare the pre-reading responses to the text in question to determine the amount of prior knowledge possessed by the student. The post-reading responses were then compared to the text to measure the amount of learning that had occurred. The findings showed that a student's prior knowledge needs to be at an appropriate level to the instructional text (moderate cosine between pretest essay and text) for effective learning to occur. Students who displayed either little prior knowledge or a high level of prior knowledge compared to the text showed little change in their response evaluation between the pre- and post-reading prompts. The highest degree of learning was exhibited when the student demonstrated some but not complete prior knowledge of the subject before reading the text. The results indicated the capability of LSA to both characterize prior knowledge and assess learning, and support the idea that learning is optimized in the context of proper foundational knowledge [30].

Assessing Knowledge

Another early study examined properties surrounding the ability of LSA to assess knowledge in the work by Wolfe et al. 1998. The semantic comparison (cosine similarity), as estimated by LSA, between certain domain specific instructional information and a student's essay on the topic was shown to be a reliable measure of a student's knowledge of the subject and a predictor of how much a student can learn from a text. Additionally it was demonstrated that all terms in a student's essay are equally important in assessment of the student's knowledge, not just the technical or seemingly pertinent words. Another notable finding suggests that some aspect of a student's knowledge can be assessed with a relatively small amount of written text from a student [24].

Essay Grading

Very early experiments using LSA in reference to essay grading started by using LSA to model subject-matter knowledge. Given only an introductory psychology textbook as input, a Latent Semantic Analysis simulated student was trained on the text and then tested using a multiple choice test that had been provided with the textbook. The simulated student exhibited performance that was better than chance, receiving a passing score on the test and showing promise in the modeling of subject-matter knowledge. Building on this early work, the use of LSA was explored for grading expository essays. Students wrote essays on a particular subject, which were then scored by both human graders as well as a prototype LSA scoring system. Several different scoring methods were applied. In the resulting analysis, all scoring methods, using multiple different instructional texts, correlated as well with the human scores as the raters' scores did with each other. Additionally, LSA showed better correlation with individual expert raters than the humans did with each other [20].

3.4 LSA Applications

While LSA is not a complete or perfect theory of meaning, it does simulate many aspects of human understanding of words and meanings of text, as shown in these early experiments. It demonstrates a major component of language learning and use. Therefore, it can be used in many applications to replace the workload of humans. Over the past two decades, since the inception of LSA and thinking of it as theory of meaning, LSA has been employed in many research projects and applications.

Many research projects and applications have used LSA technology either as a method for identifying similarities of words and passages in a large body of text based on their meaning or as new model of knowledge representation. LSA has been widely used in many information retrieval situations with much success. It has also been applied to cross-language information retrieval, indexing content in multiple languages simultaneously [13]. Data mining applications of LSA include CareerMap, a tool that matches military personnel to certain occupational tasks based on their training data, and performance assessment tools for team communication [31, 32].

The automated assessment of essays using LSA has been a major area of research and has been shown successful in the Intelligent Essay Assessor (IEA), which has been used in numerous applications to both score essays as well as provide instant feedback and analysis on essays of various sorts [5–8, 33]. IEA has also demonstrated ability to score creativity writing as well as expository essays [8].

LSA has also been incorporated as a key component of several educational systems. It has been shown that the exercise of writing a summary of informational text leads to improved comprehension and learning. Summary Street, an educational technology tool, helps students by interactively guiding the student through the process improving their written summarization, providing individualized feedback to prompt revision and enhance learning. Summary Street utilizes LSA to analyze the semantic content of the student's summary examining topical coverage and identifying redundant, relevant, and irrelevant information [34]. Another software literacy tool that has incorporated the capabilities and skills of IEA and Summary Street is WriteToLearn (http://www.writetolearn.net/). This technology tool is used in the classroom to help students with writing and reading [35]. AutoTutor is an example interactive intelligent system that tutors a student in certain topic areas. Using LSA to analyze the student's text responses to questions, LSA constructs a cognitive model of the student's response, determines the student's knowledge level, and then interacts with a student to help the student gain more understanding of the topic [36]. Other variants of this technology are being researched and tested [37].

Another area where LSA is used as cognitive model is in the analysis of team discourse. Transcriptions of spoken discourse can be examined to determine if a team is effectively accomplishing a certain mission [38]. Tools have been developed to analyze the content of team communication to characterize the topics and quality of information and give measures of situational awareness [39]. These tools were built upon KnowledgePost, a system using LSA to analyze online discussion groups or online course activity to generate summaries of the discussion and assess the thinking and contribution of individual participants [40–42]. LSA has been used in many additional educational tools which are discussed in The Handbook of LSA [43].

In the area of general text analytics recent applications of LSA include the development of the Word Maturity metric, a technique to model vocabulary acquisition which estimates the age at which different word meanings are learned providing insight to guide vocabulary development [44]. Cohesion and coherence metrics for text based on LSA analysis have been developed in the form of the Coh-Metrix system for assessing the difficulty of a written text [45]. LSA has also been used in Operations Management Research to uncover intellectual insights in unstructured data [46].

Of particular interest in reference to this chapter, a recent research project used LSA as a text mining tool to analyze open-ended responses to a questionnaire. In this application, military service men and women responded to open-ended prompts covering health issues. LSA was successfully used to evaluate the participants' responses to identify important health concerns that were perhaps not identified by the structured part of the questionnaire. Analysis of the open responses was used to identify trends, needs, and concerns [47].

4 Methodology

Using LSA as an unsupervised learning system has proven to be especially beneficial for evaluation of essays and open response items. Its automated nature overcomes the scaling issues attached to the use of human raters while providing consistent, repeatable, non-biased evaluation. The fact that LSA captures meanings, not keywords, makes a concept-based approach to assessment possible. An LSA system is easy to train, though it does require large volumes of general training text as well as some representation of domain expertise to provide an adequate foundation for semantic mapping. LSA provides many of the same capabilities as a human rater without the limitations on throughput and at less expense. The methodology used in the application of LSA toward the task of text analysis and more specifically to the evaluation of essay and open-ended response items is significantly different than that for information retrieval.

4.1 Objective

The objective in analyzing open-ended responses obviously cannot be retrieval oriented. Rather, it is an analytic task that is comparative in nature. For a given response the assessment task is to determine what sort of concepts it expresses, if any desired concepts are missing, and how the information it contains compares to other responses for the same prompt. Since specific content is not being retrieved from the collection, the LSA space is used as a semantic mapping system for concepts and the comparison of projected items. The content of the LSA space forms context or basis for understanding the meaning expressed in the response items as those items are projected into the space. It becomes an interpretive tool for text analysis. Evaluation methods using this semantic mapping system must exploit the concept identification it affords in ways that can yield useful output, whether in the form of comparative scores, feedback recommendations, or other indicators.

4.2 The Base Interpretive Space

As already described, LSA forms a model of meaning by producing a mapping system for representing the semantic associations of both individual words and multi-word pieces of text. In text analysis applications, this mapping system forms a semantic background, what we will call a base interpretive space (BIS), for processing the items to be analyzed. Because the meaning contained in the BIS is learned from the initial content used to build the space, the nature of the meanings represented in the space are completely dependent on the content provided as input in the space construction process. It is important that the BIS be well formed in order to represent the range of meanings that will be required by the application.

The mapping system provided by the BIS can not only be used for evaluation analysis, but is also useful for other applications where the data set of interest is small or narrowly focused, not providing enough material to establish a basis of meaning. The use of a BIS in these situations provides a contextual background that augments the meaning represented in the data set. While a set of 10,000 documents might seem like a large volume of data to a person, it is an insufficient amount from which to learn meaning. For any LSA application using a BIS, there are several important considerations when selecting content that will be used for its construction. These include the overall size of the training corpus, the presence of relative and distributed content within the corpus, and its overall term coverage.

4.2.1   Corpus Size

Initial research using LSA was attempted on relatively small bodies of content (2000 documents or less) with varied but often promising results. As the availability of computational power and electronic text both increased, the processing of larger corpora became feasible and improvements in the performance of LSA were observed. Poor results reported from some past research studies using LSA were based on the processing of ridiculously small corpora (sometimes as few as ten short paragraphs). Since LSA must learn the meanings of words and documents by forming associations, a sizeable amount of content is required to provide sufficient learning to mimic human understanding of language constructs. Providing too little training data gives insufficient meaning for developing the underlying mapping system. It has been suggested that a minimum of 100,000 paragraph sized passages is necessary to represent the language experience of an elementary student [16]. Our own work has observed that the association of domain specific content is significantly improved as even randomly selected background material is added to the content used to build a space [48].

Currently, the best policy is to use as much content as can be obtained and make the training corpora as large as possible. This of course must be tempered with knowledge of the available computing resources. It is still quite possible to build a corpus so large that it simply cannot be processed. It is also currently unknown if there is a point at which adding content makes no difference or in fact degrades the space. More research in this area is needed.

4.2.2   Relevant and Distributed Content

In addition to a large volume of input text, LSA also needs good quality content that provides concept information not just in the domain of interest, but also enough general content to represent language usage as a whole. LSA forms its representations of meaning from the analysis of text alone, and to build this notion of meaning it needs to have many representative textual associations both in the present and in prior knowledge of a potential user of the system [16, 20]. The size and content of the corpus that is used to construct the BIS influences whether the LSA representation is similar to an amateur, novice, or expert level of knowledge. If LSA is trained on content including highly technical texts of a particular domain, then it would behave similar to an expert in the field because the LSA representation is much more elaborate [25]. The desire is for the LSA representation to map similar items in close proximity, yet with enough separation to be distinguishable. This sensitivity is especially true in evaluation applications as the evaluation items are typically expected to be about a common body of subject matter [48]. Perhaps one of the failings of past attempts using LSA for information retrieval is the common practice of constructing a space using solely the retrieval targets for content. The BIS must include not just items in the domain but background meanings representative of the wide range of expression.

4.2.3 Term Coverage

Term coverage is an important aspect in the construction of a background space. Terms that are in an evaluation item can only be mapped if they are represented in the BIS. This, in turn, only occurs if the terms are present in the original content used in its construction. Unmapped terms cannot be assessed for meaning. It is essential to provide content that gives adequate term coverage for possible terms used in the expected evaluation items.

Because the meaning of a document is construed as the sum of the meaning of its terms, when some of those terms cannot be assigned a meaning the interpretation of the entire item is questionable. This is no different than if a human reader encountered a document with terms that were simply unknown to him or her. Depending on the number or significance of the unknown terms the entire document may be misunderstood or simply unintelligible. A human reader would at this point give some indication that they did not know what was going on. Unless this is explicitly monitored and reported, an automated system would simply return wrong answers. This very issue has led to flawed reports of poor LSA results due to missing terms for the items being examined [18].

Monitoring the term "hit-rate" as items are evaluated must be performed to flag individual items where the scoring may be questionable and to indicate whether the BIS is adequately formed for the domain of the items being considered. Too many missing terms would indicate that the base space is insufficient for evaluating the content of interest [48].

In general, obtaining sufficient term coverage is not difficult as most of the expected terms in a response set will be typical of the overall language and already in the vocabulary of a reasonably sized corpus. Depending on the type of items being evaluated, consideration must be made for the formality of the language anticipated. Working with transcribed conversational data will differ significantly from a proper essay or an open response short answer item. Adding domain specific content and specifically including content generated by the subject group in question are both effective strategies for obtaining appropriate term coverage.

4.3   Evaluation Algorithms

Various methods for evaluating essays and open responses have been explored. Several potential scoring methods have been proposed and tested: holistic scoring, target based scoring, essay to all other essay ranking, and other comparisons to an instructional text, either based on selected components of the text or as a whole. In several different tests all of these scoring methods were demonstrated to produce scores that correlated with human rater scoring of the items as well as the raters' scores did with each other. Additionally, LSA correlated better with individual expert raters than the humans did with each other [20]. Out of these methods, three primary evaluation approaches have emerged: target based scoring, near neighbor scoring, and additive analysis.

4.3.1   Target Based Scoring

Of the three primary evaluation methods, target based scoring is possibly the most intuitive approach. Target based scoring is essentially a measurement process comparing how the meaning articulated in the subject response compares to an expected meaning. This expected meaning is the target or "gold standard" to which the responses must map in close proximity [2]. This measurement can be produced quite easily in an LSA system simply by projecting both the target and subject response in the BIS and then computing the cosine similarity, or other desired distance measure, between the projection vectors. The nature of this method limits its output to the production of a single score for the subject essay.

The main challenge in using target based scoring is the establishment of the target. In some cases it is appropriate for a subject matter expert (SME) to provide a target by producing what might be considered an ideal response against which all of the subjects may be compared [8]. The disadvantage to this approach is that it emphasizes the influence of the single SME on the evaluation results. The target might be technically correct but unrealistic for the subject audience. Also, it may not be possible, or appropriate, to define a single target as the "correct" response. Other methods for constructing a target mitigate some of these issues, such as using the average, or centroid vector, of multiple high scoring responses or multiple SME constructed responses. Another method for constructing a target is to use a related instructional text as a target, in part or in whole [30]. Application of this method is limited to cases where such a related text exists.

A secondary challenge for this scoring method is the question of how to interpret the similarity measure as a score. Depending on the application, a simple pass/fail score may be easy to generate, but producing scores at a finer granularity might not be possible. The magnitude of the distance measure between the target and the subject responses will vary with the application, so relating the distance measurement to a score value requires adjustment for each new scenario.

4.3.2 Near Neighbor Scoring

Near neighbor scoring (referred to as holistic scoring in the IEA related publications) is a method for leveraging multiple pre-scored responses to guide the evaluation of subject response items. A collection of pre-scored items are projected in the BIS and used to provide a scoring reference for new subject items. To generate a score, a new subject item is projected in the BIS and the k-nearest pre-scored items are identified using a distance measure. The scores related to these items are then weighted by their distance from the subject item and then averaged to compute a new score for the subject. This scoring method has the advantage of allowing subjects with differing content to receive similar scores, reflecting the idea that a good response might be expressed in multiple ways or possibly focus on different concepts of equal merit [5, 8]. Another benefit of this method is the ability to produce scores on a continuous range.

Limiting factors to this scoring approach are the availability and quality of pre-scored data. A sufficient number of pre-scored items is required before any evaluation can be performed using this method. The scores must be reliable and broadly distributed. If the pre-scored content does not adequately represent the full range of possible scores, the scoring system will not be able to assign scores across the entire range.

4.3.3 Additive Analysis

In our own research we have developed an evaluation technique referred to as additive analysis. This is a component based technique that can be used to produce a holistic score, a component score, or for other non-scoring purposes. Leveraging the idea that a desired subject response should exhibit certain key concepts, it is possible to detect the presence of those concepts by adding content to a subject response to observe how the augmented response projection moves in the BIS. If the content was already present in the subject response, the projection should not move significantly, but if the content is absent from the response then the projection will move to the degree that the new content alters the meaning. The magnitude of the change in the projection can be used to weight scoring assignments or as shown in the case study below to identify appropriate feedback items.

This approach provides advantages over the target based or near neighbor approaches. Desired concepts can be expressed without tying them together as a single target. Analysis concepts can simply be enumerated rather than requiring them to be composed collectively. This method also allows for the detection of individual concepts in the subject response without requiring that the response item be decomposed in any fashion. Identification of the presence of key concepts can be used to guide the selection of appropriate feedback items.

4.4   Feedback Selection

Scoring is one thing, but providing feedback is another. Scoring is used primarily to indicate the measure of knowledge on a specified topic that has been expressed in a response. Feedback goes beyond simply measurement and provides teaching information for the subject. There are several different purposes for feedback:

- Improvement—supplies the user with concepts missing from their response
- Correction—identifies components of the response that are in error and detract from the score
- Reinforcement—identifies components of the response that are correct
- Recommendation—provides related information that may be of interest to the user based on their response

Feedback may be selected based on comparison to a target in the case of target based scoring, or may be selected based on the additive analysis method alone. Ideally, the selection should be driven by detection of the concepts expressed in the subject response being evaluated.

Definition of the feedback items is highly dependent on the application domain and presents a challenge in the development of the feedback system. Some applications require the construction of custom feedback items, while other domains have a readily available pool of potential feedback from which appropriate items may be selected. Construction of useful feedback often requires input from SMEs that can author the specific items desired.

5   Case Study NICHD Project

As an example of a working application that uses LSA as an unsupervised learning system for evaluating open-ended questionnaire responses, we discuss the *Practiced Driver* application. This system was developed with colleagues at Parallel Consulting, LLC (PC) as a functional prototype of an interactive Web-based driver's education system for beginning and novice drivers as part of an NIH-funded research program.

5.1   Background: Driver Training

The purpose of the Practiced Driver system is to assist novice drivers by facilitating the rehearsal of higher-order driving skills and the development of tacit knowledge for handling the demanding situations involved in operating a motor vehicle. Tacit knowledge is a different sort of knowledge than can be explicitly articulated and taught through direct instruction, such as traffic rules, but which can be gained through experiential learning. The Practiced Driver system provides this as a two part training process where novice drivers first engage in a Web-based exercise presenting them with a driving scenario and then prompting them to provide open-ended responses to questions about how they would handle the situation described in the scenario. During the exercise, the participants are immediately presented with tailored constructive feedback based on their responses suggesting areas of concern and providing them with strategies to consider in similar situations. The second part of the process involves suggesting behind the wheel practice recommendations tied to their responses to the scenario prompts and the suggested areas of concern. This is serviced though a client application that can be accessed using a smart phone or other mobile device.

5.1.1 Open-Ended Responses to Scenario Prompts

For the initial version of the Practiced Driver system, two driving scenarios were developed by PC that presented the user with realistic driving situations based on typical circumstances encountered by newly licensed drivers. The scenarios incorporated several hazards that are known to contribute to accidents involving novice drivers. After being presented with the scenario, users are prompted with a series of open-ended questions about how they would handle the scenario and their prior experience with any of the aspects of the scenario.

For each scenario, representative responses were collected from a large pool of novice and experienced drivers in a controlled data collection effort conducted by PC. After filtering responses for completeness and other inclusion criteria, a set of 471 responses for each scenario was available for use. Portions of this response set were used for training and testing the prototype system.

5.1.2 Provide Feedback Suggestions for Improvement

Unlike other applications focused on essay scoring, in this project the interest was almost completely based in the semantic content of the questionnaire responses. Scoring based on construction or readability was not needed in this application, so concerns about grammar and formation of the response items were minimal. The goal of the system was to automatically analyze each scenario response and give the user feedback consisting of tailored, relevant, and safe driving strategies. It was important to avoid making suggestions in areas where a participant already exhibited adequate knowledge, so the system had to recognize those aspects that were already present in the response.

Feedback items selected for the driving scenarios were also tied to suggested practice driving skills to be pushed to the mobile client application which would recommend targeted practice skills. These suggestions were designed to exercise aspects of driver knowledge that were indicated as needing improvement based on the scenario responses of the participant.

5.2   Construction of the Background Space

For this project it was desired to have a background space that first represented general English language usage as well as adequate representation of driving specific language and specific language usage for the teenage demographic that were the primary audience for the driving scenario questions. We experimented with a series of background data sets, noting several improvements in the grouping of related items as content was added to the BIS. In our work on this project, we also explored several cursory metrics for analyzing the characteristics of a LSA semantic space [48].

The final content set used to construct the BIS for the Practiced Driver prototype consisted of a total 112,627 documents including 185,730 unique terms. The content areas represented in the total set include 100,000 general language documents randomly selected from the RTRC corpus [49] to provide a basic foundation of language usage. To this were added 6629 driving subject documents collected from an array of websites. This content consisted of state licensure instruction manuals, driver training material, and articles about driving related topics. An additional 5056 teen-authored articles were collected from high school newspapers published on the Web in order to round out the subject audience vocabulary. After initial training and testing, the 942 scenario response items that had been collected in the initial data gathering effort were also included in the content set.

5.3   Establish Target and Feedback Items

As with all scoring and evaluation systems, the first challenge was to determine how responses would be rated. Since the LSA approach involves quantitative measurements of semantic similarity, one way to achieve this objective is to define a target against which response items could be evaluated. It was decided early in the project effort to use the responses gathered from the data collection effort to guide the selection of both response targets and feedback items. These response items were initially processed and projected into the BIS both to observe the semantic groupings of the response items and to test the performance of the BIS.

5.3.1 Human Input: The SME

To form the initial basis for validating the performance of the system, two human raters reviewed all of the responses to identify the most common strategies used in both good (safe) and poor (unsafe) responses. This information was used to establish a scoring rubric, and the responses were then hand scored on a 4-point scale based the recognition of hazards and the driving strategies employed, with 4 representing the best or "most safe" responses. Using the projection vectors for the response items in each score group, a representative centroid vector was calculated for each of the four groups and the distribution of the response vectors was compared to the score group centroids to verify that the groups were semantically separated in the space.

The safest responses (highest scoring) were selected and then reviewed by SME to identify the very best responses based on safety factors, eliminating any that exhibited strategies that could be dangerous for novice drivers. Professional driving instructors also reviewed the initial scenario response essays and selected the best responses for each scenario prompt.

Using the top rated selected safe responses, a target response was generated by calculating a centroid from the selected group of items. This had the advantage of producing a target that was informed by the SME while also being based on language usage from the participant responses in the semantic representation of the target. Additionally, using a composite target of this sort was an attempt to capture the possibility of there being multiple good responses that each contained different concepts. This target served to describe an ideal "best response" that became the standard against which all responses would be measured for sufficiency. Since score outputs were not being produced by the system, mappings for the lower score points did not need to be created.

5.3.2 Human Selected Feedback Items

While not reporting a score, the Practiced Driver system did need to return constructive feedback to the user. This feedback consisted of individual suggestion items that could be reported to the user based on the concepts they expressed in their response. Feedback snippets were developed by selecting individual sentences from the safest (highest scoring) responses covering all of the driving hazards included in the scenarios. The snippets were chosen to represent concepts or strategies that might be used to mitigate the hazards contained in the scenarios. These snippets were reviewed by SMEs to verify that the suggestions were both safe and practical for novice drivers. The snippets were edited to refine the wording so as to improve their readability for novice drivers. The final feedback database consisted of 26 snippets for scenario one and 28 snippets for scenario two.

5.4   Feedback Selection Method

In this application we selected appropriate feedback items by using an additive
analysis process in combination with a target based approach to determine which
snippets provided the most improvement based on the relative proximity of a
response to the target.

Each user response was projected into the BIS compared to the target response
mapping and then augmented with an individual feedback snippet by simply
concatenating the text and then projecting the augmented response into the BIS. This
projection of the augmented response was again compared to the target response
mapping and the change recorded. After performing this process iteratively for
each candidate feedback item, the feedback item pairing that resulted in the most
improved augmented response closest to the target was selected as a feedback item
to be returned. This process was repeated with the new augmented response to find a
possible second appropriate feedback item, and again for the selection of a possible
third feedback item (see Fig. 4). A minimum threshold of improvement was set to
eliminate the production of feedback items that offered no significant contribution.
This threshold was established through empirical testing.

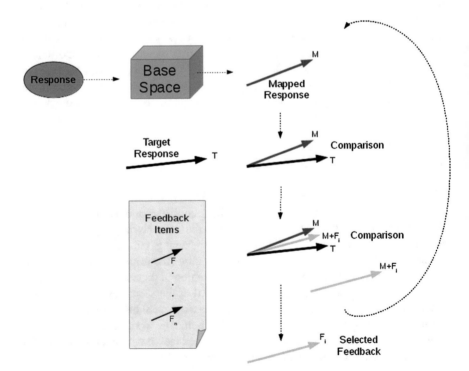

Fig. 4 Flowchart of the feedback selection process

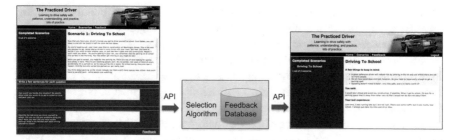

Fig. 5 The Practiced Driver prototype

5.5 *Results*

The feedback generation process was first validated using the original scenario responses as input. For all teen driver participants, this process produced at least two feedback items with a positive effect (i.e., moving the response closer to the target). The feedback produced from validation testing was reviewed by SMEs to verify that it was appropriate. Testing demonstrated that the feedback items generated were both specific to the concepts not addressed in the user's original response and not redundant with each other.

The prototype system (see Fig. 5) was fielded for demonstration and further testing purposes. The Practiced Driver system was able to interactively provide response feedback in real time for multiple users. Reviews from driving academies and individual driving instructors were all positive. The adaptive nature of the feedback generation, the fact that users will not receive a suggested strategy if they have already included in their response, was seen as a significant quality and desirable capability by the reviewers.

6 Conclusion

While essays and open-ended questionnaires are useful analysis tools, scoring and evaluating them are often problematic due to the need for human input and effort. Human raters limit the scalability of the evaluation task which prevents the use of open response items on a wider scale. Automated systems for evaluating open response items are necessary, but must be able to perform comparably to human raters in order for such systems to be accepted. LSA provides a unique unsupervised learning approach to address the need for automated evaluation. The feasibility of applying LSA to this purpose has been demonstrated by the case study and the several projects described in this chapter.

The types of assessment to be performed vary widely depending on the application domain or the need of the users. Our own current research projects using

LSA in assessment cover the range of measuring diverse and creative thinking, identifying chains of reasoning exhibited in essay responses, and monitoring situational awareness. The development of evaluation methods is ongoing and there remains much work to be done both in the refinement of evaluation algorithms and in exploiting and enhancing the unsupervised learning capabilities of LSA. Many current systems are based on measuring proximity to a defined target or targets in the semantic space, but there are areas of application where nearness to a semantic target is not the true measure of response quality or potentially not even of interest. Development of additional evaluation methods, such as additive analysis, to exploit the capabilities of LSA for analyzing open response type items will allow the extension of LSA into broader areas of application.

Acknowledgments Research reported in this publication was supported by the Eunice Kennedy Shriver National Institute Of Child Health & Human Development of the National Institutes of Health under Award Number R41HD074300. The content is solely the responsibility of the authors and does not necessarily represent the official views of the National Institutes of Health. Computations reported in this research were performed using the LSA_Toolkit™ produced by Small Bear Technologies, Inc. (http://SmallBearTechnologies.com).

References

1. Page, E.B.: The imminence of... grading essays by computer. Phi Delta Kappan **47**(5), 238–243 (1966)
2. Foltz, P.W., Laham, D., Landauer, T.K.: The intelligent essay accessor: applications to educational technology. Interactive Multimedia Electron J Comput Enhanced Learn **1**(2) (1999)
3. Page, E., Petersen, N.S.: The computer moves into essay grading: updating the ancient test. Phi Delta Kappan **76**(7), 561–565 (1995)
4. Page, E.B.: Computer grading of student prose, using modern concepts and software. J. Exp. Educ. **62**(2), 127–142 (1994)
5. Landauer, T.K., Laham, D., Foltz, P.: Automatic essay assessment. Assess. Educ. Principles Policy Pract. **10**(3), 295–308 (2003). doi:10.1080/0969594032000148154
6. Foltz, P.W., Streeter, L.A., Lochbaum, K.E., Landauer, T.K.: Implementation and applications of the intelligent essay accessor. In: Shermis, M., Burstein, J. (eds.) Handbook of Automated Essay Evaluation, pp. 68–88. Routledge, New York (2013)
7. Hearst, M.A.: The debate on automated essay grading. IEEE Intell. Syst. Appl. **15**(5), 22–37 (2000). doi:10.1109/5254.889104
8. Landauer, T.K., Laham, D., Foltz, P.W.: Automated scoring and annotation of essays with the intelligent essay accessor. In: Shermis, M.D., Burstein, J. (eds.) Automated Essay Scoring: A Cross-Disciplinary Perspective, pp. 87–112. Lawrence Erlbaum Associates, Mahwah (2003)
9. Dumais, S. T., Furnas, G., Landauer, T. K., Deerwester, S., Harshman, R.: Using latent semantic analysis to improve access to textual information. In: SIGCHI Conference on Human Factors in Computing Systems, pp. 281–285. ACM (1988)
10. Furnas, G.W., Landauer, T.K., Gomez, L.M., Dumain, S.T.: The vocabulary problem in human-system communication: an analysis and a solution. Commun. ACM **30**(11), 964–971 (1987)
11. Martin, D.I., Berry, M.W.: Latent semantic indexing. In: Bates, M.J., Maack, M.N. (eds.) Encyclopedia of Library and Information Sciences (ELIS), vol. 3, pp. 2195–3204. Taylor & Francis, Oxford (2010)

12. Dumais, S.T.: Improving the retrieval of information from external sources. Behav. Res. Methods Instrum. Comput. **23**(2), 229–236 (1991)
13. Dumias, S.T.: LSA and information retrieval: getting back to basics. In: Landauer, T.K., McNamara, D.S., Dennis, S., Kintsch, W. (eds.) Handbook of Latent Semantic Analysis. Lawrence Erlbaum Associates, Mahwah (2007)
14. Deerwester, S., Dumais, S.T., Furnas, G., Landauer, T.K., Harshman, R.: Indexing by latent semantic analysis. J. Am. Soc. Inf. Sci. **41**(6), 391–407 (1990)
15. Martin, D.I., Berry, M.W.: Mathematical foundations behind latent semantic analysis. In: Landauer, T.K., McNamara, D.S., Dennis, S., Kintsch, W. (eds.) Handbook of Latent Semantic Analysis. Lawrence Erlbaum Associates, Mahwah (2007)
16. Landauer, T.K.: LSA as a theory of meaning. In: Landauer, T.K., McNamara, D.S., Dennis, S., Kintsch, W. (eds.) Handbook of Latent Semantic Analysis. Lawrence Erlbaum Associates, Mahwah (2007)
17. Landauer, T.K., Dumais, S.T.: A solution to Plato's problem: the latent semantic analysis theory of acquisition, induction, and representation of knowledge. Psychol. Rev. **104**(2), 211–240 (1997)
18. Landauer, T.K.: On the computational basis of learning and cognition: arguments from LSA. In: Ross, N. (ed.) Psychology of Learning and Motivation, vol. 41, pp. 43–84. Elsevier, Amsterdam (2002)
19. Martin, J.C.: How does LSA work. http://smallbeartechnologies.com/lsa-technology/how-does-lsa-work/ (2012). Retrieved 19 Mar 2015
20. Landauer, T.K., Foltz, P.W., Laham, D.: An introduction to latent semantic analysis. Discourse Process. **25**(2–3), 259–284 (1998)
21. Martin, D.I., Martin, J.C., Berry, M.W., Browne, M.: Out-of-core SVD performance for document indexing. Appl. Numer. Math. **57**(11–12), 1230–1239 (2007). doi:10.1016/j.apnum.2007.01.002
22. Berry, M.W., Martin, D.: Principal component analysis for information retrieval. In: Kontoghiorghes, E. (ed.) Statistics: A Series of Textbooks and Monographs: Handbook of Parallel Computing and Statistics, pp. 399–413. Chapman & Hall/CRC, Boca Raton (2005)
23. Martin, J.C.: The LSA_Toolkit™.http://smallbeartechnologies.com/products-and-services/lsa-toolkit/ (2015). Retrieved 19 Mar 2015
24. Rehder, B., Schreiner, M.E., Wolfe, M.B.W., Laham, D., Landauer, T.K., Kintsch, W.: Using latent semantic analysis to assess knowledge: some technical considerations. Discourse Process. **25**(2–3), 337–354 (1998)
25. Foltz, P.W., Kintsch, W., Landauer, T.K.: The measurement of textual coherence with latent semantic analysis. Discourse Process. **28**(2–3), 285–307 (1998). doi:10.1080/01638539809545029
26. Landauer, T.K.: Learning and representing verbal meaning: the latent semantic analysis theory. Curr. Dir. Psychol. Sci. **7**(5), 161–164 (1998)
27. Landauer, T.K., Laham, D., Foltz, P.W.: Learning human-like knowledge by singular value decomposition: a progress report. In: Jordan, M.I., Kearns, M.J., Solla, S.A. (eds.) Advances in Neural Information Processing Systems, pp. 45–51. MIT Press, Cambridge (1998)
28. Foltz, P.W.: Latent semantic analysis for text-based research. Behav. Res. Methods Instrum. Comput. **28**(2), 197–202 (1996)
29. Kintsch, W.: The use of knowledge in discourse processing: a construction-integration model. Psychol. Rev. **95**(2), 163–182 (1988)
30. Wolfe, M.B.W., Schreiner, M.E., Rehder, B., Laham, D., Foltz, P.W., Kintsch, W., Landauer, T.K.: Learning from text: matching readers and texts by latent semantic analysis. Discourse Process. **28**(2–3), 309–336 (1998)
31. Foltz, P.W., Oberbreckling, R.J., Laham, R.D.: Analyzing job and occupational content using latent semantic analysis. In: Wilson, M.A., Bennett, W., Gibson, S.G., Alliger, G.M. (eds.) The Handbook of Work Analysis Methods, Systems, Applications and Science of Work Measurement in Organizations. Taylor & Francis Group, New York (2012)

32. Laham, D., Bennett, W., Landauer, T.K.: An LSA-based software tool for matching jobs, people, and instruction. Interact. Learn. Environ. **8**(3), 171–185 (2000). doi:10.1076/1049-4820(200012)8:3;1-D;FT171
33. Foltz, P.W., Gilliam, S., Kendall, S.: Supporting content-based feedback in online writing evaluation with LSA. Interact. Learn. Environ. **8**(2), 111–127 (2000)
34. Kintsch, E., Caccamise, D., Franzke, M., Johnson, N., Dooley, S.: Summary street: computer-guided summary writing. In: Landauer, T.K., McNamara, D.S., Dennis, S., Kintsch, W. (eds.) Handbook of Latent Semantic Analysis. Lawrence Erlbaum Associates, Mahwah (2007)
35. Landauer, T.K., Lochbaum, K.E., Dooley, S.: A new formative assessment technology for reading and writing. Theory Pract. **48**(1), 44–52 (2009). doi:10.1080/00405840802577593
36. Graesser, A., Penumatsa, P., Ventura, M., Cai, Z., Hu, X.: Using LSA in autotutor: learning through mixed-initiative dialogue in natural language. In: Landauer, T.K., McNamara, D.S., Dennis, S., Kintsch, W. (eds.) Handbook of Latent Semantic Analysis. Lawrence Erlbaum Associates, Mahwah (2007)
37. D'mello, S., Graesser, A.: AutoTutor and affective autotutor: learning by talking with cognitively and emotionally intelligent computers that talk back. ACM Trans. Interact. Intell. Syst. **2**(4) (2012). http://doi.org/10.1145/2395123.2395128
38. Foltz, P.: Automated content processing of spoken and written discourse: text coherence, essays, and team analyses. Inform. Des. J. **13**(1), 5–13 (2005)
39. Foltz, P., Lavoie, N., Oberbreckling, R., Rosenstein, M.: Automated performance assessment of teams in virtual environments. In: Schmorrow, D., Cohn, J., Nicholson, D. (eds.) The PSI Handbook of Virtual Environments for Training and Education: Developments for the Military and Beyond. Praeger Security International, Westport (2008)
40. Boyce, L., Lavoie, N., Streeter, L., Lochbaum, K., Psotka, J.: Technology as a tool for leadership development: effectiveness of automated web-based systems in facilitating tacit knowledge acquisition. Mil. Psychol. **20**(4), 271–288 (2008). doi:10.1080/08995600802345220
41. Lavoie, N., Streeter, L., Lochbaum, K., Wroblewski, D., Boyce, L.A., Krupnick, C., Psotka, J.: Automating expertise in collaborative learning environments. J. Asynchronous Learn. Netw. **14**(4), 97–119 (2010)
42. Streeter, L., Lochbaum, K., Lavoie, N., Psotka, J.: Automated tools for collaborative learning environments. In: Landauer, T.K., McNamara, D.S., Dennis, S., Kintsch, W. (eds.) Handbook of Latent Semantic Analysis. Lawrence Erlbaum Associates, Mahwah (2007)
43. Landauer, T.K., McNamara, D.S., Dennis, S., Kintsch, W. (eds.): Handbook of Latent Semantic Analysis. Lawrence Erlbaum Associates, Mahwah (2007)
44. Biemiller, A., Rosenstein, M., Sparks, R., Landauer, T.K., Foltz, P.W.: Models of vocabulary acquisition: direct tests and text-derived simulations of vocabulary growth. Sci. Stud. Read. **18**(2), 130–154 (2014). doi:10.1080/10888438.2013.821992
45. McNamara, D.S., Graesser, A.C., McCarthy, P.M., Cai, Z.: Automated Evaluation of Text and Discourse with Coh-Metrix. Cambridge University Press, Cambridge (2014)
46. Kulkarni, S.S., Apte, U., Evangelopoulos, N.: The use of latent semantic analysis in operations management research. Decis. Sci. **45**(5), 971–994 (2014). doi:10.1111/deci.12095
47. Leleu, T.D., Jacobson, I.G., Leardmann, C.A., Smith, B., Foltz, P.W., Amoroso, P.J., Smith, T.C.: Application of latent semantic analysis for open-ended responses in a large, epidemiologic study. BMC Med. Res. Methodol. **11**, 136 (2011). doi:10.1186/1471-2288-11-136
48. Martin, J.C., Martin, D.I., Lavoie, N., Parker, J.: Quantitative Metrics Assessing the Quality of a Large Hyper-dimensional Space for Latent Semantic Analysis (To Appear)
49. Lewis, D., Yang, Y., Rose, T., Li, F.: RCV1: a new benchmark collection for text categorization research. J. Mach. Learn. Res. **5**, 361–397 (2004)
50. Martin, D.I., Berry, M.W.: Text mining. In: Higham, N. (ed.) Princeton Companion to Applied Mathematics, pp. 887–891. Princeton University Press, Princeton, NJ (2015)

Mining Evolving Patterns in Dynamic Relational Networks

Rezwan Ahmed and George Karypis

Abstract Dynamic networks have recently been recognized as a powerful abstraction to model and represent the temporal changes and dynamic aspects of the data underlying many complex systems. This recognition has resulted in a burst of research activity related to modeling, analyzing, and understanding the properties, characteristics, and evolution of such dynamic networks. The focus of this growing research has been on mainly defining important recurrent structural patterns and developing algorithms for their identification. Most of these tools are not designed to identify time-persistent relational patterns or do not focus on tracking the changes of these relational patterns over time. Analysis of temporal aspects of the entity relations in these networks can provide significant insight in determining the conserved relational patterns and the evolution of such patterns over time. In this chapter we present new data mining methods for analyzing the temporal evolution of relations between entities of relational networks. A qualitative analysis of the results shows that the discovered patterns are able to capture network characteristics that can be used as features for modeling the underlying dynamic network in the context of a classification task.

1 Introduction

As the capacity to exchange and store information has soared, so has the amount and diversity of available data. This massive growth in scale, combined with the increasing complexity of the data, has spawned the need to develop efficient methods to process and extract useful information from this data. Within the research dedicated to solving this problem, a significant amount of attention has been given to develop efficient mining tools to analyze graphs and networks modeling relations between objects or entities. Due to their flexibility and availability of theoretical and applied tools for efficient analysis, networks have been used as generic model to represent the relations between various entities in diverse applications. Examples

R. Ahmed (✉) • G. Karypis
Department of Computer Science and Engineering, University of Minnesota,
Minneapolis, MN 55455, USA
e-mail: ahmed@cs.umn.edu; karypis@cs.umn.edu

© Springer International Publishing Switzerland 2016
M.E. Celebi, K. Aydin (eds.), *Unsupervised Learning Algorithms*,
DOI 10.1007/978-3-319-24211-8_17

485

of some widely studied networks include the friend-networks of popular social networking sites like Facebook, the Enron email network, co-authorship and citation networks, and protein–protein interaction networks. Until recently, the focus of the research has been on analyzing graphs and networks modeling static data. However, with the emergence of new application areas, it has become apparent that static models are inappropriate for representing the temporal changes underlying many of these systems, and that it is imperative to develop models capable of capturing the dynamic aspects of the data, as well as tools to analyze and process this data.

In recent years there has been a burst of research activity related to modeling, analyzing, and understanding the properties and characteristics of such dynamic networks. The growing research has focused on finding frequent patterns [5], clustering [8], characterizing network evolution rules [4], detecting related cliques [7, 21], finding subgraph subsequences [13], and identifying co-evolution patterns capturing attribute trends [10, 15] in dynamic networks. Although the existing techniques can detect the frequent patterns in a dynamic network, most of them are not designed to identify time-persistent relational patterns and do not focus on tracking the changes of these conserved relational patterns over time. Analysis of temporal aspects of the entity relations in these networks can provide significant insight determining the conserved relational patterns and the evolution of such patterns over time.

The objective of this chapter is to present new data mining methods that are designed to analyze and identify how the relations between the entities of relational networks evolve over time. The key motivation underlying this research is that significant insights can be gained from dynamic relational networks by analyzing the temporal evolution of their relations. Such analysis can provide evidence to the existence of, possibly unknown, coordination mechanisms by identifying the relational motifs that evolve and move in a similar and highly conserved fashion, and to the existence of external factors that are responsible for changing the stable relational patterns in these networks. Different classes of evolving relational patterns are introduced that are motivated by considering two distinct aspects of relational pattern evolution. The first is the notion of state transition and seeks to identify sets of entities whose time-persistent relations change over time and space. The second is the notion of coevolution and seeks to identify recurring sets of entities whose relations change in a consistent way over time and space.

The rest of the chapter is organized as follows. Section 2 reviews some graph-related definitions and introduces notation used throughout the chapter. Section 3 presents a new class of patterns, referred to as the evolving induced relational states (EIRS), which is designed to analyze the time-persistent relations or states between the entities of the dynamic networks. These patterns can help identify the transitions from one conserved state to the next and may provide evidence to the existence of external factors that are responsible for changing the stable relational patterns in these networks. We developed an algorithm to efficiently mine all maximal non-redundant evolution paths of the stable relational states of a dynamic network. Next in Sect. 4, we introduce a class of patterns, referred to as coevolving relational motifs (CRM), which is designed to identify recurring sets of entities whose relations change in a consistent way over time. CRMs can provide evidence to the existence of, possibly unknown, coordination mechanisms by identifying the relational motifs that evolve in a similar and highly conserved fashion. An algorithm is presented to efficiently analyze the frequent relational changes between the entities of the dynamic networks and capture all frequent coevolutions as CRMs. In Sect. 5, we define a new class of patterns built upon the concepts of CRMs, referred to as coevolving induced relational motifs (CIRM), is designed to represent patterns in which all the relations among recurring sets of nodes are captured and some of the relations undergo changes in a consistent way across different snapshots of the network. We also present an algorithm to efficiently mine all frequent coevolving induced relational motifs. In Sect. 6 we provide a qualitative analysis of the information captured by each class of the discovered patterns. For example, we show that some of the discovered CRMs capture relational changes between the nodes that are thematically different (i.e., edge labels transition between two clusters of topics that have very low similarity). Moreover, some of these patterns are able to capture network characteristics that can be used as features for modeling the underlying dynamic network. A comprehensive evaluation of the performance and scalability of all the algorithms is presented through extensive experiments using multiple dynamic networks derived from real-world datasets from various application domains. The detailed analysis of the performance and scalability results can be found in [1–3]. Section 7 summarizes the conclusions and presents some future research directions.

2 Definitions and Notation

A relational network is represented via labeled graphs. A *graph* $G = (V, E)$ consists of a set of vertices V and a set of edges E that contain pairs of vertices. A graph G is called *labeled* if there are functions χ and ψ that assign labels to the vertices and edges, respectively. That is, there is a set of vertex labels λ_v and edge labels λ_e such that $\chi : V \rightarrow \lambda_v$ and $\psi : E \rightarrow \lambda_e$. For simplicity, we will use l_u and l_v to denote the labels of vertices u and v, respectively, and use l_e to denote the label of edge e. The labels assigned to the vertices (edges) do not need to be distinct and the same label can be assigned to different vertices (edges). A graph $G' = (V', E')$ is a *subgraph* of G if $V' \subseteq V$ and $E' \subseteq E$. A subgraph $G'' = (V'', E'')$ of G is an *induced subgraph* if $\{(u, v) | (u, v) \in V'' \times V''\} \cap E = E''$.

Given a connected graph $G = (V, E)$ and a depth-first search (DFS) traversal of G, its *depth-first search tree* T is the tree formed by the forward edges of G. All nodes of G are encoded with subscripts to order them according to their discovery time. Given a DFS tree T of graph G containing n nodes, the root node is labeled as (v_0) and the last discovered node is labeled as (v_{n-1}). The *rightmost path* of the DFS tree T is the path from vertex v_0 to vertex v_{n-1}.

A *dynamic network* $\mathcal{N} = \langle G_1, G_2, \ldots, G_T \rangle$ is modeled as a finite sequence of labeled undirected graphs, where $G_t = (V, E_t)$ describes the state of the system at a discrete time interval t. Each of these labeled undirected graphs will be called a *snapshot*. Note that the set of vertices in each snapshot is assumed to be the same whereas the set of edges can change across the different snapshots. The set of vertices is referred to as the nodes of $\mathcal{N}$, denoted by $V_{\mathcal{N}}$. When nodes appear or disappear over time, the set of nodes of each snapshot is the union of all the nodes over all snapshots. The nodes across the different snapshots are numbered consistently, so that the ith node of G_k ($1 \leq k \leq T$) will always correspond to the same ith node of $\mathcal{N}$. In addition, each of these snapshots is assumed to have their own vertex- and edge-labeling functions (χ_t and ψ_t), which allows for the label of each vertex/edge to change across the snapshots.

An edge (u, v) is in a *consistent state* over a maximal time interval $s:e$ if it is present in all snapshots $G_s, \ldots, G_e$ with the same label and it is different in both G_{s-1} and G_{e+1} (assuming $s > 1$ and $e < T$). We define the *span sequence* of an edge as the sequence of maximal-length time intervals in which an edge is present in a consistent state. The span sequence of an edge will be described by a sequence of vertex labels, edge labels, and time intervals of the form $\langle (l_{u_1}, l_{e_1}, l_{v_1}, s_1:e_1), \ldots, (l_{u_n}, l_{e_n}, l_{v_n}, s_n:e_n) \rangle$, where $s_i \leq e_i$ and $e_i \leq s_{i+1}$.

A *persistent dynamic network* $\mathcal{N}^{\phi}$ is derived from a dynamic network $\mathcal{N}$ by removing all the edges from the snapshots of $\mathcal{N}$ that do not occur in a consistent state in at least ϕ consecutive snapshots, where $1 \leq \phi \leq T$. It can be seen that $\mathcal{N}^{\phi}$ can be derived from $\mathcal{N}$ by removing from the span sequence of each edge all the intervals whose length is less than ϕ.

An *injection* is defined as a function $f : A \rightarrow B$ such that $\forall a, b \in A$, if $f(a) = f(b)$, then $a = b$. A function $f : A \rightarrow B$ is a bijective function or *bijection*, *iff* $\forall b \in B$, there is a unique $a \in A$ such that $f(a) = b$. A function composition $g \circ f$ implies that for function $f : X \rightarrow Y$ and $g : Y \rightarrow Z$, then *composite* function $g \circ f : X \rightarrow Z$.

An induced subgraph that involves the same set of vertices and whose edges and their labels remain conserved across a sequence of snapshots will be referred to as the *induced relational state* (IRS). The IRS definition is illustrated in Fig. 1. The three-vertex induced subgraph consisting of the dark shaded nodes that are connected via the directed edges corresponds to an induced relational state as it remains conserved in the three consecutive snapshots. The key attribute of an IRS is that the set of vertices and edges that compose the induced subgraph must remain

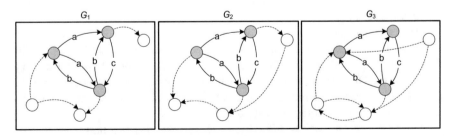

Fig. 1 Examples of relational states

the same in the consecutive snapshots. From this definition we see that an IRS corresponds to a time-conserved pattern of relations among a fixed set of entities (i.e., nodes) and as such can be thought as corresponding to a *stable* relational pattern. An IRS S_i will be denoted by the tuple $S_i = (V_i, s_i : e_i)$, where V_i is the set of vertices of the induced subgraph that persists from snapshot G_{s_i} to snapshot G_{e_i} ($s_i \leq e_i$) and it does not persist in G_{s_i-1} and G_{e_i+1} (assuming $s_i > 1$ and $E_i < T$). We will refer to the time interval $s_i : e_i$ as the *span* of S_i, and to the set of consecutive snapshots $G_{s_i}, \ldots, G_{e_i}$ as its *supporting set*. By its definition, the span of an IRS and the length of its supporting set are *maximal*. Note that for the rest of the thesis, any references to subsequences of snapshots will be assumed to be consecutive. The induced subgraph corresponding to an IRS S_i will be denoted as $g(S_i)$. A snapshot G_t supports an induced relational state S_i, if $g(S_i)$ is an induced subgraph of G_t. The size of an IRS S_i is defined in terms of the number of vertices and represented as k, where $k = |V_i|$. Based on the type of the dynamic network, a low k value may generate a large number of IRSs capturing trivial relational information, whereas a large k value may not detect any IRS.

A *relational motif* is a subgraph that occurs frequently in a single snapshot or a collection of snapshots. It is similar to the traditional definition of frequently occurring graph patterns [16]. An *embedding* of a relational motif in a snapshot is determined by performing subgraph isomorphism operation of the motif's graph pattern. A motif *occurs* in a snapshot (i.e., a snapshot supports a motif) means that there exists an embedding of the motif in to the snapshot. In order to determine, how many times a motif occurs in a collection of snapshots, we count the total number of embeddings of the motif in that collection of snapshots. In Fig. 2, the three-vertex subgraph consisting of the shaded nodes that are connected via the labeled edges corresponds to a relational motif that occurs a total of four times (twice in G_1 and once in each of G_2 and G_3). Note that the set of nodes that support the multiple occurrences of a relational motif do not need to be the same. In Fig. 2, there are three non-identical sets of nodes supporting the four occurrences, since the shaded

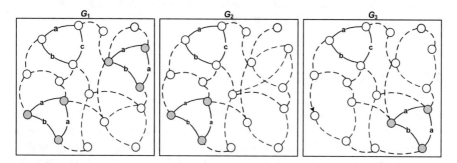

Fig. 2 Examples of relational motifs. The *shaded nodes* that are connected via the *labeled edges* correspond to a relational motif. The three-vertex subgraph consisting of the *shaded nodes* that are connected via the *labeled edges* corresponds to a relational motif that occurs a total of four times (twice in G_1 and once in each of G_2 and G_3)

node set in G_2 is same as one of the node set of G_1 (lower left one). We will use M to denote a relational motif and the underlying subgraph will be denoted by the tuple (N, A, L_N, L_A), where N is the set of nodes, A is the set of edges (arcs), and the functions L_N and L_A that assign labels to the nodes and edges, respectively.

Note that a key difference between relational states and relational motifs is that the set of nodes that support the relational state is the same in the consecutive snapshots. On the other hand, the set of nodes that support the multiple occurrences of a relational motif do not need to be the same. This difference changes the computational complexity required to identify these two types of relational patterns. Specifically, since the sets of vertices involved in the relational state remain fixed across its supporting set of snapshots, we can determine if a snapshot supports a relational state by simply checking if the relational state's graph pattern is a subgraph of that snapshot. On the other hand, in order to determine if a snapshot supports a relational motif (and how many times), we need to perform subgraph isomorphism operations (i.e., identify the embeddings of the relational motif's graph pattern), which can be expensive if the number of distinct vertex labels in the motif and/or snapshot is small.

An *induced relational motif* is an induced subgraph that occurs frequently in a single snapshot or a collection of snapshots. In order to determine if a snapshot supports an induced relational motif (and how many times), we need to perform induced subgraph isomorphism operations (i.e., identify the embeddings of the relational motif's graph pattern).

3 Mining the Evolution of Conserved Relational States

Significant insights regarding the stable relational patterns among the entities can be gained by analyzing temporal evolution of the complex entity relations. This can help identify the transitions from one conserved state to the next and may provide evidence to the existence of external factors that are responsible for changing the stable relational patterns in these networks.

3.1 Evolving Induced Relational State

An example of the type of evolving patterns in dynamic networks that the work in this section is designed to identify is illustrated in Fig. 3. This figure shows a hypothetical dynamic network consisting of 14 consecutive snapshots, each modeling the annual relations among a set of entities. The four consecutive snapshots for years 1990 through 1993 show an induced relational state S_1 that consists of nodes {a, b, e, f}. This state was evolved to the induced relational state S_2 that occurs in years 1995–1999 that contains nodes {a, b, d, e, h}. Finally, in years 2000–2003,

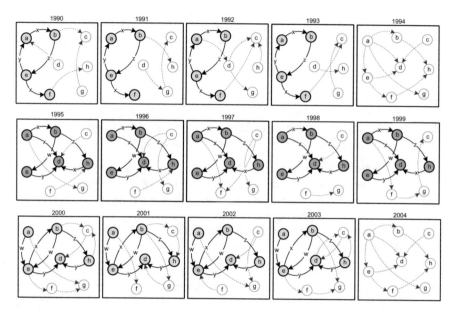

Fig. 3 An example of an evolving relational state

the induced relational state S_2 was further evolved to the induced relational state S_3 that contains the same set of nodes but has a different set of relations. Note that even though the sets of nodes involved in S_1 and S_2 are different, there is a high degree of overlap between them. Moreover, the transition from S_1 to S_2 did not happen in consecutive years, but there was a 1 year gap, as the snapshot for 1994 does not contain either S_1 or S_2. Such a sequence of induced relational states $S_1 \rightsquigarrow S_2 \rightsquigarrow S_3$ represents an instance of what we refer to as an *evolving induced relational state* (EIRS) and represents the types of patterns that the work in this section is designed to identify.

EIRSs identify entities whose relations transition through a sequence of time-persistent relational patterns and as such can provide evidence of the existence of external factors responsible for these relational changes. For example, consider a dynamic network that captures the trading patterns between a set of entities. The nodes in this *trading network* model the trading entities (e.g., countries, states, businesses, individuals) and the directed edges model the transfer of goods and their types from one entity to another. An EIRS in this trading network can potentially identify how the trading patterns change over time (e.g., addition/deletion of edges or inclusion of new trading partners) signalling the existence of significant economic, political, and environmental factors that drive such changes (see Sect. 6 for some examples of such patterns in an inter-country trading network). Similarly, in a dynamic network that captures the annual citation structure of U.S. Patents (or other scientific publications), EIRSs can identify how stable knowledge transfer or knowledge sharing sub-networks among different science areas have evolved and thus facilitate the identification of transformative science developments that changed the state of these sub-networks.

The formal definition of an EIRS is as follows.

Definition 1 (**Evolving Induced Relational State**). Given a dynamic network $\mathcal{N}$ containing T snapshots, a value ϕ ($1 \le \phi \le T$), and a value β ($0 < \beta \le 1$), an evolving induced relational state of length m is a sequence of induced relational states $\langle S_1, S_2, \ldots, S_m \rangle$ that satisfies the following constraints:

   (i) the supporting set of each induced relational state S_i contains at least ϕ consecutive snapshots in the persistent dynamic network $\mathcal{N}^\phi$ of $\mathcal{N}$,
  (ii) for each $1 \le i < m$, the first snapshot in S_{i+1}'s supporting set follows the last snapshot in S_i's supporting set,
 (iii) for each $1 \le i < m$, $g(S_i)$ is different from $g(S_{i+1})$, and
 (iv) for each $1 \le i < m$, $|V_i \cap V_{i+1}|/|V_i \cup V_{i+1}| \ge \beta$.

The value ϕ, referred to as the support of the EIRS, is used to capture the requirement that each induced relational state occurs in a sufficiently large number of consecutive snapshots and as such it represents a set of relations among the entities involved that are stable. The value β, referred to as the *inter-state similarity*, is used to enforce the minimum vertex-level similarity between the selected relational states. This ensures that the EIRS captures the relational transitions of a consistent set of vertices (i.e., they do not jump on entirely different parts of the network in successive relational states) but at the same time allows for the inclusion of new vertices and/or the elimination of existing vertices, if they are required to describe the new relational state. The inter-state similarity can be also defined in terms of the maximum number of vertices that can be different between the selected relational states. The third constraint in the above definition is used to eliminate EIRSs that contain consecutive IRSs with identical induced subgraphs. This is motivated by our desire to find EIRSs that capture changes in the time-persistent relations. However, the above definition allows for the same induced subgraph to occur multiple times in the same EIRS, as long as these occurrences do not happen one after the other (e.g., it allows for EIRSs of the form $\langle S_1, S_2, S_3 \rangle$ in which $g(S_1) = g(S_3)$).

An important aspect of the definition of an EIRS is that it is defined with respect to the persistent dynamic network $\mathcal{N}^\phi$ of $\mathcal{N}$ and not $\mathcal{N}$ itself. This is because we are interested in finding how the persistent relations among a set of entities have changed over time and as such we first eliminate the set of relations that appear for a short period of time. Note that we focus on finding the maximal EIRSs, since they represent the non-redundant set of EIRSs.

Given the above definition, the work in this section is designed to develop efficient algorithms for solving the following problem:

Problem 1 (**Maximal Evolving Induced Relational State Mining**). Given a dynamic network $\mathcal{N}$ containing T snapshots, a user defined support ϕ ($1 \le \phi \le T$), an inter-state similarity β $(0 < \beta \leq 1)$, a minimum size of k_{min} and a maximum size of k_{max} vertices per IRS, and a minimum EIRS length m_{min}, find all EIRSs such that no EIRS is a subsequence of another EIRS.

Since the set of maximal EIRSs contains all non-maximal EIRSs, the above problem will produce a succinct set of results. Also, the minimum and maximum constraints on the size of the IRSs involved is introduced to allow an application to focus on IRSs of meaningful size, whereas the minimum constraint on the EIRS length is introduced in order to eliminate short paths.

3.2 Finding Evolving Induced Relational States

The algorithm that we developed for finding all maximal EIRSs (Problem 1) follows a two-step approach. In the first step, the dynamic network $\mathcal{N}$ is transformed into its persistent dynamic network $\mathcal{N}^\phi$ and a recursive enumeration algorithm is used to identify all the IRSs $\mathcal{S}$ whose supporting set is at least ϕ in $\mathcal{N}^\phi$. The $\mathcal{N}$ to $\mathcal{N}^\phi$ transformation is done by removing spans that are less than ϕ from each edge's span sequence and then removing the edges with empty span sequences. In the second step, the set of IRSs are mined in order to identify their maximal non-redundant sequences that satisfy the constraints of EIRS's definition.

3.2.1 Step 1: Mining of Induced Relational States

The algorithm that we developed to mine all induced relational states is based on a recursive approach to enumerate all (connected) induced subgraphs of a graph that satisfy minimum and maximum size constraints. In the rest of this section we first describe the recursive algorithm to enumerate all induced subgraphs in a simple graph and then describe how we modified this approach to mine the induced relational states in a dynamic network. The enumeration algorithm was inspired by the recursive algorithm to enumerate all spanning trees [23]. Our discussion initially assumes that the graph is undirected and the necessary modifications that apply for directed graphs are described afterwards. Also, any references to induced subgraphs assumes *connected* induced subgraphs.

Induced Subgraph Enumeration

Given a graph $G = (V, E, L[E])$, let $G_i = (V_i, E_i, L[E_i])$ be an induced subgraph of G (V_i can also be empty), V_f be a subset of vertices of V satisfying $V_i \cap V_f = \emptyset$, and let $F(V_i, V_f)$ be the set of induced subgraphs of G that contain V_i and zero or more vertices from V_f. Given these definitions, the complete set of induced subgraphs of G is given by $F(\emptyset, V)$. The set $F(V_i, V_f)$ can be computed using the recurrence relation.

$$F(V_i, V_f) =$$

$$\begin{cases} V_i, & \text{if } \text{sgadj}(V_i, V_f) = \emptyset \\ & \text{or } V_f = \emptyset \\ F(\{u\}, V_f \setminus u) \cup F(\emptyset, V_f \setminus u), & \text{if } V_i = \emptyset \wedge V_f \neq \emptyset \\ \text{where } u \in V_f \\ F(V_i \cup \{u\}, V_f \setminus u) \cup F(V_i, V_f \setminus u), & \text{otherwise,} \\ \text{where } u \in \text{sgadj}(V_i, V_f) \end{cases} \qquad (1)$$

where $\text{sgadj}(V_i, V_f)$ (*subgraph-adjacent*) denotes the vertices in V_f that are adjacent to at least one of the vertices in V_i.

To show that Eq. (1) correctly generates the complete set of induced subgraphs, it is sufficient to consider the three conditions of the recurrence relation. The first condition, which corresponds to the initial condition of the recurrence relation, covers the situations in which either (1) none of the vertices in V_f are adjacent to any of the vertices in V_i and as such V_i is the only induced subgraphs that can be generated, or (2) V_f is empty and as such V_i cannot be extended further. The second condition, which covers the situation in which V_i is empty and V_f is not empty, decomposes $F(\emptyset, V_f)$ as the union of two sets of induced subgraphs based on an arbitrarily selected vertex $u \in V_f$. The first is the set of induced subgraphs that contain vertex u (corresponding to $F(\{u\}, V_f \setminus u)$) and the second is the set of induced subgraphs that do not contain u (corresponding to $F(\emptyset, V_f \setminus u)$). Since any of the induced subgraphs in $F(\emptyset, V_f)$ will either contain u or not contain u, the above decomposition covers all possible cases and it correctly generates $F(\emptyset, V_f)$. Finally, the third condition, which corresponds to the general case, decomposes $F(V_i, V_f)$ as the union of two sets of induced subgraphs based on an arbitrarily selected vertex $u \in V_f$ that is adjacent to at least one vertex in V_i. The first is the set of induced subgraphs that contain V_i and u (corresponding to $F(V_i \cup \{u\}, V_f \setminus u)$) and the second is the set of induced subgraphs that contain V_i but not u (corresponding to $F(V_i, V_f \setminus u)$). Similarly to the second condition, this decomposition covers all the cases with respect to u and it correctly generates $F(V_i, V_f)$. Also the requirement that $u \in \text{sgadj}(V_i, V_f)$ ensures that this condition enumerates only the connected induced subgraphs[1]. Since each recursive call in Eq. (1) removes a vertex from V_f, the recurrence relation will terminate due to the first condition. Finally, since the three conditions in Eq. (1) cover all possible cases, the overall recurrence relation is correct.

In addition to correctness, it can be seen that the recurrence relation of Eq. (1) does not have any overlapping sub-problems, and as such, each induced subgraph of $F(V_i, V_f)$ is generated only once, leading to an efficient approach for generating $F(V_i, V_f)$. Constraints on the minimum and maximum size of the induced subgraphs

---

[1]Given a connected (induced) subgraph g, it can be grown by adding one vertex at a time while still maintaining connectivity; e.g., an MST of g (which exists due to its connectivity) can be used to guide the order by which vertices are added.

can be easily incorporated in Eq. (1) by returning $\emptyset$ in the first condition when $|V_i|$ is less than the minimum size and not performing the recursive exploration for other two cases.

Induced Relational State Enumeration

There are two key challenges in extending the induced subgraph enumeration approach of Eq. (1) in order to enumerate the IRSs in a dynamic network. First, the addition of a vertex to an IRS is different from adding a vertex to an induced subgraph as it can result in multiple IRSs depending on the overlapping spans between the vertex being added and the original IRS. Consider an IRS $S_i = (V_i, s_i:e_i)$, a set of vertices V_f such that $V_i \cap V_f = \emptyset$, and a vertex $v \in V_f$ that is adjacent to at least one of the vertices in V_i. If v's span sequence contains multiple spans that have overlaps greater than or equal to ϕ with S_i's span, then the inclusion of v leads to multiple IRSs, each supported by different disjoint spans.

Figure 4 illustrates the vertex addition process during an IRS expansion. Figure 4a shows a simple case of vertex addition where an IRS $S_1 = (\{a, b, e\}, 1:7)$ is expanded by adding adjacent vertex c having a single span of $2:4$. The resultant IRS is $S_2 = (\{a, b, e, c\}, 2:4)$ that contains all the vertices and only the overlapping span of S_1 and c. Figure 4b shows a more complex case where the vertex h that is added to S_1 has the span sequence of $\langle 1:3, 5:9 \rangle$. In this case, the overlapping spans $1:3$ and $5:7$ form two separate IRSs $S_3 = (\{a, b, e, h\}, 1:3)$ and $S_4 =$

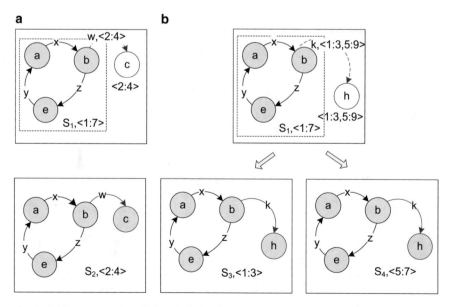

Fig. 4 Adding a vertex to an induced relational state

($\{a, b, e, h\}, 5:7$), each of which needs to be considered for future expansions in order to discover the complete set of IRSs.

Second, the concept of removing a vertex from V_f used in Eq. (1) to decompose the set of induced subgraphs needs to be re-visited so that to account for the temporal nature of dynamic networks. Failure to do so will lead to an IRS discovery algorithm that will not discover the complete set of IRSs and the set of IRSs that it discovers will be different based on the order that it chooses to add vertices in the IRS under consideration.

This is illustrated in the example of Fig. 5. The IRS $S_1 = (\{a, b\}, 0:12)$ is expanded to $S_2 = (\{a, b, c\}, 1:5)$ by adding the adjacent vertex c. In terms of Eq. (1), this corresponds to the third condition and leads to the recursive calls of $F(\{a, b, c\}, V_f \setminus c)$ and $F(\{a, b\}, V_f \setminus c)$ (i.e., expand S_2 and expand S_1). It is easy to see that the set of IRSs that will be generated from these recursions will not contain $S_4 = (\{a, b, c, d\}, 6:11)$, since it can only be generated from S_2 but its span does not overlap with S_4's span. On the other hand, if S_1 is initially expanded by adding vertex d, resulting in the IRS $S_3 = (\{a, b, d\}, 6:11)$, then the recursive calls of $F(\{a, b, d\}, V_f \setminus d)$ and $F(\{a, b\}, V_f \setminus d)$ will generate S_4 and S_2, respectively. Thus, based on the order by which vertices are selected and included in an IRS, some IRSs may be missed. Moreover, different vertex inclusion orders can potentially miss different IRSs.

To address both of these issues the algorithm that we developed for enumerating the complete set of IRSs utilizes a recursive decomposition approach that extends Eq. (1) by utilizing two key concepts. The first is the notion of the set of vertex-span tuples that can be used to grow a given IRS. Formally, given an IRS $S_i = (V_i, s_i : e_i)$

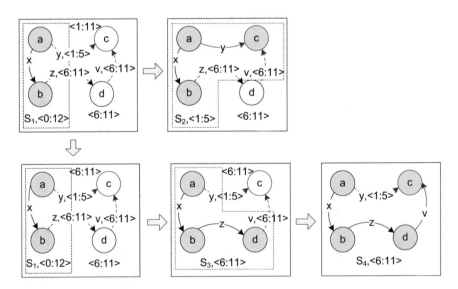

Fig. 5 Updating span sequence of a vertex

and a set of vertices V_f in $\mathcal{N}^\phi$ with $V_i \cap V_f = \emptyset$, the irsadj(S_i, V_f) is the set of vertex-span tuples of the form $(u, s_{u_j} : e_{u_j})$ such that $u \in V_f$ and $(V_i \cup \{u\}, s_{u_j} : e_{u_j})$ is an IRS whose span is at least ϕ. Note that irsadj(S_i, V_f) can contain multiple tuples for the same vertex if that vertex can extend S_i in multiple ways (each having a different span and possibly an induced subgraph with different sets of edges). The tuples in irsadj(S_i, V_f) represent possible extensions of S_i and since a vertex can occur multiple times, it allows for the generation of IRS with the same set of vertices but different spans (addressing the first challenge). To illustrate how this operation can address the first challenge, consider the example of Fig. 4b. When vertex h containing the span sequence of $\langle 1:3, 5:9 \rangle$ is added into $S_1 = (\{a, b, e\}, 1:7)$, each vertex-span tuple of h generates separate overlapping spans $1:3$ and $5:7$. This results in forming two separate IRSs $S_3 = (\{a, b, e, h\}, 1:3)$ and $S_4 = (\{a, b, e, h\}, 5:7)$. Now, S_3 and S_4 are further expanded in order to discover the complete set of IRSs.

The second is the notion of vertex-span deletion, which is used to eliminate the order dependency described earlier and generates the complete set of IRSs. The key idea is when a tuple $(u, s_{u_j} : e_{u_j})$ is added into S_i, instead of removing u from V_f, only remove the span $s_{u_j} : e_{u_j}$ from u's span sequence in V_f. Vertex u will only be removed from V_f iff after the removal of $s_{u_j} : e_{u_j}$ its span sequence becomes empty or the remaining spans have lengths that are smaller than ϕ. Formally, given V_f, a vertex $u \in V_f$, and a vertex-span tuple $(u, s_{u_j} : e_{u_j})$, the *span-deletion* operation, denoted by $V_f | (u, s_{u_j} : e_{u_j})$, updates the span sequence of u by removing the span $s_{u_j} : e_{u_j}$ from its span sequence, eliminating any of its spans that become shorter than ϕ, and eliminating u if its updated span sequence becomes empty. The span-deletion operation is the analogous operation to vertex removal of Eq. (1). To illustrate how this operation can address the second challenge, consider again the example of Fig. 5. Once c is added into S_1, the span that it used (i.e., $1:5$) is deleted from its span sequence, resulting in a new span sequence containing $\langle 6:11 \rangle$. Now the recursive call corresponding to $F(\{a, b\}, V_f | (c, 1:5))$ will be able to identify S_4 as c is still part of V_f.

Given the above definitions, the recursive approach for enumerating the complete set of IRSs can now be formally defined. Let $S_i = (V_i, s_i : e_i)$ be an IRS, V_f a set of vertices in $\mathcal{N}^\phi$ with their corresponding span-sequences in $\mathcal{N}^\phi$ such that $V_i \cap V_f = \emptyset$, and $H(S_i, V_f)$ be the set of IRSs that (1) contain V_i and zero or more vertices from V_f and (2) their span is a sub-span[2] of $s_i : e_i$. Given the above, the complete set of IRSs in $\mathcal{N}^\phi$ is given by $H((\emptyset, 1:T), V(\mathcal{N}^\phi))$. The recurrence relation for $H(S_i, V_f)$ is:

---

[2] The *sub-span* of a span corresponds to a time interval that is either identical to the span or is contained within it.

$$H(S_i, V_f) =$$

$$
\begin{cases}
S_i, & \text{if } \mathrm{irsadj}(S_i, V_f) = \emptyset \\
& \text{or } V_f = \emptyset \\[1.5ex]
H((\{u\}, s_u : e_u), V_f | (u, s_u : e_u)) \\
\quad \cup\, H((\emptyset, 1 : T), V_f | (u, s_u : e_u)), & \text{if } V_i = \emptyset \wedge V_f \neq \emptyset \quad (2) \\
\text{where } u \in V_f \text{ and } s_u : e_u \text{ is a span of } u \\[1.5ex]
H((V_i \cup \{u\}, s_u : e_u), V_f | (u, s_u : e_u)) \\
\quad \cup\, H(S_i, V_f | (u, s_u : e_u)), & \text{otherwise.} \\
\text{where } (u, s_u : e_u) \in \mathrm{irsadj}(S_i, V_f)
\end{cases}
$$

The above recurrence relation shares the same overall structure with the corresponding recurrence relation for enumerating the induced subgraphs [Eq. (1)] and its correctness can be shown in a way similar to that used for Eq. (1). To eliminate redundancy, we omit the complete proof of Eq. (2), and only focus on discussing the third condition, which represents the general case. This condition decomposes $H(S_i, V_f)$ as the union of two sets of IRSs based on an arbitrarily selected vertex-span tuple $(u, s_u : e_u) \in \mathrm{irsadj}(S_i, V_f)$. Since the span $s_u : e_u$ is a maximal overlapping span between u and S_i, the set of vertices $V_i \cup \{u\}$ with the span of $s_u : e_u$ is an IRS. With respect to vertex-span tuple $(u, s_u : e_u)$, the set of IRSs in $H(S_i, V_f)$ can belong to one of the following three groups: (i) the set of IRSs that contain u and have a span that is a sub-span of $s_u : e_u$; (ii) the set of IRSs that contain u and have a span that is disjoint with $s_u : e_u$, and (iii) the set of IRSs that do not contain u. The $H((V_i \cup \{u\}, s_u : e_u), V_f | (u, s_u : e_u))$ part of the third condition generates (i), whereas the $H(S_i, V_f | (u, s_u : e_u))$ part generates (ii) and (iii). What is missing from the above groups is the group corresponding to the set of IRSs that contain u and have a span that partially overlaps with $s_u : e_u$. The claim is that this cannot happen. Consider an IRS $S_j = (V_j, s_j : e_j) \in H(S_i, V_f)$ that contain u and without loss of generality, assume that $s_u < s_j < e_u < e_j$. Since we are dealing with induced subgraphs and stable topologies (i.e., from the definition of an IRS), the connectivity of u to the vertices in V_i remains the same during the span of $s_u : e_u$ and also during the span of $s_j : e_j$, which means that the connectivity of u to the vertices in V_i remains the same during the entire span of $s_u : e_j$. This is a contradiction, since $(u, s_u : e_u) \in \mathrm{irsadj}(S_i, V_f)$ and as such is a maximal length span of stable relations due to the fact that $(V_i \cup \{u\}, s_u : e_u)$ is an IRS and the span of an IRS is maximal. Thus, the two cases of the third condition in Eq. (2) cover all possible cases and it correctly generate the complete set of IRSs. Also, since each recursive call modifies at least the set V_f, none of the recursive calls lead to overlapping subproblems, ensuring that each IRS is only generated once.

The high-level structure of the IRS enumeration algorithm is shown in Algorithm 1. A relational state S_i, the set of available vertices V_f that are not in S_i, and a list of all identified IRSs $\mathscr{S}$ are passed to *enumerate* to incrementally grow S_i. To generate all IRSs, the enumeration process starts with enumerate($\emptyset, V_f, \mathscr{S}$) where V_f includes all vertices for $\mathscr{N}^\phi$. Lines 1–6 initiate separate enumeration for each vertex. Line 7 selects an adjacent vertex of S_i from V_f. If no adjacent vertex is found,

---

Algorithm 1 enumerate($S_i, V_f, \mathscr{S}$)

1: **if** S_i is $\emptyset$ **then**
    for each vertex v **in** V_f **do**
2:        $S_i \leftarrow$ construct a relational state using v and its span
3:        remove v from V_f
4:        enumerate($S_i, V_f, \mathscr{S}$)
5:    **return**
6: $v \leftarrow$ select an adjacent vertex of S_i from V_f
7: **if** there is no v found for expansion of S_i **then**
8:    **return**
9: $rslist \leftarrow$ detect all relational states by including v in S_i
    for each relational state s **in** $rslist$ **do**
10:    **if** s contains at least minimum number of required vertices **then**
11:        add s to $\mathscr{S}$
12: $v' \leftarrow$ update v's span sequence by removing the span of each relational state in $rslist$
13: **if** v' is not empty **then**
14:    replace v by v' in V_f
15: **else**
16:    remove v from V_f
17: enumerate($S_i, V_f, \mathscr{S}$)
18: **if** v' is not empty **then**
19:    remove v' from V_f
    for each relational state s **in** $rslist$ **do**
20:    **if** s can be expanded **then**
21:        enumerate($s, V_f, \mathscr{S}$)
22: add v to V_f
23: **return**

---

recursion terminates (Lines 8–9). Once the vertex v is identified, a list of new IRS $rslist$ is constructed by including v in S_i. This is the step (Line 10) in which multiple IRSs are generated. For each new relational state s in $rslist$ (Lines 11–13), if s satisfies the minimum size requirements, it is recorded as part of $\mathscr{S}$. Then v's span sequence is updated by removing the span of each IRS in $rslist$ and stored as v'. v is removed from V_f and v' is added to V_f (Lines 14–18). At this point (Line 19), the algorithm recursively grows S_i with updated V_f. When the recursion completes enumerating S_i, v' is removed from V_f (Lines 20–21). For each new relational state s in $rslist$ (Lines 22–24), if s can be expanded further, a new recursion is started with s being the IRS and V_f. When the recursion returns(Line 25), v is added back to V_f to restore original list of V_f. When the initial call to *enumerate* terminates, the list $\mathscr{S}$ contains all qualified IRSs.

Handling Directed Edges

To handle directed edges, we consider each direction of an edge separately, such that a directed edge $a \rightarrow b$ is listed separately from $a \leftarrow b$. The direction of an edge is stored as part of the label along with the span sequence of that edge. The

---

Algorithm 2 mpm($G^{RS}, u, t, d[], p$)

1: /* u is the current node */
2: /* t is the current time */
3: /* $d[]$ is the discovery array */
4: /* p is the current path */
5: $d[u] = t++$
6: push u into p
7: **if** $adj(u) = \emptyset$ **and** $|p| >$ minimum EIRS-length **then**
8:     record p
9: **else**
    for each node v **in** ($adj(u)$ sorted in increasing end-time order) **do**
10:       **if** $d[v] < d[u]$ **then**
11:         mpm(G^{RS}, v, t, d, p)
12: pop p

---

direction of an edge at a certain span is coded as 0, 1 or 2 to represent $a \rightarrow b$, $a \leftarrow b$ or $a \leftrightarrow b$. Note that the ordering of the vertices in an edge (a, b) are stored in increasing vertex-number order (i.e, $a < b$). Using the above representation of an edge, we determine the direction of all the edges of an IRS during its span duration.

3.2.2 Step 2: Mining of Maximal Evolution Paths

The algorithm that we developed to identify the sequence of IRSs that correspond to the maximal EIRSs is based on a modified DFS traversal of a directed acyclic graph that is referred to as the *induced relational state graph* and will be denoted by G^{RS}. G^{RS} contains a node for each of the discovered IRSs and a virtual root node r which is connected to all nodes. For each pair of IRSs $S_i = (V_i, s_i : e_i)$ and $S_j = (V_j, s_j : e_j)$, G^{RS} contains a directed edge from the node corresponding to S_i to the node corresponding to S_j iff $V_i \neq V_j$, $|V_i \cap V_j|/|V_i \cup V_j| \geq \beta$ and $e_i < s_j$ (i.e., constraints (ii)–(iv) of Definition 1).

The algorithm, referred to as *mpm* (Maximal Path Miner), uses a discovery array $d[]$ to record the discovery times of each node during traversal and all discovery times are initially set to -1. The traversal starts from the root node and proceeds to visit the rest of the nodes. Given a node u, the *mpm* algorithm selects among its adjacent nodes the node v that has the earliest end-time and $d[v] < d[u]$. The *mpm* algorithm also keeps track of the current path from the root to the node that it is currently at. If that node (i.e., node v) has no outgoing edges, then it outputs that path (i.e., u, v). The sequence of the relational states corresponding to the nodes of that path (the root node is excluded) represents an EIRS. The pseudocode is shown in Algorithm 2.

A close inspection of the *mpm* algorithm reveals that it is similar in nature to a traditional depth-first traversal with two key differences. First, the adjacent nodes of each node are visited in non-decreasing end-time order (line 10) and second, the condition on line 11 prevents the traversal of what are essentially forward edges in the depth-first tree but allows for the traversal of cross edges. To see that the *mpm* algorithm generates the complete set of maximal paths in a non-redundant fashion (i.e., each maximal path is output once), it is sufficient to consider the following. First, *mpm* without the condition on line 11 will generate all paths and each path will be generated once and it will terminate. This follows directly from the fact that G^{RS} is a directed acyclic graph. Second, the condition on line 11 eliminates paths that are contained within another path. To see this, consider the case in which while exploring the nodes adjacent to u, a vertex $v \in \text{adj}(u)$ is encountered such that $d[v] > d[u]$. The fact that $d[v] > d[u]$ indicates that vertex v was encountered from another path from u that traversed a vertex $v' \in \text{adj}(u)$ that was explored earlier. Thus, there is a path $p = u \rightarrow v' \rightsquigarrow v$ whose length is greater than one edge. As a result, the length of all paths that contain the edge $u \rightarrow v$ can be increased by replacing the edge with p. Third, the paths generated are maximal. Let $p_1 = (u_{i_1} \rightsquigarrow u_{i_j} \rightarrow u_{i_{j+1}} \rightsquigarrow u_{i_k})$ be a path discovered by *mpm* and assume that is not maximal. Without loss of generality, let $p_2 = (u_{i_1} \rightsquigarrow u_{i_j} \rightarrow u_{i_{j'}} \rightsquigarrow u_{i_{j+1}} \rightsquigarrow u_{i_k})$ be the maximal path that contains p_1 as a sub-path. Since path p_2 contains $u_{i_{j'}}$ before $u_{i_{j+1}}$, the end-time of $u_{i_{j'}}$ has to be less than the end-time of $u_{i_{j+1}}$. While exploring the nodes adjacent to u_{i_j}, the *mpm* algorithm would have selected $u_{i_{j'}}$ prior to $u_{i_{j+1}}$ (since adjacent nodes are visited in increasing end-time order) and in the course of the recursion would have visited $u_{i_{j+1}}$ from $u_{i_{j'}}$. Consequently, when $u_{i_{j+1}}$ is considered at a later point as an adjacent node of u_{i_j}, $d[u_{i_{j+1}}] > d[u_{i_j}]$ will prevent *mpm* from any further exploration. Thus, *mpm* will not generate the non-maximal path $(u_{i_1} \rightsquigarrow u_{i_j} \rightarrow u_{i_{j+1}} \rightsquigarrow u_{i_k})$.

Selective Materialization

The discussion so far assumes that G^{RS} has been fully materialized. However, this can be expensive as it requires pairwise comparisons between a large number of IRSs in order to determine if they satisfy constraints (ii)–(iv) of Definition 1. For this reason, the *mpm* algorithm materializes the portions of G^{RS} that it needs during the traversal. This allows it to reduce the rather expensive computations associated with some of the constraints by not having to visit forward edges. Moreover it utilizes the minimum EIRS length constraint to further prune the parts of G^{RS} that it needs to generate.

For a given node u, the *mpm* algorithm needs the adjacent node of u that has the earliest end-time. Let e_u be the end-time of node u. Since a node (i.e., an IRS) is required to have at least a span of length ϕ, we start u's adjacent node search among the nodes in G^{RS} that have the end-time of $e_k = e_u + \phi$. According to constraint (iv) of Definition 1, a certain minimum threshold of similarity is desired between two IRSs of an EIRS. Thus, it is sufficient to compare u with only those nodes that have at least a common vertex with u. We index the nodes of G^{RS} based on the vertices so that the similar nodes of u can be accessed by looking up all the nodes that have at least one vertex in common with u. If a node v is similar to u and $d[v] < d[u]$, we add the node to the adjacency list. If the search fails to detect any adjacent node, we initiate another search looking for nodes that have an end-time of $e_k + 1$ and continue such incremental search until an adjacent node of u is found or all possible end-times have been explored.

4   Mining the Coevolving Relational Motifs

Computational methods and tools that can efficiently and effectively analyze the temporal changes in dynamic complex relational networks enable us to gain significant insights regarding the relations between the entities and how these relations have evolved over time. Such *coevolving* patterns can provide evidence to the existence of, possibly unknown, coordination mechanisms by identifying the relational motifs that evolve in a similar and highly conserved fashion.

4.1   Coevolving Relational Motifs

Coevolving relational motifs are designed to identify the relational patterns that change in a consistent way over time. An example of this type of conservation is illustrated in Fig. 6, in the context of a hypothetical country-to-country trading network where labels represent the commodities being traded. The network for 1990 shows a simple relational motif (M_1) between pairs of nodes that occurs four times (shaded nodes and solid labeled edges). This relational motif has evolved in the network for 2000 in such a way so that in all four cases, a new motif (M_2) that includes an additional node has emerged. Finally, in the network for 2005 we see that three out of these four occurrences have evolved to a new motif (M_3) that now involves four nodes. This example shows that the initial relational motif among the four sets of nodes has changed in a fairly consistent fashion over time (i.e., it *coevolved*) and such a sequence of motifs $M_1 \rightsquigarrow M_2 \rightsquigarrow M_3$ represents an instance of a CRM.

CRMs identify consistent patterns of relational motif evolution that can provide valuable insights on the processes of the underlying networks. For example, the CRM of Fig. 6 captures the well-known phenomenon of production specialization due to economic globalization, in which the production of goods have been broken down into different components performed by different countries [11]. Similarly, CRMs in health-care networks can capture how the set of medical specialties required to treat certain medical conditions have changed over the years, in communication networks CRMs can capture the evolution of themes being discussed among groups of individuals as their lifestyles change, whereas CRMs in corporate email networks can capture how the discussion related to certain topics moves through the companies' hierarchies.

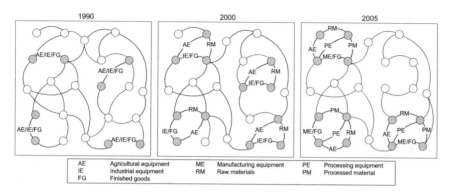

Fig. 6 An example of a coevolving relational motif in the context of a hypothetical country-to-country trading network where labels represent the commodities being traded

The formal definition of a CRM that is used in this chapter is as follows:

Definition 2. A CRM of length m is a tuple $\{N, \langle M_1, \ldots, M_m \rangle\}$, where N is a set of vertices, m is the number of relational motifs, and each relational motif $M_j = (N_j, A_j, L_{N_j}, L_{A_j})$ is defined over a subset of vertices N if there is an injection ξ_j from N_j to N.

4.1.1 CRM Embedding/Occurrence

An m-length CRM occurs in a dynamic network $\mathcal{N}$ whose node set is V, if there is a sequence of n snapshots $\langle G_{i_1}, G_{i_2}, \ldots, G_{i_n} \rangle$, where $m \leq n$, and a subset of vertices B of V (i.e., $B \subseteq V$) such that:

(1) there is a bijection ξ from N to B
(2) the injection $\xi \circ \xi_j$ is an embedding of M_j in G_{i_j}
(3) there is no embedding of M_j via the injection $\xi \circ \xi_j$ in $G_{i_{j+1}}$ or no embedding of M_{j+1} via the injection $\xi \circ \xi_{j+1}$ in G_{i_j}.

4.1.2 CRM Constraints

A CRM needs to satisfy the following constraints:

(1) it occurs in $\mathcal{N}$ at least ϕ times,
(2) each occurrence of the CRM in $\mathcal{N}$ uses a non-identical set of nodes,
(3) $M_j \neq M_{j+1}$, and
(4) $|N_j| \geq \beta |N|$ where $0 < \beta \leq 1$.

Note that the third condition of the CRM embedding in the above definition is designed to ensure that for each pair of successive motifs at least one of them is not supported by the snapshot-nodes pair that supported the other motif. This is done to ensure that there is a relational change between the nodes associated with those embeddings in each others snapshot.

The purpose of the CRM constraints in Definition 2 is as follows: First, a CRM *occurs* in the dynamic network when there exists an embedding of that CRM (i.e., an embedding of the evolving motif sequence) in to the dynamic network. Thus, the number of times a CRM occurs in a dynamic network can be used as a metric to evaluate the significance of a certain evolving pattern. For example, the number of occurrences of the CRM in Fig. 6 is 3. The parameter ϕ is used to eliminate sequences of evolving motifs that are not frequent enough to indicate the existence of an underlying process driving these changes. Second, two occurrences of a CRM are considered to use non-identical sets of nodes if their motifs use distinct embeddings in the snapshots that support them. This constraint is used to eliminate CRMs that converge to the same sets of nodes for a particular motif (i.e., CRMs sharing embeddings). The third constraint of Definition 2 limits the discovered CRMs to only an evolving sequence of motifs and not a sequence that remains the same. Finally, the parameter β is used to control the degree of change between the sets of nodes involved in each motif of a CRM and enforces a minimum node overlap among all motifs of CRM. Note that the frequent dynamic subgraphs introduced by Borgwardt et al. [5] correspond to CRMs in which the snapshots supporting each set of nodes are restricted to be consecutive and $\beta = 1$.

An edge of a CRM captures the relational changes between two nodes over time. Given a pair of vertices u and v of a CRM, the edge (u, v) may change label over time or may not be present in all motifs of the CRM. The number of edges in a CRM can be thought as the union of all edges between the nodes of the CRM. For example, an edge of the CRM captured in Fig. 6 is $AE/IE/FG \rightsquigarrow IE/FG \rightsquigarrow ME/FG$. Also, the total number of edges of the CRM is 5, since the last motif contains five edges between the four shaded nodes and all edges of previous motifs are earlier versions of the edges existing in the last motif.

In this chapter we focus on developing an efficient algorithm to mine a subclass of the CRMs, such that in addition to the conditions mentioned in Definition 2, the motifs that make up the CRM, also share at least one edge that itself is a CRM. Formally, we focus on identifying the CRMs $c = \{N_c, \langle M_1, \ldots, M_m \rangle\}$ that contain at least a pair of vertices $\{u, v\} \in N_C$ such that each induced subgraph M_i' of M_i on $\{u, v\}$ is connected and $x = \{\{u, v\}, \langle M_1', \ldots, M_m' \rangle\}$ is a CRM. A CRM like x that contains only one edge and two vertices will be called an *anchor*. We focus our CRM enumeration problem around the anchors, since they ensure that the CRM's motifs contain at least a pair of nodes in common irrespective of the specified overlap constraint that evolves in a conserved way. It also characterizes how the network around these core set of entities coevolved with them. In addition, we enforced the anchor constraint to simplify the enumeration problem and handle the exponential complexity of the CRM enumeration in various dynamic networks. We will refer to the class of CRMs that contain an anchor as *anchored CRMs*. For the rest of the discussion, any references to a CRM will assume it is an anchored CRM.

Given the above definition, the work in this chapter is designed to develop efficient algorithm for solving the following problem:

Problem 2. Given a dynamic network $\mathcal{N}$ containing T snapshots, a user defined minimum support ϕ $(1 \leq \phi)$, a minimum number of edges k_{min} per CRM, and a minimum number of motifs m_{min} per CRM, find all CRMs such that the motifs that make up the CRM, also share an anchor CRM.

A CRM that meets the requirements specified in Problem 2 is referred to as a *frequent* CRM and it is *valid* if it also satisfies the minimum node overlap constraint (Definition 2(iv)).

4.2 Finding Coevolving Relational Motifs

A consequent of the way anchored CRMs are defined is that the number of motifs that they contain is exactly the same as the number of motifs that exist in the anchor(s) that they contain. As a result, the CRMs can be identified by starting from all the available anchors and try to grow them by repeatedly adding edges as long as the newly derived CRMs satisfy the constraints and have exactly the same number of motifs as the anchor. Since a CRM can contain more than one anchor, this approach may identify redundant CRMs by generating the same CRM from multiple anchors. Therefore, the challenge is to design a strategy that is complete and non-redundant. To achieve this, we develop an approach that generates each CRM from a unique anchor. Given a CRM, the anchor from which it will be generated is referred to as its *seed anchor*.

The algorithm that we developed, named CRMminer, for finding all non-redundant valid CRMs (Problem 2) initially identifies all frequent anchors and then performs a depth-first exploration of each anchor pattern space along with a canonical labeling that is derived by extending the ideas of the minimum DFS code [24] to the case of CRMs for redundancy elimination. We impose frequency-based constraints by stopping any further exploration of a CRM when the pattern does not occur at least ϕ times in the dynamic network $\mathcal{N}$.

4.2.1 CRM Representation

A CRM $c = \{N, \langle M_1, \ldots, M_m \rangle\}$ is represented as a graph $G_c = (N, E_c)$, such that an edge $(u, v) \in E_c$ is a five-item tuple $(u, v, l_u, l_{u,v}, l_v)$, where $u, v \in N$, the vectors l_u and l_v contain the vertex labels and $l_{u,v}$ contains the edge labels of all motifs. If the CRM consists of m motifs, then $l_u = \langle l_{u_1}, \ldots, l_{u_m} \rangle$, $l_v = \langle l_{v_1}, \ldots, l_{v_m} \rangle$ and $l_{u,v} = \langle l_{u_1,v_1}, \ldots, l_{u_m,v_m} \rangle$. The kth entry in each vector l_u, $l_{u,v}$, and l_v records the connectivity information among the vertices of the kth motif (M_k). If an edge (u, v) is part of motif M_k, then the kth entry of l_u, l_v, and $l_{u,v}$ are set to the labels of u and v vertices, and the label of the (u, v) edge, respectively. If both vertices u and v are part of motif M_k, but the (u, v) edge is not, or at least one of the vertices u or v is not part of the M_k motif (i.e., no (u, v) edge is possible), then ω is inserted at the kth entry of the $l_{u,v}$ to capture the disconnected state. Similarly, if u or v does not have any incident edges in the M_k motif (i.e., the vertices are not present in that motif), then ω is added as the vertex label at the kth entry of l_u or l_v. Note that the value of ω is lexicographically greater than the maximum edge and vertex label.

This representation is illustrated in Fig. 7. The CRM consists of three motifs $\langle M_1, M_2, M_3 \rangle$ and represents relations among vertices $N=\{v_0, v_1, v_2, v_3\}$ using five edges. The edge between (v_0, v_1) exists in all three motifs capturing changes in relation as the edge label changes from $AE/IE/FG \;\leadsto\; IE/FG \;\leadsto\; ME/FG$. It is represented as $l_{v_0}=\langle BEL, BEL, BEL \rangle$, $l_{v_1}=\langle NTH, NTH, NTH \rangle$, and $l_{v_0,v_1}=\langle AE/IE/FG, IE/FG, ME/FG \rangle$. The next edge between (v_1, v_2) appears in two motifs and the label vectors are represented as $l_{v_1}=\langle NTH, NTH, NTH \rangle$, $l_{v_2}=\langle \omega, GDR, GDR \rangle$, and $l_{v_0,v_1}=\langle \omega, RM, PM \rangle$. Following similar process, we can represent edges (v_2, v_0), (v_2, v_3), and (v_3, v_0).

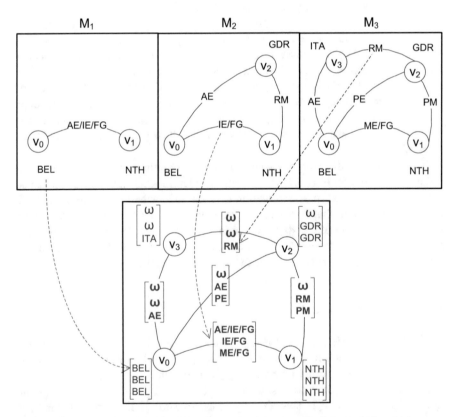

Fig. 7 A CRM representation. The CRM c consists of three motifs $\langle M_1, M_2, M_3 \rangle$ and represents relations among vertices $N = \{v_0, v_1, v_2, v_3\}$ using five edges. G_c (*bottom graph*) shows the CRM representation capturing vertex and edge label vectors

4.2.2 Mining Anchors

The search for CRMs is initiated by locating the frequent anchors that satisfy the CRM definition and the restrictions defined in Problem 2. This is done as following: Given a dynamic network $\mathcal{N}$, we sort all the vertices and edges by their label frequency and remove all infrequent vertices and edges. Remaining vertices and edges are relabeled in decreasing frequency. We determine the span sequences of each edge and list every edge's span sequence if that sequence contains at least a span with an edge label that is different from the rest of the spans. At this point, we use the sequential pattern mining technique prefixSpan [19] to determine all frequent span sequences. Since the frequent sequences can be partial sequences of the original input span sequences, it is not guaranteed that they all contain consecutive spans with different labels. Thus, the frequent sequences that contain different consecutive spans in terms of label are considered as the anchors. The number of spans in a frequent span sequence corresponds to the total number of motifs in the anchor.

4.2.3 CRM Enumeration

Given an anchor c, we generate the set of desired CRMs by growing the size of the current CRM one edge at a time following a depth-first approach. To ensure that each CRM is generated only once in the depth-first exploration, we use an approach similar to the gSpan algorithm [24], which we have extended for the problem of CRM mining.

gSpan explores the frequent pattern lattice in a depth-first fashion. The pattern lattice is represented as a hierarchical search space where each node corresponds to a connected frequent pattern, the highest node being an empty pattern (i.e., a single vertex), the next level nodes represent one-edge patterns, and so on. The nth level nodes, which represent n-edge patterns, contain one more edge than the corresponding $(n-1)$ level nodes. To ensure that each frequent pattern in this lattice is visited exactly once, gSpan's exploration amounts to visiting the nodes of the lattice (i.e., frequent patterns) by traversing a set of edges that form a spanning tree of the lattice. This spanning tree is defined by assigning to each node of the lattice a canonical label, called the minimum DFS code. A DFS code of a graph is a unique label that is formed based on the sequence of edges added to that node during a depth-first exploration. The minimum DFS code is the DFS code that is lexicographically the smallest. Given this canonical labeling, the set of lattice edges that are used to form the spanning tree correspond to the edges between two successive nodes of the lattice (parent and child) such that the minimum DFS code of the child can be obtained by simply appending the extra edge to the minimum DFS code of the parent. For example, given a DFS code $\alpha = \langle a_0, a_1, \cdots, a_m \rangle$, a valid child DFS code is $\gamma = \langle a_0, a_1, \ldots, a_m, b \rangle$, where b is the new edge. This spanning tree guarantees that each node has a unique parent and all nodes are connected [24]. To efficiently grow a node, gSpan generates the child nodes by only adding those edges that originate from the vertices on the rightmost path of the DFS-tree representation of the parent node. It then checks whether the resulting DFS code of the child node corresponds to the minimum DFS code. Construction of the child nodes generated by adding other edges (i.e., not from the rightmost path) is skipped, since such child nodes will never contain a DFS code that corresponds to the minimum DFS code.

In order to apply the ideas introduced by gSpan to the problem of efficiently mining CRMs, we need to develop approaches for (1) representing the DFS code of a CRM, (2) ordering the DFS codes of a CRM using the DFS lexicographic ordering, (3) representing the minimum DFS code of a CRM to use as the canonical label, and (4) extending a DFS code of a CRM by adding an edge. Once properly defined, the correctness and completeness of frequent CRM enumeration follows directly from the corresponding proofs of gSpan.

DFS Code of a CRM

In order to derive a DFS code of a CRM, we need to develop a way of ordering the edges. Given a CRM c, represented as a graph $G_c = (N, E_c)$, we perform a depth-first search in G_c to build a DFS tree T_c. The vertices (N) are assigned subscripts from 0 to $n - 1$ for $|N| = n$ according to their discovery time. The edges (E_c) are grouped into two sets: the forward edge set contains all edges in the DFS tree and denoted as $E_{T_c,fw} = \{(v_i, v_j) \in E_c \mid i < j\}$, and the backward edge set contains all edges which are not in the DFS tree and denoted as $E_{T_c,bw} = \{(v_i, v_j) \in E_c \mid i > j\}$. Let us denote a partial order on $E_{T_c,fw}$ as $\prec_{T_c,fw}$, a partial order on $E_{T_c,bw}$ as $\prec_{T_c,bw}$, and a partial order on E_c as $\prec_{T_c,bw+fw}$. Given two edges $e_1 = (v_{i_1}, v_{j_1})$ and $e_2 = (v_{i_2}, v_{j_2})$, the partial order relations are defined as:

(a) $\forall e_1, e_2 \in E_{T_c,fw}$, if $j_1 < j_2$, then $e_1 \prec_{T_c,fw} e_2$.
(b) $\forall e_1, e_2 \in E_{T_c,bw}$, if $i_1 < i_2$ or $(i_1 = i_2$ and $j_1 < j_2)$, then $e_1 \prec_{T_c,bw} e_2$.
(c) $\forall e_1 \in E_{T_c,bw}$ and $\forall e_2 \in E_{T_c,fw}$, if $i_1 < j_2$, then $e_1 \prec_{T_c,bw+fw} e_2$.
(d) $\forall e_1 \in E_{T_c,fw}$ and $\forall e_2 \in E_{T_c,bw}$, if $j_1 \leq i_2$, then $e_1 \prec_{T_c,bw+fw} e_2$.

The combination of the three partial orders defined above enforces a linear order $\prec_{T_c,E_c}$ on E_c.

Given this linear order $\prec_{T_c,E_c}$, we can order all edges in G_c and construct an edge sequence to form a DFS code of a CRM, denoted as $code(c, T_c)$. An edge of the DFS code of a CRM is represented similar to the CRM edge definition and uses a 5-tuple representation $(i, j, l_i, l_{i,j}, l_j)$, where i and j are the DFS subscripts (i.e., the discovery time) of the vertices, and l_i, l_j, and $l_{i,j}$ are the label vectors of the vertices and edge, respectively. The kth entry in each vector l_i, l_j, and $l_{i,j}$ contains the labels of vertices and edges of motif M_k.

For example, Fig. 8 presents two CRMs and their corresponding DFS codes. The DFS codes are listed below to show the sequence of edges and their differences (i.e.,

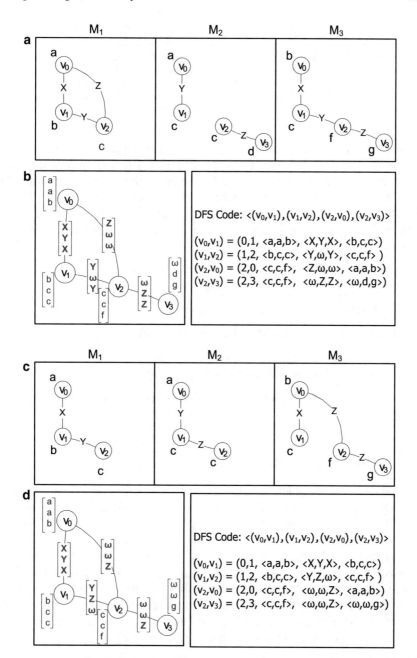

Fig. 8 DFS code for two CRMs represented as a sequence of edges (v_0, v_1), (v_1, v_2), (v_2, v_0), and (v_2, v_3). (**a**) Presents CRM C_1 consisting of three motifs, four vertices, and four edges. (**b**) Presents G_{C_1} and the corresponding DFS code for CRM C_1. (**c**) Presents CRM C_2 consisting of three motifs, four vertices, and four edges. (**d**) Presents G_{C_2} and the corresponding DFS code for CRM C_2

CRM C_1—Fig. 8a	CRM C_2—Fig. 8c
$\langle(0,1,\langle a,a,b\rangle,\langle X,Y,X\rangle,\langle b,c,c\rangle),$	$\langle(0,1,\langle a,a,b\rangle,\langle X,Y,X\rangle,\langle b,c,c\rangle),$
$(1,2,\langle b,c,c\rangle,\langle Y,\omega,Y\rangle,\langle c,c,f\rangle),$	$(1,2,\langle b,c,c\rangle,\langle Y,Z,\omega\rangle,\langle c,c,f\rangle),$
$(2,0,\langle c,c,f\rangle,\langle Z,\omega,\omega\rangle,\langle a,a,b\rangle),$	$(2,0,\langle c,c,f\rangle,\langle \omega,\omega,Z\rangle,\langle a,a,b\rangle),$
$(2,3,\langle c,c,f\rangle,\langle \omega,Z,Z\rangle,\langle \omega,d,g\rangle))$	$(2,3,\langle c,c,f\rangle,\langle \omega,\omega,Z\rangle,\langle \omega,d,g\rangle))$

the edge labels): Note that the kth entry of a vertex and edge label vector of a DFS code is filled with ω if the corresponding vertex or edge is not present in motif M_k.

To maintain the edge ordering of a DFS code similar to gSpan, our algorithm follows the DFS code's neighborhood restriction rules [24]. Given a DFS code of a CRM code(c,T_c), assume two neighboring edges $a_k = (i_k,j_k,l_{i_k},l_{i_kj_k},l_{j_k})$ and $a_{k+1} = (i_{k+1},j_{k+1},l_{i_{k+1}},l_{i_{k+1}j_{k+1}},l_{j_{k+1}})$. Then, a_k and a_{k+1} must meet the following rules:

(a) If $a_k \in E_{T_c,bw}$, then one of the following must be true:

    (1) when $a_{k+1} \in E_{T_c,fw}$, $i_{k+1} \leq i_k$ and $j_{k+1} = i_{k+1}$.
    (2) when $a_{k+1} \in E_{T_c,bw}$, $i_{k+1} = i_k$ and $j_k < j_{k+1}$.

(b) If $a_k \in E_{T_c,fw}$, then one of the following must be true:

    (1) when $a_{k+1} \in E_{T_c,fw}$, $i_{k+1} \leq j_k$ and $j_{k+1} = j_k + 1$.
    (2) when $a_{k+1} \in E_{T_c,bw}$, $i_{k+1} = j_k$ and $j_{k+1} < i_k$.

Note that the above stated rules ensure that the edge (v_2, v_0) appears before the edge (v_2, v_3) in Fig. 8b, d.

DFS Lexicographic Ordering

To establish a canonical labeling system for a CRM, CRMminer defines the DFS lexicographical ordering based on the CRM's DFS code definition. The linear ordering is defined as follows. Let two DFS codes of a CRM consisting of m motifs be $\alpha = \text{code}(c_\alpha, T_\alpha) = \langle e_\alpha^0, e_\alpha^1, \ldots, e_\alpha^p\rangle$ and $\delta = \text{code}(c_\delta, T_\delta) = \langle e_\delta^0, e_\delta^1, \ldots, e_\delta^q\rangle$, where p and q are the number of edges in α and δ, and $0 \leq p, q$. Let the forward and backward edge set for T_α and T_δ be $E_{\alpha,fw}$, $E_{\alpha,bw}$, $E_{\delta,fw}$, and $E_{\delta,bw}$, respectively. Also, let $e_\alpha^x = (i_\alpha^x, j_\alpha^x, l_{i_\alpha^x}, l_{i_\alpha^x j_\alpha^x}, l_{j_\alpha^x})$ and $e_\delta^y = (i_\delta^y, j_\delta^y, l_{i_\delta^y}, l_{i_\delta^y j_\delta^y}, l_{j_\delta^y})$ be two edges, one in each of the α and δ DFS codes. We define a check $PA(e)$ to determine whether a CRM edge is a potential anchor edge that contains no ω label and changes between consecutive motifs with respect to its edge/vertex labels. Now, $e_\alpha^x < e_\delta^y$, if any of the following is true:

(1) $F_\omega(l_{i_\alpha^x j_\alpha^x}) \leq F_\omega(l_{i_\delta^y j_\delta^y})$, where $F_\omega(A)$ is the number of ω's in vector A, or
(2) $PA(e_\alpha^x) = \text{true}$ and $PA(e_\delta^y) = \text{false}$, or
(3) $e_\alpha^x \in E_{\alpha,bw}$ and $e_\delta^y \in E_{\delta,fw}$, or
(4) $e_\alpha^x \in E_{\alpha,bw}$, $e_\delta^y \in E_{\delta,bw}$, and $j_\alpha^x < j_\delta^y$, or

(5) $e_\alpha^x \in E_{\alpha,bw}$, $e_\delta^y \in E_{\delta,bw}$, $f_\alpha^x = f_\delta^y$ and $l_{i_\alpha^x f_\alpha^x} < l_{i_\delta^y f_\delta^y}$, or

(6) $e_\alpha^x \in E_{\alpha,fw}$, $e_\delta^y \in E_{\delta,fw}$, and $i_\delta^y < i_\alpha^x$, or

(7) $e_\alpha^x \in E_{\alpha,fw}$, $e_\delta^y \in E_{\delta,fw}$, $i_\alpha^x = i_\delta^y$ and $l_{i_\alpha^x} < l_{i_\delta^y}$, or

(8) $e_\alpha^x \in E_{\alpha,fw}$, $e_\delta^y \in E_{\delta,fw}$, $i_\alpha^x = i_\delta^y$, $l_{i_\alpha^x} = l_{i_\delta^y}$ and $l_{i_\alpha^x f_\alpha^x} < l_{i_\delta^y f_\delta^y}$, or

(9) $e_\alpha^x \in E_{\alpha,fw}$, $e_\delta^y \in E_{\delta,fw}$, $i_\alpha^x = i_\delta^y$, $l_{i_\alpha^x} = l_{i_\delta^y}$, $l_{i_\alpha^x f_\alpha^x} = l_{i_\delta^y f_\delta^y}$, and $l_{f_\alpha^x} < l_{f_\delta^y}$.

Based on the above definitions, we derive the following conditions to compare two DFS codes of a CRM. We define $\alpha \leq \delta$, iff any of the following conditions is true:

(a) $\exists t, 0 \leq t \leq \min(p, q)$, $e_\alpha^k = e_\delta^k$ for $k < t$, and $e_\alpha^t < e_\delta^t$, or
(b) $e_\alpha^k = e_\delta^k$ for $0 \leq k \leq p$ and $p \leq q$.

Note that the label vectors (i.e., $l_{i_\alpha^x f_\alpha^x}$ and $l_{i_\delta^y f_\delta^y}$) are compared lexicographically. The DFS lexicographical ordering ranks edges with label vectors containing no ω labels higher than the edges which does. To define the relation between ω and valid vertex/edge label, the value of ω is set to a lexicographically higher value than the maximum edge and vertex label. Furthermore, we rank an anchor edge over an edge that remains same. This is important as we show later in Sect. 4.2.5.

In order to provide a detailed example of the DFS lexicographical ordering, let us compare the DFS codes of CRMs C_1 and C_2 presented in Fig. 8 following the rules presented above. For both the DFS codes, the first edge (v_0, v_1) is the same. In case of the second edge (v_1, v_2), both the DFS codes contain the same vertex label vectors. However, the edge label vectors are different and the edge label vector $\langle Y, Z, \omega \rangle$ of DFS code 2 is smaller than $\langle Y, \omega, Y \rangle$ DFS code 1. Thus, DFS code 2 is lexicographically smaller than DFS code 1.

Minimum DFS Code

A CRM can be represented by different DFS trees resulting in different DFS codes. To define a canonical label for a CRM, we select the minimum DFS code according to the DFS lexicographic order, represented by min code(c). To efficiently determine the minimum DFS code of a CRM, we start from the smallest edge according to the DFS lexicographic order. To grow the code by adding another edge, we select all adjacent edges following the rightmost extension rules and select the smallest one according to the DFS lexicographic order. Similar to simple graph isomorphism, given two CRMs c and c', c is isomorphic to c' if and only if min code(c) = min code(c'). Thus, by searching frequent minimum DFS codes, we can identify the corresponding frequent CRMs.

Pattern Lattice Growth

The difference between a simple graph pattern and a CRM is the vertex/edge label representation. One contains a single label and the other contains a label vector (i.e., sequence of labels including an empty label ω). The growth of a simple graph by one edge results in a number of simple graphs based on the number of unique labels of the frequent edges on the rightmost path. However, a CRM extended by an edge can generate large number of CRMs, since these are formed based on the combination of the label vectors of the frequent edges on the rightmost path.

In Fig. 9, we illustrated one edge growth of a simple graph and a CRM according to rightmost extension. Both the simple graph and the CRM are expanded by adding a frequent edge (v_2, v_3). In case of the simple graph (Fig. 9a), the original DFS code consisted sequence of two edges (v_0, v_1) and (v_1, v_2) represented as $(0, 1, a, X, b)$ and $(1, 2, b, Y, c)$. After the one edge extension, the new edge $(2, 3, c, Z, d)$ connected the new vertex v_3 to the rightmost vertex v_2. When the CRM is considered (Fig. 9c), the original DFS code consisted of the same sequence of edges $\{(v_0, v_1), (v_1, v_2)\}$ represented as: $(0, 1, \langle a, a, b \rangle, \langle X, Y, X \rangle, \langle b, c, c \rangle)$, $(1, 2, \langle b, c, c \rangle, \langle Y, \omega, Y \rangle, \langle c, \omega, f \rangle)$. Based on the combination of the label vectors of vertices v_2 and v_3, and edge (v_2, v_3), we have the following options for the (v_2, v_3) edge: $(2, 3, \langle c, c, f \rangle, \langle \omega, Z, Z \rangle, \langle \omega, d, g \rangle)$, $(2, 3, \langle c, c, f \rangle, \langle \omega, \omega, Z \rangle, \langle \omega, \omega, g \rangle)$, and $(2, 3, \langle c, c, f \rangle, \langle \omega, Z, \omega \rangle, \langle \omega, d, \omega \rangle)$. Note that we allow a CRM to

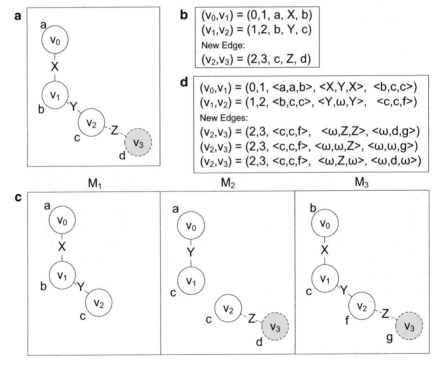

Fig. 9 Adding an edge according to rightmost extension rules. (**a**) Extending a simple graph, (**b**) DFS code of the simple graph, (**c**) extending a CRM, and (**d**) DFS code of the CRM

grow by an edge that may not be present or frequent in all motifs of that CRM to ensure complete set of the results.

To efficiently determine the frequent candidate edges during the rightmost extension, we apply the sequential pattern mining technique [19]. Even though there can be significantly more number of child CRMs from one edge extension of a CRM than a simple graph, the extended lexicographic ordering is able to order all candidates and enable us to perform pre-order search on the pattern lattice. Since traversal of the pattern lattice of a CRM remains similar to a simple graph, it allows CRMminer to prune the pattern with non-minimum DFS codes and their descendants similar to gSpan without impacting the completeness of the results.

4.2.4 CRMminer Algorithm

The high-level structure of CRMminer is shown in Algorithm 3. It first finds all frequent anchors according to Sect. 4.2.2. After locating all embeddings of an anchor c, the algorithm grows it recursively by adding one frequent adjacent edge at a time. The recursive function *ExpandCRM* first checks the support of c to prune any infrequent expansion. If c meets the minimum size (k_{min}) and overlap requirement (β), it is added to the $\mathscr{C}$ to be recorded as a valid CRM. It terminates expansion when c reaches the maximum size (k_{max}). To grow c further, the algorithm collects all adjacent edges of c in $\mathscr{N}$ following the DFS rightmost extension rules in Sect. 4.2.3. Each candidate edge e is added to c to construct its child c'. To prevent redundancy and ensure completeness, it only grows c' if it represents the canonical label of c', i.e., the minimum DFS code of c'.

Algorithm 3 CRMminer($\mathscr{N}, \mathscr{C}$)

1: $C_{anchor} \leftarrow$ find all frequent anchors from $\mathscr{N}$
    for each c **in** C_{anchor} **do**
2:    $embd_c \leftarrow$ all embeddings of c in $\mathscr{N}$
3:    *ExpandCRM*($\mathscr{N}, \mathscr{C}, c, embd_c$)
4: **return**

4.2.5 Algorithm Completeness

To eliminate redundancy during the CRM expansion process, we use minimum DFS code as the canonical label and construct the pattern lattice to ensure that every node (i.e., a CRM) is connected to a unique parent and grown via single edge addition. This process ensures that each potential CRM is only explored once. To ensure that all discovered CRMs contain at least one anchor, we show how we can identify an anchor of a valid CRM. Given a valid CRM c and its canonical label (i.e., the minimum DFS code) min code(c) $= \langle e_1, e_2, \ldots, e_k \rangle$, where e_i is an

---

Algorithm 4 ExpandCRM($\mathcal{N}$, $\mathcal{C}$, c, $embd_c$)

1: **if** $support(c, embd_c) < \phi$ **then**
2:     **return**
3: **if** $size(c) \geq k_{min}$ **then**
4:     **if** $overlap(c) \geq \beta$ **then**
5:         $\mathcal{C} \leftarrow \mathcal{C} \cup c$
6:     **if** $size(c) = k_{max}$ **then**
7:         **return**
8: $E \leftarrow$ all frequent adjacent edges of c in $\mathcal{N}$
    for each edge candidate e in E **do**
9:     $c' \leftarrow$ add e to c
10:     **if** $c' = CanonicalLabel(c')$ **then**
11:         $embd_{c'} \leftarrow$ all embeddings of c' in $\mathcal{N}$
12:         ExpandCRM($\mathcal{N}$, $\mathcal{C}$, c', $embd_{c'}$)
13: **return**

---

edge in the lexicographically ordered edge sequence. We claim that the first edge (e_1) of the canonical label of a CRM (c) is an anchor of that CRM. To prove this claim, assume e_1 is not an anchor and e_x is an anchor, where $1 < x \leq k$. Given the graph in CRM c, we can construct a DFS code for c that starts with e_x as the first edge. Assume, the new DFS code is represented as $code(c) = \langle e_x, e_{p_1}, \ldots, e_{p_{k-1}} \rangle$, where $\langle e_{p_1}, \ldots, e_{p_{k-1}} \rangle$ is an edge sequence containing a permutation of the edges $\{e_1, \ldots, e_k\} \setminus \{e_x\}$. Since e_1 is not an anchor, it either contains some ω labels or its labels remain same at least for some consecutive motifs of c. Based on the DFS lexicographic ordering and e_x being an anchor, $e_x < e_1$. Hence, we can state that $\langle e_x, e_{p_1}, \ldots, e_{p_{k-1}} \rangle < \langle e_1, e_2, \ldots, e_k \rangle$. This is a contradiction, since $\langle e_1, e_2, \ldots, e_k \rangle$ is the minimum DFS code of c. Thus, e_1 is an anchor of CRM c. Given that the first edge is an anchor, then CRMminer will generate that CRM by starting from the anchor and then following its rightmost extension rule to add the rest of the edges one by one.

4.2.6 Search Space Pruning

One of the challenges that any graph mining algorithm needs to handle is the exponential growth of the search space during enumeration. Traditionally, user specified constraints are used to prune the search space. To ensure discovery of the complete set of patterns, the pruning constraints need to have the anti-monotonicity property (i.e., a specific measure of a graph can never exceeds the measure of its subgraphs). For CRMminer, we use both support measure and minimum overlap constraints to prune the search space.

Minimum Support (ϕ)

To efficiently search patterns in a single large graph using a minimum support constraint, the support measure needs to guarantee the anti-monotonicity property. Bringmann [6] presented the minimum image based support measure to prune the search space in a single large graph. Given a pattern $p = (V_p, E_p, L_p)$ and a single large graph $G = (V_G, E_G, L_G)$, this measure identifies the vertex in p which is mapped to the least number of unique vertices in G and uses this number as the frequency of p in G. To formally define the support measure, let each subgraph g of G that is isomorphic to p be defined as an *occurrence* of p in G. For each occurrence g, there is a function $\varphi : V_p \to V_G$ that maps the nodes of p to the nodes in G such that (1) $\forall v \in V_p \Rightarrow L_p(v) = L_G(\varphi(v))$ and (2) $\forall (u, v) \in E_p \Rightarrow (\varphi(u), \varphi(v)) \in E_G$. The minimum image based support of a pattern p in G is defined as:

$$\sigma(p, G) = \min_{v \in V_p} | \ \{\varphi_i(v) : \ \varphi_i \text{ is an occurrence of } p \text{ in } G\} \ |. \tag{3}$$

This minimum image based support is anti-monotonic [6].

We adopted the minimum image based support measure to calculate the minimum support of a CRM in a dynamic network. As defined in Sect. 2, a dynamic network $\mathcal{N}$ can be represented as a single large graph where the nodes $\mathcal{N}$ are considered as the vertices of the large graph. Hence, it is possible to calculate the least number of unique vertices of the dynamic network that are mapped to a particular vertex of a CRM. Given a CRM $c = \{N_c, \langle M_1, M_2, \ldots, M_m \rangle\}$ in a dynamic network $\mathcal{N} = \{V_{\mathcal{N}}, \langle G_0, G_1, \ldots, G_T \rangle\}$ where $m \leq T$, the minimum image based support of c is defined as:

$$\sigma(c, \mathcal{N}) = \min_{v \in V_c} | \ \{\varphi_i(v) : \varphi_i \text{ is an occurrence of } c \text{ in } \mathcal{N}\} \ | . \tag{4}$$

Similar to the support measure of a pattern in a single large graph, by selecting the support of the vertex in c that has the least number of unique mapping in $\mathcal{N}$, we maintain the anti-monotonicity property.

Recall from Sect. 4.2.3 that the frequent candidate edges are identified using frequent sequence mining. Since we use the minimum image based support measure, the frequent edges detected by the sequence mining tool may not have sufficient support when calculated using such measure. Thus, we compute the minimum image based support for all candidate edges to only consider the edges that ensures the CRM extension to have sufficient minimum image based support.

Minimum Overlap (β)

Each motif of a CRM needs to contain at least a minimum percentage of the nodes from all the nodes of the CRM. This minimum node overlap threshold (defined as β in Sect. 4) controls the degree of change that is allowed between the sets of nodes in each motif of a CRM. Given a CRM $c = \{N_c, \langle M_1, M_2, \ldots, M_m \rangle\}$ containing m motifs, the *minimum overlap* of a CRM is defined as:

$$\rho(c) = \frac{\min_{1 \leq i \leq m} \{|V_{M_i}|\}}{|N_c|}, \tag{5}$$

where V_{M_i} is the set of nodes in motif M_i. Even though the minimum overlap constraint is a reasonable approach to ensure that the motifs that make the CRM are coherent, it is not anti-monotonic [26]. Thus, to generate a complete set of CRMs that meet user specified thresholds of support and overlap, we cannot prune CRMs that do not satisfy this constraint as CRMs derived from it can satisfy the constraint.

For example, let us assume the minimum overlap threshold is 60 % in Fig. 10. In step (1), motif M_1 contains two out of three nodes of the CRM (i.e., the minimum). Therefore, the minimum overlap at step (1) (2/3 > 60 %) is valid. We added vertex v_3 by including edge (v_2, V_3) to the CRM at step (2). This dropped the minimum overlap (2/4) below the threshold. However, in step (3), inclusion of edge (v_3, v_0) adds vertex v_3 to motif M_1. This increases the minimum overlap to be 3/4 and makes the CRM valid again. Hence, we need to enumerate all CRMs that meet the support threshold and then search the output space for CRMs that meet the minimum overlap requirement.

To improve the performance, we developed an approximate version of our algorithm, named CRMminer$_x$, that discovers a subset of the valid CRMs (i.e., meet the constraints of Definition 2) by pruning the pattern lattice using the minimum overlap threshold. We first check whether a CRM meets the overlap threshold. If it does, we continue the enumeration process. If it does not, then we check whether any of the patterns at the previous level of the pattern lattice from which the current

Fig. 10 Example of a CRM growth when the minimum overlap constraint is not anti-monotonic. Assume the minimum overlap constraint is 60 %

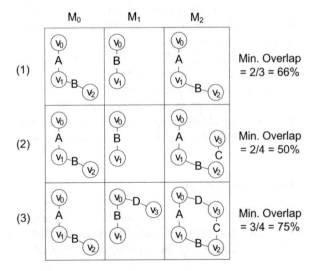

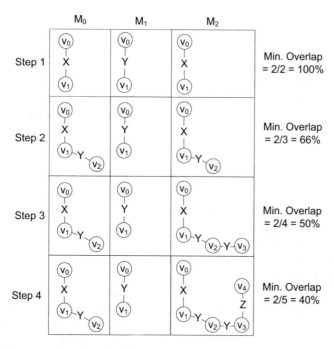

Fig. 11 Minimum overlap calculation based on (6) is used for search space pruning during the CRM enumeration process. Assuming $\beta = 60\%$, this CRM enumeration terminates at step 4

pattern was derived, referred to as parent CRMs, meet the overlap threshold. If at least one does, we continue enumeration. If none of the parent CRMs meet the overlap threshold, we prune that CRM.

To approximately calculate the minimum overlap of the parent CRMs, we do not generate all possible parent patterns. The parent is defined to contain one less node than the current CRM; thus, we remove a node $v \in N_c$. In such case, the parent pattern will contain $(|N_c| - 1)$ nodes and each motif M_i may contain $(|V_{M_i}| - 1)$ nodes if $v \in V_{M_i}$ or $(|V_{M_i}|)$ nodes if $v \notin V_{M_i}$. For at least one parent CRM to meet the overlap threshold, we consider the best case when $v \notin V_{M_i}$. Therefore, the minimum overlap threshold of a parent CRM c_p of the CRM $c = \{N_c, \langle M_1, M_2, \ldots, M_m\rangle\}$ is defined as:

$$\rho(c_p) = \frac{\min_{1 \leq i \leq m} \{|V_{M_i}|\}}{|N_c| - 1}. \tag{6}$$

In Fig. 11, we illustrate the minimum overlap calculation during search space pruning. Assume the user specified $\beta = 60\%$. We start with the anchor at step 1 when the minimum overlap threshold is 100%. Next the edge (v_1, v_2) is added for motif M_0 and M_2 and the minimum overlap based on motif M_1 is $(2/3) = 66\%$. Since the β threshold is met, we continue the enumeration process. At step 3, an edge (v_2, v_3) is added to motif M_2. Since motif M_1 contains the lowest number of nodes, the minimum overlap is $(2/4) = 50\%$, which does not meet the β threshold. At this point, we check whether any of the parent CRM meets the β threshold and the minimum threshold for its parent is $(2/(4-1)) = 66\% > \beta$. Thus, we continue the enumeration by adding an edge (v_3, v_4) to motif M_2 at step 4. The minimum overlap is $(2/5) = 40\%$ and its parent overlap threshold is $(2/(5-1)) = 50\%$. Both thresholds are lower than β, hence we stop enumerating this CRM any further.

5 Mining the Coevolving Induced Relational Motifs

Based on our work on CRMs, we realized that to understand how the relations between groups of entities have changed over time, we need to take into account all their relations. For example, the analysis of a co-authorship network to identify the popular topics of collaboration among authors can help us understand the current scientific research trends. By exploring how a group of authors collaborate over the years and observing how their relations change, we may understand the developmental path of their research area. By thoroughly analyzing the relations among different groups of authors, we may find some common factors that encourage them to collaborate. Patterns that include all relations among a set of entities are well suited to address such problem.

5.1 Coevolving Induced Relational Motifs

Coevolving Induced Relational Motifs (CIRMs) are designed to capture frequent patterns that include all relations among a set of entities (i.e., induced) at a certain time in the dynamic network and change in a consistent way over time. To illustrate this type of patterns consider the network of Fig. 12. The network for 1990 shows an induced relational motif (M_1) between pairs of nodes that occurs four times (shaded nodes and solid labeled edges). Three out of the four occurrences have evolved into a new motif (M_2) that includes an additional node in the network for 2000. Finally, in 2005 we see a new motif (M_3) that now involves four nodes and occurs two times. This example shows that the initial relational motif has changed in a fairly consistent fashion over time (i.e., it *coevolved*) and such a sequence of motifs $M_1 \rightsquigarrow M_2 \rightsquigarrow M_3$ that captures all relations among a set of entities represents an instance of a CIRM whose frequency is two (determined by M_3).

The formal definition of a CIRM that is used in this section is as follows:

Definition 3. A CIRM of length m is a tuple $\{N, \langle M_1, \ldots, M_m \rangle\}$, where N is a set of vertices and each $M_j = (N_j, A_j)$ is an **induced** relational motif defined over a subset of the vertices of N that satisfies the following constraints:

  (i) it occurs at least ϕ times,

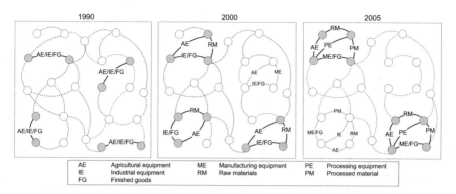

Fig. 12 An example of a coevolving induced relational motif in the context of a hypothetical country-to-country trading network where labels represent the commodities being traded. Assume the minimum support threshold (ϕ) for the CIRM is 2

ii) each occurrence uses a non-identical set of nodes,

(iii) $M_j \neq M_{j+1}$, and

(iv) $|N_j| \geq \beta |N|$ where $0 < \beta \leq 1$.

An induced relational motif M_j is defined over a subset of vertices N if there is an injection ξ_j from N_j to N. An m-length CIRM *occurs* in a dynamic network whose node set is V if there is a sequence of m snapshots $\langle G_{i_1}, G_{i_2}, \ldots, G_{i_m} \rangle$ and a subset of vertices B of V (i.e., $B \subseteq V$) such that:

(1) there is a bijection ξ from N to B
(2) the injection $\xi \circ \xi_j$ is an embedding of M_j in G_{i_j}
(3) there is no embedding of M_j via the injection $\xi \circ \xi_j$ in $G_{i_{j+1}}$ or no embedding of M_{j+1} via the injection $\xi \circ \xi_{j+1}$ in G_{i_j}.

The parameter ϕ is used to eliminate sequences of evolving motifs that are not frequent enough. Whereas the parameter β is used to control the degree of change between the sets of nodes involved in each motif of a CIRM and enforces a minimum node overlap among all motifs of CIRM.

In this section, we focus on identifying a subclass of CIRMs, called anchored CIRMs, such that in addition to the conditions of Definition 3, all motifs of the CIRM share at least one edge (anchor) that itself is a CIRM (i.e., the anchor is an evolving edge). This restriction ensures that all motifs of a CIRM contain at least a common pair of nodes and captures how these core set of entities coevolved.

Note that due to the above restriction, the number of motifs in a CIRM will be exactly the same as the number of motifs (i.e., edge spans) in its anchor. In some cases this will fail to identify evolving patterns that started from an anchor and then experience multiple relational changes between any two non-anchor nodes within the span of a motif. To address the above, we also identify a special class of CIRMs, referred to as *CIRM split extensions*, that have additional motifs than the anchor. A pattern is a CIRM split extension if:

(a) all of its motifs share an edge that satisfies properties (i), (ii), and (iv) of Definition 3, and
(b) for each maximal run of edge-spans $(x_1, x_2, \ldots, x_k)$ with the same label, there is another edge-span in the network that starts at the first snapshot of x_1 and ends at the last snapshot of x_k such that this edge-span is supported by the snapshots starting at the first snapshot that supports x_1 and ends at the last snapshot that supports x_k.

Given the above definition, the work in this section is designed to develop an efficient algorithm for solving the following problem:

Problem 3. Given a dynamic network $\mathcal{N}$ containing T snapshots, a user defined minimum support ϕ $(1 \leq \phi)$, a minimum number of vertices k_{min} per CIRM, and a minimum number of motifs m_{min} per CIRM, find all frequent anchored CIRMs and all CIRM split extensions.

A CIRM that meets the requirements specified in Problem 3 is referred to as a *frequent* CIRM and it is *valid* if it also satisfies the minimum node overlap constraint [Definition 3(iv)].

5.2 Coevolving Induced Relational Motif Mining

We developed an algorithm for solving Problem 3, called CIRMminer that follows a depth-first exploration approach from each anchor and identifies in a non-redundant fashion the complete set of CIRMs that occur at least ϕ times. For the rest of the discussion, any references to a relational motif or a motif will assume it is an induced relational motif. We represented CIRMs similar to CRMs in Sect. 4.2.1.

5.2.1 Mining Anchors

The search for CIRMs is initiated by locating the frequent anchors that satisfy the CIRM definition and the restrictions defined in Problem 3. This process is illustrated in Fig. 13. Given a dynamic network $\mathcal{N}$, we sort all the vertices and edges by their label frequency and remove all infrequent vertices. The remaining vertices and all edges are relabeled in decreasing frequency. We determine the span sequences of each edge and collect every edge's span sequence if that sequence contains at least a span with an edge label that is different from the rest of the spans. At this point, we use the sequential pattern mining technique prefixSpan [19] to determine all frequent span subsequences. Each of the frequent span subsequences is supported by a group of node pairs where the edge between each pair of nodes evolve in a consistent way over time. Since the frequent subsequences can be partial sequences of the original input span sequences, it is not guaranteed that they all contain consecutive spans with different labels. Thus, the frequent subsequences that

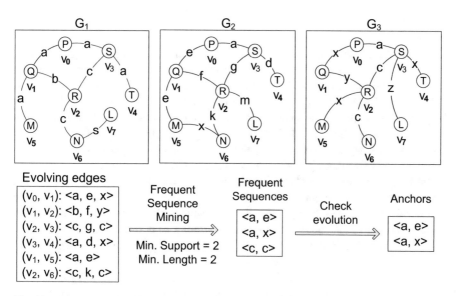

Fig. 13 The process of mining anchors from the network $\langle G_1, G_2, G_3 \rangle$. Since all vertex labels remained consistent over time, we listed the edge label sequences as the span sequence of the evolving edges

contain different consecutive spans in terms of label are considered as the anchors. The number of spans in a frequent span sequence corresponds to the total number of motifs in the anchor.

5.2.2 CIRM Enumeration

Given an anchor, CIRMminer generates the set of desired CIRMs by growing the size of the current CIRM one vertex at a time following a depth-first approach. The vertex-based CIRM growth approach was selected in order to ensure that each motif of the CIRM contains all edges among the vertices of that motif (i.e., it is an induced subgraph). To ensure that each CIRM is generated only once in the depth-first exploration, it uses an approach similar to gSpan [24], which we have extended for the problem of CIRM mining. gSpan performs a depth-first exploration of the pattern lattice avoiding redundant visits to the same node using a canonical labeling scheme, called minimum DFS code, and traverses a set of edges that form a spanning tree of the lattice. In order to apply the ideas introduced by gSpan to efficiently mine CIRMs, we defined the minimum DFS code of a CIRM following the definition of the minimum DFS code of a CRM [3]. We use the minimum DFS code of a CIRM as the canonical label. Due to the space constraint, we omit the discussion on minimum DFS code of a CIRM and encourage readers to refer to [3]. Once properly defined, the correctness and completeness of frequent CIRM enumeration follows directly from the corresponding proofs of gSpan.

CIRM Growth

The CIRM enumeration process follows the rightmost extension rule [24] to select candidates for the next expansion and discards all CIRM extensions that do not contain a minimum DFS code. This is done by searching through all embeddings of the CIRM to identify the adjacent vertices that (1) connect to the nodes on the rightmost path and (2) the span of the vertices overlaps the span of the CIRM. For each adjacent vertex, CIRMminer goes through all embeddings of the CIRM to collect the sets of edges that connect that vertex with all existing vertices within the CIRM's span. To identify a set of edges for an embedding, it traverses through all snapshots within each motif's span and selects all maximal sets of edges connecting that vertex with the other CIRM vertices and remain in a consistent state. Each maximal set of edges along with the associated span is assigned a unique ID. The vertex and edge labels and the corresponding span determine the ID. Given these IDs, the algorithm then represents the set of edges resulting from a particular embedding of a CIRM as a sequence of IDs.

Figure 14 presents an example of generating a sequence of IDs during the process of CIRM growth. The CIRM consists of three motifs $\langle M_1, M_2, M_3 \rangle$, three vertices (shaded nodes), and two edges (solid labeled edges). An embedding of the CIRM is shown where the vertices are v_0, v_1, and v_2, and the edges are (v_0, v_1) and (v_1, v_2). We omitted vertex labels to form a simple example. To grow the CIRM by adding a new vertex, CIRMminer selects the adjacent vertex v_3 for this embedding. Hence, it needs to consider the set of edges that connects v_3 to the existing vertices v_0, v_1, and v_2. By analyzing the overlapping spans of the edges (v_0, v_3), (v_1, v_3), and (v_2, v_3), it identifies five different segments such that each one contains a maximal set of edges in a consistent state. As a result, these maximal sets are assigned unique IDs I_1 through I_5 where each ID contains a set of edges and a specific span. The set of edges is represented as: $\langle I_1, I_2, I_3, I_4, I_5 \rangle$.

CIRMminer represents the collected sets of edges from all embeddings as a collection of ID sequences and applies frequent sequence mining technique [19] to find all frequent ID sequences. Each frequent ID sequence is then considered as a frequent set of edges associated with a candidate vertex for the next CIRM extension. For example, if the subsequence $\langle I_1, I_2, I_5 \rangle$ from Fig. 14 is frequent, then the following set of edges is considered for vertex v_3: $(0, 3, \langle P, P, P \rangle, \langle d, h, w \rangle, \langle S, S, S \rangle)$, $(1, 3, \langle Q, Q, Q \rangle, \langle k, \omega, \omega \rangle, \langle S, S, S \rangle)$, $(2, 3, \langle R, R, R \rangle, \langle c, g, z \rangle, \langle S, S, S \rangle)$.

It is possible that a frequent ID sequence contains multiple IDs that belong to a particular motif of the CIRM. For example, if an ID sequence $\langle I_1, I_2, I_4, I_5 \rangle$ is frequent (in Fig. 14), both IDs I_2 and I_4 contain the span that belongs to motif M_2. To identify CIRMs according to Definition 3, it needs to divide these set of edges as multiple candidate sets where each set contains only one of the overlapping spans for each motif to match the total number of motifs of the original CIRM. To find the CIRM split extensions, the algorithm considers all such set of edges as valid extensions. This inclusion leads to identifying a super set of anchored CIRMs. Some of the CIRM split extensions may violate constraint (iii) of Definition 3 (i.e., $M_j \neq M_{j+1}$). It discards such CIRM split extensions as a post-processing step. Note that

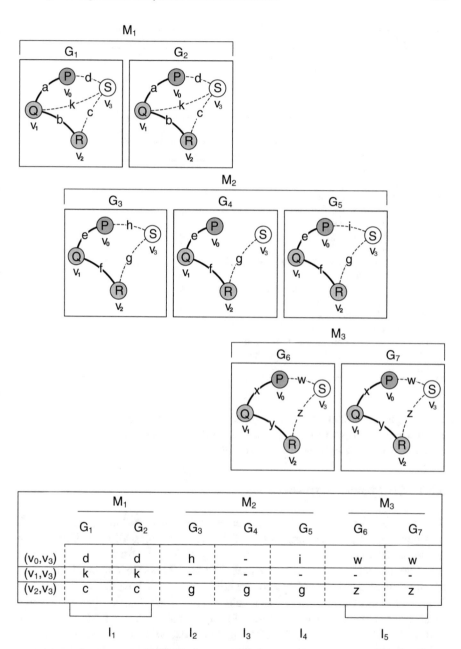

Fig. 14 Generating ID sequences from a set of edges. *Shaded vertices* and *solid line edges* are part of the existing CIRMs. Vertex v_3 is considered as the candidate vertex. For this example, since the vertex labels remain same in all snapshots, the *edges* are represented using only edge labels

---

Algorithm 5 CIRMminer($\mathcal{N}$, $\mathcal{C}$)

1: $C_{anchor} \leftarrow$ find all frequent anchors from $\mathcal{N}$
    for each c **in** C_{anchor} **do**
2:     $e_c \leftarrow$ all embeddings of c in $\mathcal{N}$
3:     *ExpandCIRM*($\mathcal{N}$, $\mathcal{C}$, c, e_c)
4: **return**

Algorithm 6 ExpandCIRM($\mathcal{N}$, $\mathcal{C}$, c, e_c)

1: **if** $support(c, e_c) < \phi$ **then**
2:     **return**
3: **if** $size(c) \geq k_{min}$ **then**
4:     **if** $overlap(c) \geq \beta$ **then**
5:         $\mathcal{C} \leftarrow \mathcal{C} \cup c$
6:     **if** $size(c) = k_{max}$ **then**
7:         **return**
8: $V \leftarrow$ all frequent adjacent nodes of c in $\mathcal{N}$
    for each candidate v **in** V **do**
9:     $S \leftarrow$ find all frequent edge sets that connect v to c
        for each edge set s **in** S **do**
10:     $c' \leftarrow$ add s to c
11:     **if** $c' = CanonicalLabel(c')$ **then**
12:         $e_{c'} \leftarrow$ all embeddings of c' in $\mathcal{N}$
13:         ExpandCIRM($\mathcal{N}$, $\mathcal{C}$, c', $e_{c'}$)
14: **return**

---

each frequent candidate set of edges is added to the CIRM following the rightmost extension rules to determine the exact edge order ensuring that the minimum DFS code check can be performed on the extended CIRM.

5.2.3 CIRMminer Algorithm

The high-level structure of CIRMminer is shown in Algorithm 5. It first finds all frequent anchors according to Sect. 5.2.1. After locating all embeddings of an anchor c, the algorithm grows it recursively by adding one frequent adjacent node at a time. The recursive function *ExpandCIRM* first checks the support of c to prune any infrequent expansion. If c meets the minimum size and overlap requirement, it is added to the $\mathcal{C}$ to be recorded as a valid CIRM. It terminates expansion when c reaches the maximum size. To grow c further, the algorithm collects all adjacent nodes of c in $\mathcal{N}$ following the DFS rightmost extension rules. For each of the candidate node $v \in V$, it finds all frequent edge sets that connect the new node v to the existing nodes of c according to Sect. 5.2.2. Each of the candidate edge set s is added to c to construct its child c'. To prevent redundancy and ensure completeness, it only grows c' if it represents the canonical label of c', i.e., the minimum DFS code of c'.

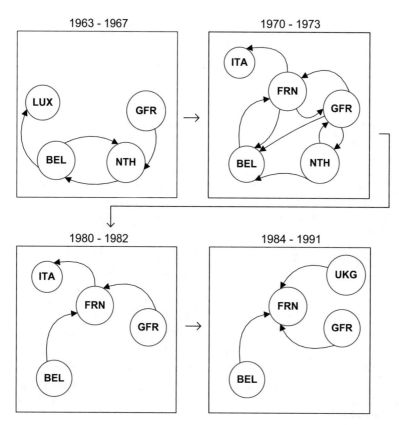

Fig. 15 An EIRS capturing a trade relation between EU countries. The nodes in the figure are *LUX*=Luxembourg, *BEL*=Belgium, *GFR*=German Federal Republic, *NTH*=Netherlands, *ITA*=Italy, and *UKG*=United Kingdom

6 Qualitative Analysis and Applications

In this section we present some of the evolving patterns that were discovered by our algorithms in order to illustrate the type of information that can be extracted from different dynamic networks.

6.1 Analysis of a Trade Network

This trade network dataset models the yearly export and import relations of 192 countries from 1948 to 2000 based on the Expanded Trade and GDP Data [12]. The nodes model the trading countries and the direct edges model the export or import activity between two countries for a certain year. The snapshots of the dynamic

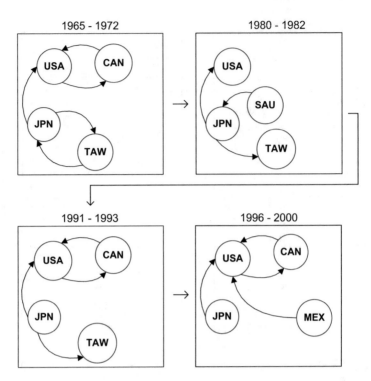

Fig. 16 An EIRS capturing a trade relation of USA. The nodes in the figure are *USA*=United States of America, *CAN*=Canada, *JPN*=Japan, *TAW*=Taiwan, *SAU*=Saudi Arabia, and *MEX*=Mexico

network that we created correspond to the trade network of each year, leading to a dynamic network consisting of $2000 - 1948 = 53$ snapshots. If the export amount from country A to country B in a given year is more than 10 % of the total export amount of A and total import amount of B for that year, a directed edge $A \rightarrow B$ is added to that year's trade graph.

To illustrate the type of information that can be extracted from this dynamic network, we will focus on how stable relations among the entities changed over time. We present some of the EIRSs that were discovered by our algorithm. In Fig. 15 we present an EIRS generated from the trade network capturing trade relations between some of the European countries over 30 years period and the chosen $\phi=3$. The EIRS mainly captures trade relations between Belgium, Netherlands, Germany, and France. The other countries, such as Luxembourg, Italy, and United Kingdom, participate for a partial period of time. Based on the illustration, initially (during 1963–1967) Belgium and Netherlands were strong trade partners as they exported and imported from each other. The period 1970–1973 shows that the countries were heavily trading between each other. By evaluating the historical events, political and economic situation of that period, we could find the cause of higher trade activity. The periods 1980–1982 and 1984–1991 capture how France's trade relations with

Belgium and Germany became one sided as France only imported from those countries. The cause of such changes could be that France was exporting to other countries or Belgium and Germany decided to import from some other countries. In Fig. 16 we present another EIRS generated from the trade network capturing a stable trade relation of USA with other countries over 35 years period. We notice that USA and Canada have strong trade relations over a long period of time. Even though the strong tie in trading seems obvious due to the geographical co-location of the countries, it is interesting that the algorithm could discover such relation from the historical data. The EIRS also captures steady relation between the USA and Japan.

6.2 Analysis of a Co-authorship Network

This network models the yearly co-authorship relations from 1958 to 2012 based on the DBLP Computer Science Bibliography Database [18]. The snapshots correspond to the co-authorship network of each year, leading to a dynamic network consisting of $2012 - 1958 = 55$ snapshots. The nodes model the authors of the publications and the undirected edges model the collaboration between two authors at a certain year. To assign edge labels, we cluster the publication titles into 50 thematically related groups and use the cluster number as the labels. This representation contained $1,057,524$ nodes and on average $72,202$ edges per snapshot.

In order to illustrate the information that can be gathered from this dynamic network, we focus on identifying the coevolution of the relational entities. We present some of the CRMs that were discovered by our algorithm analyzing the *DBLP* dataset. The yearly co-authorship relations among the authors are divided into 50 clusters based on the title of the papers and we use the most frequent words that belong to a cluster to describe the topic it represents. To rank the discovered CRMs, we use the cosine similarity between the centroids of the clusters, referred to as *topic similarity*, that ranges from 0.02 to 0.52. For each CRM, we determine a score by calculating the average topic similarity based on all topic transitions (i.e., edge label changes) between two consecutive motifs of the CRM. The CRMs containing the least score ranks the highest. This ranking is designed to capture the frequent co-authorship relational changes that are thematically the most different.

Two of the high-ranked CRMs are shown in Fig. 17. The first CRM shows the periodic changes in research topics represented as 8 and 41 and the topic similarity between these topics is 0.22. The CRM captures the periodic transitions of the relations as author a and b collaborate with other authors c, d, and e over the time. The second CRM shows the periodic changes in research topics represented as 29 and 26 and the topic similarity between these topics is 0.20. The CRM captures similar the periodic transitions of the relations as author a and b collaborate with other authors c, d, and e over the time.

6.3 Analysis of a Multivariate Time-Series Dataset

This is a cell culture bioprocess dataset [17], referred to as the *GT* dataset that tracks the dynamics of various process parameters at every minute over 11 days period for 247 production runs. To represent this data as a dynamic network, we computed 47 correlation matrices for 14 of the process parameters using a sliding window of 12 hours interval with 50 % overlap. To construct a network snapshot based on a correlation matrix, we use each parameter as a vertex and two parameters/vertices are connected with an edge labeled as positive/negative if their correlation is above or below a certain threshold. Data from each run forms a small graph based on the correlation matrix at a certain time interval and the union of the graphs from all 247 runs at a certain time interval represents the snapshot. This dataset contains 47 snapshots where each snapshot contains 3458 nodes and on average 1776 edges.

As it is difficult to understand the relations between the nodes (i.e., parameters) without sufficient domain knowledge in Bio-Chemical processes, we decided to analyze the *GT* dataset based on discovered CRMs/CIRMs by using them as features to discriminate the production runs. For the *GT* dataset, the discovered CRMs/CIRMs display the dynamics of various process parameters during the cell culture process. Out of the 247 production runs included in the *GT* dataset, based on the quality of the yields, 48 of the runs are labeled as *Good*, 48 of the runs are labeled as *Bad*, and the remaining were not labeled. To understand the class distribution (i.e., Good or Bad) of the discovered CRMs and CIRMs, we analyzed the embeddings of the discovered CRMs and CIRMs from three different experiments using ϕ of 70, 60 and 50. Since we have 96 labeled runs, we do not count the embeddings that belong to unlabeled runs. Figure 18 shows the class distribution of the embeddings for the discovered CRMs and CIRMs. It shows that the CRMs/CIRMs were present mostly as part of the high yield runs, since more than 75 % of the embeddings belong to the Good class. The class representation is consistent for different support parameters. These results suggest that there is a consistency of what makes some thing good but runs can go bad for many reasons. Note that using such information, if we can determine the low yield runs at the early stage of experiment, we can terminate the experiment and save a lot of resources. In addition, the ratio of the embeddings supporting the Good class remains consistent between CRMs (Good c) and CIRMs (Good ic). Even though there are fewer CIRMs detected compared to CRMs, the information captured within the discovered CIRMs represents the characteristics of the underlying dynamic network as well as the CRMs.

7   Conclusion and Future Research Directions

In this chapter we presented several algorithms that can efficiently and effectively analyze the changes in dynamic relational networks. The new classes of dynamic patterns enable the identification of hidden coordination mechanisms underlying

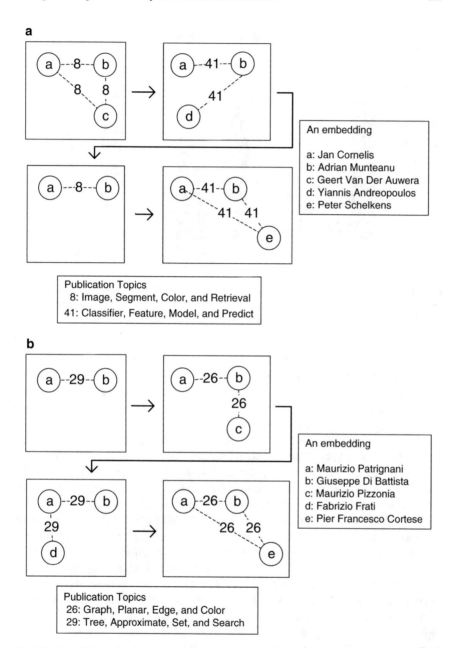

Fig. 17 Two CRMs capturing co-authorship patterns. The *edge labels* represent the domain or subject of the publications the authors were involved together. The *vertices* are labeled to show the changes in relations between the nodes. These CRMs are collected using $\phi = 90$ and $\beta = 0.60$

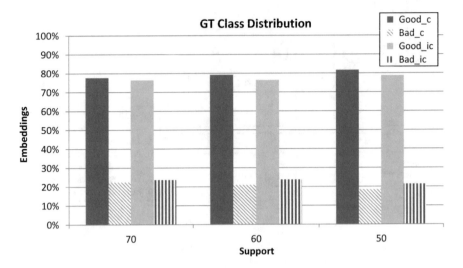

Fig. 18 A distribution of the CRM and CIRM embeddings. The Good class represents the production runs with high yield and the Bad class represents with poor yield. CIRMs were collected using $\phi = (70, 60$ and $50)$, $\beta = 0.50$, $m_{min} = 3$, $k_{min} = 3$, and $k_{max} = 8$

the networks, provide information on the recurrence and the stability of its relational patterns, and improve the ability to predict the relations and their changes in these networks. Specifically, the qualitative analysis of each class of patterns has shown the information captured by these patterns about the underlying networks and proven to be useful for building models.

There are a number of ways we can extend the existing definitions of EIRS and CRM/CIRM to address other real-world pattern mining problems. First, combine the notions of state transition and coevolution into a single pattern class in order to derive patterns that capture the coevolution of recurring stable relations. Second, relax the assumption that the various occurrences of the relational states and motifs match perfectly in terms of node and edge labels. Two general approaches can be designed to allow for flexibility in matching the types of relations and entities. The first utilizes an application-specific similarity function along with a user-supplied similarity threshold γ and the second is based on employing clustering techniques on the edge and node labels. The similarity-function based approach provides control over the degree of approximation that is allowed during the mining of evolving relational patterns. The clustering approach can be used to replace the complex edge and node labels by two distinct sets of categorical labels corresponding to the cluster number that the edges and nodes belong to.

Third, consider dynamic relational networks modeling real-life datasets that contain noisy data and missing links. Two different approaches and their combination can be considered for addressing such problems. The first solution focuses on the modeling phase of the underlying datasets. One approach is to temporally smooth the sequence of snapshots in order to eliminate noise that is localized in time. Another approach can be to eliminate transient relations from the snapshots all together and focus the analysis on only the relations that have some degree of persistence. The second solution focuses on the pattern mining phase that allows for missing edges and nodes during the subgraph occurrence operations (i.e., the subgraph test in the case of relational states and the subgraph isomorphism in the case of relational motifs). The degree of allowed match tolerance can be controlled by a user-supplied parameter.

Fourth, utilize CRMminer/CIRMminer algorithms to mine discriminative dynamic relational patterns, whose presence or absence can help us classify a dynamic network. By combining pattern frequency and discriminative measures, it is shown that discriminative frequent patterns are very effective for classification [9]. There are several algorithms, such as LEAP [25], CORK [22], GraphSig [20], and LTS [14] that mine discriminative frequent patterns in static graph. To efficiently mine discriminative dynamic relational patterns, we need to focus on addressing two problems. The first is to design an efficient algorithm to select discriminative features among a large number of frequent dynamic patterns (i.e., CRMs/CIRMs) and to define a context aware prediction function that can measure the discriminative potential of a dynamic pattern. The second is to use both frequency and the discriminative potential function during dynamic pattern enumeration for pruning the exponential search space.

References

1. Ahmed, R., Karypis, G.: Algorithms for mining the evolution of conserved relational states in dynamic networks. Knowledge and Information Systems, 33(3):1–28 (2012)
2. Ahmed, R., Karypis, G.: Mining coevolving induced relational motifs in dynamic networks. In: SDM Networks, the 2nd Workshop on Mining Graphs and Networks (2015)
3. Ahmed, R., Karypis, G.: Algorithms for Mining the Coevolving Relational Motifs in Dynamic Networks. In: ACM Trans. Knowl. Discov. Data, 10(1), ACM, NY (2015)
4. Berlingerio, M., Bonchi, F., Bringmann, B., Gionis, A.: Mining graph evolution rules. In: Machine Learning pp. 115–130. Springer, Berlin/Heidelberg (2009)
5. Borgwardt, K.M., Kriegel, H.P., Wackersreuther, P.: Pattern mining in frequent dynamic subgraphs. In: IEEE ICDM, pp. 818–822 (2006)
6. Bringmann, B., Nijssen, S.: What is frequent in a single graph? In: Advances in Knowledge Discovery and Data Mining, pp. 858–863. Springer, Berlin/Heidelberg (2008)
7. Cerf, L., Nguyen, T., Boulicaut, J.: Discovering relevant cross-graph cliques in dynamic networks. In: Foundations of Intelligent Systems, pp. 513–522. Springer, Berlin/Heidelberg (2009)
8. Chakrabarti, D., Kumar, R., Tomkins, A.: Evolutionary clustering. In: ACM KDD, pp. 554–560 (2006)
9. Cheng, H., Yan, X., Han, J., Hsu, C.W.: Discriminative frequent pattern analysis for effective classification. In: IEEE 23rd International Conference on Data Engineering, 2007. ICDE 2007, pp. 716–725. IEEE, Istanbul (2007)
10. Desmier, E., Plantevit, M., Robardet, C., Boulicaut, J.F.: Cohesive co-evolution patterns in dynamic attributed graphs. In: Discovery Science, pp. 110–124. Springer, Heidelberg (2012)
11. Friedman, T.L.: The World Is Flat: A Brief History of the Twenty-First Century. Farrar, Straus & Giroux, New York (2005)

12. Gleditsch, K.S.: Expanded trade and GDP data. In: J. Confl. Resolut., **46**(5), pp. 712–724 (2002)
13. Inokuchi, A., Washio, T.: Mining frequent graph sequence patterns induced by vertices. In: Proc. of 10th SDM, pp. 466–477 (2010)
14. Jin, N., Wang, W.: LTS: Discriminative subgraph mining by learning from search history In: 2011 IEEE 27th International Conference on Data Engineering (ICDE), pp. 207–218. IEEE,Washington (2011)
15. Jin, R., McCallen, S., Almaas, E.: Trend motif: A graph mining approach for analysis of dynamic complex networks. In: IEEE ICDM, pp. 541–546 (2007)
16. Kuramochi, M., Karypis, G.: Finding frequent patterns in a large sparse graph. In: Data Mining and Knowledge Discovery, **11**(3), pp. 243–271, Springer US (2005)
17. Le, H., Kabbur, S., Pollastrini, L., Sun, Z., Mills, K., Johnson, K., Karypis, G., Hu, W.S.: Multivariate analysis of cell culture bioprocess data–lactate consumption as process indicator. In: Journal of biotechnology, 162, pp. 210–223 (2012)
18. Ley, M.: Dblp, computer science bibliography. Website, http://www.informatik.uni-trier.de/~ley/ (2008)
19. Pei, J., Han, J., Mortazavi-Asl, B., Pinto, H., Chen, Q., Dayal, U., Hsu, M.C.: PrefixSpan: Mining Sequential Patterns Efficiently by Prefix-Projected Pattern Growth. In: Proceedings IEEE International Conference on Data Engineering (ICDE), pp. 215–224 (2001)
20. Ranu, S., Singh, A.K.: Graphsig: a scalable approach to mining significant subgraphs in large graph databases. In: IEEE 25th International Conference on Data Engineering, 2009. ICDE'09, pp. 844–855. IEEE, Shanghai (2009)
21. Robardet, C.: Constraint-based pattern mining in dynamic graphs. In: IEEE ICDM, pp. 950–955 (2009)
22. Thoma, M., Cheng, H., Gretton, A., Han, J., Kriegel, H.P., Smola, A.J., Song, L., Philip, S.Y., Yan, X., Borgwardt, K.M.: Near-optimal supervised feature selection among frequent subgraphs. In: SDM, pp. 1076–1087. SIAM, Sparks (2009)
23. West, D.B.: Introduction to Graph Theory. Prentice Hall, Upper Saddle River (2001)
24. Yan, X., Han, J.: gSpan: graph-based substructure pattern mining. In: IEEE ICDM, pp. 721–724 (2002)
25. Yan, X., Cheng, H., Han, J., Yu, P.S.: Mining significant graph patterns by leap search. In: Proceedings of the 2008 ACM SIGMOD international conference on Management of data, pp. 433–444. ACM, New York (2008)
26. Zhu, F., Yan, X., Han, J., Philip, S.Y.: gPrune: a constraint pushing framework for graph pattern mining. In: Advances in Knowledge Discovery and Data Mining, pp. 388–400. Springer, Heidelberg (2007)

Probabilistically Grounded Unsupervised Training of Neural Networks

Edmondo Trentin and Marco Bongini

Abstract The chapter is a survey of probabilistic interpretations of artificial neural networks (ANN) along with the corresponding unsupervised learning algorithms. ANNs for estimating probability density functions (pdf) are reviewed first, including parametric estimation via constrained radial basis functions and nonparametric estimation via multilayer perceptrons. The approaches overcome the limitations of traditional statistical estimation methods, possibly leading to improved pdf models. The focus is then moved from pdf estimation to online neural clustering, relying on maximum-likelihood training. Finally, extension of the techniques to the unsupervised training of generative probabilistic hybrid paradigms for sequences of random observations is discussed.

1 Introduction

Alongside with traditional statistical techniques, learning machines have long been applied to pattern classification, nonlinear regression, and clustering. Despite such a popularity, practitioners are usually not concerned with (nor, even aware of) any probabilistic interpretation of the machines they use. Nonetheless, mathematically grounded interpretation of learnable models turns out to be fruitful in terms of better understanding the nature of the models, as well as of better fitting the (probabilistic) laws characterizing the data samples at hand. In fact, theorems confirm that, under rather loose conditions, artificial neural networks (ANNs) can be trained as optimal estimates of Bayes posterior probabilities. Whilst estimation of probabilistic quantities via ANNs is feasible due to the simplicity of satisfying the probability constraints, probabilistic-grounded unsupervised training of connectionist models (e.g., estimation of probability density functions (pdf), feature extraction/reduction, data clustering) may be much harder. In density estimation, for instance, the ANN must cope with the very nature of the unknown pdf underlying the unlabeled data sample, i.e., a nonparametric function which may possibly take any non-negative, unbounded value, and whose integral over the feature space shall equal 1.

E. Trentin (✉) • M. Bongini
DIISM, University of Siena, Siena, Italy
e-mail: trentin@dii.unisi.it; bongini@dii.unisi.it

© Springer International Publishing Switzerland 2016

533

M.E. Celebi, K. Aydin (eds.), *Unsupervised Learning Algorithms*,
DOI 10.1007/978-3-319-24211-8_18

Yet, neural models of pdf could fit, and potentially improve, the advantages of parametric and nonparametric statistical estimation techniques. Moreover, these estimates could be valuable for enforcing statistical paradigms for sequential and structured pattern recognition, namely hidden Markov models (HMM) and some probabilistic graphical models, which rely on the so-called emission pdfs (that, in turn, are usually modeled under strong parametric assumptions). Furthermore, some relevant instances of clustering and feature reduction algorithms stem (implicitly or explicitly) from a proper probabilistic description (i.e., the pdf) of the data.

This chapter surveys probabilistic interpretations of ANNs, moving the focus from the traditional supervised learning of posterior probabilities to unsupervised learning of pdfs, clustering, and density estimation for sequence processing. The approaches are expected to overcome the major limitations of traditional statistical methods, possibly leading to improved models. Unsupervised density estimation is reviewed first (Sect. 2). A constrained maximum-likelihood algorithm for unsupervised training of radial basis functions (RBF) networks is handed out in Sect. 2.1, resulting in a connectionist parametric pdf estimation tool. A survey of techniques for nonparametric density estimation via multilayer perceptrons is then reviewed in Sect. 2.2. Starting from pdf estimation and the maximum-likelihood criterion, it is then feasible to move to a popular online neural algorithm for clustering unsupervised samples of data (Sect. 3), and even to probabilistic modeling of sequences of random observation drawn from an unknown pdf, by means of a hybrid ANN/hidden Markov model paradigm (reviewed in Sect. 4). Conclusions are drawn in Sect. 5.

2 Unsupervised Estimation of Probability Density Functions

One major topic in pattern recognition is the problem of estimating pdf [11]. Albeit popular, parametric techniques (e.g., maximum likelihood for Gaussian mixtures) rely on an arbitrary assumption on the form of the underlying, unknown distribution. Nonparametric techniques (e.g., k_n-nearest neighbors [11]) remove this assumption and attempt a direct estimation of the pdf from a data sample. The Parzen Window (PW) is one of the most popular nonparametric approaches to pdf estimation, relying on a combination of local window functions centered on the patterns of the training sample [11]. Although effective, PW suffers from several limitations, including: (1) the estimate is not expressed in a compact functional form (i.e., a probability law), but it is a sum of as many local windows as the size of the sample; (2) the local nature of the window functions tends to yield a fragmented model, which is basically "memory based" and (by definition) is prone to overfitting; (3) the whole training sample has to be kept always in memory in order to compute the estimate of the pdf over any new (test) patterns, resulting in a high complexity of the technique in space and time; (4) the form of the window function chosen has a deep influence on the eventual form of the estimated model, unless an asymptotic case (i.e., infinite sample) is considered; (5) the PW model heavily depends on the choice of an initial width of the local region of the feature space where the windows are centered.

ANNs are, in principle, an alternative family of nonparametric models [18]. Given the "universal approximation" property [4] of certain ANN families (multilayer perceptrons [37] and radial basis function networks [4]), they might be a suitable model for any given (continuous) form of data distributions. While ANNs are intensively used for estimating probabilities (e.g., posterior probabilities in classification tasks [4]), they have not been intensively exploited so far for estimating pdfs. One of the main rationales behind this fact is that connectionist modeling of probabilities is easily (and somewhat heuristically) achieved by standard supervised backpropagation [37], once 0/1 target outputs are defined for the training data [4] (as pointed out in Sect. 1). Moreover, it is also simple to introduce constraints within the model that ensure the ANN may be interpreted in probabilistic terms, e.g., using sigmoid output activations (that range in the (0, 1) interval), along with a softmax-like mechanism [2] which ensures that all the outputs sum to 1. Learning a pdf, on the contrary, is an unsupervised and far less obvious task.

In the following, we discuss the connectionist approaches to the problem of pdf estimation. The former generates within a maximum-likelihood parametric estimation technique under specific assumptions on the form of the underlying pdf, namely a Gaussian-based RBF network. It is presented in Sect. 2.1 and has its point of strength in the overwhelming simplicity. More generic (and recent) nonparametric frameworks are reviewed in Sect. 2.2, relying on multilayer perceptrons.

2.1 Estimating pdfs via Constrained RBFs

Let $\mathscr{T} = \{\mathbf{x}_1, \ldots, \mathbf{x}_N\}$ be a random sample of N d-dimensional continuous-valued observations, identically and independently drawn (iid) from the unknown pdf $p(\mathbf{x})$. A radial basis function (RBF)-like network [4] can be used to obtain a model $\hat{p}(\mathbf{x}|\boldsymbol{\theta}_{\hat{p}})$ of the unknown pdf $p(\mathbf{x})$, where $\boldsymbol{\theta}_{\hat{p}}$ is the parameter vector of the RBF. Namely, $\boldsymbol{\theta}_{\hat{p}}$ includes the RBF hidden-to-output weights, and the parameters characterizing the Gaussian kernels. In order to ensure that a pdf is obtained (i.e., non-negative outputs, and a unit integral over the whole definition domain), constraints have to be placed on the hidden-to-output connection weights of the RBF (assuming that normalized Gaussian kernels are used).

Relying on the iid assumption, maximum-likelihood (ML) estimation of the RBF parameters $\boldsymbol{\theta}_{\hat{p}}$ given $\mathscr{T}$ is accomplished maximizing the quantity

$$p(\mathscr{T}|\boldsymbol{\theta}_{\hat{p}}) = \prod_{i=1}^{N} \hat{p}(\mathbf{x}_i|\boldsymbol{\theta}_{\hat{p}}). \tag{1}$$

A hill-climbing algorithm for maximizing $p(\mathscr{T}|\boldsymbol{\theta}_{\hat{p}})$ w.r.t. $\boldsymbol{\theta}_{\hat{p}}$ is obtained in two steps. (1) *Initialization*: start with some initial, e.g., random, assignment of values to the RBF parameters.[1] (2) *Gradient-ascent*: repeatedly apply a learning rule in the form $\Delta\boldsymbol{\theta}_{\hat{p}} = \eta\nabla_{\boldsymbol{\theta}_{\hat{p}}}\{\prod_{i=1}^{N}\hat{p}(\mathbf{x}_i|\boldsymbol{\theta}_{\hat{p}})\}$ with $\eta \in \Re^+$. This is a batch learning rule. In practice, neural network learning may be simplified, yet even improved, with the adoption of an online training scheme that prescribes $\Delta\boldsymbol{\theta}_{\hat{p}} = \eta\nabla_{\boldsymbol{\theta}_{\hat{p}}}\{\hat{p}(\mathbf{x}|\boldsymbol{\theta}_{\hat{p}})\}$ upon presentation of each individual training observation $\mathbf{x}$. The gradient is calculated w.r.t. two distinct families of adaptive parameters.

(1) Mixing parameters $c_1,\ldots,c_n$, i.e., the hidden-to-output weights of the RBF network. A constrained ML estimation process is required to ensure that $c_j \in (0,1)$ for $j = 1,\ldots,n$, and that $\sum_{j=1}^{n} c_j = 1$. To satisfy the constraints we introduce n latent parameters $\gamma_1,\ldots,\gamma_n$, which are unconstrained, and we let

$$c_i = \frac{\varsigma(\gamma_i)}{\sum_{j=1}^{n}\varsigma(\gamma_j)}, i = 1,\ldots,n \tag{2}$$

where $\varsigma(x) = 1/(1 + e^{-x})$. Each γ_i is then treated as an unknown parameter to be estimated via ML.

(2) d-dimensional mean vector $\boldsymbol{\mu}_i$ and $d \times d$ covariance matrix Σ_i for each of the Gaussian kernels $K_i(\mathbf{x}) = G(\mathbf{x};\boldsymbol{\mu}_i,\Sigma_i)$, $i = 1,\ldots,n$ of the RBF, where $G(\mathbf{x};\boldsymbol{\mu}_i,\Sigma_i)$ denotes a multivariate Gaussian pdf having mean vector $\boldsymbol{\mu}_i$, covariance matrix Σ_i, and evaluated over the random vector $\mathbf{x}$. A common (yet effective) simplification is to consider diagonal covariance matrices, i.e., independence[2] among the components of the input vector $\mathbf{x}$. This assumption leads to the following three major consequences: (1) modeling properties are not affected significantly, according to [26]; (2) generalization capabilities of the overall model may turn out to be improved, since the number of free parameters is reduced; (3) i-th multivariate kernel K_i may be expressed in the form of a product of d univariate Gaussian pdfs as:

$$K_i(\mathbf{x}) = \prod_{j=1}^{d}\frac{1}{\sqrt{2\pi}\sigma_{ij}}\exp\left\{-\frac{1}{2}\left(\frac{x_j - \mu_{ij}}{\sigma_{ij}}\right)^2\right\} \tag{3}$$

i.e., the free parameters to be estimated are the means μ_{ij} and the standard deviations σ_{ij}, for each kernel $i = 1,\ldots,n$ and for each component $j = 1,\ldots,d$ of the input space. Recapitulating, the RBF system equation can thus be expressed in the concise form $\hat{p}(\mathbf{x}|\boldsymbol{\theta}_{\hat{p}}) = \sum_{i=1}^{n} c_i G(\mathbf{x};\boldsymbol{\mu}_i,\Sigma_i)$.

An explicit form for $\Delta\boldsymbol{\theta}_{\hat{p}}$ is now obtained by calculating the partial derivatives of $\hat{p}(\mathbf{x}|\boldsymbol{\theta}_{\hat{p}})$ w.r.t. the two families of free parameters in the model. For a generic mixing parameter $c_i, i = 1,\ldots,n$, from Eq. (2) and since $\hat{p}(\mathbf{x}||\boldsymbol{\theta}_{\hat{p}}) = \sum_{k=1}^{n} c_k K_k(\mathbf{x})$ we have

---

[1] RBF initialization can be accomplished using more structured approaches. See, for instance, [10].

[2] More precisely, diagonal covariance matrices entail uncorrelated components of the random vector. Uncorrelatedness implies independence only when the components have a joint Gaussian distribution.

$$\frac{\partial \hat{p}(\mathbf{x}|\boldsymbol{\theta}_{\hat{p}})}{\partial \gamma_i} = \sum_{j=1}^{n} \frac{\partial \hat{p}(\mathbf{x}|\boldsymbol{\theta}_{\hat{p}})}{\partial c_j} \frac{\partial c_j}{\partial \gamma_i}$$

$$= \sum_{j=1}^{n} K_j(\mathbf{x}) \frac{\partial}{\partial \gamma_i} \left(\frac{\varsigma(\gamma_j)}{\sum_{k=1}^{n} \varsigma(\gamma_k)} \right)$$

$$= K_i(\mathbf{x}) \left\{ \frac{\varsigma'(\gamma_i) \sum_k \varsigma(\gamma_k) - \varsigma(\gamma_i)\varsigma'(\gamma_i)}{[\sum_k \varsigma(\gamma_k)]^2} \right\} + \sum_{j \neq i} K_j(\mathbf{x}) \left\{ \frac{-\varsigma(\gamma_j)\varsigma'(\gamma_i)}{[\sum_k \varsigma(\gamma_k)]^2} \right\}$$

$$= K_i(\mathbf{x}) \frac{\varsigma'(\gamma_i)}{\sum_k \varsigma(\gamma_k)} - \sum_j K_j(\mathbf{x}) \frac{\varsigma(\gamma_j)\varsigma'(\gamma_i)}{[\sum_k \varsigma(\gamma_k)]^2}$$

$$= K_i(\mathbf{x}) \frac{\varsigma'(\gamma_i)}{\sum_k \varsigma(\gamma_k)} - \left\{ \sum_j c_j K_j(\mathbf{x}) \right\} \frac{\varsigma'(\gamma_i)}{\sum_k \varsigma(\gamma_k)}$$

$$= \frac{\varsigma'(\gamma_i)}{\sum_k \varsigma(\gamma_k)} \left\{ K_i(\mathbf{x}) - \hat{p}(\mathbf{x}||\boldsymbol{\theta}_{\hat{p}}) \right\}. \tag{4}$$

As for the means μ_{ij} and the standard deviations σ_{ij} we proceed as follows. Let θ_{ij} denote the free parameter, i.e., μ_{ij} or σ_{ij}, to be estimated. It is seen that $\frac{\partial \hat{p}(\mathbf{x}|\boldsymbol{\theta}_{\hat{p}})}{\partial \theta_{ij}} = c_i \frac{\partial K_i(\mathbf{x})}{\partial \theta_{ij}}$, where the calculation of $\frac{\partial K_i(\mathbf{x})}{\partial \theta_{ij}}$ is accomplished as follows. First, let us observe that for any real-valued, differentiable function $f(.)$ this property holds true: $\frac{\partial f(.)}{\partial x} = f(.) \frac{\partial \log[f(.)]}{\partial x}$. Thence, from Eq. (3) we can write

$$\frac{\partial K_i(\mathbf{x})}{\partial \theta_{ij}} = K_i(\mathbf{x}) \frac{\partial \log K_i(\mathbf{x})}{\partial \theta_{ij}} \tag{5}$$

$$= K_i(\mathbf{x}) \frac{\partial}{\partial \theta_{ij}} \sum_{k=1}^{d} \left\{ -\frac{1}{2} \left[\log(2\pi\sigma_{ik}^2) + \left(\frac{x_k - \mu_{ik}}{\sigma_{ik}} \right)^2 \right] \right\}.$$

For the means, i.e., $\theta_{ij} = \mu_{ij}$, Eq. (5) yields $\frac{\partial K_i(\mathbf{x})}{\partial \mu_{ij}} = K_i(\mathbf{x}) \frac{x_j - \mu_{ij}}{\sigma_{ij}^2}$. For the covariances, i.e., $\theta_{ij} = \sigma_{ij}$, Eq. (5) takes the form $\frac{\partial K_i(\mathbf{x})}{\partial \sigma_{ij}} = K_i(\mathbf{x}) \frac{\partial}{\partial \sigma_{ij}} \left\{ -\frac{1}{2} \log(2\pi\sigma_{ij}^2) - \frac{1}{2} \left(\frac{x_j - \mu_{ij}}{\sigma_{ij}} \right)^2 \right\}$, that is $\frac{\partial K_i(\mathbf{x})}{\partial \sigma_{ij}} = \frac{K_i(\mathbf{x})}{\sigma_{ij}} \left\{ \left(\frac{x_j - \mu_{ij}}{\sigma_{ij}} \right)^2 - 1 \right\}$ which completes the calculation of $\Delta \boldsymbol{\theta}_{\hat{p}}$. The training algorithm is guaranteed to increase the likelihood up to a (possibly local) maximum. Due to the universal approximation properties of RBFs [4] it is seen that, under the same mild conditions assumed therein, the resulting ANN is a universal model of any continuous and bounded pdf of multivariate observations.

2.2 Estimating pdfs via Multilayer Perceptrons

As we say, RBFs realize linear combinations of Gaussian kernels. All in all, the gradient-ascent ML solution handed out in the previous section reduces to a neural implementation of the classic ML parametric estimation for a Gaussian mixture, thence it is seldom palatable.

The search for a suitable, nonparametric ANN-based pdf estimator shall rather rely on a multilayer perceptron (MLP) [4], due to its flexibility, its nice generalization capabilities, and the very nature of its activation functions (yielding a mixture of sigmoids) which pose themselves as a significant alternative to Gaussian-like kernels. Since regular, supervised BP cannot be applied, variations on the theme of MLP training in the unsupervised framework are sought.

The most straightforward idea is to apply ML gradient-ascent training to the MLP, as with RBFs. If $\mathscr{T} = \{\mathbf{x}_1, \ldots, \mathbf{x}_n\}$ is the sample of the i.i.d. data drawn from the unknown pdf $p(.)$, and if we write Ω to denote the ordered set of the MLP parameters (weights and biases), then the likelihood of the model given the sample is $p(\mathscr{T} \mid \Omega) = \prod_{i=1}^{n} p(\mathbf{x}_i \mid \Omega)$, where $p(\mathbf{x}_i \mid \Omega)$ is the MLP output for input $\mathbf{x}_i$. As in the previous section, gradient ascent prescribes the modification Δw of a generic parameter $w \in \Omega$ according to $\Delta w = \eta \partial \prod_{i=1}^{n} p(\mathbf{x}_i \mid \Omega)/\partial w$, where $\eta \in \Re^+$ is the learning rate. Unfortunately, since an unconstrained mixture of sigmoids is not a pdf, training the MLP this way will only lead to an unbounded increase in the value of its weights (hence, of the integral of the function the MLP realizes), yielding a degenerate solution. This phenomenon is known as the "divergence problem" [42]. To the contrary of RBFs, where Kolmogorov's axioms of probability were satisfied by means of simple constraints on the mixing parameters, no simple workarounds are available for constraining the MLP to satisfy the "integral equals 1" requirement that any proper pdf shall meet. Alternative approaches to MLP training for pdf estimation were thus proposed in the literature.

Commonsense suggests that a solution to the divergence problem lies in controlling the growth of the integral of the function $\phi(.)$ realized by the MLP during training. Roughly speaking, the MLP output might be normalized (at the end of each training iteration) by the (approximated) integral of $\phi(.)$, computed via any numerical integration technique. The general mathematical formalization of this idea is proposed in [28]. The latter assumes a maximum log-likelihood criterion where the output activation function of the MLP is the exponential, ensuring the correct range for a pdf, namely $(0, +\infty)$. The bias of the output unit plays basically the role of the normalization factor, that is the integral of the function $\exp(\phi(.))$ realized by the MLP over a given compact set $X \subset \Re^d$. It turns out that the calculations of the gradient of the criterion w.r.t. the MLP parameters Ω involve the computation of two separate integrals (namely, $\int_X \exp(\phi(\mathbf{x}, \Omega))d\mathbf{x}$, and $\int_X \exp(\phi(\mathbf{x}, \Omega))\nabla_\Omega \phi(\mathbf{x}, \Omega)d\mathbf{x}$), on a parameter-by-parameter basis (since the latter integral requires the knowledge of $\nabla_\Omega \phi(\mathbf{x}, \Omega)$ for all—possibly backpropagated—MLP weights and biases), at each step of the BP procedure.

Albeit elegant, this approach presents a few, severe drawbacks. First of all, no algorithmic solution to the effective computation of the integrals is given in [28]. This is pivotal, since numeric integration (to be accomplished at each training iteration and for each training pattern) raises a critical trade-off (in terms of tightness of the integration grid) between numerical stability and computational complexity, especially in multivariate scenarios. Also the output exponential function may be troublesome from the numerical stability standpoint. In fact, the value of its derivatives (involved in the BP procedure) spans over a different numeric range w.r.t. the derivatives of the logistic functions used in the hidden layer. This renders the selection of a common, viable learning rate unlikely at best. Moreover, most real-world, multivariate pdfs have extremely close-to-zero values over all $\mathfrak{R}^d$. That is to say, the expected MLP output occurs at the far left-tail of the exp(.) function, where derivatives have (quasi-)null numerical values. Since (due to the BP mechanism) the derivatives of the output activation are involved in the computation of the gradients throughout the whole MLP, this results in null updates Δw for all the parameters in the MLP (i.e., learning simply gets stuck). These are possibly some of the reasons why the approach was validated empirically only on a very simple, univariate problem (requiring 5000 data points for training) [28]. Further theoretical developments of this technique are discussed in [29], which makes explicit the adoption of a minimum complexity estimation scheme. The analysis proposed in [29] (limited to one-hidden-layer MLPs over the unit interval in $\mathfrak{R}^d$), albeit mathematically valuable, is not instantiated in the form of an algorithm. In particular, the usual problem of the computation of the normalizing integral is not given any practical solution.

As a consequence, alternative approaches were proposed (e.g., [25, 44]), shifting the focus from a direct estimation of the pdf to the estimation of the corresponding cumulative distribution function (cdf). After properly training the MLP model $\phi(.)$ of the cdf, the pdf can be recovered by applying differentiation to $\phi(.)$ (i.e., indirect estimation is accomplished). The idea is sound, since cdfs usually have a (mixture of) sigmoid(s)-like form which fits the nature of the activation functions. Above all, the requirements that $\phi(.)$ has to satisfy to be interpretable as a proper cdf (namely, that it ranges between 0 and 1, and that it is monotonically nondecreasing) appear to be more easily met than the corresponding constraint on pdf models (i.e., the unit integral).

In [44] (where a kernel-based solution is presented) the empirical cdf is obtained from the data sample, and it is basically used for creating target outputs for each training pattern, such that any (differentiable) supervised regression model can then be applied. Of course the empirical cdf results in a rough, discontinuous approximation of the true cdf, but experiments presented in [44] over univariate, simple tasks involving Gaussian pdfs highlight the approach may be viable to some extent. In [25] an MLP is trained to estimate the cdf $F(x)$ of a univariate pdf $p(x)$ using a stochastic approach. Relying on the fact that if x is a random variable distributed according to $p(x)$ then the corresponding random variable $y = F(x)$ has a uniform distribution over $(0, 1)$, target outputs for the BP algorithm are generated as follows. A random sample of "targets" $\{y_1, \ldots, y_n\}$ is drawn from this uniform distribution. Since cdfs are intrinsically monotonically increasing, both the original data sample $\{x_1, \ldots, x_n\}$ and the targets $\{y_1, \ldots, y_n\}$ are sorted in a nondecreasing order (s.t. $x_k \leq x_{k+1}$ and $y_k \leq y_{k+1}$ for $k = 1, \ldots, n-1$). A supervised training set $\{(x_1, y_1), \ldots, (x_n, y_n)\}$ is thus obtained, and BP is applied. A penalty term is added to the squared error criterion function for BP, too, enforcing the monotonicity of the resulting ANN. Resampling of the targets is then accomplished at regular periods of t iterations of BP. The method is theoretically proofed to exhibit a noticeable convergence rate to the true distribution as the number of training patterns increases [25].

Unfortunately, there are drawbacks to the cdf-based approaches. First of all, they require differentiation of the network output $\phi(x)$ w.r.t. its input x in order to recover the estimated value of the pdf sought. This entails two consequences. First, a differentiation method has to be applied at test time, once training has been accomplished (this requires a backpropagation of partial derivatives throughout the multilayer ANN over each new test pattern), increasing complexity at test time and possibly affecting the numerical quality of the results. Second (and much more troubling), a good approximation of the cdf may not necessarily translate into a similarly good estimate of its derivative. In fact, a small squared error between $\phi(.)$ and $F(.)$ does not mean that $\phi(.)$ is free from steep fluctuation that imply huge, rapidly changing values of the derivative. Negative values of $\frac{\partial \phi(x)}{\partial x}$ may occasionally occur, since a linear combination of logistics is not necessarily monotonically increasing, even if the technique proposed in [25] is applied (actually, the penalty term used in [25] involves a pseudo-Lagrange multiplier which enforces monotonicity but does not guarantee it). Another problem of cdf-based algorithms is that they naturally apply to univariate cases, whilst extension to multivariate pdfs is far less realistic. For this reason, [25] proposes also a multivariate deterministic variant of the estimation technique. Targets y for BP are created by means of an empirical estimation of the multivariate cdf over $\mathbf{x} \in \mathfrak{R}^d$ as $y = \frac{1}{n} \sum_{i=1}^{n} g(\mathbf{x} - \mathbf{x}_i)$, where $g(x_1, \ldots, x_d) = 1$ if $x_j \geq 0$ for all $j = 1, \ldots, d$, otherwise $g(x_1, \ldots, x_d) = 0$, and where $\{\mathbf{x}_1, \ldots, \mathbf{x}_n\}$ is the training sample. It is clear that this approximation of the cdf requires a huge number of training patterns over high-dimensional feature spaces, in order to ensure a significant coverage of $\mathfrak{R}^d$. This "sampling problem" is somewhat related to the computation of the normalization integral of the pdf involved in the ML approaches reviewed above. The application of differentiation techniques for retrieving the multivariate pdf sought is subject to the same drawbacks we mentioned in the univariate case, but it becomes even more troublesome in severely high-dimensional spaces (requiring approximated, numerical techniques), as pointed out in [25].

Finally, we present a neural-based algorithm for unsupervised, nonparametric density estimation that was recently introduced in [41]. The approach overcomes the limitations of statistical (either parametric or nonparametric) techniques, as well the drawbacks of the aforementioned MLP-based approaches, leading to improved pdf models. The technique is related to cdf-based methods, insofar that a nonparametric statistical model is used to create target output for backpropagation (BP) training [4]. The algorithm is introduced by reviewing the basic concepts of the traditional

Parzen window (PW) estimation [11]. Let us consider a pdf $p(\mathbf{x})$, defined over a real-valued, d-dimensional feature space. The probability that a pattern $\mathbf{x}' \in \mathcal{R}^d$, drawn from $p(\mathbf{x})$, falls in a certain region R of the feature space is $P = \int_R p(\mathbf{x})d\mathbf{x}$. Let then $\mathcal{T} = \{\mathbf{x}_1, \ldots, \mathbf{x}_n\}$ be an unsupervised sample of n patterns, identically and independently distributed (i.i.d.) according to $p(\mathbf{x})$. If k_n patterns in $\mathcal{T}$ fall within R, an empirical estimate of P can be obtained as $P \simeq k_n/n$. If $p(\mathbf{x})$ is continuous and R is small enough to prevent $p(\mathbf{x})$ from varying its value over R in a significant manner, we are also allowed to write $\int_R p(\mathbf{x})d\mathbf{x} \simeq p(\mathbf{x}')V$, where $\mathbf{x}' \in R$, and V is the volume of region R. As a consequence of the discussion, we can obtain an estimated value of the pdf $p(\mathbf{x})$ over pattern $\mathbf{x}'$ as:

$$p(\mathbf{x}') \simeq \frac{k_n/n}{V_n} \tag{6}$$

where V_n denotes the volume of region R_n (i.e., the choice of the region width is explicitly written as a function of n), assuming that smaller regions around $\mathbf{x}'$ are considered as the sample size n increases. This is expected to allow Eq. (6) to yield improved estimates of $p(\mathbf{x})$, i.e., to converge to the exact value of $p(\mathbf{x}')$ as n (hence, also k_n) tends to infinity (a discussion of the asymptotic behavior of nonparametric models of this kind can be found in [11]).

The basic instance of the PW technique assumes that R_n is a hypercube having edge h_n, such that $V_n = h_n^d$. The edge h_n is usually defined as a function of n as $h_n = h_1/\sqrt{n}$, in order to ensure a correct asymptotic behavior. The value h_1 has to be chosen empirically, and it heavily affects the resulting model. The formalization of the idea requires to define a unit-hypercube window function in the form

$$\varphi(\mathbf{y}) = \begin{cases} 1 & \text{if } |\, y_j \,| \leq 1/2, j = 1, \ldots, d \\ 0 & \text{otherwise} \end{cases} \tag{7}$$

such that $\varphi(\frac{\mathbf{x}'-\mathbf{x}}{h_n})$ has value 1 iff $\mathbf{x}'$ falls within the d-dimensional hyper-cubic region R_n centered in $\mathbf{x}$ and having edge h_n. This implies that $k_n = \sum_{i=1}^n \varphi(\frac{\mathbf{x}'-\mathbf{x}_i}{h_n})$. Using this expression, from Eq. (6) we can write

$$p(\mathbf{x}') \simeq \frac{1}{n} \sum_{i=1}^n \frac{1}{V_n} \varphi(\frac{\mathbf{x}'-\mathbf{x}_i}{h_n}) \tag{8}$$

which is the PW estimate of $p(\mathbf{x}')$ from the sample $\mathcal{T}$. The model is usually refined by considering smoother window functions $\varphi(.)$, instead of hypercubes, e.g., standard Gaussian kernels with zero mean and unit covariance matrix.

Let us now consider an MLP that we wish to train in order to learn a model of the probability law $p(\mathbf{x})$ from the unsupervised data set $\mathcal{T}$. The idea is to use the PW model as a target output for the ANN, and to apply standard BP to learn the ANN connection weights. An unbiased variant of this idea is proposed, according to the following unsupervised algorithm (expressed in pseudo-code):

```
Input:  𝒯 = {x₁,...,xₙ},  h₁.
Output: p̃(.) // the neural estimate of p(.)
1. Let hₙ = h₁/√n
2. Let Vₙ = hₙᵈ
3. For i=1 to n do // loop over 𝒯
3.1          Let 𝒯ᵢ = 𝒯 \ {xᵢ}
3.2          Let yᵢ = 1/n Σₓ∈𝒯ᵢ 1/Vₙ φ(xᵢ-x/hₙ) // target output
4. Let 𝒮 = {(xᵢ,yᵢ) | i = 1,...,n} // supervised training set
5. Train the ANN via backpropagation over 𝒮
6. Let p̃(.) be equal to the function computed by the
ANN
7. Return p̃(.)
```

Since the MLP output is assumed to be an estimate of a pdf, it must be non-negative. This is granted once standard sigmoids (in the form $y = \frac{1}{1+e^{-x}}$) are used in the output layer. Standard sigmoids range in the $(0, 1)$ interval, while pdfs may take any positive value. For this reason, sigmoids with adaptive amplitude λ (i.e., in the form $y = \frac{\lambda}{1+e^{-x}}$), as described in [40], should be used. A direct alternative is using linear output activation functions, forcing negative outputs to zero once training is completed. Nevertheless, as in the popular k_n-nearest neighbor technique [11], the resulting MLP is not necessarily a pdf (in general, the integral of $p̃(.)$ over the feature space is not 1).

There are two major aspects of the algorithm that shall be clearly pointed out. First, the PW generation of target outputs (steps 3–3.2) is unbiased. Computation of the target for i-th input pattern x_i does not involve x_i in the underlying PW model. This is crucial in smoothing the local nature of PW. In practice, the target (estimated pdf value) over x_i is determined by the concentration of patterns in the sample (different from x_i) that occurs in the surroundings of x_i. In particular, if an isolated pattern (i.e., an outlier) is considered, its exclusion from the PW model turns out to yield a close-to-zero target value. This phenomenon is evident along the possible tails of certain distributions, and it is observed in the experiments described in [41].

A second relevant aspect of the algorithm is that it trains the MLP only over the locations (in the feature space) of the patterns belonging to the original sample. At first glance, a different approach could look more promising: once the PW model has been estimated from $\mathcal{T}$, generate a huge supervised training set by covering the input interval in a "uniform" manner, and by evaluating target outputs via the PW model. A more homogeneous and exhaustive coverage of the feature space would be expected, as well as a more precise ANN approximation of the PW. As a matter of fact, training the ANN this way reduces its generalization capabilities, resulting in a more "nervous" surface of the estimated pdf, since the PW model has a natural tendency to yield unreliable estimates over regions of the feature space that are not covered by the training sample (again, refer to the experimental demonstration in [41]).

It is immediately seen that, in spite of the simplicity of its training algorithm, eventually the MLP is expected to overcome most of the PW limitations listed above. The experiments reported in [41] highlight that, in addition, the MLP may turn out to be more accurate than the PW estimate.

3   From pdf Estimation to Online Neural Clustering

A connectionist approach to online unsupervised clustering relying on ML parametric techniques is represented by competitive neural networks (CNN) [19]. CNNs are one-layer feed-forward models whose dynamics forms the basis for several other unsupervised connectionist architectures. These models are called *competitive* because each unit competes with all the others for the classification of each input pattern: the latter is indeed assigned to the *winner unit*, which is the closest (according to a given distance measure) to the input itself, i.e., the most representative within a certain set of *prototypes*. This is strictly related to the concept of *clustering* [11], where each unit represents the centroid (or mean, or codeword) of one of the clusters. The propagation of the input vector through the network consists in a projection of the pattern onto the weights of the connections entering each output unit. Connection weights are assumed to represent the components of the corresponding centroid. The aim of the forward propagation is to establish the closest centroid to the input vector, i.e., the winner unit. During training, a simple weight update rule is used to move the winner centroid toward the novel pattern, so that the components of the centroid represent a sort of moving average of the input patterns belonging to the corresponding cluster. This is basically an online version of some partitioning clustering algorithms [11]. In the following, we will derive it as a consequence of maximum-likelihood estimation of the parameters of a mixture of Gaussians under certain assumptions. Of course, the calculations are related to those we carried out in Sect. 2.1 for RBF-based ML estimation of pdfs.

Let us consider the training set $\mathcal{T} = \{\mathbf{x}_k \mid k = 1, \ldots, N\}$, where N input samples $\mathbf{x}_1, \ldots, \mathbf{x}_N$ are supposed to be drawn from the *finite mixture* density

$$p(\mathbf{x} \mid \Theta) = \sum_{i=1}^{C} \Pi_i p_i(\mathbf{x} \mid \boldsymbol{\theta}_i) \tag{9}$$

where the parametric form of the component densities $p_i(), i = 1, \ldots, C$ is assumed to be known, as well as the *mixing parameters* $\Pi_1, \ldots, \Pi_C$ (*a priori* probabilities of the components) and $\Theta = (\boldsymbol{\theta}_1, \ldots, \boldsymbol{\theta}_C)$ is the vector of all parameters associated with each component density. In the present setup we want to use the unlabeled training samples to estimate the parameters Θ. In classical pattern recognition this is an unsupervised parametric estimation problem. The following discussion will introduce a connectionist approach to the same problem, leading to the formulation of an unsupervised learning rule for CNN that is consistent (under certain assumptions) with the above-mentioned statistical estimation.

Assuming the samples are independently drawn from $p(\mathbf{x} \mid \Theta)$, the likelihood of the observed data $\mathscr{T}$ given a certain choice of parameters Θ can be written as:

$$p(\mathscr{T} \mid \Theta) = \prod_{j=1}^{N} p(\mathbf{x}_j \mid \Theta). \tag{10}$$

Maximum-likelihood estimation techniques search for the parameters Θ that maximize expression (10), or equivalently the *log-likelihood*, as:

$$l(\Theta) = \log p(\mathscr{T} \mid \Theta) = \sum_{j=1}^{N} \log p(\mathbf{x}_j \mid \Theta). \tag{11}$$

Since the logarithm is a monotonic increasing function, parameters that maximize expressions (10) and (11) are the same. If a certain parameter vector Θ' maximizes $l(\Theta)$ then it has to satisfy the following necessary, but not sufficient, condition:

$$\nabla_{\boldsymbol{\theta}'_i} l(\Theta') = \mathbf{0} \; i = 1, \ldots, C \tag{12}$$

where the operator $\nabla_{\boldsymbol{\theta}'_i}$ denotes the gradient vector computed with respect to the parameters $\boldsymbol{\theta}'_i$, and $\mathbf{0}$ is the vector with all components equal to zero. In other words, we are looking for the zeros of the gradient of the log-likelihood. Substituting Eq. (9) into Eq. (11) and setting the latter equal to zero we can write:

$$\begin{aligned}
\nabla_{\boldsymbol{\theta}'_i} l(\Theta') &= \sum_{j=1}^{N} \nabla_{\boldsymbol{\theta}'_i} \log p(\mathbf{x}_j \mid \Theta') \\
&= \sum_{j=1}^{N} \nabla_{\boldsymbol{\theta}'_i} \log \sum_{i=1}^{C} \Pi_i p_i(\mathbf{x}_j \mid \boldsymbol{\theta}'_i) \\
&= \sum_{j=1}^{N} \frac{1}{p(\mathbf{x}_j \mid \Theta')} \nabla_{\boldsymbol{\theta}'_i} \Pi_i p_i(\mathbf{x}_j \mid \boldsymbol{\theta}'_i) \\
&= \mathbf{0}.
\end{aligned} \tag{13}$$

Using Bayes' Theorem we have:

$$P(i \mid \mathbf{x}_j, \boldsymbol{\theta}'_i) = \frac{\Pi_i p_i(\mathbf{x}_j \mid \boldsymbol{\theta}'_i)}{p(\mathbf{x}_j \mid \Theta)} \tag{14}$$

where $P(i \mid \mathbf{x}_j, \boldsymbol{\theta}'_i)$ is the *a posteriori* probability of class i given the observation $\mathbf{x}_j$ and the parameters $\boldsymbol{\theta}'_i$. Equation (13) can thus be rewritten as:

$$\nabla_{\boldsymbol{\theta}'_i} l(\Theta') = \sum_{j=1}^{N} \frac{P(i \mid \mathbf{x}_j, \boldsymbol{\theta}'_i)}{\Pi_i p_i(\mathbf{x}_j \mid \boldsymbol{\theta}'_i)} \nabla_{\boldsymbol{\theta}'_i} \Pi_i p_i(\mathbf{x}_j \mid \boldsymbol{\theta}'_i)$$

$$= \sum_{j=1}^{N} P(i \mid \mathbf{x}_j, \boldsymbol{\theta}'_i) \nabla_{\boldsymbol{\theta}'_i} \log \Pi_i p_i(\mathbf{x}_j \mid \boldsymbol{\theta}'_i)$$

$$= \mathbf{0}. \tag{15}$$

In the following, we will concentrate our attention on the case in which the component densities of the mixture are multivariate normal distributions. In this case

$$p_i(\mathbf{x}_j \mid \boldsymbol{\theta}_i) = (2\pi)^{-\frac{d}{2}} \mid \Sigma_i \mid^{-\frac{1}{2}} e^{\{-\frac{1}{2}(\mathbf{x}_j - \mu_i)^t \Sigma_i^{-1}(\mathbf{x}_j - \mu_i)\}}, \tag{16}$$

where d is the dimension of the feature space, t denotes the transposition of a matrix, and the parameters to be estimated for the i-th component density are its mean vector and its covariance matrix, that is to say:

$$\boldsymbol{\theta}_i = (\mu_i, \Sigma_i). \tag{17}$$

Substituting Eq. (16) into Eq. (15) we can write the gradient of the log-likelihood for the case of normal component densities as:

$$\nabla_{\boldsymbol{\theta}_i} l(\Theta) = \sum_{j=1}^{N} P(i \mid \mathbf{x}_j, \boldsymbol{\theta}_i) \nabla_{\boldsymbol{\theta}_i} \{ \log \Pi_i (2\pi)^{-\frac{d}{2}} \mid \Sigma_i \mid^{-\frac{1}{2}}$$

$$- \frac{1}{2} (\mathbf{x}_j - \mu_i)^t \Sigma_i^{-1} (\mathbf{x}_j - \mu_i) \}. \tag{18}$$

Suppose now that the only unknown parameters to be estimated are the mean vectors of the Gaussian distributions, e.g., the covariances are supposed to be known in advance. There are practical situations, for example, in data clustering, in which the estimation of the means is sufficient; furthermore, this assumption simplifies the following mathematics, but the extension to the more general case of unknown covariances is rather simple. By setting $\Theta = (\mu_1, \ldots, \mu_C)$, Eq. (18) reduces to:

$$\nabla_{\boldsymbol{\theta}_i} l(\Theta) = \sum_{j=1}^{N} \{ P(i \mid \mathbf{x}_j, \mu_i)$$

$$\times \nabla_{\mu_i} [-\frac{1}{2} (\mathbf{x}_j - \mu_i)^t \Sigma_i^{-1} (\mathbf{x}_j - \mu_i)] \} \tag{19}$$

$$= \sum_{j=1}^{N} P(i \mid \mathbf{x}_j, \mu_i) \Sigma_i^{-1} (\mathbf{x}_j - \mu_i).$$

Again, we are looking for the parameters $\Theta' = (\mu'_1, \ldots, \mu'_C)$ that maximize the log-likelihood, i.e., that correspond to a zero of its gradient. From Eq' (20), setting $\nabla_{\theta_i} l(\Theta') = 0$ allows us to write:

$$\sum_{j=1}^{N} P(i \mid \mathbf{x}_j, \mu'_i)\mathbf{x}_j = \sum_{j=1}^{N} P(i \mid \mathbf{x}_j, \mu'_i)\mu'_i \tag{20}$$

which finally leads to the following central equation:

$$\mu'_i = \frac{\sum_{j=1}^{N} P(i|\mathbf{x}_j, \mu'_i)\mathbf{x}_j}{\sum_{j=1}^{N} P(i|\mathbf{x}_j, \mu'_i)} \tag{21}$$

which shows that the maximum-likelihood estimate for the i-th mean vector is a weighted average of the samples of the training set, where each sample gives a contribution that is proportional to the *estimated* probability of i-th class given the sample itself. Equation (21) cannot be explicitly solved, but it can be written in quite an interesting and practical iterative form by making explicit the dependence of the current estimate on the number t of iterative steps that have already been accomplished:

$$\mu'_i(t + 1) = \frac{\sum_{j=1}^{N} P(i|\mathbf{x}_j, \mu'_i(t))\mathbf{x}_j}{\sum_{j=1}^{N} P(i|\mathbf{x}_j, \mu'_i(t))} \tag{22}$$

where $\mu'_i(t)$ denotes the estimate obtained at t-th iterative step, and the corresponding value is actually used to compute the new estimate at $(t+1)$-th step. Considering the fact that $P(i|\mathbf{x}_j, \mu'_i(t))$ is large when the Mahalanobis distance between $\mathbf{x}_j$ and $\mu'_i(t)$ is small, it is reasonable to estimate an approximation of $P(i|\mathbf{x}_j, \mu'_i(t))$ in the following way:

$$P(i|\mathbf{x}_j, \mu'_i(t)) \approx \begin{cases} 1 \text{ if } dist(\mathbf{x}_j, \mu'_i(t)) = \min_{l=1,\ldots,C} dist(\mathbf{x}_j, \mu'_l(t)) \\ 0 \text{ otherwise} \end{cases} \tag{23}$$

for a given distance measure dist() (Mahalanobis distance should be used, but Euclidean distance is usually an effective choice), so expression (22) reduces to:

$$\mu'_i(t + 1) = \frac{1}{n_i(t)} \sum_{k=1}^{n_i(t)} \mathbf{x}_k^{(i)} \tag{24}$$

where $n_i(t)$ is the number of training patterns estimated to belong to class i at step t, i.e., the number of patterns for which $P(i|\mathbf{x}_j, \mu'_i(t)) = 1$ according to Eq. (23), and $\mathbf{x}_k^{(i)}, k = 1, \ldots, n_i(t)$ is the k-th of these patterns. Note that Eq. (24) is the *k-means* clustering algorithm [11]. To switch to an incremental form for Eq. (24), consider what happens when, after the i-th mean vector has been calculated at step t over $n_i(t)$ patterns, a new pattern $\mathbf{x}^{(i)}_{n_i(t)+1}$ has to be assigned to component i. This occurs when $P(i|\mathbf{x}^{(i)}_{n_i(t)+1}, \boldsymbol{\mu}'_i(t))$ given by relation (23) equals 1, while $P(i|\mathbf{x}^{(i)}_{n_i(t)+1}, \boldsymbol{\mu}'_i(t-1))$ is 0. According to Eq. (24), the new mean vector is:

$$\boldsymbol{\mu}'_i(t+1) = \frac{1}{n_i(t)+1} \left(\sum_{k=1}^{n_i(t)} \mathbf{x}^{(i)}_k + \mathbf{x}^{(i)}_{n_i(t)+1} \right)$$

$$= \frac{n_i(t)}{n_i(t)+1} \boldsymbol{\mu}'_i(t) + \frac{1}{n_i(t)+1} \mathbf{x}^{(i)}_{n_i(t)+1}$$

$$= \boldsymbol{\mu}'_i(t) + \frac{1}{n_i(t)+1} (\mathbf{x}^{(i)}_{n_i(t)+1} - \boldsymbol{\mu}'_i(t)). \tag{25}$$

We can thus estimate the new mean vector after presentation of the $n_i + 1$-th pattern in the following way:

$$\boldsymbol{\mu}'_i(t+1) = \boldsymbol{\mu}'_i(t) + \delta_{t+1}(\mathbf{x}^{(i)}_{n_i(t)+1} - \boldsymbol{\mu}'_i(t)) \tag{26}$$

where $\delta_{t+1} = \frac{1}{n_i(t)+1}$ is a vanishing quantity. Using a fixed, small value η instead of δ_{t+1} and representing each component of the i-th mean vector $\boldsymbol{\mu}_i = (\mu_{i1}, \ldots, \mu_{id})$ with the weights $w_{i1}, \ldots, w_{id}$ of the corresponding connections in the competitive neural network model, the following weight update (learning) rule is obtained:

$$\Delta w_{ij} = \eta(x_j - w_{ij}) \tag{27}$$

where w_{ij} is the weight of the connection between input unit j and output unit i (the winner unit), η is the learning rate, and x_j is the j-th component of the pattern, of class i, presented to the network. When a new pattern is fed into the network, its distance is computed with respect to all the output units—using the connection weights as components of the corresponding mean vectors—and, according to relation (23), it is assigned to the nearest unit (mixture component). The latter is referred to as the *winner* unit. Weight update, or *learning*, is then accomplished by modifying the connection weights of the winner unit by a direct application of Eq. (27). The weights of the other units are left unchanged. This is a typical *winner-take-all* approach, where the units are in competition for novel input patterns. This is the rationale for using the name CNN. The way used here to derive the learning rule makes it clear that CNN can be seen as an online version of the popular *k-means* clustering algorithm [11].

The same mathematical formulation can be extended to the case of unknown covariance matrices for the component densities of the mixture, and even for the case of unknown class prior probabilities (the mixing parameters). A very good review is presented in [11]. The extension provides us iterative formulas similar to

Eq. (22), for estimating the means and the other unknown parameters. This means that an online (incremental) connectionist version of these estimates could also be derived. Finally, in addition to their online ever-adapting mechanism, CNNs differ from traditional clustering algorithms insofar that any growing–pruning algorithm for adapting the architecture of the ANN [19] can be used alongside with the aforementioned learning rule, resulting in a clustering scheme capable of discovering spontaneously a suitable number C of clusters for the data at hand (i.e., the user is no longer required to fix C arbitrarily in advance).

4 Maximum-Likelihood Modeling of Sequences

In several relevant applications (e.g., automatic speech recognition, handwritten text recognition, bioinformatics) the patterns to be classified are not static vectors described in a real-valued feature space, but sequences Y of observations, $Y = y_1, \ldots, y_T$. These sequences may have a dramatic length (from a few hundreds to millions), which may be different from sequence to sequence. Furthermore, it may happen that an input sequence is made up of subsequences (where boundaries between pairs of adjacent subsequences are not known in advance) which may be drawn individually from different probability distributions. Modeling of these probabilistic distributions is therefore of the utmost relevance to the aforementioned applications, and it is the subject of the following sections.

4.1 Motivation: Beyond Hidden Markov Models and Recurrent ANNs

Let $p(Y)$ denote the overall probability density of any given random observation sequence Y. Roughly speaking, $p(.)$ describes the statistics of sequences of parameterized observations (random "feature vectors") in the feature space. HMM [20, 36] are far the most popular (parametric) model for probability densities defined over sequences (note that a review of HMMs in out of the scope of the present chapter: the interested reader is invited to refer to [36]). Although HMMs are effective approaches, allowing for good modeling under many circumstances, they suffer from some limitations too (see [7]). For instance, the classical HMMs rely on strong assumptions on the statistical properties of the phenomenon at hand: the stochastic processes involved are modeled by first-order Markov chains, and the parametric form of the pdf that represent the emission probabilities associated with all states is heavily constraining. In addition, correlations among the input features are lost in practical implementations, since the covariance matrix of the Gaussian components (used to model the emission probabilities [36]) is usually assumed to be diagonal (in order to reduce the complexity of the machine, and to reduce numerical stability problems). Finally, HMMs are basically memory-based models (a Gaussian component is placed over dense locations of the training data) which is unlikely to generalize—due also to the lack of any regularization techniques. Reduced generalization capabilities are particularly evident when the HMMs are applied to noisy tasks (e.g., speech processing under severe noise conditions).

For these reasons, starting from the late 1980s many researchers began to use ANNs for sequence processing, namely time-delay ANNs [19, 45, 46] and, more recently, recurrent neural networks (RNN). RNNs provide a powerful extension of feed-forward connectionist models by allowing to introduce connections between arbitrary pairs of units, independently from their position within the topology of the network. Self-recurrent loops of a unit onto itself, as well as backward connections to previous layers, or lateral links between units belonging to the same layer are all allowed. RNNs behave like dynamical systems. Once fed with an input, the recurrent connections are responsible for an evolution in time of the internal state of the network. RNNs are particularly suited for sequence processing, due to their ability to keep an internal trace, or memory, of the past. This memory is combined with the current input to provide a context-dependent output. Several RNN architectures were proposed in literature [12, 21, 31], along with a variety of training algorithms, mostly based on gradient-descent techniques. Among the latter ones, particularly remarkable are Recurrent Back Propagation [34], Back-Propagation for Sequences (BPS) [16], Real Time Recurrent Learning [48, 49], Time-dependent Recurrent Back-Propagation [33, 38, 47], and the most popular *Back-Propagation Through Time* [27, 37].

In spite of their ability to classify short-time sequences, ANNs have mostly failed, so far, as a general framework for sequence processing. This is mainly due to the lack of ability to model long-term dependencies in RNNs. The theoretical motivations underlying this problem were well analyzed by Bengio et al. [1]. In the early 1990s this fact led to the idea of combining HMMs and ANNs within a single, novel model, broadly known as *hybrid HMM/ANN* [3, 8, 13, 17, 24, 30, 32, 39]. This direction turned out to be effective, exploiting the long-term modeling capabilities of HMMs and the universal, nonparametric nature of ANNs. The most popular ANN/HMM hybrid system is the classic Bourlard and Morgan's paradigm [6–8, 30], based on a maximum-a-posteriori HMM framework where an MLP is applied in order to estimate the posterior probability of the HMM states. It is a discriminative, supervised learning machine. A fully generative, unsupervised ANN/HMM hybrid approach was later proposed in [42], where the ANN outputs are expected to model the emission probabilities [36] (i.e., pdfs) of the HMM within a maximum-likelihood global optimization scheme. It is reviewed in the next section.

4.2   Unsupervised ML Learning in Generative ANN/HMM Hybrids

The approach revolves around the idea of preserving Bourlard and Morgan's architecture (an MLP having an output unit for each state of an underlying HMM), along with a novel ML training algorithm which may jointly optimize the parameters of the ANN and of the HMM. In so doing, the ANN is expected to model pdfs (the emission probabilities) instead of posteriors. As we observed in Sect. 2, this is a nontrivial task and requires ad-hoc techniques. The model described in this section was introduced in [42] first.

The machine relies on an HMM topology, including standard *initial* probabilities π and *transition* probabilities $\mathbf{a} = [a_{ij}]$ estimated by means of the *Baum–Welch* algorithm [36], while the *emission* probabilities $\mathbf{b(y)}$ [36] are estimated by an ANN. An output unit of the ANN holds for each of the states in the HMM, with the understanding that i-th output value $o_i(t)$ represents the emission probability $b_{i,t}$ for the corresponding (i-th) state, evaluated over current observation $\mathbf{y}_t$. In the following, we refer to this ANN with the symbol ψ. Once trained, the machine can be applied to new (e.g., test) sequences by relying on the usual Viterbi algorithm [20, 36]. Learning rules for connection weights and for neuron biases are calculated according to gradient-ascent to maximize a global criterion function, namely the *likelihood* of the model given the observation sequences (ML criterion). The global criterion function to be maximized by the model during training is the *likelihood* L of the observations given the model, that is[3] [2, 35]

$$L = \sum_{i \in \mathscr{F}} \alpha_{i,T}. \tag{28}$$

The sum is extended to the set $\mathscr{F}$ of all possible *final* states [2] within the HMM corresponding to the current training sequence. This HMM is supposed to involve Q states, and T is the length of the current observation sequence $Y = \mathbf{y}_1, \ldots, \mathbf{y}_T$. The *forward* terms $\alpha_{i,t} = P(q_{i,t}, \mathbf{y}_1, \ldots, \mathbf{y}_t)$ and the *backward* terms $\beta_{i,t} = P(\mathbf{y}_{t+1}, \ldots, \mathbf{y}_T | q_{i,t})$ for state i at time t can be computed recursively as follows [35]:

$$\alpha_{i,t} = b_{i,t} \sum_j a_{ji} \alpha_{j,t-1} \tag{29}$$

and

$$\beta_{i,t} = \sum_j b_{j,t+1} a_{ij} \beta_{j,t+1} \tag{30}$$

---

[3]A standard notation is used in the following to refer to quantities involved in HMM training (see [36]).

where a_{ij} denotes the transition probability from i-th state to j-th state, $b_{i,t}$ denotes emission probability associated with i-th state over t-th observation $\mathbf{y}_t$, and the sums are extended to all possible states within the HMM. The initialization of the forward probabilities is accomplished as in standard HMMs [35], whereas the backward terms at time T are initialized in a slightly different manner, namely:

$$\beta_{i,T} = \begin{cases} 1 \text{ if } i \in \mathscr{F} \\ 0 \text{ otherwise.} \end{cases} \tag{31}$$

Assuming that an ANN may represent the emission pdfs in a proper manner, and given a generic weight w of the ANN, hill-climbing gradient-ascent over C prescribes a learning rule of the kind:

$$\Delta w = \eta \frac{\partial L}{\partial w} \tag{32}$$

where η is the *learning rate*. In the following, we derive a learning rule for the ANN aimed at maximizing the likelihood of the observations, which is the usual criterion adopted to train HMMs. Eventually, the learning rule can be applied in parallel with the Baum–Welch algorithm, limiting the latter to the ML estimation of those quantities (namely the initial and transition probabilities in the underlying HMM) that do not explicitly depend on the ANN weights.

Let us observe, after [2], that the following property can be easily shown to hold true by straightforwardly taking the partial derivatives of the left- and right-hand sides of Eq. (29) with respect to $b_{i,t}$:

$$\frac{\partial \alpha_{i,t}}{\partial b_{i,t}} = \frac{\alpha_{i,t}}{b_{i,t}}. \tag{33}$$

In addition, borrowing the scheme proposed by Bridle [9] and Bengio [2], the following theorem can be proved to hold true (see [42]): $\frac{\partial L}{\partial \alpha_{i,t}} = \beta_{i,t}$, for each $i = 1, \ldots, Q$ and for each $t = 1, \ldots, T$. Given this theorem and Eq. (33), repeatedly applying the chain rule we can expand $\frac{\partial L}{\partial w}$ by writing:

$$\begin{aligned} \frac{\partial L}{\partial w} &= \sum_i \sum_t \frac{\partial L}{\partial b_{i,t}} \frac{\partial b_{i,t}}{\partial w} \\ &= \sum_i \sum_t \frac{\partial L}{\partial \alpha_{i,t}} \frac{\partial \alpha_{i,t}}{\partial b_{i,t}} \frac{\partial b_{i,t}}{\partial w} \\ &= \sum_i \sum_t \beta_{i,t} \frac{\alpha_{i,t}}{b_{i,t}} \frac{\partial b_{i,t}}{\partial w} \end{aligned} \tag{34}$$

where the sums are extended over all states $i = 1, \ldots, Q$ of the HMM, and to all $t = 1, \ldots, T$, respectively. From now on, attention is focused on the calculation of $\frac{\partial b_{i,t}}{\partial w}$, where $b_{i,t}$ is the output from i-th unit of the ANN at time t. Let us consider a multilayer feed-forward network (e.g., an MLP) the j-th output of which, computed over t-th input observation $\mathbf{y}_t$, is expected to be a nonparametric estimate of the emission probability $b_{j,t}$ associated with j-th state of the HMM at time t. An activation function $f_j(x_j(t))$, either linear or nonlinear, is attached to each unit j of the ANN, where $x_j(t)$ denotes input to the unit itself at time t. The corresponding output $o_j(t)$ is given by $o_j(t) = f_j(x_j(t))$. The net is assumed to have n layers $\mathcal{L}_0, \mathcal{L}_1, \ldots, \mathcal{L}_n$, where $\mathcal{L}_0$ is the input layer and is not counted, and $\mathcal{L}_n$ is the output layer. For notational convenience we write $i \in \mathcal{L}_k$ to denote the index of i-th unit in layer $\mathcal{L}_k$.

When considering the case of a generic weight w_{jk} between the k-th unit in layer $\mathcal{L}_{n-1}$ and j-th unit in the output layer, we can write:

$$\frac{\partial b_{j,t}}{\partial w_{jk}} = \frac{\partial f_j(x_j(t))}{\partial w_{jk}}$$

$$= f'_j(x_j(t)) \frac{\partial \sum_{i \in \mathcal{L}_{n-1}} w_{ji} o_i(t)}{\partial w_{jk}}$$

$$= f'_j(x_j(t)) o_k(t). \tag{35}$$

By defining the quantity[4]

$$\delta_i(j, t) = \begin{cases} f'_j(x_j(t)) & \text{if } i = j \\ 0 & \text{otherwise} \end{cases} \tag{36}$$

for each $i \in \mathcal{L}_n$, we can rewrite Eq. (35) in the compact form:

$$\frac{\partial b_{j,t}}{\partial w_{jk}} = \delta_j(j, t) o_k(t). \tag{37}$$

Consider now the case of connection weights in the first hidden layer. Let w_{kl} be a generic weight between l-th unit in layer $\mathcal{L}_{n-2}$ and k-th unit in layer $\mathcal{L}_{n-1}$.

$$\frac{\partial b_{j,t}}{\partial w_{kl}} = \frac{\partial f_j(x_j(t))}{\partial w_{kl}}$$

$$= f'_j(x_j(t)) \sum_{i \in \mathcal{L}_{n-1}} \frac{\partial w_{ji} o_i(t)}{\partial w_{kl}}$$

---

[4]Dependency on the specific HMM state j under consideration and on current time t is written explicitly, since this will turn out to be useful in the final formulation of the learning rule.

$$= f_j'(x_j(t)) \frac{\partial w_{jk} o_k(t)}{\partial o_k(t)} \frac{\partial o_k(t)}{\partial w_{kl}}$$

$$= f_j'(x_j(t)) w_{jk} \frac{\partial f_k(x_k(t))}{\partial w_{kl}}$$

$$= f_j'(x_j(t)) w_{jk} f_k'(x_k(t)) \frac{\partial \sum_{i \in \mathscr{L}_{n-2}} w_{ki} o_i(t)}{\partial w_{kl}}$$

$$= f_j'(x_j(t)) w_{jk} f_k'(x_k(t)) o_l(t). \tag{38}$$

Similarly to definition (36), we introduce the quantity:

$$\delta_k(j,t) = f_k'(x_k(t)) \sum_{i \in \mathscr{L}_n} w_{ik} \delta_i(j,t) \tag{39}$$

for each $k \in \mathscr{L}_{n-1}$, which allows us to rewrite Eq. (38) in the form:

$$\frac{\partial b_{j,t}}{\partial w_{kl}} = \delta_k(j,t) o_l(t) \tag{40}$$

which is formally identical to Eq. (37).

Finally, we carry out the calculations for a generic, hypothetic weight w_{lm} between m-th unit in layer $\mathscr{L}_{n-3}$ and l-th unit in layer $\mathscr{L}_{n-2}$, but the same results will also apply to subsequent layers ($\mathscr{L}_{n-4}, \mathscr{L}_{n-5}, \ldots, \mathscr{L}_0$):

$$\frac{\partial b_{j,t}}{\partial w_{lm}} = \frac{\partial f_j(x_j(t))}{\partial w_{lm}}$$

$$= f_j'(x_j(t)) \sum_{i \in \mathscr{L}_{n-1}} \frac{\partial w_{ji} o_i(t)}{\partial w_{lm}}$$

$$= f_j'(x_j(t)) \sum_{i \in \mathscr{L}_{n-1}} \frac{\partial w_{ji} o_i(t)}{\partial o_i(t)} \frac{\partial o_i(t)}{\partial w_{lm}}$$

$$= f_j'(x_j(t)) \sum_{i \in \mathscr{L}_{n-1}} w_{ji} \frac{\partial o_i(t)}{\partial w_{lm}}. \tag{41}$$

Efforts now concentrate on the development of the term $\frac{\partial o_i(t)}{\partial w_{lm}}$. This is accomplished by writing:

$$\frac{\partial o_i(t)}{\partial w_{lm}} = \frac{\partial f_i(x_i(t))}{\partial w_{lm}}$$

$$= f_i'(x_i(t)) \sum_{k \in \mathscr{L}_{n-2}} \frac{\partial w_{ik} o_k(t)}{\partial w_{lm}}$$

$$= f_i'(x_i(t)) \frac{\partial w_{il} o_l(t)}{\partial o_l(t)} \frac{\partial o_l(t)}{\partial w_{lm}}$$

$$= f_i'(x_i(t)) w_{il} \frac{\partial f_l(x_l(t))}{\partial w_{lm}}$$

$$= f_i'(x_i(t)) w_{il} f_l'(x_l(t)) \frac{\partial x_l(t)}{\partial w_{lm}}$$

$$= f_i'(x_i(t)) w_{il} f_l'(x_l(t)) \sum_{k \in \mathscr{L}_{n-3}} \frac{\partial w_{lk} o_k(t)}{\partial w_{lm}}$$

$$= f_i'(x_i(t)) w_{il} f_l'(x_l(t)) o_m(t) \tag{42}$$

which can be substituted into Eq. (41) obtaining:

$$\frac{\partial b_{j,t}}{\partial w_{lm}} = f_j'(x_j(t)) \sum_{i \in \mathscr{L}_{n-1}} w_{ji} f_i'(x_i(t)) w_{il} f_l'(x_l(t)) o_m(t). \tag{43}$$

Again, as in (36) and (39), for each $l \in \mathscr{L}_{n-2}$ we define the quantity[5]:

$$\delta_l(j, t) = f_l'(x_l(t)) \sum_{i \in \mathscr{L}_{n-1}} w_{il} \delta_i(j, t) \tag{44}$$

and rewrite Eq. (43) in the familiar, compact form:

$$\frac{\partial b_{j,t}}{\partial w_{lm}} = \delta_l(j, t) o_m(t) \tag{45}$$

that is formally identical to Eqs. (37) and (40).

Using the above developments to expand Eq. (34) and substituting it into Eq. (32), the latter can now be restated in the form of a general learning rule for a generic weight w_{jk} of the network, by writing:

$$\Delta w_{jk} = \eta \sum_{i=1}^{Q} \sum_{t=1}^{T} \beta_{i,t} \frac{\alpha_{i,t}}{b_{i,t}} \delta_j(i, t) o_k(t) \tag{46}$$

where the term $\delta_j(i, t)$ is computed via Eq. (36), Eqs. (39) or (44), according to the fact that the layer under consideration is the output layer, the first hidden layer, or one of the other hidden layers, respectively.

Experimental results reported in [42] (where several workaround are presented to prevent occurrences of the divergence problem) show a dramatic improvement in

---

[5]For lower layers the sum in Eq. (44) must be accomplished over the units belonging to the immediately upper layer.

terms of sequence pdf modeling capability over traditional HMMs. The improvement is particularly significant in sequence processing tasks which involve noisy data [43]. This is likely to be due to: (1) the flexibility of the nonparametric modeling over the parametric assumptions made in standard HMM; and (2) the generalization capabilities of ANNs, in contrast with the generalization problems of traditional HMMs.

5 Conclusions

Traditionally, unsupervised learning in ANNs revolves around methods that do not take the probabilistic nature of phenomena into explicit account. Examples are the auto-associative MLP, Hopfield network, Kohonen's self-organizing maps, etc. [19]. To the contrary, the chapter focuses on learning machines that undergo a proper probabilistic interpretation, and whose training algorithms exploit (and respect) the probability laws that underlie the data samples at hand. In so doing, estimates of probabilistic quantities (that are typically involved in statistical analysis, pattern classification, and clustering) are obtained, providing the practitioner with robust tools, alternative to statistical approaches. The flexibility and the generalization capabilities of ANNs yield less constrained, more general models of pdfs than traditional statistical estimates do. Indeed, empirical evidence reported in the literature shows that ANN-based estimates may improve performance significantly over their statistical counterparts. As well, traditional parametric estimation methods may be used to obtain unsupervised learning rules for ANNs that result in an online, ever-adaptive version of popular partitive clustering algorithms, or that can learn the emission probabilities of an underlying HMM.

In most cases, the algorithms exploit the maximum-likelihood criterion, which replaces the minimum squared error (MSE) criterion typical of standard back-propagation. ML forms the basis for completely unsupervised training algorithms, suitable for RBF-based pdf estimation, CNN-based clustering, and ANN/HMM-based estimation of the probability density of sequences of random vectors. Nonetheless, the community resorted also to regular BP over the MSE loss functional for tackling the divergence problem that may occur in ML training of unconstrained MLPs. To this end, synthetic target outputs for BP are generated automatically from the unsupervised training data by relying on statistical estimates of pdfs or cdfs.

Our expectations for future scenarios embrace the following issues. First and foremost, an exact solution to the divergence problem via an algorithmically constrained MLP that may well undergo ML training. Then, extension of the ANN-based probabilistic models for HMMs to broader families of probabilistic graphical models that operate under a Markovian assumption (e.g., Markov random fields [22], conditional random fields [23], hybrid random fields [14], and dynamic random fields [5]). Finally, just like hybrid statistical models can be learned from samples of sequential data, robust neural estimates of probabilistic laws underlying structured data (like random graphs [15]) are sought.

References

1. Bengio, Y., Simard, P., Frasconi, P.: Learning long-term dependencies with gradient descent is difficult. IEEE Trans. Neural Netw. **5**(2), 157–166 (1994). Special issue on Recurrent Neural Networks, March 94
2. Bengio, Y.: Neural Networks for Speech and Sequence Recognition. International Thomson Computer, London (1996)
3. Bengio, Y., De Mori, R., Flammia, G., Kompe, R.: Global optimization of a neural network-hidden Markov model hybrid. IEEE Trans. Neural Netw. **3**(2), 252–259 (1992)
4. Bishop, C.M.: Neural Networks for Pattern Recognition. Oxford University Press, Oxford (1995)
5. Bongini, M., Trentin, E.: Towards a novel probabilistic graphical model of sequential data: a solution to the problem of structure learning and an empirical evaluation. In: Artificial Neural Networks in Pattern Recognition - 5th INNS IAPR TC 3 GIRPR Workshop, ANNPR 2012, pp. 82–92. Trento, Italy, September 17–19, 2012. Proceedings (2012)
6. Bourlard, H., Morgan, N.: Continuous speech recognition by connectionist statistical methods. IEEE Trans. Neural Netw. **4**(6), 893–909 (1993)
7. Bourlard, H., Morgan, N.: Connectionist Speech Recognition. A Hybrid Approach. The Kluwer International Series in Engineering and Computer Science, vol. 247. Kluwer Academic, Boston (1994)
8. Bourlard, H., Wellekens, C.: Links between hidden Markov models and multilayer perceptrons. IEEE Trans. Pattern Anal. Mach. Intell. **12**, 1167–1178 (1990)
9. Bridle, J.: Alphanets: a recurrent 'neural' network architecture with a hidden Markov model interpretation. Speech Comm. **9**(1), 83–92 (1990)
10. Celebi, M.E., Kingravi, H.A., Vela, P.A.: A comparative study of efficient initialization methods for the k-means clustering algorithm. Expert Syst. Appl. **40**(1), 200–210 (2013)
11. Duda, R.O., Hart, P.E.: Pattern Classification and Scene Analysis. Wiley, New York (1973)
12. Elman, J.: Finding structure in time. Cogn. Sci. **14**, 179–211 (1990)
13. Franzini, M., Lee, K., Waibel, A.: Connectionist Viterbi training: a new hybrid method for continuous speech recognition. In: International Conference on Acoustics, Speech and Signal Processing, pp. 425–428. Albuquerque, NM (1990)
14. Freno, A., Trentin, E.: Hybrid Random Fields – A Scalable Approach to Structure and Parameter Learning in Probabilistic Graphical Models. Springer, Berlin (2011)
15. Gilbert, E.N.: Random graphs. Ann. Math. Stat. **30**(4), 1141–1144 (1959)
16. Gori, M., Bengio, Y., De Mori, R.: BPS: a learning algorithm for capturing the dynamical nature of speech. In: Proceedings of the International Joint Conference on Neural Networks, pp. 643–644. IEEE, New York/Washington D.C. (1989)
17. Haffner, P., Franzini, M., Waibel, A.: Integrating time alignment and neural networks for high performance continuous speech recognition. In: International Conference on Acoustics, Speech and Signal Processing, pp. 105–108. Toronto (1991)
18. Haykin, S.: Neural Networks (A Comprehensive Foundation). Macmillan, New York (1994)
19. Hertz, J., Krogh, A., Palmer, R.: Introduction to the Theory of Neural Computation. Addison Wesley, Massachusetts (1991)
20. Huang, X.D., Ariki, Y., Jack, M.: Hidden Markov Models for Speech Recognition. Edinburgh University Press, Edinburgh (1990)

21. Jordan, M.: Serial order: A parallel, distributed processing approach. In: Elman, J., Rumelhart, D. (eds.) Advances in Connectionist Theory: Speech. Lawrence Erlbaum, Hillsdale (1989)
22. Kindermann, R., Snell, J.L.: Markov Random Fields and Their Applications. American Mathematical Society, Providence (1980)
23. Lafferty, J., McCallum, A., Pereira, F.: Conditional random fields: probabilistic models for segmenting and labeling sequence data. In: Proceeding of the 18th International Conference on Machine Learning, pp. 282–289. Morgan Kaufmann, San Francisco (2001)
24. Levin, E.: Word recognition using hidden control neural architecture. In: International Conference on Acoustics, Speech and Signal Processing, pp. 433–436. Albuquerque, NM (1990)
25. Magdon-Ismail, M., Atiya, A.: Density estimation and random variate generation using multilayer networks. IEEE Trans. Neural Netw. 13(3), 497–520 (2002)
26. McLachlan, G., Basford, K. (eds.): Mixture Models: Inference and Applications to Clustering. Marcel Dekker, New York (1988)
27. Minsky, M., Papert, S.: Perceptrons. MIT, Cambridge (1969)
28. Modha, D.S., Fainman, Y.: A learning law for density estimation. IEEE Trans. Neural Netw. 5(3), 519–23 (1994)
29. Modha, D.S., Masry, E.: Rate of convergence in density estimation using neural networks. Neural Comput. 8, 1107–1122 (1996)
30. Morgan, N., Bourlard, H.: Continuous speech recognition using multilayer perceptrons with hidden Markov models. In: International Conference on Acoustics, Speech and Signal Processing, pp. 413–416. Albuquerque, NM (1990)
31. Mozer, M.C.: Neural net architectures for temporal sequence processing. In: Weigend, A., Gershenfeld, N. (eds.) Predicting the Future and Understanding the Past, pp. 243–264. Addison-Wesley, Redwood City (1993)
32. Niles, L., Silverman, H.: Combining hidden Markov models and neural network classifiers. In: International Conference on Acoustics, Speech and Signal Processing, pp. 417–420. Albuquerque, NM (1990)
33. Pearlmutter, B.: Learning state space trajectories in recurrent neural networks. Neural Comput. 1, 263–269 (1989)
34. Pineda, F.: Recurrent back-propagation and the dynamical approach to adaptive neural computation. Neural Comput. 1, 161–172 (1989)
35. Rabiner, L., Juang, B.H.: Fundamentals of Speech Recognition. Prentice Hall, Englewood Cliffs (1993)
36. Rabiner, L.R.: A tutorial on hidden Markov models and selected applications in speech recognition. Proc. IEEE 77(2), 257–286 (1989)
37. Rumelhart, D., Hinton, G., Williams, R.: Learning internal representations by error propagation. In: Rumelhart, D., McClelland, J. (eds.) Parallel Distributed Processing, vol. 1, chap. 8, pp. 318–362. MIT, Cambridge (1986)
38. Sato, M.: A real time learning algorithm for recurrent analog neural networks. Biol. Cybern. 62, 237–241 (1990)
39. Tebelskis, J., Waibel, A., Petek, B., Schmidbauer, O.: Continuous speech recognition using linked predictive networks. In: Lippman, R.P., Moody, R., Touretzky, D.S. (eds.) Advances in Neural Information Processing Systems, vol. 3, pp. 199–205. Morgan Kaufmann, San Mateo/Denver (1991)
40. Trentin, E.: Networks with trainable amplitude of activation functions. Neural Netw. 14(4–5), 471–493 (2001)
41. Trentin, E.: Simple and effective connectionist nonparametric estimation of probability density functions. In: Artificial Neural Networks in Pattern Recognition, Second IAPR Workshop, ANNPR 2006, pp. 1–10. Ulm, Germany, August 31–September 2, 2006. Proceedings (2006)
42. Trentin, E., Gori, M.: Robust combination of neural networks and hidden Markov models for speech recognition. IEEE Trans. Neural Netw. 14(6), 1519–1531 (2003)
43. Trentin, E., Gori, M.: Inversion-based nonlinear adaptation of noisy acoustic parameters for a neural/HMM speech recognizer. Neurocomputing 70(1–3), 398–408 (2006)

44. Vapnik, V.N., Mukherjee, S.: Support vector method for multivariate density estimation. In: Advances in Neural Information Processing Systems, pp. 659–665. MIT, Cambridge (2000)
45. Waibel, A.: Modular construction of time-delay neural networks for speech recognition. Neural Comput. **1**, 39–46 (1989)
46. Waibel, A., Hanazawa, T., Hinton, G., Shikano, K., Lang, K.: Phoneme recognition using time-delay neural networks. IEEE Trans. Acoust. Speech Signal Process. **37**, 328–339 (1989)
47. Werbos, P.: Generalization of backpropagation with application to a recurrent gas market model. Neural Netw. **1**, 339–356 (1988)
48. Williams, R., Zipser, D.: Experimental analysis of the real-time recurrent learning algorithm. Connect. Sci. **1**, 87–111 (1989)
49. Williams, R., Zipser, D.: A learning algorithm for continually running fully recurrent neural networks. Neural Comput. **1**, 270–280 (1989)

Printed in the United States
By Bookmasters